The Population Ecology and Conservation of *Charadrius* Plovers

Studies in Avian Biology Editorial Board
Studies in Avian Biology

A Publication of The American Ornithological Society

www.crcpress.com/browse/series/crcstdavibio

Studies in Avian Biology is a series of works founded and published by the Cooper Ornithological Society in 1978, and published by The American Ornithological Society since 2017. Volumes in the series address current topics in ornithology and can be organized as monographs or multiauthored collections of chapters. Authors are invited to contact the Series Editor to discuss project proposals and guidelines for preparation of manuscripts.

The Population Ecology and Conservation of *Charadrius* Plovers

Edited by
Mark A. Colwell and Susan M. Haig

CRC Press is an imprint of the
Taylor & Francis Group, an **informa** business

CRC Press
Taylor & Francis Group
6000 Broken Sound Parkway NW, Suite 300
Boca Raton, FL 33487-2742

First issued in paperback 2020

ISBN-13: 978-1-4987-5582-5 (hbk)
ISBN-13: 978-0-367-72997-4 (pbk)

Visit the Taylor & Francis Web site at
http://www.taylorandfrancis.com

and the CRC Press Web site at
http://www.crcpress.com

CONTENTS

ACKNOWLEDGMENTS

Production of this book on the ecology and conservation of *Charadrius* plovers was conceived as a means to gather summaries of long-term data from plover researchers across the globe in order to draw from their hundreds of years of combined experience and to address pressing conservation questions and issues of the day. Inspiration for our work on plovers was garnered during the time we spent as graduate students of Lewis Oring and with lots of help along the way from David Lank. We dedicate this book to Lew and Dov with our profound gratitude.

We sincerely thank the chapter authors for their hard work and patience over the past 2 years. We further thank the many reviewers who took the time to provide comments on each chapter. Ram Papish provided the artwork for the cover and throughout the book. Finally, we are grateful to Kate Huyvaert for shepherding the book through the publication process and for CRC Press for their work on book production.

MARK A. COLWELL
SUSAN M. HAIG

EDITORS

Mark Colwell is a professor in the Wildlife Department at Humboldt State University, where he has taught Ornithology and Shorebird Ecology for 30 years. At HSU, he has been recognized for his scholarly and teaching contributions. He has published 90 papers and one book, mostly on topics related to ecology and conservation of shorebirds.

Susan Haig is a senior scientist (emerita) at the U.S. Geological Survey Forest and Rangeland Ecosystem Science Center, professor of Wildlife Ecology at Oregon State University and a research associate of the Smithsonian Institution. Her work on small population biology and conservation genetics focused on shorebirds with an emphasis on Piping Plovers. For many years, she lead the U.S. Fish and Wildlife Service Piping Plover recovery team in the Great Lakes and Northern Great Plains. For this work, she was awarded the Miller Medal by the American Ornithologists' Union and a Distinguished Service Award from the U.S. Department of the Interior.

CONTRIBUTORS

DANIEL H. CATLIN
Department of Fish and Wildlife Conservation
Virginia Tech
Blacksburg, Virginia

MARK A. COLWELL
Wildlife Department
Humboldt State University
Arcata, California

JESSE R. CONKLIN
Conservation Ecology Group
Groningen Institute for Evolutionary Life Sciences
University of Groningen
Groningen, The Netherlands

STEPHEN J. DINSMORE
Department of Natural Resource Ecology and Management
Iowa State University
Ames, Iowa

NATALIE DOS REMEDIOS
Department of Animal and Plant Sciences
University of Sheffield
Sheffield, United Kingdom

LUKE J. EBERHART-PHILLIPS
Behavioral Genetics and Evolutionary Ecology Research Group
Max Planck Institute for Ornithology
Seewiesen, Germany

JAMES D. FRASER
Department of Fish and Wildlife Conservation
Virginia Tech
Blacksburg, Virginia

SUSAN M. HAIG
Forest and Rangeland Ecosystem Science Center
United States Geological Survey
Corvallis, Oregon

CLEMENS KÜPPER
Behavioral Genetics and Evolutionary Ecology Research Group
Max Planck Institute for Ornithology
Seewiesen, Germany

ERICA NOL
Department of Biology
Trent University
Peterborough, Ontario, Canada

GARY W. PAGE
Pacific Coast and Central Valley Group
Point Blue Conservation Science
Petaluma, California

LYNNE E. STENZEL
Pacific Coast and Central Valley Group
Point Blue Conservation Science
Petaluma, California

MICHAEL A. WESTON
Centre for Integrative Ecology
Deakin University
Burwood, Victoria, Australia

Ram Papish

CHAPTER ONE

An Overview of the World's Plovers*

Mark A. Colwell and Susan M. Haig

Abstract. Plovers of the genus Charadrius and their close allies are a diverse group, numbering 40 species, many with subspecies. They breed on all continents except Antarctica, in open, sparsely vegetated habitats of tundra and grasslands, and along shores of oceans, rivers, and inland lakes. Most are migratory, especially those breeding in arctic and temperate regions; others are partial migrants or sedentary. On migration, they are poorly studied and do not always correspond to the typical shorebird (i.e., sandpiper) pattern characterized by dense flocks concentrating at a few staging areas. Their foraging ecologies are rather uniform in that all species search visually for prey using a "run-stop-peck" maneuver. Breeding birds defend nesting and foraging territories while nonbreeding birds forage in loose flocks, which may stem from individuals minimizing interference with conspecifics while enhancing benefits of shared vigilance for predators. In breeding, they are conservative, laying two to four eggs at daily or longer intervals; replacement clutches are common, especially in species with prolonged breeding seasons. Precocial young hatch after comparatively long incubation that is correlated with development of neural centers associated with vision. Their mating systems are a mix of social monogamy and biparental care, with frequent sequential polygamy, especially in temperate and tropical taxa that breed for extended periods. Population sizes vary over several orders of magnitude; several species are highly endangered. Other species are abundant and widely distributed, although their populations may also be in decline. Regardless of their status, most plovers occupy habitats throughout the year that put them at conservation risk owing to anthropogenic factors including climate change, human disturbance, habitat loss, and predation. In this book, we draw from the expertise of an international group of researchers to outline the ecologies, behaviors, and challenges of plovers throughout the annual cycle so that decision makers can be most successful in their endeavors to conserve and manage populations.

Keywords: Charadrius, conservation, ecology, management, plover, shorebird, migratory connectivity, annual cycle.

Many of the 40 plover species (see Table 1.1 for Latin names) and their close relatives are widespread and well known to humans throughout history and today. The Latin name for the plover family, subfamily, and genus derives from the Greek word χαραδριός, which

* Mark A. Colwell and Susan M. Haig. 2019. An Overview of the World's Plovers. Pp. 3–15 in M.A. Colwell and S.M. Haig (editors). The Population Ecology and Conservation of Charadrius Plovers Studies in Avian Biology (no. 52), CRC Press, Boca Raton, FL.

TABLE 1.1
An overview of plover species[a] and subspecies included in this book.

Common name	Latin name/subspecies	Population size[b]	Population status[b]
S. Red-breasted Dotterel	*C. obscurus*	250	Endangered
N. Red-breasted Dotterel	*C. aquilonius*	2,175	Near Threatened
Lesser Sand-Plover	*C. m. mongolus*	310,000–390,000	Least Concern
	C. m. pamirensis		
	C. m. atrifrons		
	C. m. schaeferi		
	C. m. stegmanni		
Greater Sand-Plover	*C. l. leschenaultii*	180,000–360,000	Least concern
	C. l. columbinus		
	C. l. scythicus		
Caspian Plover	*C. asiaticus*	10,000–100,000	Least concern
Collared Plover	*C. collaris*	<10,000	Least concern
Puna Plover	*C. alticola*		Least concern
Two-banded Plover	*C. falklandicus*	10,000–100,000	Least concern
Double-banded Plover	*C. bicinctus*	50,000	Least concern
Kittlitz's Plover	*C. pecuarius*	475,000	Least concern
Red-capped Plover	*C. ruficapillus*	95,000	Least concern
Malay Plover	*C. peronii*	<10,000	Near threatened
Kentish Plover	*C. a. alexandrinus*		Least concern
	C. a. seebohmi	5,000–10,000	
	C. a. nihonensis	10,000	
White-faced Plover	*C. dealbatus*		Data deficient
Snowy Plover	*C. n. nivosus*	25,869	Near threatened
	C. n. occidentalis		
Javan Plover	*C. javanicus*	6,000	Near threatened
Wilson's Plover	*C. w. wilsonia*		Least concern
	C. w. beldingi		
	C. w. cinnamominus		
	C. w. crassirostris		
Common Ringed Plover	*C. h. hiaticula*	360,000–1,300,000	Least concern
	C. h. tundrae		
Semipalmated Plover	*C. semipalmatus*	>150,000	Least concern
Long-billed Plover	*C. placidus*	1,000–25,000	Least concern
Piping Plover	*C. m. melodus*	3,362	Threatened/ endangered
	C. m. circumcinctus	2,361	Endangered/ threatened
Black-banded (Madagascar) Plover	*C. thoracicus*	1,800–2,300	Vulnerable
Little Ringed Plover	*C. d. dubius*	280,000–530,000	Least concern
	C. d. curonicus		
	C. d. jerdoni		

(Continued)

TABLE 1.1 (*Continued*)

An overview of plover species[a] and subspecies included in this book.

Common name	Latin name/subspecies	Population size[b]	Population status[b]
African Three-banded Plover	*C. tricollaris*	70,000–140,000	Least concern
Madagascar Three-banded Plover	*C. bifrontatus*	10,000–30,000	Least concern
Forbes's Plover	*C. forbesi*	100,000(?)	Least concern
White-fronted Plover	*C. m. marginatus*		Least concern
	C. m. mechowi		
	C. m. arenaceus		
	C. m. tenellus		
Chestnut-banded Plover	*C. p. pallidus*	11,000–16,000	Near threatened
	C. p. venustus	6,500	
Killdeer	*C. v. vociferus*	1,000,000	Least concern
	C. v. ternominatus		
	C. v. peruvianus		
Mountain Plover	*C. montanus*	11,000–14,000	Near threatened
Oriental Plover	*C. veredus*	160,000	Least concern
Eurasian Dotterel	*C. morinellus*	50,000–220,000	Least concern
St Helena Plover	*C. sanctaehelenae*	<500	Critically endangered
Rufous-chested Plover	*C. modestus*	33,000–63,000	Least concern
Red-kneed Dotterel	*Erythrogonys cinctus*	26,000	Least concern
Hooded Plover	*Thinornis cucullatus*	~10,000	Vulnerable
Shore Plover	*T. novaeseelandiae*	~300	Endangered
Black-fronted Dotterel	*Elseyornis melanops*	15,500	Least concern
Inland Dotterel	*Peltohyas australis*	14,000	Least concern
Wrybill	*Anarhynchus frontalis*	4,500–5,000	Vulnerable

[a] Clements et al. (2014); del Hoyo et al. (2016) Handbook of Bird of the World Alive. www.hbw.com/node/467298; accessed 9 September 2016.
[b] IUCN (2016).

derives from kharadra (meaning ravine) and refers to the bird's use of river valley habitats. In Greek history, Plutarch wrote of the healing nature of the plover's gaze: "...brought into the presence of any sick person, the bird foretells his recovery or his death by looking toward or away from him" (Wolf 1912). The salubrious gaze of the bird, however, runs counter to the effect that humans have had on plovers worldwide. That is, several plovers rank among the world's most endangered species, and even widespread and common plovers are in decline (Sanzenbacher and Haig 2001). This book is motivated by a long-standing interest in the ecology and conservation of plovers, as exemplified by the accumulated publishing records of its coauthors. Collectively, we have spent over 325 years studying plovers and have published 264 scientific papers addressing their evolution, ecology, and conservation. Our efforts have been focused, however, on a small subset of the clade (Figure 12.1, Chapter Twelve, this volume). Each of us is motivated by a sincere interest in insuring that future generations benefit from the opportunity to observe, marvel at, and study the diverse and wondrous facets of these easily observed shorebirds.

FOSSIL HISTORY AND EVOLUTIONARY RELATIONSHIPS

Shorebirds have been a phylogenetic enigma since Linnaeus (1758) first produced a modern classification (Livezey 2010). Feduccia (1999) portrayed the ancestral bird that survived the K/T [Cretaceous (K) and Tertiary (T) time periods] extinction event as a shorebird; perhaps this was a Stone-Curlew (*Burhinus* spp.), a species which may be a close relative of plovers, depending on the phylogeny one uses (Livezey 2010). There is, however, limited fossil evidence to inform and enlighten our understanding of birds in general and shorebirds specifically (Mayr 2009). Hence, discussions of the evolutionary history of shorebirds rely on molecular data, which suggests a Neogene origin for the order Charadriiformes (suborders Charadrii, Scolopaci, and Lari) estimated at 10–15 million years ago (Jetz et al. 2012). However, after that, there has been no agreement as to the origin of shorebirds, let alone plovers (but see Chapter Two, this volume, for more details).

Today, the world's shorebirds (or waders) include roughly 215 species divided unevenly among 14 families (Colwell 2010). The two most diverse families are the sandpipers (88 species; Scolopacidae) and plovers (65 species; Charadriidae). Interestingly, the shorebirds continue to present a phylogenetic "problem," which seems to be largely ignored by most scientists, conservationists, and the public. Birders, field guides (e.g., Hayman et al. 1986), and conservationists (e.g., Colwell 2010) alike treat them as a unified group, while students of modern bird taxonomy recognize that shorebirds are a diverse "mix" of clades (Livezey 2010). Increasing evidence indicates that the plovers and sandpipers likely represent separate evolutionary lineages with uncertain affinities to alcids, jaegers, gulls, and terns, which are either sandwiched between them or sit as a parallel group. Morphological (Livezey 2010) and molecular (e.g., van Tuinen et al. 2004, Baker et al. 2007) evidence often yields conflicting patterns as to the true affinities of Charadriidae in the order Charadriiformes. In this book, we focus on the subfamily Charadriinae, which includes the collared and ringed plovers (genus Charadrius), along with a suite of taxa of uncertain phylogenetic affinities (e.g., genera *Thinornus, Anarhynchus, Eudromius, Elseyornis, Peltohyas*; Table 1.1).

SALIENT FEATURES OF THE CHARADRIINAE

The Charadrius plovers and allies are united in appearance by features that belie their similar ecologies. They are small, ranging in mass from 25 to 120 g; more than half weigh less than 50 g. Sexes are similar in size and other morphological measures, although males tend to have brighter ornamentation (see below). Precocial, nidifugous chicks hatch after comparatively long incubation periods of 25 or more days (Nol 1986). Young of subtropical species (e.g., Kittlitz's Plover) grow more slowly, have longer pre-fledging intervals, and lower energy expenditure compared with northern taxa (Tjorve et al. 2008). This has been attributed to conditions of lower food availability in the subtropics compared with seasonally productive northern climes.

As a group, the Charadrius plovers possess 11 primaries, 8–14 secondaries, and 12 rectrices (Prater et al. 1977). The contour feathers of the body are mostly light (i.e., white) below, sometimes suffused with reddish brown on the breast; this contrasts with browns or grays above. In many species, alternate plumages of males are darkly feathered on the breast, ear coverts, crown, or forehead; females are typically a muted version of this pattern (Graul 1973). This sexual dimorphism is more exaggerated in polygamous taxa (Argüelles-Ticó et al. 2016). Additionally, dimorphism may be correlated with sex differences in parental care in which brightly plumaged males incubate at night when visual predators are less likely to detect them on the nest (Ekanayake et al. 2015).

Molt follows a complex alternate strategy (*sensu* Howell et al. 2010). This pattern suggests that today's plovers evolved from a migratory ancestor (Howell 2010), which is supported by a recent phylogenetic analysis (dos Remedios et al. 2015; Chapter Two, this volume) but contradicted by another (Joseph et al. 1999). Surprisingly, little has been published on the details of molts and plumages in any one species. For example, the pre-alternate molt in the Kentish Plover has been described as beginning in some birds in November (Prater et al. 1977, Cramp and Simmons 1981). In the closely related Snowy Plover, the pre-alternate molt has been suggested to begin between February and April (Page et al. 2009). However, observations of marked Snowy Plovers (in a partial migrant population in which breeding spans 6 months) show that some males begin molting much earlier, in October (DeJoannis 2016). A detailed study of a marked Hooded Plover population showed that primary (i.e., prebasic) molt lasted 7 months and overlapped completely with the breeding period, which is rare among birds (Rogers et al. 2014). Perhaps even more interesting was the plasticity of molt progression in Hooded

Plovers, with individuals shifting between a "fast" and "slow" sequence within one molt. The reasons for these shifts in molt speed are unresolved for this species but may include integration of molt with varying energetic demands of successful and failed reproductive attempts (Rogers et al. 2014).

Plovers forage visually using a "run-stop-peck" maneuver, which is consider the ancestral condition in shorebirds (Barbosa and Moreno 1999). This stereotypic foraging behavior requires a well-developed sense of vision. Nol (1986) showed that plovers (Charadriidae; including some Charadrius) had longer incubation periods than comparable-sized (i.e., gauged by egg volume) sandpipers (Scolopacidae). She interpreted longer incubation in plovers as necessary for the development of complex neural tissue (i.e., optic tectum) associated with visual foraging behavior compared with neural tissue necessary for tactile feeding in sandpipers. In tandem with an emphasis on visual foraging, most plovers have short, straight bills (Barbosa and Moreno 1999). The exception is New Zealand's Wrybill, the only bird with an asymmetric bill (in this case curved to the right at an angle of 15–22°, Johnsgard 1981). The plover bill is characterized by a "basal hinge and bone spring kinesis," which enables a doubly rhynchokinetic flexion outward along points of the upper and lower mandible (Zusi 1984, pp. 3–6). Detailed descriptions of the digestive system are not available to offer insight regarding the manner in which plovers digest their prey. There is treatment, however, of adaptations of the excretory system, especially for species that occupy hypersaline habitats for portions of the annual cycle. For example, Staaland (1967) reported on the differences in mass of the nasal, or salt, gland for a variety of Charadriiformes that varied in their use of fresh water and marine habitats. The Common Ringed Plover had a comparatively large gland, which may predispose it to greater flexibility in a world in which wetland salinization is increasing (Rubega and Robinson 1998). On the other hand, inland Killdeer appear to be tied to fresh water year round (Plissner et al. 2000).

SPECIES OVERVIEW

This book addresses the ecologies and conservation of 40 plover species (Table 1.1). It includes several recent taxonomic changes in which subspecies were elevated to species; we have included at least one taxon, the White-faced Plover, that has a controversial recent history regarding its status as a species. In other cases, recent phylogenetic analyses have given rise to the splitting and renaming of species. For example, the New Zealand Dotterel was split into the Southern and Northern Red-breasted dotterels (Barth et al. 2013), differences between Kentish and Snowy plovers were resolved (Küpper et al. 2009), and the three-banded plovers of Madagascar and continental Africa were recognized as distinct species. While debate continues regarding some of these decisions (see Chapter Two, this volume), we largely adhered to the latest "hypothesis" regarding phylogenetic relationships (dos Remedios et al. 2015). To put current knowledge of the clade into perspective, we note that Johnsgard's (1981) detailed and inclusive tome covering shorebirds of the world identified 33 species in the group. Notable absences from his survey were the Puna and Forbes's plovers. In summary, the Charadrius plovers represent a diverse group with a phylogenetic history that continues to spawn debate (dos Remedios et al. 2015).

The 40 plover species are even more diverse when one considers subspecies (Table 1.1). Twelve species (30%) include recognized subspecies. A simple Spearman's rank correlation between range size and number of subspecies ($r_s = 0.49$; $P = 0.002$) suggests that several widely distributed plovers (average subspecies = 3.0 ± 1.0) may require further attention to consider whether taxonomic revision is necessary to include "new" taxa. As mentioned earlier, a controversy remains over the unresolved status of the White-faced Plover, which may be a morph of the Kentish Plover (Chapter Two, this volume). Furthermore, such decisions will affect the conservation status of those taxa with restricted ranges (Stattersfield et al. 1998) because isolated taxa will, *ceteris paribus*, have smaller population sizes.

POPULATION BIOLOGY

We have organized the book around the concept of examining migratory connectivity throughout the annual cycle (Figure 1.1), emphasizing key facets that affect vital rates of productivity and survivorship. Our goal is to summarize the current state of knowledge regarding plover biology

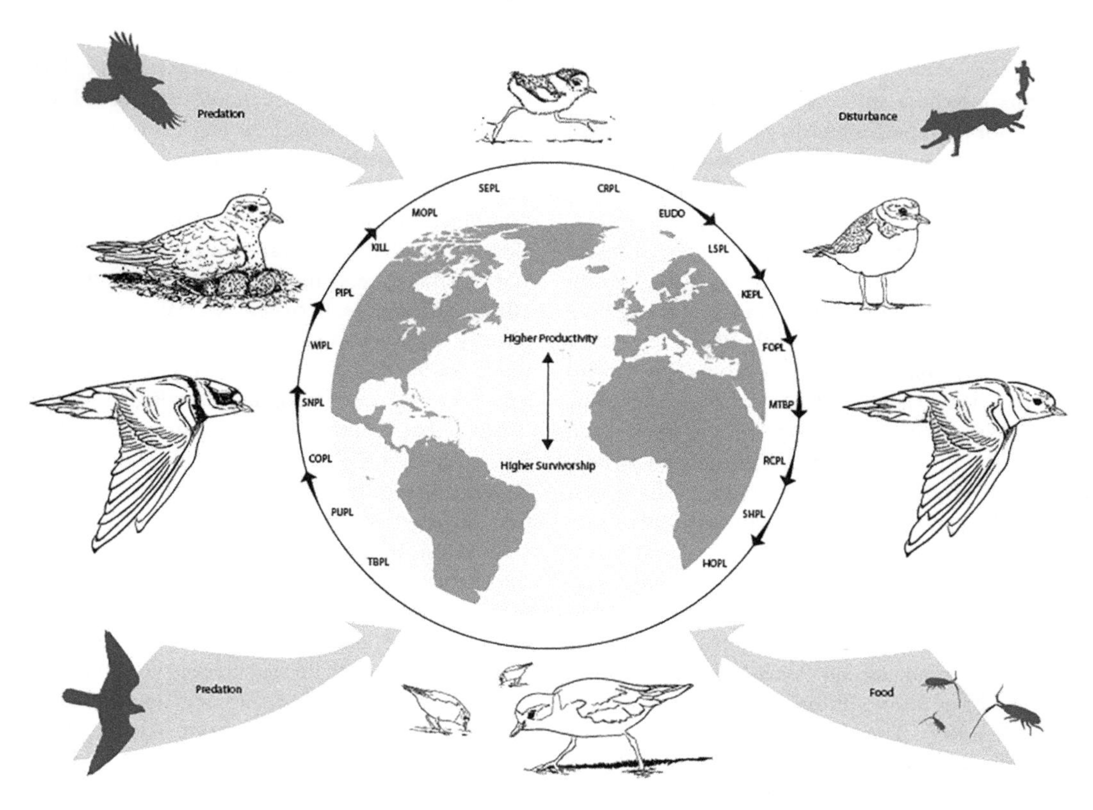

Figure 1.1. The annual cycle (e.g., molt, migration, breeding, nonbreeding) and geographic distribution of representative species of Charadrius plovers. Productivity (e.g., clutch size) is greater in migratory species breeding at northern latitudes, whereas survivorship is higher among sedentary taxa occupying southern regions. Natural and anthropogenic factors, represented by gray arrows, affect the demography of species and are concerns of management and conservation actions aimed at increasing population size.

in order to stimulate research and enhance conservation. At its foundation, effective conservation requires knowledge of a species' demography (Chapter Ten, this volume). Plovers are long-lived, especially nonmigratory taxa; annual survival estimates often vary around 75%, although in some years mortality can be high. In most species, individuals first breed as yearlings. The reproductive rates of plovers are low, with small clutches; replacement clutches are common, however, especially in populations breeding in temperate or tropical regions where nest loss is common and breeding seasons are long (Chapter Five, this volume). Overall, productivity of populations is highly variable. Breeding-site fidelity tends to be high, although philopatry is often low. Genetic diversity within populations is generally high, whereas genetic differences among populations are low (Chapter Two, this volume).

BREEDING BIOLOGY AND MATING SYSTEMS

Plovers exhibit diverse mating systems (Oring 1982, 1986, Székely et al. 2006; Chapter Four, this volume). Many species are socially monogamous with males and females sharing parental care of eggs and occasionally chicks (e.g., Killdeer); however, uniparental care of broods by males predominates. Some temperate-breeding species exhibit sequential polygamy, with females typically abandoning males to care for chicks while they pursue additional mating opportunities (e.g., Kentish Plover, Snowy Plover). In the Eurasian Dotterel, social polyandry is flexible and, in one location, assumes an exceptional form among birds, akin to lekking. In Scotland, females competed for access to mates at a predictable site, identified as an "arena." Females acquired a mate, laid a clutch to be tended by the male, after which females pursued second mates (Owens et al. 1994). The causes of this substantial variation in mating system among and within species have traditionally been explained as emerging from biased sex ratios at hatch, which are exacerbated by greater female mortality as chicks (Székely et al. 2006, Saunders and Cuthbert 2015), juveniles (Stenzel et al. 2007), and adults (Stenzel et al. 2011). However, Eberhart-Phillips et al. (2017) found hatching sex ratios (in plovers and other vertebrates) are parous and follow Fisher's expectation of equal sex allocation. The deviation from parity arises after hatch during sex-biased juvenile survival. When combined with protracted breeding seasons of many species, the results are operational sex ratios (*sensu* Emlen and Oring 1977) that favor female desertion and polygamy. Evidence indicates that extra-pair fertilizations are rare (Wallander et al. 2001) or uncommon (Küpper et al. 2003, Owens et al. 1995, Maher et al. 2017), which may arise owing to the risk of desertion by males investing more in parental care than females. A summary of field observations suggests that nest parasitism is opportunistic in most species (Yom-Tov 2001), perhaps owing to the short "window of opportunity" for effectively parasitizing a host given that clutch sizes are small.

In many ways, plovers are conservative breeders. Modal clutch sizes never exceed four eggs, although many species lay three or, less commonly, two eggs (Walters 1984, Colwell 2006; Chapter Five, this volume). Interspecific variation in clutch size and egg-laying intervals correlate with breeding latitude: boreal and arctic species tend to lay four eggs, with daily intervals between successive eggs in a clutch; temperate and tropical taxa lay smaller clutches with longer intervals between eggs (Colwell 2006). Reproductive failure is common but highly variable among years; females readily replace failed clutches if time permits in the breeding season. Individual clutches represent a large percentage of female body mass (59 ± 16%, $n = 11$), which suggests that the costs of reproductive effort may affect sex differences in adult mortality (Chapters Four and Ten, this volume; but see Colwell et al. 2013). Adults tend precocial, nidifugous chicks by brooding them, leading them from danger, and by distracting and occasionally attacking predators (Gochfeld 1984); young feed themselves (Walters 1984).

THE NONBREEDING SEASON AND MIGRATORY CONNECTIVITY

Understanding the location, movement, and population mixing (i.e., the migratory connectivity, Marra et al. 2006) of birds throughout the annual cycle is critical to successful conservation planning. Plover migration varies, as expected from the breadth of their breeding latitudes (Chapter Seven, this volume). Northern species (e.g., Common Ringed Plover, Semipalmated Plover) are migratory, whereas tropical taxa (e.g., Madagascar Plover, St. Helena Plover) are typically sedentary; temperate species (e.g., Snowy Plover) are partial migrants, whereas some species appear "irruptive" in their movements. This variation has generated contrasting views regarding the biogeographic origin of the clade and evolution of migration (i.e., Joseph et al. 1999, dos Remedios et al. 2015). Details of any species' migration ecology are, however, surprisingly poorly known. For instance, very little is published for most species on (1) timing of passage of the sexes, as well as individuals of different age classes, (2) staging behavior (e.g., duration of stay) and condition (e.g., mass gain) at individual sites, (3) concentrations that allow assessment of percentage of flyway populations at particular sites (Iwamura et al. 2016), and (4) their routes. Even less is known about where individual populations winter and spend much of the year, whether they mix with individuals from other populations, and where they go after the postbreeding season. Do they return to previous natal or breeding sites or do they follow individuals from other populations?

These data deficiencies (see Figure 12.1, Chapter Twelve, this volume) have consequences for conservation strategies applied to plovers, especially if knowledge derived from closely related species is misapplied. For example, the Ramsar Convention and Western Hemisphere Shorebird Reserve Network recognize the importance of a wetland complex based on the percentage of a flyway population that occurs at a site (e.g., San Francisco Bay, CA; Page et al. 1999). This approach was first articulated for shorebirds by Myers et al. (1987), who emphasized the importance of protecting especially vulnerable wetlands where a large percentage of a flyway population concentrated during migration. This strategy is well supported by empirical data (e.g., Red Knot, *Calidris canutus rufa*, Niles et al. 2008) and modeling for sandpipers

(Iwamura et al. 2016). However, plovers do not appear to routinely concentrate in large flocks in the manner typical of sandpipers, although there are large mixed flocks of post- and prebreeding plovers on sandflats throughout Prairie Canada, the Gulf of Mexico, and the Caribbean (SMH pers. obs.). Hence, strategies to protect plover populations during the nonbreeding season should differ from sandpipers.

For plovers, the nonbreeding season (i.e., not migrating or breeding) is an interval of varying duration when individuals spend the majority of their time foraging, roosting, and evading predators (Chapters Eight and Nine, this volume). At this time, most species occur in loose flocks, although there is one report of a massive aggregation associated with a concentrated food source during drought in northwestern Australia (Piersma and Hassell 2010). Social organization is occasionally characterized by territoriality (Colwell 2006; Chapter Eight, this volume). In many other species, the typical flock of feeding plovers consists of individuals widely separated from conspecifics.

The characteristic "run-stop-peck" foraging maneuver and reliance on visual cues for acquiring prey (Barbosa and Moreno 1999) predisposes plovers to social organization and dispersion patterns that differ profoundly from those of sandpipers. Compared with tactile-feeding sandpipers, these behaviors are also associated with a tendency to form smaller, less dense flocks (Barbosa 1995), in which individuals are less prone to interference while foraging (Goss-Custard 1976, Stinson 1980). The activity budgets of plovers wintering in the tropics vary greatly (McNeil et al. 1992; Chapter Eight, this volume). At least 14 species foraged day and night, with several species more likely to forage at night under conditions of bright moonlight. Whether this preponderance of nocturnal foraging is driven by a preference to avoid predation (e.g., by diurnal raptors) or arising from increased activity patterns of invertebrate prey remains unresolved, although Eberhart-Phillips (2016) found Killdeer foraging varied with the lunar cycle. In support of the latter, the intake rate of adult Kentish Plovers was 3.7 times greater at night than during the day (Kuwae 2007). While suggestive that benefits of increased intake rate and lower predation risk at night exceed costs, a formal assessment of trade-offs is lacking. From a conservation perspective, it is essential to determine the extent to which individuals can make up for lowered intake rates by feeding at night or day, especially if limited by human disturbance (Chapter Eleven, this volume).

Throughout the year, plovers occupy open, sparsely vegetated habitats (Chapter Six, this volume). The amount and quality of these habitats may have strong impacts on individual behaviors (e.g., vigilance), which affect fitness (e.g., survival, reproductive success) and translate into population growth (Chapter Eight, this volume). Since plovers feed year round using vision, it may predispose them to correlated behaviors at other times of the annual cycle. For example, the alert and evasive maneuvers of foraging plovers in response to an approaching raptor are similar to their tendency to quietly slip off a nest on the approach of a nest predator or humans (Chapter Eleven, this volume).

CONSERVATION

Plovers face many threats, almost all of which are exacerbated by human actions. Climate change (Chapter Three, this volume) is likely to reduce the amount and quality of plover habitat during all phases of the annual cycle (Gieder et al. 2014, Iwamura et al. 2016). In the arctic, the degree of warming is expected to be greater than temperate environs (Taylor et al. 2013), with consequences for the extent, structure, and composition of tundra flora that comprise breeding habitats of several species (Tape et al. 2006). In coastal areas, plovers contend with increased sea levels and altered water chemistry which threaten plover food resources, whereas inland populations face increased desertification and salinity of wetlands, which threaten the survivorship of adults and juveniles. Encroaching human populations in coastal and agricultural areas will increase human disturbance, with negative consequences for plover fitness. Native and invasive predators (Chapter Six, this volume) threaten several island endemics, whereas chronic disturbance to widespread or common plovers is also problematic for the maintenance of viable populations. Disturbance by humans (Chapter Eleven, this volume) poses a threat to plovers in a variety of habitats, especially when they breed in areas favored by humans for recreation or agriculture. The seriousness of each of these threats is rivaled by the challenges of solutions.

TABLE 1.2

Ecological and anthropogenic factors affecting conservation status of plovers.

IUCN status[a]	Predation	Habitat loss	Disturbance	Climate change	Examples
Endangered (3)	Compromises vital rates of evolutionarily naïve (insular) taxa	Restricted range exacerbates habitat loss and lowers carrying capacity	Human activity lowers vital rates in habitat valued for recreation	Sea level rise restricts extent of habitat	S. Red-breasted Dotterel Shore Plover St. Helena Plover
Threatened (7)	Abundant synanthropic predators compromise vital rates, especially productivity	Agricultural expansion, altered river hydrology, and coastal reclamation negatively affect habitat suitability	Human populations concentrated in coastal habitats	Habitat quality and extent limited by sea level rise, and warming and drying of interior habitats	N. Red-breasted Dotterel Piping Plover Snowy Plover Mountain Plover Malay Plover Javan Plover Chestnut-banded Plover
Vulnerable (3)	Native and introduced predators compromise vital rates	Altered river hydrology affects habitat quality	Human recreational activity affects vital rates	Sea level rise limits extent of habitat	Hooded Plover Wrybill Black-banded (Madagascar) Plover
Least concern (26)	Variable effects on productivity and survival	Remote and large breeding range diminishes impact of habitat loss	Human activity increased in areas of high human activity	Limited migration bottlenecks lessen effect of sea level rise and habitat loss during nonbreeding season	Sand plovers Common Ringed Plover Semipalmated Plover Kittlitz's Plover

[a] Number of species listed in the category according to IUCN Red-list database (15 December 2017).

To gain a broader perspective on the conservation status of the Charadriinae, we used the International Union for Conservation of Nature (IUCN) database, supplemented with information derived from North American conservation plans (Brown et al. 2001, Partners in Flight online) to categorize plovers by conservation status (Table 1.2). We then discuss the conservation threats posed for plovers throughout the annual cycle (Figure 1.1). The 40 species of plover lie along continua describing population size and geographic range (Figure 12.2, Chapter Twelve, this volume).

Endangered

A common feature of the three species in this category is their restricted island range. Curiously, IUCN recently changed the status of the St. Helena Plover to Vulnerable based on a small increase in population size. We question this change given the species' small population and restricted range, which will require constant management to alleviate the threat of introduced predators and maintain the pasture habitats it favors. In New Zealand, introduced predators pose similar conservation challenges for the Shore Plover, and Southern and Northern Red-breasted dotterels.

Threatened/Vulnerable

Most of the ten species in these (combined) IUCN categories have declining populations estimated at less than 10,000 and restricted breeding ranges. At some point in the annual cycle, each of the ten species occupy temperate or tropical habitats, which places them in conflict with humans in coastal and inland habitats that will continue to be compromised by direct loss of habitat and its degradation via human disturbance and climate change.

Least Concern

IUCN places most (n = 26; 65%) species in this category. Not surprisingly, larger population sizes (>100,000) and broad geographic ranges characterize many taxa in this group; three northern species with large populations occupy expansive breeding ranges in tundra. Further, Palearctic taxa such as Kentish and Common Ringed plovers, and Greater and Lesser sand-plovers breed across extensive regions, whereas their nonbreeding ranges are often coastal. IUCN population trend data suggest declines for ten of these species, with many (n = 12) species of unknown status.

SUMMARY

As this chapter and the following contributions detail, plovers provide a model system for the comparative study of behavior and ecology. This stems from their accessibility and ease of study, and inherent variation in life history traits and behavioral strategies. While seemingly similar, plovers provide a global perspective on a broad spectrum of conservation challenges facing birds in general and many other life forms. We hope readers of this compendium are stimulated to pursue additional study of a fascinating group of birds, with the goal of improving conservation measures for the benefit of plovers, and other species, including humankind.

ACKNOWLEDGMENTS

We thank Luke Eberhart-Phillips and Carrie Phillips for reviewing this chapter. Funding was partially provided by the USGS Forest and Rangeland Ecosystem Science Center. Any use of trade, product, or firm names is for descriptive purposes only and does not imply endorsement by the U.S. Government.

LITERATURE CITED

Argüelles-Ticó, A., C. Küpper, R. N. Kelsh, A. Kostalányi, T. Székely, and R. E. van Dijk. 2016. Geographic variation in breeding system and environment predicts melanin-based plumage ornamentation of male and female Kentish plovers. *Behavioral Ecology and Sociobiology* 70:49–60.

Baker, A. J., S. L. Pereira, and T. A. Paton. 2007. Phylogenetic relationships and divergence times of Charadriiformes genera: Multigene evidence for the Cretaceous origin of at least 15 clades of shorebirds. *Biology Letters* 3:205–209.

Barbosa, A. 1995. Foraging strategies and their influence on scanning and flocking behavior of waders. *Journal of Avian Biology* 26:182–186.

Barbosa, A., and E. Moreno. 1999. Evolution of foraging strategies in shorebirds: An ecomorphological approach. *Auk* 116:712–725.

Barth, J. M. I., M. Matschiner, and B. C. Robertson. 2013. Phylogenetic position and subspecies divergence of the endangered New Zealand Dotterel (*Charadrius obscurus*). *PLoS One* 8:e78068.

Brown, S., C. Hickey, B. Harrington, and R. Gill (editors). 2001. *United States Shorebird Conservation Plan*. Manomet Center for Conservation Science, Manomet, MA.

Clements, J. F., T. S. Schulenberg, M. J. Iliff, D. Roberson, T. A. Fredericks, B. L. Sullivan, and C. L. Wood. 2014. The eBird/Clements checklist of birds of the world: Version 6.9.

Colwell, M. A. 2006. Egg-laying intervals in shorebirds. *Wader Study Group Bulletin* 111:50–59.

Colwell, M. A. 2010. *Shorebird Ecology, Conservation, and Management*. University of California Press, Berkeley, CA.

Colwell, M. A., W. J. Pearson, L. J. Eberhart-Phillips, and S. J. Dinsmore. 2013. Apparent survival of Snowy Plovers (*Charadrius nivosus*) varies with reproductive effort and year and between sexes. *Auk* 130:725–732.

Cramp, S. and K. E. L. Simmons (editors). 1981. *The Birds of the Western Palearctic*, vol. 3. Oxford University Press, Oxford, UK.

DeJoannis, A. 2016. Molt in individuals: A description of prealternate molt phenology in a population of Snowy Plovers in Humboldt County, California. M.Sc. thesis, Humboldt State University, Arcata, CA.

Del Hoyo, J. et al. 2016. *Handbook of the Birds of the World Alive*. Lynx Edicions, Barcelona.

dos Remedios, N. D., P. L. M. Lee, T. Burke, T. Székely, and C. Küpper. 2015. North or south? Phylogenetic and biogeographic origins of a globally distributed clade. *Molecular Phylogenetics and Evolution* 89:151–159.

Eberhart-Phillips, L. J. 2016. Dancing in the moonlight: Evidence that Killdeer foraging behaviour varies with the lunar cycle. *Journal of Ornithology* 158:253–262.

Eberhart-Phillips, L. J., C. Küpper, T. E. X. Miller, M. Cruz-López, K. Maher, N. dos Remedios, M. A. Stoffel, J. I. Hoffman, O. Krüger, and T. Székely. 2017. Sex-specific early survival drives adult sex ratio bias in snowy plovers and impacts mating system and population growth. *Proceedings of the National Academy of Sciences USA* 114:E5474–E5481.

Ekanayake, K. B., M. A. Weston, D. G. Nimmo, G. S. Maguire, J. A. Endler, and C. Küpper. 2015. The bright incubate at night: Sexual dichromatism and adaptive incubation division in an open-nesting shorebird. *Proceedings of the Royal Society B*. 282:20143026.

Emlen, S. T., and L. W. Oring. 1977. Ecology, sexual selection, and the evolution of mating systems. *Science* 197:215–223.

Feduccia, A. 1999. *The Origin and Evolution of Birds*. Yale University Press, New Haven, CT.

Gieder, K. D., S. M. Karpanty, J. D. Fraser, D. H. Catlin, B. T. Gutierrez, N. G. Plant, A. M. Turecek, and E. R. Thieler. 2014. A Bayesian network approach to predicting nest presence of the federally threatened piping plover (*Charadrius melodus*) using barrier island features. *Ecological Modelling* 276:38–50.

Gochfeld, M. 1984. Antipredator behavior: Aggressive and distraction displays of shorebirds, pp. 289–377 in J. Burger and B. L. Olla (editors), *Shorebirds: Breeding Behavior and Populations*. Plenum Press, New York.

Goss-Custard, J. D. 1976. Variation in dispersion of Redshank *Tringa totanus* on their wintering grounds. *Ibis* 118:257–263.

Graul, W. D. 1973. Possible functions of head and breast markings in Charadriinae. *Wilson Bulletin* 85:60–70.

Haig, S.M., S.P. Murphy, J.D. Matthews, I. Arismendi, and M. Safeeq. 2019. Climate-altered wetlands challenge waterbird use and migratory connectivity in arid landscapes. Scientific Reports 9: 4666.

Hayman, P., J. Marchant, and T. Prater. 1986. *Shorebirds: An Identification Guide to Waders of the World*. Houghton Mifflin Co., New York, NY.

Howell, S. N. G. 2010. *Molt of North American Birds*. Houghton Mifflin Harcourt Publishing Co., New York.

Iwamura, T., H. P. Possingham, I. Chadès, C. Minton, N. J. Murray, D. I. Rogers, E. A. Treml, and R. A. Fuller. 2016. Migratory connectivity magnifies the consequences of habitat loss from sea-level rise for shorebird populations. *Proceedings Royal Society B* 280:20130325.

Jetz, W., G. H. Thomas, J. B. Joy, K. Hartmann and A. O. Mooers. 2012. The global diversity of birds. *Nature* 491:444–448.

Johnsgard, P. A. 1981. *The Plovers, Sandpipers, and Snipes of the World*. University of Nebraska Press, Lincoln, NE.

Joseph, L., E. P. Lessa, and L. Christidis. 1999. Phylogeny and biogeography in the evolution of migration: Shorebirds of the *Charadrius* complex. *Journal of Biogeography* 26:329–342.

Küpper, C., J. Kis, A. Kosztolányi, T. Székely, I. C. Cuthill, and D. Blomqvist. 2003. Genetic mating system and timing of extra-pair fertilizations in the Kentish plover. *Behavioral Ecology and Sociobiology* 57:32–39.

Küpper, C., J. Augustin, A. Kosztolányi, T. Burke, J. Figuerola, and T. Székely. 2009. Kentish versus Snowy plover: Phenotypic and genetic analyses of *Charadrius alexandrinus* reveal divergence of Eurasian and American subspecies. *Auk* 126:839–852.

Kuwae, T. 2007. Diurnal and nocturnal feeding rate in Kentish plovers *Charadrius alexandrinus* on an intertidal flat as recorded by telescopic video systems. *Marine Biology* 151:663–673.

Linnaeus, C. 1758. *Systema naturae per regna tria naturae*. 10th ed. Rev. 2 vol. L. Salmii, Homiiae.

Livezey, B. C. 2010. Phylogenetics of modern shorebirds (Charadriiformes) based on phenotypic evidence: Analysis and discussion. *Zoological Journal of the Linnean Society* 160:567–618.

Maher, K. H., L. J. Eberhart-Phillips, A. Kosztolányi, N. D. Remedios, M. C. Carmona-Isunza, M. Cruz-López, S. Zefania, J. S. Clair, M. Alrashidi, M. A. Weston, M. A. Serrano-Meneses, O. Krüger, J. I. Hoffman, T. Székely, T. Burke, and C. Küpper. 2017. High fidelity: Extra-pair fertilisations in eight Charadrius plover species are not associated with parental relatedness or social mating system. *Journal of Avian Biology* 48:1–11.

Marra, P. P., D. R. Norris, S. M. Haig, M. Webster, and J. A. Royle. 2006. *Migratory Connectivity*. Cambridge University Press, Cambridge, UK.

Mayr, G. 2009. *Paleogene Fossil Birds*. Springer-Verlag, Heidelberg, Germany.

McNeil, R., P. Drapeau, and J. D. Goss-Custard. 1992. The occurrence and adaptive significance of nocturnal habits in waterfowl. *Biological Review* 67:381–419.

Myers, J. P., R. I. G. Morrison, P. Z. Antas, B. A. Harrington, T. E. Lovejoy, M. Sallaberry, S. E. Senner, and A. Tarak. 1987. Conservation strategy for migratory species. *American Scientist* 75:19–26.

Niles, L. J., H. P. Sitters, A. D. Dey, P. W. Atkinson, A. J. Baker, K. A. Bennett, R. Carmona, K. E. Clark, N. A. Clark, C. Espoz, P. M. González, B. A. Harrington, D. E. Hernández, K. S. Kalasz, R. G. Lathrop, R. N. Matus, C. D. T. Minton, R. I. G. Morrison, M. K. Peck, W. Pitts, R. A. Robinson, and I. L. Serrano. 2008. Status of the Red Knot (*Calidris canutus rufa*) in the western Hemisphere. *Studies in Avian Biology*, vol. 8. Cooper Ornithological Society, Allen Press, Lawrence, KS.

Nol, E. 1986. Incubation period and foraging technique in shorebirds. *American Naturalist* 128:115–119.

Oring, L. W. 1982. Avian mating systems, pp. 1–92 in D. S. Farner, J. R. King, and K. C. Parkes (editors), *Avian Biology*, vol. 6. Academic Press, New York.

Oring, L. W. 1986. Avian polyandry, pp. 309–351 in R. F. Johnston (editor), *Current Ornithology*, vol. 3. Plenum Press, New York, NY.

Owens, I. P. F., T. Burke, and D. B. A. Thompson. 1994. Extraordinary sex roles in the Eurasian Dotterel: Female mating arenas, female-female competition, and female mate choice. *American Naturalist* 144:76–100.

Owens, I. P. F., A. Dixon, T. Burke, and D. B. A. Thompson. 1995. Strategic paternity assurance in the sex-role reversed Eurasian Dotterel (*Charadrius morinellus*): Behavioral and genetic evidence. *Behavioral Ecology* 6:14–21.

Page, G. W., L. E. Stenzel, and J. E. Kjelmyr. 1999. Overview of shorebird abundance and distribution in wetlands of the Pacific coast of the contiguous United States. *Condor* 101:461–471.

Page, G. W., L. E. Stenzel, J. S. Warriner, J. C. Warriner, and P. W. Paton. 2009. Snowy Plover (Charadrius nivosus). In A. Poole and F. Gill (editors), *The Birds of North America, No. 154*. The Academy of Natural Sciences, Philadelphia, PA.

Piersma, T., and C. Hassell. 2010. Record numbers of grasshopper-eating waders (Oriental Pratincole, Oriental Plover, Little Curlew) on coastal west-Kimberley grasslands of NW Australia in mid-February 2010. *Wader Study Group Bulletin* 117:103–108.

Plissner, J. H., L. W. Oring, and S. M. Haig. 2000. Space use of Killdeer at a Great Basin breeding area. *Journal Wildlife Management* 64:421–429.

Prater, A. J., J. H. Marchant, and J. Vuorinen. 1977. *Guide to the Identification and Ageing of Holarctic Waders*. British Trust for Ornithology, Beech Grove, UK.

Rogers, K., D. I. Rogers, and M. A. Weston. 2014. Prolonged and flexible primary moult overlaps extensively with breeding in beach-nesting Hooded Plovers *Thinornis rubricollis*. *Ibis* 156:840–849.

Rubega, M. A., and J. A. Robinson. 1998. Water salinization and shorebirds: Emerging issues. *International Wader Studies* 9:45–54.

Sanzenbacher, P. M., and S. M. Haig. 2001. Killdeer population trends in North America. *Journal of Field Ornithology* 72:160–169.

Saunders, S. P., and F. J. Cuthbert. 2015. Chick mortality leads to male-biased sex ratios in endangered Great Lakes Piping Plovers. *Journal of Field Ornithology* 86:103–114.

Staaland, H. 1967. Anatomical and physiological adaptations of the nasal glands in Charadriiformes birds. *Comparative Biochemistry and Physiology* 23:933–944.

Stattersfield, A. J., M. J. Crosby, A. J. Long, and D. C. Wege. 1998. *Endemic Bird Areas of the World*. BirdLife Conservation Series No. 7. BirdLife International, Cambridge, UK.

Stinson, C. H. 1980. Flocking and predator avoidance: Models of flocking and observations on the spatial dispersion of foraging wintering shorebirds (Charadrii). *Oikos* 34:35–43.

Stenzel, L. E., G. W. Page, J. C. Warriner, J. S. Warriner, D. E. George, C. R. Eyster, B. A. Ramer, and K. K. Neuman. 2007. Survival and natal dispersal of juvenile Snowy Plovers (*Charadrius alexandrinus*) in central coastal California. *Auk* 124:1023–136.

Stenzel, L. E., G. W. Page, J. C. Warriner, J. S. Warriner, K. K. Neuman, D. E. George, C. R. Eyster, and F. C. Bidstrup. 2011. Male-biased sex ratio, survival, mating opportunity and annual productivity in the Snowy Plover *Charadrius alexandrinus*. *Ibis* 153:312–322.

Székely, T., G. Thomas, and I. C. Cuthill. 2006. Sexual conflict, ecology, and breeding systems in shorebirds. *Bioscience* 56:801–808.

Tape, K., M. Sturm, and C. Racine. 2006. The evidence for shrub expansion in northern Alaska and the Pan-Arctic. *Global Change Biology* 12:686–702.

Taylor, P. C., M. Cai, A. Hu, J. Meehl, W. Washington, and G. J. Zhang. 2013. A decomposition of feedback contributions to polar warming amplification. *Journal of Climate* 26:7023–7043.

Tjorve, K. M., L. G. Underhill, and G. H. Visser. 2008. The energetic implications of precocial development for three shorebird species breeding in a warm environment. *Ibis* 150:125–138.

Van Tuinen, M., D. Waterhouse, and G. J. Dyke. 2004. Avian molecular systematics on the rebound: A fresh look at modern shorebird phylogeny. *Journal of Avian Biology* 35:191–194.

Wallander, J., D. Blomqvist, and J. T. Lifjeld. 2001. Genetic and social monogamy – does it occur without mate guarding in the ringed plover? *Ethology* 107:561–572.

Walters, J. R. 1984. The evolution of parental behavior and clutch size in shorebirds, pp. 243–287 in J. Burger and B. L. Olla (editors), *Shorebirds: Breeding Behavior and Populations*. Plenum Press, New York.

Wolf, S. E. 1912. *The Greek Romances in Elizabethan Prose Fiction*. Columbia University Press, New York.

Yom-Tov, Y. 2001. An updated list and some comments on the occurrence of intraspecific nest parasitism in birds. *Ibis* 143:133–143.

Zusi, R. L. 1984. *A Functional and Evolutionary Analysis of Rhynchokinesis in Birds*. Smithsonian Contributions to Zoology No. 395. Smithsonian Institution Press, Washington, DC.

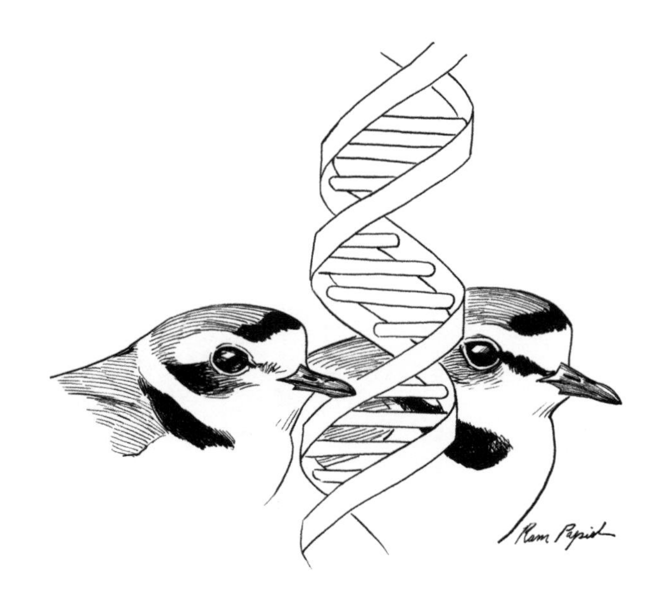
Ram Papish

CHAPTER TWO

Defining Species and Populations*

MOLECULAR GENETIC STUDIES IN PLOVERS

Clemens Küpper and Natalie dos Remedios

Abstract. Molecular genetic studies shed light on the processes and constraints that led to present species distributions. DNA sequences have become instrumental in constructing a modern phylogeny of plovers with subsequent comparative analyses revealing that plover ancestors evolved in the Northern hemisphere. Detailed molecular investigations have helped to clarify taxonomic relationships and species delimitations of controversial plover taxa. The results of molecular studies have practical implications for conservation management, helping to delineate species and populations. For example, an analysis of molecular characters identified substantial differences between the threatened American Snowy Plover (*Charadrius nivosus*) and the Eurasian Kentish Plover (*C. alexandrinus*) leading to restoration of full species status for Snowy plovers. Yet taxonomic analyses are ongoing and phylogenetic studies constantly bring up new questions and challenges. Whether *Charadrius* plovers form a monophyletic clade or some species are more closely related to lapwings (*Vanellus* spp.) is currently under debate. Some species such as Kentish/White-faced Plover (*C. alexandrinus*) and Greater Sand-Plover (*C. leschenaultii*) may harbor further cryptic species and are awaiting a more detailed investigation. At the population level, genetic analyses show patterns of high gene flow and a general lack of genetic structure over large geographic distances within many plover species. The extent of gene flow and lack of population differentiation across entire continental distributions shown by some plover species is remarkable even in comparison with other highly dispersive birds. Variation in the patterns of gene flow among plover species is explained by four characteristics: (1) habitat specialization, (2) mating behavior and breeding dispersal, (3) colonization of oceanic islands, and (4) past population fluctuations due to climatic changes. Surprisingly, the role of seasonal migration in maintaining gene flow appears to be of less importance in plovers. Methodological improvements, such as high throughput sequencing, offer great promise for the future. New molecular studies are likely to resolve many open questions on the past and present status of plover populations that are of interest to conservationists, ecologists, and evolutionary biologists alike.

Keywords: conservation genetics, gene flow, genetic structure, phylogeny, phylogeography, population differentiation, population structure.

* Clemens Küpper and Natalie dos Remedios. 2019. Defining Species and Populations. Pp. 17–43 in M. A. Colwell and S. M. Haig (editors). The Population Ecology and Conservation of *Charadrius* Plovers Studies in Avian Biology (no. 52), CRC Press, Boca Raton, FL.

Genetics and the use of molecular markers are now at the very heart of recent advances in our understanding of biogeography, evolution, and ecology, as well as taxonomic classification and conservation law. For plovers, molecular approaches have been chiefly implemented to investigate relationships between species and populations, phylogeography, and the genetic health and diversity of populations, with a view toward improving conservation strategies (Haig et al. 2011). Assessments of genetic health may include identifying changes in genetic diversity, estimating past and present population size, and assessing demographic parameters such as sex ratios, reproductive success, and survivorship. Surprisingly, some of the most theoretical molecular studies often have the most applied and/or legal implications (e.g., estimates of effective population size; Haig and D'Elia 2010). Genetic studies are integral to modern taxonomic classification of plover species and subspecies, which were once largely based on analysis of morphological characters (dos Remedios et al. 2015, Livezey 2010). Hence, understanding the fundamentals of taxonomy and population genetics is imperative for everyone from behavioral ecologists to law enforcement agents.

When embarking on a genetic study the first question is, what genetic markers or loci should be used. DNA regions differ in their variability and, depending on the research question, some markers will be more appropriate than others (Box 2.1). Most studies of plovers to date have focused on a relatively small number of genetic markers. Using more markers provides higher confidence in the results and reduces the stochastic effects that come with single markers. Theoretically, there should be no shortage of genetic markers. Although smaller than most other vertebrate genomes, a bird genome still contains, on average, 1.3 billion base pairs and ~26,000 genes (Gregory 2005, Warren et al. 2017). The challenge is that it remains expensive to decipher and to compare entire genomes. The ongoing

BOX 2.1 Commonly Used DNA Markers in Ecological and Evolutionary Research

Genetic markers are regions of DNA variation that provide information on genetic differentiation and/or diversity at the individual, population, or species levels. Choosing the right genetic markers is the first requirement in any molecular analysis. Variation at a locus is introduced by mutations that accumulate over time. With reproductive isolation between organisms, these mutations ultimately result in divergent evolution of DNA sequences. Loci differ in their variability with some DNA regions being more constrained (conserved) than others. Phylogenetic analyses that aim to shed light on relationships among species require relatively slowly evolving (highly conserved) markers and introns (non-protein-coding regions) of conserved genes. In contrast, analyses of closely related populations or individuals require faster evolving markers with high mutation rates such as microsatellites or single-nucleotide polymorphisms (SNPs).

Markers must either be developed *de novo* for the species studied, or existing markers previously developed in related species may be utilized after optimization (Sunnucks 2000). Ideally, all markers should be selectively neutral, that is, particular mutations in these regions confer no evolutionary advantage or disadvantage and hence are not under any form of selection. Neutrality is a requirement of most phylogenetic and population genetic models. Individual markers are highly stochastic in this regard and therefore the most reliable results are obtained through analysis of multiple loci. The number of markers analyzed is usually constrained by the time and resources available, and often a trade-off exists between the number of markers and number of samples to be analyzed.

The organization of bird genomes is generally highly similar among species (Dawson et al. 2007). This has enabled the identification of a number of markers for analyzing genetic variation across many species, including shorebirds (Küpper et al. 2008). However, even these cross-species markers require thorough testing and optimization before use, as levels of DNA variation will differ between species and loci.

(Continued)

BOX 2.1 (*Continued*) Commonly Used DNA Markers in Ecological and Evolutionary Research

Below we outline the most important genetic markers currently used in population genetic and phylogenetic studies.

Conserved Nuclear Loci

Regions of protein-coding genes (usually introns) or noncoding ultraconserved elements that are highly similar across multiple species. These markers have a relatively low mutation rate and are very popular for phylogenetic analyses and taxonomic studies for evaluating relatedness at the species and higher-order levels.

Mitochondrial DNA

Mitochondrial genomes originated early in the history of multicellular life on earth and derive from the genomes of bacteria that were once engulfed by ancestral eukaryotic cells. These circular DNA molecules contain only about 40 genes and have a length of around 17,000 bp. Mitochondrial DNA is usually inherited through the maternal line and has a high mutation rate suitable for studying variation at the population or species levels. Once the workhorse of phylogeographic analyses, its popularity is now waning, though it is still commonly used for species identification through DNA barcoding.

Microsatellites

Also called short tandem repeats, these are hypervariable stretches of DNA with a repeated short 2–4 bp motif such as GT or AGTC. Alleles differ in their number of motif repeats due to errors in DNA replication introduced by slippage of the DNA polymerase enzyme in these regions. Their hypervariability makes these markers suitable for identification of individuals and studies at the population level.

Single-Nucleotide Polymorphisms (SNPs)

SNPs are single variable base positions in DNA sequence. Recent technological advancements have enabled rapid, automated analysis of thousands of SNPs located across the whole genome. Most commonly, SNPs are bi-allelic, each allele featuring one of two possible nucleotide bases, often with a more common (major) and a less common (minor) allele. The variability of SNPs is lower than that of microsatellites such that about ten times more markers are needed to achieve the same resolution. SNPs are increasingly popular and are replacing microsatellite markers more and more in population and individual identification studies. One frequently used method for genotyping SNPs in non-model organisms is RAD-Sequencing (Baird et al. 2008, Davey et al. 2011), which does not require a separate marker development step.

Transposons

Transposons are mobile DNA elements that can cut and paste (or copy and paste) themselves into different positions in the genome. They are noncoding sequences and usually considered as genomic parasites or selfish DNA, although some transposons may play an important role in genome function and evolution. Several different classes of transposons exist, and all classes combined make up a considerable amount of the entire genome (between 4% and 10% in birds, Kapusta and Suh 2017). As the probability of repeated loss/gain of transposons at orthologous sites is low, the presence or absence of transposons can be used as characters in phylogenetic studies providing a binary matrix that can be efficiently analyzed (Suh et al. 2015).

initiative B10K (http://b10k.genomics.cn/) aims to provide genome maps for all extant bird species. These maps will provide information on where specific genes (or genetic markers) are located and, in this way, aid marker discovery and development. B10K will greatly enhance the possibilities of genetic surveys and analyses because the laborious stage of finding enough useful markers for many study species will become redundant.

Even without whole genome information, molecular studies have already made significant contributions to solving questions concerning plovers. For example, phylogenetic studies have provided new insights on the geographic origin of the group. The Charadrius plovers are currently distributed across all continents except Antarctica, with an approximately equal number of species occupying the Northern and Southern hemispheres. Early researchers debated the hemispheric origin based on the identification of phenotypic traits that were considered more likely to represent the ancestral condition. For example, the black breast bands, black lores (Graul 1973), and four egg clutches (Maclean 1972) of several northern species were suggested to be more primitive, with reduced plumage markings and smaller clutch sizes evolving after colonization of the Southern hemisphere. Alternatively, the distribution of species at the tips of southern landmasses were considered as evidence for an Antarctic origin and plovers were thought to have inhabited the continent at a time when it was not ice-covered (Vaughan 1980).

An early molecular analysis pointed toward a Southern hemisphere origin of plovers (Joseph et al. 1999). However, a recent analysis with more thorough taxon sampling identified the Northern hemisphere as the ancestral home of the genus (Figure 2.1). The latter analysis, based on multiple nuclear and mitochondrial genes, suggested that the earliest plovers radiated out of the Palearctic or Nearctic realms, reaching the Southern hemisphere through multiple independent colonizations (dos Remedios et al. 2015).

PHYLOGENY

Analyses of morphological and other phenotypic traits were once the most common means of taxonomic classification and were used for determining evolutionary relationships among species. Today, phylogenies are predominantly based on genetic characters. DNA sequences are widely seen as the most reliable source for building phylogenetic trees and determining evolutionary relationships. Phenotypic traits can be unreliable for two principal reasons. First, phenotypic traits evolve under the influence of natural and sexual selection. Different lineages that face similar environmental conditions are more likely to evolve similar traits by convergent or parallel evolution (Nixon and Carpenter 2012). In contrast, much of the genome is evolving neutrally, meaning that DNA sequences are less likely to be influenced by selective pressures and genetic variation is more likely to be shaped by phylogenetic history. Second, changes in phenotypic characters can occur very rapidly, especially when species first diverge and organisms face new environmental challenges; for example, when they colonize new areas and/or niches. In contrast, neutral DNA sequence changes tend to occur at a more constant rate over time, making genetic data more reliable for determining the time since two groups last shared a common ancestor (Hillis 1987, Wake et al. 2011).

Position of Charadriiformes among Other Bird Orders

Determining the taxonomic position of the order Charadriiformes (shorebirds, gulls, alcids, and allies) has been described as "chronically challenging" even when full genome sequences are considered (Jarvis et al. 2014). Initial attempts to date the origin of the Charadriiformes (Baker et al. 2007) are now considered erroneous due to unreliable fossil evidence used in molecular clock calibrations (Mayr 2011). Phylogenetic analyses usually involve a trade-off between the number of taxa versus the number of characters involved. Translated into genetic analyses, the trade-off is between the number of loci or base pairs and the number of taxa that can be sampled. The debate among phylogeneticists over which approach is more informative is ongoing. Including 9,993 avian species but with limited genetic information, Jetz et al. (2012) suggested that the Charadriiform order is a distinct lineage, with a common ancestor that diverged from other extant groups prior to the Cretaceous-Paleogene (K-Pg) mass extinction event, which occurred 66 million years ago. A more recent estimate based on 259 ultraconserved elements (Box 2.1; Prum et al. 2015) and 198 species of living birds placed the

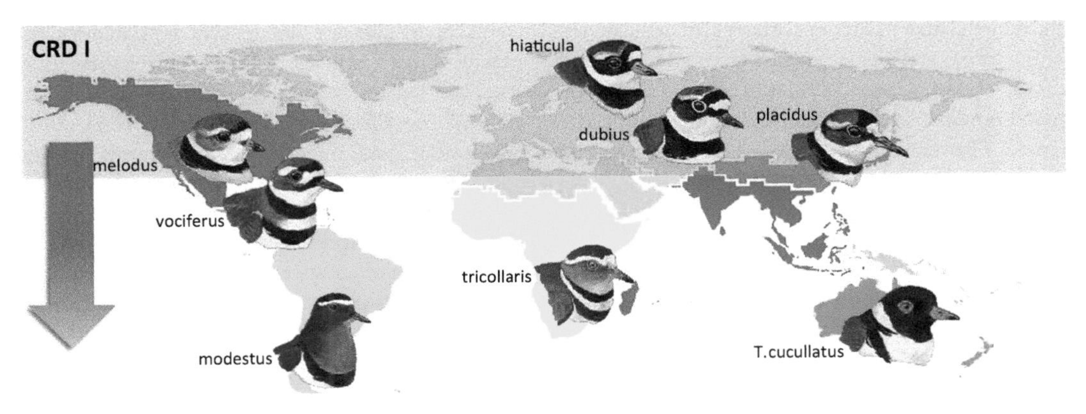

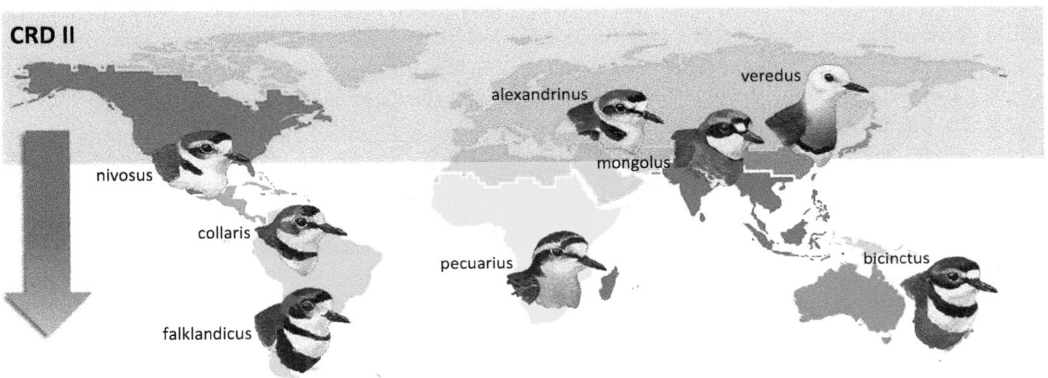

Figure 2.1. Biogeographic origins of the genus *Charadrius*. Representative species that highlight the geographic and phenotypic diversity of plovers are shown. Colored areas on the map represent different biogeographic realms. The gray-shaded regions on the Northern hemisphere represent the ancestral origins of both *CRD I* and *CRD II* clades. Colonization of the Southern hemisphere took place more recently in the evolutionary history of the genus, following southwards dispersal.

Charadriiformes alongside all other extant water birds, Aequorlitornithes, within a single clade. This study identified the Mirandornithes (flamingos and grebes) as the closest sister taxon of the Charadriiformes with a common ancestor that existed during the Paleocene (66–56 million years ago). Other clades within the extant water birds included Eurypgimorphae (kagu, sunbittern, and tropicbirds) and Aequornithes (penguins, herons, storks, and allies). Analyzing whole genome sequences from 48 birds representing all orders, Jarvis et al. (2014) also placed the Charadriiform origin during the Paleocene period but identified the Gruiformes (cranes, crakes, rails, and allies) as the closest sister group. A similar conclusion was derived from an analysis of mobile DNA elements called transposons (Box 2.1; Suh et al. 2015). As is apparent, the field of avian phylogenetics is currently very dynamic, and several phylogenetic hypotheses are under debate. One major challenge for the field is that, following the K-Pg mass extinction event, rapid species divergence occurred over a relatively short time. Such rapid radiation makes it extremely difficult to resolve the precise order of divergence between taxa (Degnan and Rosenberg 2006, Corl and Ellegren 2013, Suh et al. 2015). Consequently, the debate will probably continue for some time. Until a conclusion is reached, it is safe to say that the order Charadriiformes belongs to the Neoaves superorder, a group that includes nearly all extant bird species (Prum et al. 2015).

Relationships within Charadriiformes

Among the 385 extant Charadriiform species are 215 shorebirds (Jetz et al. 2012, Gill and Donsker 2015). The taxonomic division between shorebirds and their non-shorebird allies—the gulls, terns, and alcids—stems from the 19th century, when Gadow (1892) proposed two clades based on analysis of 40 characters of "organic systems." This classification has stuck despite recent phylogenetic updates such that the shorebirds are now considered a paraphyletic group, meaning that not all descendants of their last common ancestor are included. Within shorebirds, three main clades (or suborders) exist: Charadrii (plovers, oystercatchers, avocets, and allies), Scolopaci (sandpipers, phalaropes, jacanas, and allies), and Lari (pratincoles, coursers, and allies) (van Tuinen et al. 2004, Baker et al. 2007, Fain and Houde 2007). The full phylogenetic clade Lari includes both "shorebird" and "non-shorebird" taxa such as gulls and terns.

Within the Charadrii, plovers belong to the family Charadriidae, which includes the genera Charadrius (plovers), *Vanellus* (lapwings), and *Pluvialis* (Golden- and Black-bellied plovers), as well as several other genera, each with just one or two members (namely *Anarhynchus*, *Elseyornis*, *Oreopholus*, *Peltohyas*, *Phegornis*, and *Thinornis*; see the **Taxonomic Clarifications** section, this chapter). The classification of all plover genera within a monophyletic clade follows traditional theories, and has been confirmed by DNA–DNA hybridization analysis and phylogenetic analysis of eight gene regions (Sibley and Ahlquist 1990, Baker et al. 2012). Additionally, this analysis settled the debate over the erroneous placement of *Pluvialis* plovers outside the family Charadriidae, which had been suggested following analyses of a smaller number of gene regions (Baker et al. 2007, Fain and Houde 2007). It is worth noting that the terms "plover," "dotterel," and "lapwing" were coined prior to modern taxonomic synthesis. As such, species including the Crab Plover (*Dromas ardeola*), Egyptian Plover (*Pluvianus aegyptius*), and Magellanic Plover (*Pluvianellus socialis*) are not true plovers and belong outside the family Charadriidae, within Dromadidae, Pluvianidae, and Pluvianellidae, respectively.

Molecular clock dating offers the potential to better understand the timing of diversification within clades. This has proven useful in dating the origins of diverse animal taxa including birds (Ho and Duchene 2014, Lavinia et al. 2016). However, precise date estimates require reliable fossils from verified ancestral members of the study species, as well as accurate carbon dating (Parham et al. 2012). Such data are not readily available for the extant plover species. Use of unreliable fossil data has led to erroneous estimates for the origins of the Charadriiformes (Baker et al. 2007, Mayr 2011), and, due to this controversy, the best estimates on the origins of the group have been gained by extrapolation from the dating of closely related taxa. Such reports place the divergence of ancestral *Charadrius* species from the ancestors of genera *Haematopus* (oystercatchers) and *Recurvirostra* (avocets) at the beginning of the Neogene period, approximately 23 million years ago (Prum et al. 2015). The diversification of the genus Charadrius may, therefore, should be placed within the Neogene period, a time of rapid radiation when 36 bird lineages emerged within 10–15 million years (Jarvis et al. 2014).

Species Relationships within Charadrius

The study of species-level relationships within the genus *Charadrius* began over 250 years ago, with taxonomic studies based on morphometric and anatomical analyses (Linnaeus 1758). While Linneaus' approach was useful, the fine details of the evolutionary history of the genus remained unresolved until the emergence of molecular techniques. These techniques have advanced from DNA–DNA hybridization (Sibley and Ahlquist 1990), allozyme analysis (Christian et al. 1992), and analysis of mitochondrial DNA alone (Joseph et al. 1999), to analysis of multiple nuclear and mitochondrial gene regions (Barth et al. 2013, dos Remedios et al. 2015). Though analyses of partial datasets led to uncertainty in early phylogenetic estimates, greater resolution has now emerged following the first species-level phylogeny including nearly all species, based on multiple genes (dos Remedios et al. 2015). The latest taxonomies that include genetic data suggest the existence of two major Charadrius (CRD) clades, *CRD I* and *CRD II* (Figure 2.2), the common ancestors of which diverged early in the evolutionary history of the genus (Barth et al. 2013, dos Remedios et al. 2015). Within these two clades, six minor clades of sister species have further been identified and this classification is supported by phylogenetic, geographic, and phenotypic data.

Phylogenetic relationships among the *Charadrius* plovers, however, are not without controversy as exemplified by recently published taxonomic

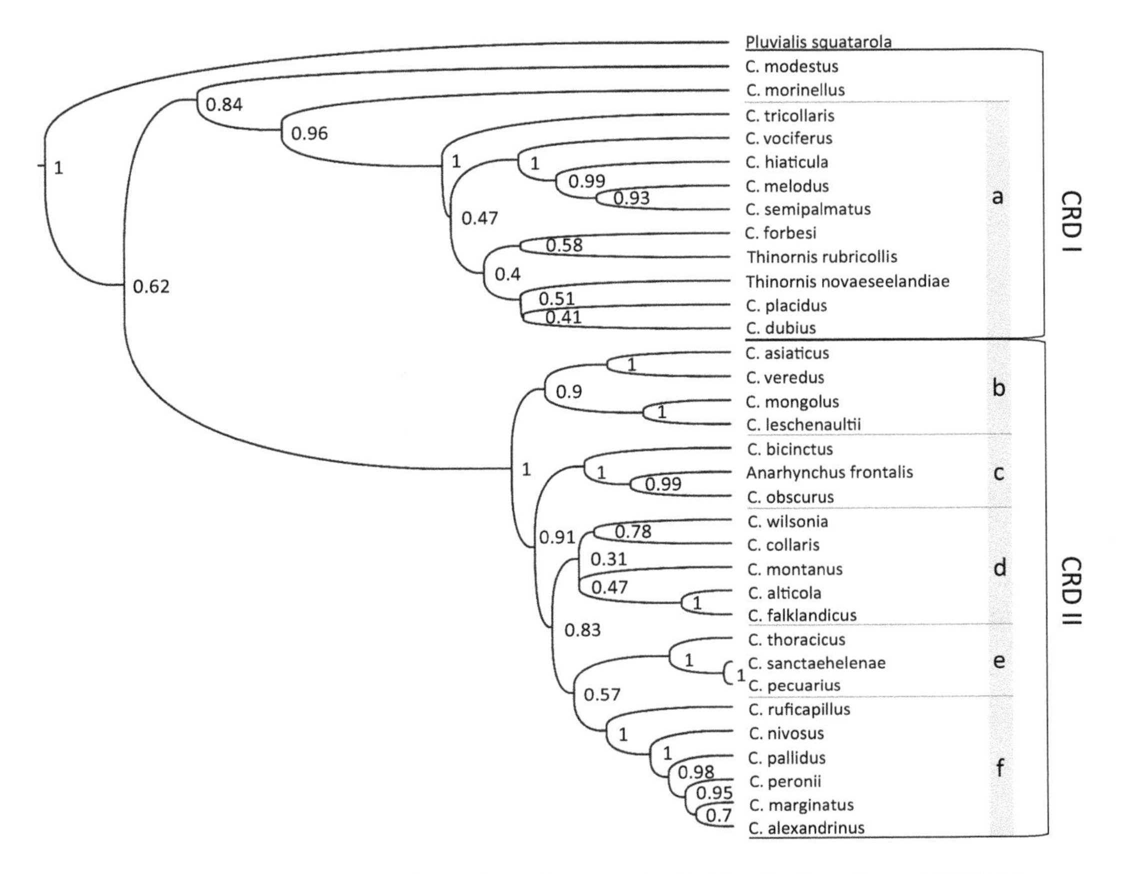

Figure 2.2. Phylogeny of plovers based on four nuclear and two mitochondrial loci (dos Remedios et al. 2015). Values at each node indicate associated posterior probabilities determined by a Bayesian phylogenetic analysis. Minor clades within CRD I and CRD II are labeled a–f. CRD I includes **clade a**: the widespread ringed plovers (*C. hiaticula* and *C. dubius*), two African species (*C. tricollaris* and *C. forbesi*), three Nearctic/Neotropical species (*C. vociferus*, *C. melodus*, and *C. semipalmatus*), and one Asian species (*C. placidus*). CRD I also includes two species distinct from **clade a**, that appear to have diverged early in the history of the group, namely *C. modestus* and *C. morinellus*. Within CRD II, greater resolution has been attained and five minor clades have been described: **clade b**, four Asian red-breasted species (*C. asiaticus*, *C. veredus*, *C. leschenaultia*, and *C. mongolus*); **clade c**, two Oceanian species (*C. bicinctus* and *C. obscurus*); **clade d**, five American (Nearctic and Neotropical) species (*C. alticola*, *C. collaris*, *C. falklandicus*, *C. montanus*, and *C. wilsonia*); **clade e**, three African species, including one widespread species (*C. pecuarius*) and two endemic island species (*C. sanctaehelenae* and *C. thoracicus*); and lastly **clade f**, seven species distributed across all continents but Antarctica (Küpper et al. 2009; *C. alexandrinus*, *C. nivosus*, *C. marginatus*, *C. pallidus*, *C. peronii*, *C. ruficapillus* and also, though not analyzed to date, *C. javanicus*, a species previously considered conspecific with *C. alexandrinus*; Cramp and Simmons 1983, Iqbal et al. 2013).

papers. Barth et al. (2013) analyzed sequence data for the genus *Charadrius* alongside four *Vanellus* species including the Southern Lapwing (*V. chilensis*), Masked Lapwing (*V. miles*), Andean Lapwing (*V. resplendens*), and Northern Lapwing (*V. vanellus*), and suggested that the genus is paraphyletic based on observation of *Vanellus* spp. occurring within the *Charadrius* tree topology. Their results suggested CRD II to be more closely related to *Vanellus* spp. than to CRD I, a claim that requires further investigation. Barth et al.'s (2013) study was based on a relatively small number of genes with a large proportion of sequence data missing from analysis. As such, the results are likely to have been strongly influenced by the presence of sequence insertions or deletions, something that often leads to erroneous interpretations (Lemmon et al. 2009, Roure et al. 2013). Topologies based on phenotypic characters alone are dramatically different from those based on genetic data. Livezey's (2010) phylogeny did not support the CRD I and CRD II clades; instead, he positioned several CRD I species within the *CRD II* clade, as sister species to the Wilson's Plover (*C. wilsonia*) and Double-banded Plover (*C. bicinctus*). Additionally, Livezey (2010) placed the Kentish Plover (*C. alexandrinus*), Snowy Plover (*C. nivosus*), and Malay Plover (*C. peronii*) away from their genetic sister species, the Chestnut-banded Plover (*C. pallidus*) and White-fronted Plover (*C. marginatus*), and instead alongside the genetically clearly distinct Lesser Sand-plover (*C. mongolus*) and Greater Sand-plover (*C. leschenaultia*).

Phylogenetic methods and molecular techniques continue to advance, and genome-wide sequence comparisons offer the potential for greater resolution across the *Charadrius* tree. Such analyses are likely to be particularly informative concerning species-level relationships within *CRD I* and also concerning the order of divergence between minor clades which are not well supported based on present phylogenetic estimates (dos Remedios et al. 2015).

TAXONOMIC CLARIFICATIONS

Comparisons of molecular characters have helped to settle many previous taxonomic debates. Recent studies have suggested updates to taxonomic classification regarding the genus *Charadrius* that deserve further attention. Species for particular consideration are discussed below.

Genera Anarhynchus, Erythrogonys, Oreopholus, Peltohyas, Phegornis, *and* Thinornis

Phylogenetic analyses based on mitochondrial and nuclear DNA (Barth et al. 2013, dos Remedios et al. 2015) indicated that inclusion within the genus *Charadrius* is warranted for the Wrybill (*Anarhynchus frontalis*), Black-fronted Dotterel (*Elseyornis melanops*), Shore Plover (*Thinornis novaeseelandiae*), and Hooded Plover (*T. rubricollis*). One study additionally highlighted three species that may have diverged early in the history of *CRD I* and *CRD II* (Barth et al. 2013), namely the Red-kneed Dotterel (*Erythrogonys cinctus*), Tawny-throated Dotterel (*Oreopholus ruficollis*), and Diademed Plover (*Phegornis mitchellii*). These species may also deserve placement within the genus *Charadrius*, but further phylogenetic attention will be necessary to confirm their taxonomic classification, as well as that of the Inland Dotterel (*Peltohyas australis*), which has not yet been the subject of a molecular genetic analysis.

Kentish and Snowy Plovers

For nearly a century, the Kentish and Snowy plovers were lumped into one species, despite their distributions on different continents (Küpper et al. 2009). A reevaluation based on mitochondrial DNA and nuclear microsatellite markers, as well as phenotypic data such as body size and chick plumage, confirmed that the Kentish and Snowy plovers are two distinct species (Küpper et al. 2009). This new classification has been adopted by the International Ornithological Congress and American Ornithologists' Union (Chesser et al. 2011).

White-faced Plover

The White-faced Plover has a convoluted taxonomic history of species splitting and finally lumping with its close relative, the Kentish Plover (Kennerley et al. 2008). Phenotypically, White-faced Plovers are characterized by pale plumage and white, rather than black, lores during the breeding season in comparison to Kentish Plovers (Kennerley et al. 2008). However, an analysis of mitochondrial and nuclear markers that included well-characterized museum collection samples failed to find enough support to raise the White-faced Plover to full species status (Rheindt et al. 2011). Instead, the authors suggested that the White-faced Plover should be merely considered as a separate subspecies of the Kentish Plover in East Asia. Nevertheless, based on morphological differences, the taxon has been elevated to species level by some taxonomic authorities under the scientific name of *C. dealbatus* (del Hoyo et al. 2016), though not others (Clements et al. 2015, Gill and Donsker et al. 2015). This case highlights the potential disparity between phenotype and genotype that can lead to taxonomic controversy. In the case of the White-faced Plover, great phenotypic divergence may be the result of a small number of genetic changes (Rheindt et al. 2011). Whether these differences are enough to justify a species or subspecies status is an ongoing debate.

Javan Plover

The Javan Plover (*C. javanicus*), a *Near Threatened* plover endemic to Indonesia (Birdlife International 2017), was historically classified as conspecific with the Kentish Plover (Hoogerwerf

1966). It is currently afforded full species status (Cramp and Simmons 1983, del Hoyo et al. 2016). Similar to the case of the White-faced Plover, the Javan Plover is phenotypically distinct from the Kentish Plover due to its dark brown rather than black coloring on lores and breast patches (Iqbal et al. 2013). A study based on a short mitochondrial region found that Javan Plovers form a distinct taxon from Kentish Plovers, but the detected sequence divergence (1.2%) was only moderate and, hence, may only warrant subspecies status for the Javan Plover (Ashari and Astuti 2017). However, because the study used only a single mitochondrial marker, a comprehensive assessment of the species' status should wait until further genetic evidence has been evaluated.

Three-banded Plovers and Forbes's Plover

The African and Madagascar Three-banded plovers have long been classified as subspecies (*C. tricollaris tricollaris* and *C. t. bifrontatus*; Sibley and Monroe 1990) but have recently been proposed as separate species (del Hoyo and Collar 2014). Genetic analysis based on a limited sample size has indicated a strong similarity between the two subspecies, placing questions over whether levels of genetic divergence are sufficient for species status (dos Remedios et al. in press). Further evidence is, therefore, required before an official split is warranted. In contrast, strong genetic differentiation has been identified between the Three-banded Plover and Forbes's Plover (*C. forbesi*) of West Africa, despite high phenotypic similarity, confirming that their present status as distinct species is justified (dos Remedios et al. 2015).

POPULATION AND CONSERVATION GENETICS OF PLOVERS

Long-term monitoring of wild populations provides detailed information on recent changes in population size. However, those attempting to manage and sustain population monitoring for years or decades can appreciate the amount of effort needed to keep these projects running. Genetic studies can complement field efforts, especially in cases where population monitoring is not feasible. Major demographic changes, including recent population declines, can be detected with the help of DNA markers. Moreover, genetic analyses can determine population size changes far back in the evolutionary history of a species, even pre-dating the existence of modern humans (Nadachowska-Brzyska et al. 2015).

Monitoring genetic diversity can be used to assess the genetic health of taxa of conservation concern. This helps to identify populations that are genetically impoverished and require more direct conservation actions (Allendorf et al. 2010). Mitochondrial DNA markers are frequently used to survey the genetic diversity of populations or species. Mitochondrial diversity currently provides the only available information on genetic diversity for most plovers as well as many other species. Although mitochondrial diversity does not necessarily predict population size for larger populations (Bazin et al. 2006, Nabholz et al. 2009), it correlates well with population size in small populations (Jackson et al. 2013).

We collated available sequence data for the mitochondrial *cytochrome c oxidase subunit 1 (COI)* locus to compare genetic diversity among plover species (Table 2.1). The COI locus is commonly used for species identification via DNA barcoding (Hebert et al. 2004). Through the "Barcoding of Life" initiative (https://www.ncbi.nlm.nih.gov/genbank/barcode/), genetic data from this marker are now available for a large number of taxa. We downloaded available sequences from Genbank's freely accessible nucleotide database (http://www.ncbi.nlm.nih.gov/nuccore/) and proceeded with our analyses with sequences that were at least 400 bp long, adding our own unpublished sequences, to calculate comparable summary indices of mitochondrial diversity for species grouped by conservation status (Table 2.1). We only included species if at least three individuals had been sampled.

The ability to detect genetic diversity increases with sample size. Accordingly, we found that mitochondrial diversity, measured as the number of haplotypes, nucleotide and haplotype diversity, was positively related to the number of individuals sampled per species but not to the length of the sequence. No diversity was found in eight species but in five of these, only data from three individuals were available. Nevertheless, because sample sizes for *Least Concern* species and species of conservation concern were broadly similar, we were able to carry out a cursory exploration of the relationship between mitochondrial diversity and conservation status. Five out of 19 species of *Least Concern* (26%) compared to four out of nine species of conservation concern (44%) exhibited no

TABLE 2.1

Genetic diversity estimated from the mitochondrial COI locus for 28 plover species grouped by conservation status (IUCN 2015).

Species Conservation status	N	Fragment length (bp)	No. of haplotypes	π	h
Least Concern					
Collared Plover	6	694	1	0	0
Eurasian Dotterel	6	681	1	0	0
Greater Sand-plover	5	618	3	0.0188	0.70
Kentish Plover	14	626	3	0.0008	0.48
Killdeer	8	593	2	0.0004	0.25
Little Ringed Plover	16	619	4	0.0022	0.44
Pacific Golden Plover	12	597	4	0.0008	0.56
Puna Plover	8	404	2	0.0006	0.25
Common Ringed Plover	7	611	2	0.0014	0.29
Rufous-chested Dotterel	5	533	3	0.0023	0.70
Semipalmated Plover	8	612	2	0.0004	0.25
Wilson's Plover	5	600	2	0.0020	0.60
Lesser Sand-plover	3	694	1	0	0
Caspian Plover	3	429	2	0.0015	0.67
Kittlitz's Plover	14	656	2	0.0006	0.36
Madagascar three-banded Plover	5	515	2	0.0007	0.40
Forbes's Plover	3	429	1	0	0
Red-capped Plover	3	625	1	0	0
White-fronted Plover	7	567	3	0.0030	0.52
Near Threatened					
Mountain Plover	5	626	1	0	0
Piping Plover	9	692	2	0.0003	0.22
Snowy Plover	6	609	2	0.0010	0.60
Chestnut-banded Plover	14	623	3	0.0180	0.58
Southern Red-breasted Dotterel	3	422	2	0.006	0.67
Vulnerable					
Madagascar Plover	4	626	2	0.0011	0.67
Hooded Plover	3	623	1	0	0
Endangered					
Shore Plover	3	566	1	0	0
Critically Endangered					
St. Helena Plover	4	626	1	0	0

N, number of sequences; π, nucleotide diversity; h, haplotype diversity.

genetic diversity (Table 2.1), albeit this difference was not significant (chi-square test: $\chi^2_1 = 0.28$, $P = 0.60$). Similarly, we did not detect a significant difference in average nucleotide diversity (Wilcoxon rank sum test: $W_{27} = 99$, $P = 0.52$) nor in haplotype diversity (Wilcoxon rank sum test: $W_{27} = 92.5$, $P = 0.74$) between these two groups. Instead, average nucleotide diversity was nominally lower for species of *Least Concern* than those of conservation concern whereas the values were very similar for haplotype diversity.

At the other extreme, we found unusually high mitochondrial diversity in the Greater Sand-plover and Chestnut-banded Plover. Nucleotide diversity was higher by an order of magnitude for these two species than for species with similar conservation status (Table 2.1). In both cases, elevated genetic diversity is most likely the result of highly divergent populations, as both species have disjunct geographic distributions. Sequences of Chestnut-banded Plovers included two subspecies, the eastern African subspecies *C. p. venustus* and southern African subspecies *C. p. pallidus*. As the Chestnut-banded Plover is considered *Near Threatened*, the differences in mitochondrial DNA sequences should be used as a starting point for a rigorous taxonomic review to evaluate whether assigning species status to each subspecies is warranted (dos Remedios et al. 2017). The Greater Sand-plover sequences included samples from the nominate subspecies *leschenaultii* and others sampled in Africa and India during wintering without subspecies information. As with the Chestnut-banded Plover, it is likely that at least two subspecies were represented among the samples. Interestingly though, the two genetic lineages of the Greater Sand-plover exhibited roughly the same genetic distance to each other (proportion of nucleotide differences = 0.032) as each of them exhibited to the Lesser Sand-plover (proportion of nucleotide differences = 0.035), suggesting that both lineages could in fact represent distinct species. Taken together, it is possible that both species represent additional examples of cryptic speciation, similar to that described for the Kentish and Snowy plovers. Detailed genetic and phenotypic investigations are needed to critically review the current taxonomic status of these two plover species.

Evolutionary biologists acknowledge that speciation is a highly dynamic process that operates on a continuum. In contrast, the term "species" implies a categorical system; typically, we refer to two species as "A" and "B," considering these as endpoints of the speciation process. However, the reality is more complex and attempting to define a species is not always straightforward. This is because populations and species are constantly evolving, diverging, and/or hybridizing. Further, the exact delimitations are often challenging and controversial as, in practice, it is difficult to establish exactly when two diverging sister populations become two separate species. Nevertheless, a correct taxonomy and the delineation of species and populations are important for conservation policies and management.

More than a dozen species concepts have been defined based on biological, ecological, phylogenetic, or evolutionary differences between taxa (DeQueiroz 2007, Sangster 2014) and the question of which species concept to adopt can have huge ramifications for conservation (Haig et al. 2011). Unfortunately, the underlying assumptions of most species concepts are rarely tested (Edwards et al. 2005). For example, the biological species concept requires proof that two species are reproductively isolated under natural conditions. Proof for this is rarely provided, especially when the species/populations are allopatric. Instead, it is often assumed that allopatric populations that are phenotypically similar interbreed freely. This assumption itself is questionable, however, and rarely tested until detailed genetic analyses are conducted. This has led to calls to reverse the principle of proof, for example, elevating allopatric island populations, by default, to species status until counterevidence is presented (Gill 2014). When adopting this strategy though, a danger of oversplitting and unnecessarily multiplying the number of species exists, which impedes conservation management actions such as translocations and may lead to diversion and misdirection of limited conservation funds (Haig and D'Elia 2010).

The challenges and complexities associated with species delineation are unlikely to be resolved in the near future. Yet avian taxonomy is constantly being updated to take into account new research evidence. When taxonomic committees evaluate species statuses and decide whether or not to split a species, pragmatism rules. No obvious preference for a particular species concept exists, rather the differences between two potential species have to be diagnosable and reproducible (Sangster 2014). This is facilitated by attempts to define a minimal set of quantitative criteria necessary to delimit species. Tobias et al. (2010) suggested a scoring system for assessing objectively whether quantitative differences between allopatric populations warrant a species split. This system was developed based on observed differences in quantitative traits between undisputed sympatric species and true species pairs in well-established European bird species (Tobias et al. 2010). Applying such a method has great potential for improving the taxonomy of poorly researched species, especially for the many species residing in tropical countries for which genetic assessments are yet to be conducted.

Even more contentious is the concept of subspecies (Zink 2004, Winker 2010, Haig et al. 2011). Subspecies were traditionally defined based on phenotypic differences in geographically distinct populations, but they are now increasingly assessed using molecular markers. The concept of the "evolutionarily significant unit" (ESU) puts a strong emphasis on separation during the evolutionary history of sister taxa (Ryder 1986). ESUs are chiefly defined based on differences at (neutral) genetic markers although distinctive locally adapted phenotypic characteristics, such as life history traits, and geographic separation, are also frequently used in their delimitation (Waples 1998). So far, mitochondrial DNA has featured prominently in delineating ESUs, especially when ESUs are narrowly understood as being separate monophyletic evolutionary lineages (Moritz 1994). Mitochondrial DNA analysis works well for highly differentiated populations that have been isolated for a long time. In these cases, mitochondrial DNA can provide a glimpse of past restrictions to gene flow between related ESUs. In contrast, "management units" (Moritz 1994) are distinct populations that are identified based on differences in allele frequencies at nuclear genetic markers such as microsatellites or SNPs. Management units are commonly considered as subunits of ESUs (Höglund 2009).

Conservation genetic concepts are firmly implemented in practical conservation. For the U.S. Endangered Species Act, species, subspecies, and distinct population segments are used as the basis of practical management and legislation (Haig and D'Elia 2010). A distinct population segment is a discrete population with significant importance for the survival of a taxon (Haig and D'Elia 2010). In contrast to defining ESUs, no genetic data are necessary for defining distinct population segments. However, genetic differences can help to confirm that a population segment is indeed reproductively isolated. ESUs provide the current gold standard for defining distinct population segments. Threatened distinct population segments receive conservation priority, especially if their existence is considered of high importance for the persistence of the species (Haig and D'Elia 2010). An up-to-date taxonomy is the scientific base required for streamlining the legal process that precedes the listing of threatened population segments.

CONSERVATION GENETIC CASE STUDIES OF PLOVERS

Few plover species have been subjected to full genetic investigations across their entire range. Here we discuss the results of a number of case studies carried out on plover species of conservation concern: the Piping Plover (*C. melodus*), Snowy Plover, Mountain Plover (*C. montanus*), Southern and Northern Red-breasted Dotterel (*C. obscurus* and *C. aquilonius*), and Madagascar Plover (*C. thoracicus*).

Piping Plover

The Piping Plover is a migratory species, with a disjunct distribution that includes the Atlantic coast of Canada and the United States, and the Northern Great Plains and Great Lakes areas (Elliott-Smith and Haig 2004). Two subspecies (*C. m. melodus* and *C. m. circumcinctus*) have been described based on differences in breast band patterns and geographic distributions. Both subspecies are listed as *Endangered* throughout their range, and population monitoring over several decades has documented a complex pattern of range expansions and contractions (Miller et al. 2010). Several molecular genetic studies have been conducted to inform conservation management decisions. Initial molecular studies based on protein variants (allozymes) failed to detect differences between the two subspecies (Haig and Oring 1988). However, more recent studies showed that subspecies delineation is warranted based on mitochondrial and microsatellite differences, with each subspecies constituting an ESU (Miller et al. 2010). Within each ESU, no clear evidence for further genetic structure was apparent, although, particularly within the Atlantic subspecies, elevated population divergence, evident through high F_{ST}-values in pairwise comparisons among sites, suggested the emergence of genetic structure. A reanalysis of the microsatellite data using a novel algorithm suggests that Canadian and U.S. Atlantic populations may have started to diverge (D'Urban Jackson et al. 2017). Populations of the interior ESU (subspecies *circumcinctus*) showed clear signs of recent bottlenecks, a result that was consistent with historical count records (Miller et al. 2010). Genetic analyses also helped to determine the origin of Piping Plovers wintering in the Bahamas, showing that 95% of the wintering population belonged to the Atlantic *melodus* subspecies (Gratto-Trevor et al. 2016).

Snowy Plover

Snowy Plovers inhabit the shores of brackish lakes, sandy ocean beaches, and coastal and interior salt flats in the Americas. Throughout the distribution of the nominate subspecies *nivosus*, interior populations are relatively large, whereas coastal populations are threatened and require active management to prevent further decline (Thomas et al. 2012, Eberhart-Phillips et al. 2015a, Cruz-López et al. 2017). Coastal populations exhibit variable mating and brood care behavior with a large proportion of broods being cared for by the male alone (Warriner et al. 1986, Carmona-Isunza et al. 2015, Cruz-López et al. 2017, Chapter Five, this volume). Snowy Plovers have high levels of breeding dispersal, with nests belonging to the same female documented at locations more than 1,100 km apart within a single season (Stenzel et al. 1994). Despite this extraordinary mobility, demographic data from intensive banding and resighting efforts provide little evidence for connectivity between inland and coastal populations. Based on mitochondrial and microsatellite markers, North American populations were at first thought to lack genetic structure (Funk et al. 2007, Küpper et al. 2009) meaning that the current justification for Pacific, Atlantic, and interior ESUs or management units would have to be based largely on phenotypic differences between populations. However, a reanalysis of population differentiation based on microsatellite data, including a larger sample size and an improved algorithm for detecting more fine-scale structure, showed that Snowy Plovers on the Atlantic coast do indeed represent a distinct genetic cluster (D'Urban Jackson et al. 2017) and may need to be considered as a separate management unit.

Mountain Plover

The population size and breeding distribution of the Mountain Plover, an endemic of the North American grasslands, are decreasing as a result of human-induced habitat changes at breeding and wintering grounds (Knopf and Wunder 2006). Mountain Plovers are characterized by high adult site fidelity and natal philopatry (Dinsmore et al. 2003, Knopf and Wunder 2006). Increasing habitat fragmentation is therefore expected to have driven population differentiation and a loss of genetic diversity. Contrary to these expectations, studies on Mountain Plovers breeding in the United States reported high genetic diversity, based on mitochondrial and microsatellite markers, with no indication of population differentiation (Oyler-McCance et al. 2005, 2008). This lack of genetic structure suggests that sufficient contemporary gene flow does exist, perhaps driven by dispersing juvenile Mountain Plovers (Oyler-McCance et al. 2005, 2008). The current size of the U.S. population is estimated at 5,000–11,000 individuals, meaning that a moderate population decline is unlikely to be detected with mitochondrial markers. Instead, the observed pattern of genetic diversity was consistent with a past population expansion. Such an expansion may have taken place, for example, after the last glacial maximum, an event that had a strong impact on other North American grassland species (Oyler-McCance et al. 2005).

Southern and Northern Red-breasted Dotterels

Southern and Northern Red-breasted Dotterels breed at widely separated locations; the Northern Red-breasted Dotterel inhabits the northern part of New Zealand's North Island, whereas the Southern Red-breasted Dotterel inhabits the southern part of the South Island and Stewart Island. The two dotterels differ in their behavior and morphology and were previously considered as two subspecies of New Zealand Dotterel (Herbert et al. 1993, Dowding 1999, Barth et al. 2013). The current disjunct breeding distributions of the two taxa are the result of severe population declines and extinctions following human settlement by Europeans and the introduction of new predators (Dowding 1999). An initial study examining allozyme variation found no difference between the two dotterels (Herbert et al. 1993). More recent genetic analyses using more variable markers (two mitochondrial and one nuclear intron marker) identified moderate genetic differentiation (Barth et al. 2013). Both populations harbor unique mitochondrial haplotypes; however, no genetic differences were detected when comparing nuclear intron sequences. This result is not unexpected since the intronic nuclear marker evolves slowly and its resolution may be too low to detect existing genetic differences, which are likely to have accumulated relatively recently (Barth et al. 2013).

Based on a combination of phenotypic and genetic characters, and following the species delimitation criteria suggested by Tobias et al. (2010), the dotterels were recently split into separate species by taxonomic authorities (del Hoyo et al. 2016), a split acknowledged by the International Union for the Conservation of Nature (IUCN). Though only moderate genetic differentiation is present, the geographic separation and reproductive isolation between the two dotterels mean they are likely to gradually diverge further, and this will eventually be reflected in stronger genetic differences. Nevertheless, the elevation to species level has not yet been adopted by other taxonomic committees including Clements et al. (2015) and Gill and Donsker (2015), who still treat the two dotterels as subspecies.

Madagascar Plover

The Madagascar Plover is an endemic habitat specialist that occupies sparsely vegetated coastal wetlands including the shorelines of lakes and salt marshes (Long et al. 2008). An analysis based on microsatellite markers indicated little genetic diversity (Eberhart-Phillips et al. 2015b), as expected for an island endemic with an estimated population size of only 3,500 individuals (Long et al. 2008). Along the coast, gradual genetic changes typical of isolation-by-distance (the correlation between genetic and geographic distance) were observed, leading to two distinct genetic clusters in the north and south of Madagascar. In comparison, genetic structure was much weaker among populations of two common habitat generalists, the White-fronted Plover and Kittlitz's Plover (*C. pecuarius*), which are frequently sympatric with the Madagascar Plover (Eberhart-Phillips et al. 2015b). The authors suggested that the Madagascar Plover's strong habitat specialization contributed to its genetic differentiation because a lack of suitable habitat nearby is likely to reduce dispersal.

The examples above, particularly those involving the Red-breasted Dotterel, Mountain and Snowy plovers, highlight the difficulties involved in attempting to define ESUs in species with high gene flow. Even in light of large geographic distributions and increasing habitat fragmentation, it remains challenging to delineate ESUs with traditional genetic markers for highly dispersive taxonomic groups such as the plovers. As little as one migrant per generation can be enough to prevent populations from diverging (Mills and Allendorf 1996) and this figure is commonly exceeded in many highly dispersive species. Additionally, choosing the right genetic markers is critical for detecting low levels of genetic differentiation. Using a handful of neutral markers will detect the most obvious taxonomic errors at the species and genus levels, but this approach usually fails to resolve existing adaptive variation between populations, especially if gene flow ceased only recently, or if speciation occurs despite ongoing gene flow.

PHYLOGEOGRAPHY: ECOLOGICAL AND BEHAVIORAL CORRELATES OF GENETIC STRUCTURE AND DIVERSITY

Phylogeography describes evolutionary patterns that account for the geographic distribution of individuals or species based on genetic data (Avise 2000). Broadly interpreted, phylogeography also includes many facets of biogeography and population biology (Hickerson et al. 2010). As mentioned above, few physical barriers exist that a dispersing plover cannot overcome, and their superior dispersal capabilities are reflected in exceptionally high levels of gene flow. Kentish Plovers show a complete lack of genetic structure among populations distributed over the continental landmasses of Europe, Asia, and Africa based on a robust genetic marker set of highly variable microsatellite markers and mitochondrial sequences (Küpper et al. 2012). This means that, based on their genetic profiles, Kentish Plovers breeding in Portugal or Spain are not distinguishable from those breeding in China. Similarly, Snowy Plovers sampled across continental North America, and Kittlitz's Plovers sampled across mainland Africa, exhibit very low levels of genetic differentiation over large geographic distances (Funk et al. 2007, dos Remedios et al. in press). Populations of these species are highly genetically homogenous, with high gene flow maintained over large distances, despite a patchy distribution. The (occasional) complete lack of genetic structure over large geographic distances is unusual even for birds. Below, we discuss a number of ecological and behavioral characteristics that may affect patterns of genetic differentiation within plover species, namely (1) migratory connectivity, (2) habitat requirements, (3) mating systems, and (4) past climate oscillations (Figure 2.3).

Migratory Connectivity

As is often the case for shorebirds, many plover species are migratory (Chapter Seven, this volume) and may travel thousands of kilometers to their wintering habitats. Migratory bird species exhibit high annual dispersal and are less likely to return to their place of birth to breed than nonmigratory birds (Paradis et al. 1998). Because of their high migratory connectivity, that is, a strong link between wintering and breeding populations across the species range, these species are expected to show lower levels of population differentiation than nonmigratory species.

Whether migratory connectivity alone can explain the differences in genetic structure among plovers is doubtful (D'Urban Jackson et al. 2017). Some migratory species, such as the Mountain Plover, show no sign of genetic structure (Oyler-McCance et al. 2008), whereas others, such as the Piping Plover, have moderate genetic structure (Miller et al. 2010). Similarly, levels of genetic differentiation vary among resident species. Two resident plover species breeding in Madagascar show moderate (White-fronted Plover) to high (Madagascar Plover) levels of genetic structure, whereas one species, the Kittlitz's Plover, does not (Eberhart-Phillips et al. 2015b, dos Remedios et al. in press). Partially migratory species, such as the Kentish and Snowy plovers with resident and migratory populations mixing during the winter, show very low genetic differentiation on a continental scale. In general, adults faithfully return to breed at the same site each year (Foppen et al. 2006, Pearson and Colwell 2014). However, occasionally migrants may remain at their wintering sites and breed with local residents. In particular, juveniles, who have not yet bred elsewhere, may stay at or near the wintering grounds when conditions for breeding appear promising. Such birds would help to maintain high gene flow and migratory connectivity that prevents genetic structure from developing. Yet, observations of banded migrants breeding at wintering sites are missing (Stenzel et al. 2007). Greater banding and resighting efforts at overwintering areas are needed to address this question.

In some sandpipers, females mate with multiple males and a significant proportion of matings may occur during migration (van Rhijn 1985, Lank et al. 2002). Females may then either nest near the mating site or migrate further, perhaps fertilizing eggs with stored sperm from earlier matings (Oring et al. 1992) after they have reached suitable breeding grounds. Such a spatial separation of mating and nesting promotes gene flow and prevents the development of genetic structure between populations. However, it is unlikely that matings during migration will explain much of the low genetic structure observed in plovers because the mating behavior of plovers differs from that of polygamous sandpipers. In plovers,

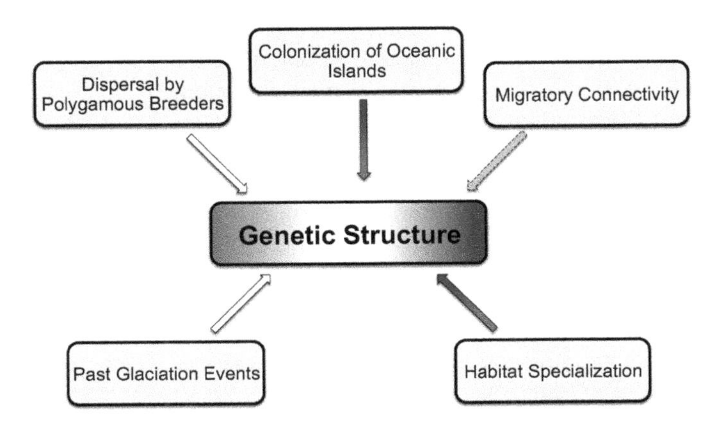

Figure 2.3. Ecological factors promoting or preventing population divergence in plovers. Increases in genetic structure will lead to population divergence and often culminate in speciation. Factors promoting population divergence are indicated by bicolored arrows, and those reducing it are indicated by uniform white arrows.

pairs are usually formed after arrival at the breeding grounds (Carmona-Isunza et al. 2015). The pair bond remains beyond copulation; the male and female typically stay together and jointly care for the offspring, at least until hatching of the chicks (Chapter Four, this volume). In addition, copulations outside the pair bond do not appear to be a major feature of plover breeding strategies because extra-pair fertilizations are rare (Küpper et al. 2004, Maher et al. 2017).

Habitat Requirements

Variation in habitat requirements may lead to differences in genetic structure between species. Species with specific habitat requirements are more prone to undergo population bottlenecks or to suffer from local extinctions, if their preferred habitat becomes unavailable (Julliard et al. 2004, Li et al. 2014). Greater genetic structure is therefore expected in habitat specialists than in generalists. One way to examine this relationship is by comparing genetic differentiation among species that occupy different habitats. Inland wetlands are relatively common and widespread, although conditions at these sites are often unpredictable and highly variable. Inland wetlands often provide a large diversity of prey and are therefore suitable for habitat generalists. In contrast, coastal habitats provide more predictable conditions but are more patchily distributed. Therefore, coastal wetlands are more likely to appeal to habitat specialists. Supporting a link between habitat requirement and genetic variation, a comparative analysis suggested that migratory shorebirds that overwinter in coastal wetlands tend to harbor more subspecies than those that overwinter inland (Kraaijeveld 2008). For example, the widespread Ruff (*Philomachus pugnax*) breeds at high latitudes all over Eurasia from Scandinavia to Chukotka. Ruffs overwinter largely inland and their populations are genetically uniform; there is no genetic differentiation across their breeding distribution (Verkuil et al. 2012), and the species is not divided into subspecies. In contrast, for the coastal overwintering Red Knot (*Calidris canutus*), which also breeds in the Arctic tundra, six subspecies are recognized. An alternative explanation proposes that species overwintering in coastal areas follow more rigid migratory pathways than those overwintering inland. Subsequently, inland overwintering species have more flexible movement patterns, leading to higher migratory connectivity (Kraaijeveld 2008).

Habitat specialization may also lead to higher genetic differentiation among resident populations. This is because habitat generalists and specialists may differ in their opportunities for dispersal (Zayed et al. 2005), with specialists having fewer opportunities to move into suitable habitat than generalists do. Support for this theory is provided by a comparison of population differentiation for three plover species breeding in Madagascar (Eberhart-Phillips et al. 2015b). The Madagascar Plover, a coastal habitat specialist, showed the highest genetic structure, whereas genetic structure was only moderate among White-fronted Plovers, a species with less stringent habitat requirements. In contrast, genetic structure was completely absent among Kittlitz's Plovers from Madagascar, who are habitat generalists. Anecdotal evidence suggests that dispersal does indeed differ between the three species: Kittlitz's Plovers may disperse much further than the other two species and marked individuals have been observed at sites more than 100 km from their ringing site, whereas the other two species have not been resighted more than 15 km from their original capture site (Zefania and Székely 2013).

Mating Systems

Sexual selection is a process whereby traits such as colorful plumage ornaments offer certain individuals an advantage over other members of the same sex when competing for mates. The strength of sexual selection varies with mating system. For example, males in polygynous and females in polyandrous species are under stronger sexual selection than their counterparts in monogamous species. Less well understood is the impact of sexual selection on population divergence and speciation. Whether sexual selection promotes or inhibits speciation is the topic of an ongoing debate (West-Eberhard 1983, Mitra et al. 1996, Morrow et al. 2003, Servedio and Bürger 2014).

According to the "engine of speciation" hypothesis (West-Eberhard 1983), sexual selection increases prezygotic isolation through rapid evolution of divergent female preferences in geographically distinct populations (allopatry) or among individuals in the same location (sympatry). In support of this hypothesis, a comparison of bird taxa showed higher species richness within promiscuous clades than monogamous clades (Mitra et al. 1996). However, other comparative studies and population genetic models have questioned the role of sexual selection as a general accelerator of speciation (Morrow et al. 2003, Servedio and Bürger 2014). A comparative study in plovers using genetic differentiation as a precursor of speciation suggested that polygamy inhibits population divergence (D'Urban Jackson et al. 2017). In this study, genetic differentiation measured at microsatellite markers was compared across the continental distributions of ten plover species. Polygamous species showed lower genetic structure than monogamous species. For monogamous species, more geographically distant populations were also moderately more genetically differentiated, whereas this pattern was absent in polygamous species. A consistent pattern was found across all shorebird species using the number of subspecies as a proxy for diversification: polygamous shorebird species generally have fewer subspecies than monogamous species (D'Urban Jackson et al. 2017). Breeding dispersal is likely to be the main driver of these differences because sequentially polygamous plovers can move large geographic distances between breeding attempts (Stenzel et al. 1994) and, in this way, will homogenize the gene pool (Figure 2.4). This was supported by a within-species analysis that found female-biased gene flow in the polyandrous Kentish Plover, suggesting that polyandrous females contribute more to gene flow than dispersing males (Küpper et al. 2012). In contrast, monogamous plovers disperse very little meaning that genetic structure is more likely to emerge in these species (Figure 2.4). Taken together, these results suggest that sexual selection slows down speciation and hence acts more as a brake than an engine of speciation in shorebirds.

Past Climate Oscillations

The climate cycles of the Pleistocene repeatedly transformed the landscape of high-latitude regions in line with oscillations between cold glacial and warmer interglacial periods. Pronounced global

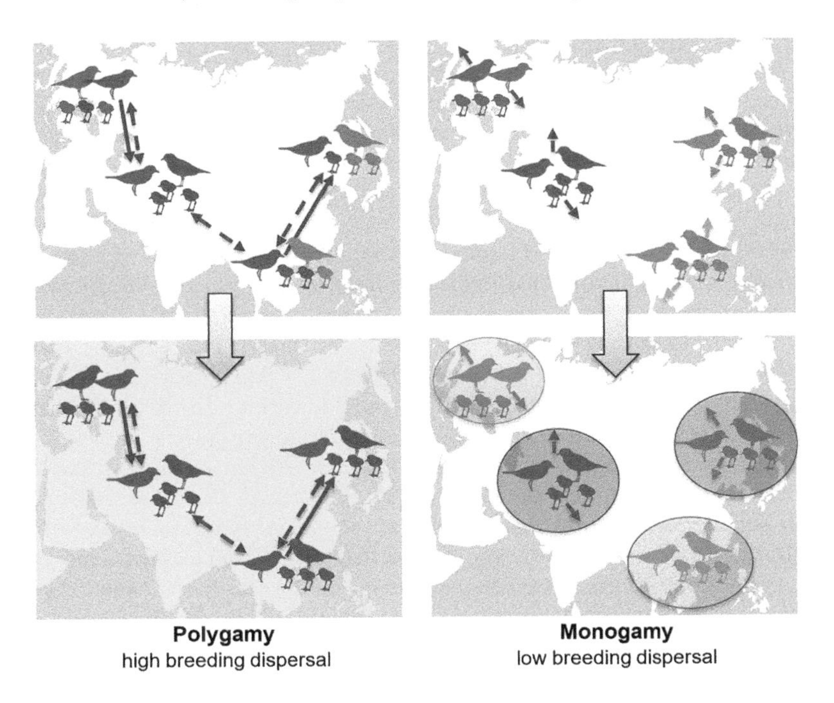

Figure 2.4. Breeding dispersal is tied to mating systems and will affect genetic differentiation and speciation (D'Urban Jackson et al. 2017). Top left: Sequentially polygamous individuals (red) will disperse after a successful mating attempt to acquire a new mate (breeding dispersal indicated by solid line). Their polygamous offspring will show the same behavior (indicated by broken lines). Bottom left: Over time, this will lead to high gene flow and a lack of genetic differentiation and inhibit speciation. Top right: Territorial, monogamous species with high site and mate fidelity. Here, low breeding dispersal leads to genetic differentiation over time. Bottom right: Populations have diverged and new allopatric species arise faster under monogamy than polygamy.

cooling led to the spawning of continental glaciers into Europe, Asia, and North America, which then retreated during warmer periods. These changes had profound implications for Holarctic species, requiring range shifts to southern refugia during glacial times, and these shifts were often associated with population bottlenecks. When temperatures increased, again contractions were followed by repeated colonization of higher latitude habitats and population expansions during interglacial periods (Avise and Walker 1998). This frequently led to population divergence culminating ultimately in allopatric speciation, for example, when populations were isolated in multiple small refugia for a long period of time (Lovette 2005).

Climatic oscillations are likely to have been important for the radiation of plover species and their colonization of the world given that the ancestor of the Charadrius plovers evolved in the Northern hemisphere (dos Remedios et al. 2015). Population contractions and expansions associated with climate changes can leave characteristic signatures across genetic markers, as highlighted in studies of sandpipers breeding at high latitudes (Wenink et al. 1994, Buehler and Baker 2005, Höglund et al. 2009). In plovers, similar data from high-latitude breeders are scarce because in-depth genetic analyses have largely been restricted to temperate or tropical species. Nevertheless, some evidence does point toward climatic oscillations impacting genetic diversity and differentiation (Thies et al. 2018). For example, in Mountain Plovers, the diversity of mitochondrial haplotypes was consistent with population expansion during the Pleistocene (Oyler-McCance et al. 2005). The two subspecies of Piping Plover may have evolved after isolation in glacial refugia (Miller et al. 2010). In Snowy Plovers, recent population changes may have overridden any such signal, though the low genetic structure observed in this species is also consistent with population and range expansion after the last glaciation (Funk et al. 2007, Küpper et al. 2009). No population expansion was detected in the continental Eurasian Kentish Plover population (Küpper et al. 2012), although only a small part of the current species range would have been affected directly by past glaciation events.

Colonization of Oceanic Islands

Despite their formidable dispersal abilities, some physical barriers to gene flow exist even for plovers. Long stretches of sea appear particularly effective in preventing or inhibiting gene flow within plover species. Populations across island archipelagos are often genetically differentiated which may eventually lead to allopatric speciation (Küpper et al. 2012, dos Remedios et al. in press, Almalki et al. 2017). Isolation in allopatry, on continents or on islands separated by water, has also lead to the emergence of closely related sister species such as the Kentish Plover superspecies complex (Küpper et al. 2009, dos Remedios et al. 2015). Examining patterns of genetic differentiation within Eurasian and African Kentish plovers, Küpper et al. (2012) found that despite a lack of genetic structure across the continental landmass, island populations were genetically distinct from mainland populations. This genetic differentiation increased linearly with distance from the mainland and was particularly strong for remote island archipelagos such as the Azores and Cape Verde Islands. Island populations were also morphologically differentiated, suggesting local adaptation (Almalki et al. 2017). In line with earlier comparative studies (Frankham 1997), small island populations of Kentish Plovers exhibited lower genetic diversity, reflected in lower mitochondrial sequence diversity and lower microsatellite heterozygosity than the large undifferentiated mainland population (Küpper et al. 2012). Results from African species such as the Kittlitz's Plover, Three-banded Plover, and White-fronted Plover, sampled in Madagascar and mainland Africa, support the hypothesis that islands are important for genetic differentiation and hence may facilitate speciation in allopatry (dos Remedios et al. in press). Two endemic island species, St. Helena and Madagascar plovers, descend from a common ancestor shared with Kittlitz's Plovers and evolved into new species at their current locations (dos Remedios et al. in press, dos Remedios et al. 2015). Similarly, Caribbean Snowy Plovers located across several island archipelagos are genetically differentiated from mainland Snowy Plovers (Funk et al. 2007). Lastly, in New Zealand, separation of populations by the Cook Strait between North and South Islands is likely to have facilitated the divergence of Northern and Southern Red-breasted Dotterels (Barth et al. 2013).

FUTURE PROSPECTS

Genetic analyses have already helped to improve our understanding of plover biology, primarily by clarifying taxonomy and assessing the status of subspecies and populations of conservation concern. Yet several facets of the current taxonomy of plovers remain unresolved. For example, it has been proposed that the *Charadrius* genus should be divided into two genera: *Charadrius* and *Anarhynchus*. The genus *Anarhynchus* currently contains only a single species, the Wrybill. If adopted, *Anarhynchus* would be adopted as genus name for diverse species such as the Caspian Plover (*C. asiaticus*), Oriental Plover (*C. veredus*), Lesser and Greater Sand-plover, Kittlitz's Plover, and Kentish Plover, among others (Sangster et al. 2016). This split was proposed after phylogenetic results showed a number of *Vanellus* taxa (lapwings) appearing within the *Charadrius* tree, suggesting the genus may not be monophyletic (Barth et al. 2013). We disagree with the proposed genus split. Rather, we argue that the suggested phylogenetic topology was most likely caused by missing data or the stochastic effects of analyzing too few genetic markers. Instead of jumping to poorly supported conclusions, we recommend further analyses using additional genetic markers, or ideally whole genome sequences, to resolve the true affinities of the plover clade.

Occasionally, the available genetic data highlight taxonomic inconsistencies that have not yet been fully recognized by taxonomists. Apparent cryptic speciation and the existence of a third phylogenetic sister group to the Lesser and Greater Sand-plover, as indicated by our comparison of available mitochondrial sequences (Table 2.1), provides an example of this. Such cases require more attention, and we suggest more rigorous analyses, ideally based on samples from across the entire geographic distribution of each species. Museum specimens may be used for such an analysis, although, in our experience, the DNA obtained from museum skin is of inferior quality and therefore makes the analysis more time-consuming and expensive. With appropriately preserved tissue or blood samples, a larger set of genetic markers is available for running a more diverse and robust set of phylogenetic and population genetic analyses. Nevertheless, museum skin samples provide an important resource for addressing a range of questions including, for example, comparing historic and present genetic diversity and gene flow for species where population changes are suspected.

Assessing the validity of subspecies and examining population connectivity is of increasing importance, particularly for planning conservation management. Molecular analyses can provide complementary evidence to assess alongside phenotypic differences in geographically separated populations. Genetic data can highlight evolutionary divergence and can indicate whether subspecies are indeed reproductively isolated. As early results in the genetic comparison of the White-faced Plover and Kentish Plover show, phenotypic differences do not necessarily reflect strong genetic differences between putative species (Rheindt et al. 2011). In the case of the White-faced Plover, this is not likely to be the end of the story. Instead, more markers are required to assess the extent of hybridization and introgression that led to the observed phenotypic differences between subspecies.

The rapid decrease in cost of DNA sequencing holds great promise, especially for species with high gene flow such as plovers. Generally, more genetic markers mean more statistical power and more robust conclusions. We expect that high-throughput sequencing, once applied to plovers, will provide far more detailed insights on population differentiation and genetic structure at a fine spatial scale. This will help to critically test the extent to which geographically separated populations are connected, helping to answer open questions on how past and present demographic events shape genetic structure. Generating these insights with traditional tracking activities such as mark-recapture studies is not practically feasible. Analyzing more genetic markers will have a highly practical use for conservation management. In particular, for species with low genetic structure, several thousand loci are often needed to confirm the extent of gene flow between populations or to identify the origin of immigrants. The early stages of genetic differentiation can be much more readily detected with a higher number of genetic markers. Given the technological advances made over the past decade (Davey et al. 2011) such studies are now feasible for all avian species.

Determining population connectivity can also be aided by other types of molecular analyses. Analysis of stable isotopes in feather or other tissues can provide information on the geographic origin of migrants (Hobson 1999). This is possible because the isotopic composition of many elements changes with latitude, food webs, and habitat, leaving characteristic signatures in tissues synthesized in different locations, such as newly molted feathers. Stable isotope analysis often includes several elements, with an additive effect for increasing spatial resolution. Such analysis has been used to determine the wintering locations of migratory shorebirds (Franks et al. 2012, Catry et al. 2016). However, it requires extensive sampling at wintering grounds for calibration and at sites which are often far apart geographically. Even with a good sample size, classification success for assigning known individuals blindly to their respective origin is sometimes only moderate (Rocque et al. 2006). One underutilized approach is combining stable isotope and genetic markers into an integrative analysis. These two marker systems complement each other very well because genetic differentiation often shows a longitudinal gradient whereas stable isotope variation follows a latitudinal gradient (Boulet et al. 2006).

With full genome sequence data, past population size changes can be inferred by sequencing a single individual from each population of interest. A comparative study using full genome sequences from 38 bird species, including the Killdeer, revealed changes in effective population size between the last 10 million and 10,000 years (Nadachowska-Brzyska et al. 2015). The results indicated that the effective population size of many species declined dramatically during the last glacial period. The Killdeer was one of the species most affected. Genetic evidence revealed that its effective population size once reached over 4 million, suggesting it was more common and perhaps more widely distributed than today, but numbers plummeted to fewer than 30,000 individuals during the glacial period. Today, the population is estimated at approximately 2,000,000 individuals; however, the species is in decline once again (Andres et al. 2012).

The field of evolutionary genetics and genomics is advancing rapidly and opening exciting new research avenues. Plovers have the potential to offer a number of suitable model systems, for example, due to their unusual diversity of mating systems (Chapter Four, this volume). Mating system variation is related to genetic diversity and effective population size (Nunney 1993). To examine this relationship empirically, patterns of genetic diversity and differentiation can be compared between autosomal and sex chromosome-linked markers (Corl and Ellegren 2012, Verkuil et al. 2014). Similar analyses can also be used to determine population ecological parameters that are difficult to measure in the field, for example, to establish whether sex-biased dispersal exists (Küpper et al. 2012).

New genomic studies in birds also provide novel insights on how phenotypic variations in species have evolved. Recent studies in Ruffs provided a rare glimpse into the underlying genetics of social behavior (Küpper et al. 2016, Lamichhaney et al. 2016). The studies revealed that a simple genetic aberration, an inversion at chromosome 11 that comprises less than 0.5% of the genome, is responsible for the morphological and behavioral differences between the three fixed reproductive morphs. Such simple genetic changes are emerging as one of the principal genetic differences for many phenotypic polymorphisms within species and may also be responsible for behavioral differences among plovers.

CONCLUSIONS

Population genetic and phylogeographic patterns indicate that in comparison with other birds, plovers exhibit high levels of population connectivity. This is not surprising given their high dispersal capabilities and mobility. Nevertheless, plovers breeding on islands are often genetically differentiated from their mainland counterparts. Studies of genetic structure have been conducted on species from diverse geographic areas, including both Northern and Southern hemispheres and tropical and temperate regions. Past climatic events, habitat specialization, and mating behavior contribute to the variation in genetic structure observed among species. In particular, high breeding dispersal in sequentially polygamous species contributes to an observed lack of genetic differentiation in polygamous plovers (Küpper et al. 2012, D'Urban Jackson et al. 2017), meaning that polygamous males and females are important in maintaining gene flow over large geographic distances. Somewhat surprisingly, though, migratory behavior does not seem to explain much of the variation in genetic structure.

Genetic studies have already provided many insights into the evolutionary processes that have shaped plovers. Advances in sequencing methodology hold great promise for further groundbreaking results in the near future. In our view, molecular studies provide useful complementary analyses to existing ecological studies. We strongly advocate a sensible biological interpretation of genetic results grounded on a solid knowledge of the natural history of a species. Field studies examining behavior and ecology should, therefore, precede or accompany genetic studies where feasible. The results of such multifaceted studies are particularly useful for assisting the design of future genetic analyses in order to maximize impact. A mere collection of samples coupled with a standardized minimal genetic analysis will ignore important features of the species' biology that have shaped genetic patterns, and may ultimately lead to false conclusions or misguided recommendations for conservation management. In contrast, knowledge of natural history enables the correct interpretation of molecular results, helping to put these results into a macrobiological context and opening new, exciting research avenues.

ACKNOWLEDGMENTS

We thank Josie D'Urban Jackson for initial discussion of ideas for this chapter. Mark Colwell, Susan Haig, Kevin Winker, and one anonymous reviewer provided valuable and constructive comments on previous drafts of this manuscript.

LITERATURE CITED

Allendorf, F. W., P. A. Hohenlohe, and G. Luikart. 2010. Genomics and the future of conservation genetics. *Nature Reviews Genetics* 11:697–709.

Almalki, M., K. Kupán, M. C. Carmona-Isunza, P. Lopez, A. Veiga, A. Kosztolányi, T. Székely, and C. Küpper. 2017. Morphological and genetic differentiation among Kentish Plover *Charadrius alexandrinus* populations in Macaronesia. *Ardeola* 64:3–16.

Andres, B. A., P. A. Smith, R. G. Morrison, C. L. Gratto-Trevor, S. C. Brown, and C. A. Friis. 2012. Population estimates of North American shorebirds, 2012. *Wader Study Group Bulletin* 119:178–194.

Ashari, H., and D. Astuti. 2017. Study on phylogenetic status of Javan Plover bird (*Charadrius*, Charadriidae, Charadriiformes) through DNA barcoding analysis. *Biosaintifika* 9:49–57.

Avise, J. C. 2000. *Phylogeography: The History and Formation of Species*. Harvard University Press, Cambridge, MA.

Avise, J. C., and D. Walker. 1998. Pleistocene phylogeographic effects on avian populations and the speciation process. *Proceedings of the Royal Society B* 265:457–463.

Baird, N. A., P. D. Etter, T. S. Atwood, M. C. Currey A. L. Shiver, Z. A. Lewis, E. U. Selker, W. A. Cresko, and E. A. Johnson. 2008. Rapid SNP discovery and genetic mapping using sequenced RAD markers. *PLoS One* 3:e3376.

Baker, A. J., S. L. Pereira, and T. A. Paton. 2007. Phylogenetic relationships and divergence times of Charadriiformes genera: Multigene evidence for the Cretaceous origin of at least 14 clades of shorebirds. *Biology Letters* 3:205–210.

Baker, A. J., Y. Yatsenko, and E. S. Tavares. 2012. Eight independent nuclear genes support monophyly of the plovers: The role of mutational variance in gene trees. *Molecular Phylogenetics and Evolution* 65:631–641.

Barth, J. M. I., M. Matschiner, and B. C. Robertson. 2013. Phylogenetic position and subspecies divergence of the endangered New Zealand Dotterel (*Charadrius obscurus*). *PLoS One* 8:e78068.

Bazin, E., S. Glémin, and N. Galtier. 2006. Population size does not influence mitochondrial genetic diversity in animals. *Science* 312:570–572.

BirdLife International [online]. 2017. *Charadrius javanicus*. The IUCN Red List of Threatened Species 2017: e.T22693839A118306149. <www.iucnredlist.org/details/22693839/0> (7 December 2017).

Boulet, M., H. L. Gibbs, and K. A. Hobson. 2006. Integrated analysis of genetic, stable isotope, and banding data reveal migratory connectivity and flyways in the northern yellow warbler (*Dendroica petechia*; *aestiva* group). *Ornithological Monographs* 61:29–78.

Buehler, D. M., and A. J. Baker. 2005. Population divergence times and historical demography in Red Knots and Dunlins. *Condor* 107:497–513.

Carmona-Isunza, M. C., C. Küpper, M. A. Serrano-Meneses, and T. Székely. 2015. Courtship behavior differs between monogamous and polygamous plovers. *Behavioral Ecology and Sociobiology* 69:2035–2042.

Catry, T., P. M. Lourenço, R. J. Lopes, P. Bocher, C. Carneiro, J. A. Alves, P. Delaporte, S. Bearhop, T. Piersma, and J. P. Granadeiro. 2016. Use of stable isotope fingerprints to assign wintering origin and trace shorebird movements along the East Atlantic Flyway. *Basic and Applied Ecology* 17:177–187.

Chesser, R. T., R. C. Banks, F. K. Barker, C. Cicero, J. L. Dunn, A. W. Kratter, I. J. Lovette, P. C. Rasmussen, J. V. Remsen, J. D. Rising, D. F. Stotz, and K. Winker. 2011. Fifty-second supplement to the American Ornithologists' Union Check-List of North American Birds. *Auk* 128:600–613.

Christian, P. D., L. Christidis, and R. Schodde. 1992. Biochemical systematics of the Australian dotterels and plovers (Charadriiformes, Charadriidae). *Australian Journal of Zoology* 40:225–233.

Clements, J. F., T. S. Schulenberg, M. J. Iliff, D. Roberson, T. A. Fredericks, B. L. Sullivan, and C. L. Wood. [online] 2015. The eBird/Clements checklist of birds of the world: v2015. <www.birds.cornell.edu/clementschecklist/download/> (5 July 2016).

Corl, A., and H. Ellegren. 2012. The genomic signature of sexual selection in the genetic diversity of the sex chromosomes and autosomes. *Evolution* 66:2138–2149.

Corl, A., and H. Ellegren. 2013. Sampling strategies for species trees: The effects on phylogenetic inference of the number of genes, number of individuals, and whether loci are mitochondrial, sex-linked, or autosomal. *Molecular Phylogenetics and Evolution* 67:358–366.

Cramp, S., and K. E. L. Simmons. 1983. *Birds of the Western Palearctic*. Oxford University Press, Oxford, UK.

Cruz-López, M., L. J. Eberhart-Phillips, G. Fernández, R. Beamonte-Barrientos, T. Székely, M. A. Serrano-Meneses, and C. Küpper. 2017. The plight of a plover: Viability of an important Snowy Plover population with flexible brood care in Mexico. *Biological Conservation* 209:440–448.

Davey, J. W., P. A. Hohenlohe, P. D. Etter, J. Q. Boone, J. M. Catchen, and M. L. Blaxter. 2011. Genome-wide genetic marker discovery and genotyping using next-generation sequencing. *Nature Reviews Genetics* 12:499–510.

Dawson, D. A., M. Åkesson, T. Burke, J. M. Pemberton, J. Slate, and B. Hansson. 2007. Gene order and recombination rate in homologous chromosome regions of the chicken and a passerine bird. *Molecular Biology and Evolution* 24:1537–1552.

Degnan, J. H., and N. A. Rosenberg. 2006. Discordance of species trees with their most likely gene trees. *PLoS Genetics* 2:e68.

del Hoyo, J., A. Elliott, J. Sargatal, D. A. Christie, and E. de Juana (editors). [online]. 2016. *Handbook of the Birds of the World Alive*. Lynx Edicions, Barcelona, Spain. <www.hbw.com> (5 July 2016).

del Hoyo, J., and N. J. Collar. 2014. *HBW and Birdlife International Illustrated Checklist of the Birds of the World*. Volume 1. Non-passerines. Lynx Edicions and Birdlife International, Barcelona, Spain and Cambridge, UK.

DeQueiroz, K. 2007. Species concepts and species delimitation. *Systematic Biology* 56:879–886.

Dinsmore, S. J., G. C. White, and F. L. Knopf. 2003. Annual survival and population estimates of Mountain Plovers in southern Phillips County, Montana. *Ecological Applications* 13:1013–1026.

dos Remedios, N., C. Küpper, T. Burke, T. Székely, N. Baker, W. Versfeld, and P. L. M. Lee. 2017. Genetic isolation in an endemic African habitat specialist. *Ibis* 159:792–802.

dos Remedios, N., C. Küpper, T. Székely, S. Zefania, F. Burns, M. Bolton, and P. M. L. Lee (in press) Genetic structure among *Charadrius* plovers on the African continent and the islands of Madagascar and St Helena. *Ibis*. doi: 10.1111/ibi.12694

dos Remedios, N., P. L. M. Lee, T. Burke, T. Székely, and C. Küpper. 2015. North or south? Phylogenetic and biogeographic origins of a globally distributed avian clade. *Molecular Phylogenetics and Evolution* 89:151–159.

Dowding, J. E. 1999. Past distribution and decline of the New Zealand Dotterel (*Charadrius obscurus*) in the South Island of New Zealand. *Notornis* 46:167–180.

D'Urban Jackson, J., N. dos Remedios, K. H. Maher, S. Zefania, S. Haig, S. Oyler-McCance, D. Blomqvist, T. Burke, M. W. Bruford, T. Székely, and C. Küpper. 2017. Polygamy slows down population divergence in shorebirds. *Evolution* 71:1313–1326.

Eberhart-Phillips, L. J., B. R. Hudgens, and M. A. Colwell. 2015a. Spatial synchrony of a threatened shorebird: Regional roles of climate, dispersal and management. *Bird Conservation International* 26:119–135.

Eberhart-Phillips, L. J., J. I. Hoffman, E. G. Brede, S. Zefania, M. J. Kamrad, T. Székely, and M. W. Bruford. 2015b. Contrasting genetic diversity and population structure among three sympatric Madagascan shorebirds: Parallels with rarity, endemism, and dispersal. *Ecology and Evolution* 5:997–1010.

Edwards, S. V., S. B. Kingan, J. D. Calkins, C. N. Balakrishnan, W. B. Jennings, W. J. Swanson, and M. D. Sorenson. 2005. Speciation in birds: Genes, geography, and sexual selection. *Proceedings of the National Academy of Sciences USA* 102:6550–6557.

Elliott-Smith, E., and S. M. Haig [online]. 2004. Piping Plover (*Charadrius melodus*), *The Birds of North America Online* (A. Poole, Ed.). Cornell Lab of Ornithology, Ithaca, NY (July 2016).

Fain, M. G., and P. Houde. 2007. Multilocus perspectives on the monophyly and phylogeny of the order Charadriiformes (Aves). *BMC Evolutionary Biology* 7:35.

Foppen, R. P. B., F. A. Majoor, F. J. Willems, P. L. Meininger, G. C. van Houwelingen, and P. A. Wolf. 2006. Survival and emigration rates in Kentish *Charadrius alexandrinus* and ringed plovers *Ch. hiaticula* in the Delta area, SW-Netherlands. *Ardea* 94:159–173.

Funk, W. C., T. D. Mullins, and S. M. Haig. 2007. Conservation genetics of Snowy Plovers (*Charadrius alexandrinus*) in the Western Hemisphere: Population genetic structure and delineation of subspecies. *Conservation Genetics* 8:1287–1309.

Franks, S. E., D. R. Norris, T. K. Kyser, G. Fernandez, B. Schwarz, R. Carmona, M. A. Colwell, J. C. Sandoval, A. Dondua, H. R. Gates, B. Haase, D. J. Hodkinson, A. Jiménez, R. B. Lanctot, B. Ortego, B. K. Sandercock, F. Sanders, J. Y. Takekawa, N. Warnock, R. C. Ydenberg, and D. B. Lank. 2012. Range-wide patterns of migratory connectivity in the Western Sandpiper *Calidris mauri*. *Journal of Avian Biology* 43:155–167.

Frankham, R. 1997. Do island populations have less genetic variation than mainland populations? *Heredity* 78:311–327.

Gadow, H. 1892. On the classification of birds. *Proceedings of the Zoological Society of London* 1892:229–256.

Gill, F. B. 2014. Species taxonomy of birds: Which null hypothesis? *Auk* 131:150–161.

Gill, F. B., and D. Donsker [online]. 2015. IOC World Bird List (v 5.2). doi: 10.14344/IOC.

Gratto-Trevor, C., S. M. Haig, M. P. Miller, T. D. Mullins, S. Maddock, E. Roche, and P. Moore. 2016. Breeding sites and winter site fidelity of Piping Plovers wintering in The Bahamas, a previously unknown major wintering area. *Journal of Field Ornithology* 87:29–41.

Graul, W. D. 1973. Possible functions of head and breast markings in Charadriinae. *Wilson Bulletin* 85:60–70.

Gregory, T. R. [online] 2005. Animal Genome Size Database. <www.genomesize.com> (1 June 2017).

Haig, S. M., and J. D'Elia. 2010. Avian subspecies and the US endangered species act. *Ornithological Monographs* 67:24–34.

Haig, S. M., and L. W. Oring. 1988. Genetic differentiation of Piping Plovers across North America. *Auk* 105:260–267.

Haig, S. M., W. M. Bronaugh, R. S. Crowhurst, J. D'Elia, C. A. Eagles-Smith, C. W. Epps, B. Knaus, M. P. Miller, M. L. Moses, S. Oyler-McCance, W. D. Robinson, and B. Sidlauskas. 2011. Genetic applications in avian conservation. *Auk* 128:205–229.

Hebert, P. D., M. Y. Stoeckle, T. S. Zemlak, and C. M. Francis. 2004. Identification of birds through DNA barcodes. *PLoS Biology* 2:e312.

Herbert J. M., J. E. Dowding, and C. H. Daugherty. 1993. *Conservation Advisory Science Notes No. 41: Genetic Variation and Systematics of the New Zealand Dotterel*. Department of Conservation, Wellington, New Zealand.

Hickerson, M. J., B. C. Carstens, J. Cavender-Bares, K. A. Crandall, C. H. Graham, J. B. Johnson, L. Rissler, P. F. Victoriano, and A. D. Yoder. 2010. Phylogeography's past, present, and future: 10 years after. *Molecular Phylogenetics and Evolution* 54:291–301.

Hillis, D. M. 1987. Molecular versus morphological approaches to systematics. *Annual Review of Ecological Systematics* 18:23–42.

Ho, S. Y. W., and S. Duchene. 2014. Molecular-clock methods for estimating evolutionary rates and timescales. *Molecular Ecology* 23:5947–5965.

Hobson, K. A. 1999. Tracing origins and migration of wildlife using stable isotopes: A review. *Oecologia* 120:314–326.

Höglund, J. 2009. *Evolutionary Conservation Genetics*. Oxford University Press, Oxford, UK.

Höglund, J., T. Johansson, A. Beintema, and H. Schekkerman. 2009. Phylogeography of the Black-tailed Godwit *Limosa limosa*: Substructuring revealed by mtDNA control region sequences. *Journal of Ornithology* 150:45–53.

Hoogerwerf, A. 1966. On the validity of *Charadrius alexandrinus javanicus* Chasen and the occurrence of *Charadrius alexandrinus ruficapillus* Temm. And of *Charadrius peronii* Schl. on Java and in New Guinea. *Philippine Journal of Science* 95:209–214.

Iqbal, M., I. Taufiqurrahman, K. Yordan, and B. van Balan. 2013. The distribution, abundance and conservation status of the Javan Plover *Charadrius javanicus*. *Wader Study Group Bulletin* 120:1–5.

IUCN [online]. 2015. The IUCN Red List of Threatened Species. Version 2015-4. <www.iucnredlist.org.>

Jackson, H., B. J. Morgan, and J. J. Groombridge. 2013. How closely do measures of mitochondrial DNA control region diversity reflect recent trajectories of population decline in birds? *Conservation Genetics* 14:1291–1296.

Jarvis, E. D., S. Mirarab, A. J. Aberer, B. Li, P. Houde, C. Li, S. Y. W. Ho, B. C. Faircloth, B. Nabholz, J. T. Howard, A. Suh, C. C. Weber, R. R. da Fonseca, J. Li, F. Zhang, H. Li, L. Zhou, N. Narula, L. Liu, G. Ganapathy, B. Boussau, M. S. Bayzid, V. Zavidovych, S. Subramanian, T. Gabaldon, S. Capella-Gutierrez, J. Huerta-Cepas, B. Rekepalli, K. Munch, M. Schierup, B. Lindow, W. C. Warren, D. Ray, R. E. Green, M. W. Bruford, X. Zhan, A. Dixon, S. Li, N. Li, Y. Huang, E. P. Derryberry, M. F. Bertelsen, F. H. Sheldon, R. T. Brumfield, C. V. Mello, P. V. Lovell, M. Wirthlin, M. P. C. Schneider, F. Prosdocimi, J. A. Samaniego, A. M. V. Velazquez, A. Alfaro-Nunez, P. F. Campos, B. Petersen, T. Sicheritz-Ponten, A. Pas, T. Bailey, P. Scofield, M. Bunce, D. M. Lambert, Q. Zhou, P.

Perelman, A. C. Driskell, B. Shapiro, Z. Xiong, Y. Zeng, S. Liu, Z. Li, B. Liu, K. Wu, J. Xiao, X. Yinqi, Q. Zheng, Y. Zhang, H. Yang, J. Wang, L. Smeds, F. E. Rheindt, M. Braun, J. Fjeldsa, L. Orlando, F. K. Barker, K. A. Jonsson, W. Johnson, K. P. Koepfli, S. O'brien, D. Haussler, O. A. Ryder, C. Rahbek, E. Willerslev, G. R. Graves, T. C. Glenn, J. Mccormack, D. Burt, H. Ellegren, P. Alstrom, S. V. Edwards, A. Stamatakis, D. P. Mindell, J. Cracraft, E. L. Braun, T. Warnow, W. Jun, M. T. P. Gilbert, and G. Zhang. 2014. Whole-genome analyses resolve early branches in the Tree of Life of modern birds. *Science* 346:1320–331.

Jetz, W., G. H. Thomas, J. B. Joy, K. Hartmann, and A. O. Mooers. 2012. The global diversity of birds in space and time. *Nature* 491:444–448.

Joseph, L., E. P. Lessa, and L. Christidis. 1999. Phylogeny and biogeography in the evolution of migration: Shorebirds of the *Charadrius* complex. *Journal of Biogeography* 26:329–342.

Julliard, R., Jiguet, F., and D. Couvet. 2004. Common birds facing global changes: What makes a species at risk? *Global Change Biology* 10:48–154.

Kapusta, A., and A. Suh. 2017. Evolution of bird genomes—a transposon's-eye view. *Annals of the New York Academy of Sciences* 1389:164–185.

Kennerley, P. R., D. N. Bakewell, and P. D. Round. 2008. Rediscovery of a long-lost *Charadrius* plover from South-East Asia. *Forktail* 24:63–79.

Knopf, F. L., and M. B. Wunder. [online] 2006. Mountain Plover (*Charadrius montanus*), *The Birds of North America Online* (A. Poole, Ed.), Cornell Lab of Ornithology, Ithaca, NY. (Accessed July 2016).

Kraaijeveld, K. 2008. Non-breeding habitat preference affects ecological speciation in migratory waders. *Naturwissenschaften* 95:347–354.

Küpper, C., J. Augustin, A. Kosztolányi, T. Burke, J. Figuerola, and T. Székely. 2009. Kentish versus Snowy Plover: Phenotypic and genetic analyses of *Charadrius alexandrinus* reveal divergence of Eurasian and American subspecies. *Auk* 126:839–852.

Küpper, C., J. Kis, A. Kosztolányi, T. Székely, I. C. Cuthill, and D. Blomqvist. 2004. Genetic mating system and timing of extra-pair fertilizations in the Kentish Plover. *Behavioral Ecology and Sociobiology* 57:32–39.

Küpper, C., M. Stocks, J. E. Risse, N. dos Remedios, L. L. Farrell, S. B. McRae, T. C. Morgan, N. Karlionova, P. Pinchuk, Y. I. Verkuil, A. S. Kitaysky, J. C. Wingfield, T. Piersma, K. Zeng, J. Slate, M. Blaxter, D. B. Lank, and T. Burke. 2016. A supergene determines highly divergent male reproductive morphs in the Ruff. *Nature Genetics* 48:79–83.

Küpper, C., S. V. Edwards, A. Kosztolányi, M. Alrashidi, T. Burke, P. Herrmann, A. Argüelles-Tico, J. A. Amat, M. Amezian, A. Rocha, H. Hötker, A. Ivanov, J. Chernicko, and T. Székely. 2012. High gene flow on a continental scale in the polyandrous Kentish Plover *Charadrius alexandrinus*. *Molecular Ecology* 21:5864–5879.

Küpper, C., T. Burke, T. Székely, and D. A. Dawson. 2008. Enhanced cross-species utility of conserved microsatellite markers in shorebirds. *BMC Genomics* 9:502.

Lamichhaney, S., G. Fan, F. Widemo, U. Gunnarsson, D. Schwochow Thalmann, M. P. Hoeppner, S. Kerje, U. Gustafson, C. Shi, H. Zhang, W. Chen, X. Liang, L. Huang, J. Wang, E. Liang, Q. Wu, S. M. Y. Lee, X. Xu, J. Höglund, X. Liu, and L. Andersson. 2016. Structural genomic changes underlie alternative reproductive strategies in the Ruff (*Philomachus pugnax*). *Nature Genetics* 48:84–88.

Lank, D. B., C. M. Smith, O. Hanotte, A. Ohtonen, S. Bailey, and T. Burke. 2002. High frequency of polyandry in a lek mating system. *Behavioral Ecology* 13:209–215.

Lavinia, P. D., K. C. R. Kerr, P. L. Tubaro, P. D. N. Hebert, and D. A. Lijtmaer. 2016. Calibrating the molecular clock beyond cytochrome b: Assessing the evolutionary rate of COI in birds. *Journal of Avian Biology* 47:84–91.

Lemmon, A. R., J. M. Brown, K. Stanger-Hall, and E. M. Lemmon. 2009. The effect of ambiguous data on phylogenetic estimates obtained by maximum likelihood and Bayesian inference. *Systematic Biology* 58:130–145.

Li, S., R. Jovelin, T. Yoshiga, R. Tanaka, and A. D. Cutter. 2014. Specialist versus generalist life histories and nucleotide diversity in *Caenorhabditis* nematodes. *Proceedings of the Royal Society B* 281:20132858.

Linnaeus, C. 1758–1759. *Systema naturae per regna tria naturae*. 10th ed. Rev. Volume 2. (L. Salmii Homiiae, Ed.) Impensis Direct. Laurentii Salvii (in Latin). https://doi.org/10.5962/bhl.title.542

Livezey, B. C. 2010. Phylogenetics of modern shorebirds (Charadriiformes) based on phenotypic evidence: Analysis and discussion. *Zoological Journal of the Linnean Society* 160:567–618.

Long, P. R., S. Zefania, R. H. ffrench-Constant, and T. Székely. 2008. Estimating the population size of an endangered shorebird, the Madagascar Plover, using a habitat suitability model. *Animal Conservation* 11:118–127.

Lovette, I. J. 2005. Glacial cycles and the tempo of avian speciation. *Trends in Ecology and Evolution* 20:57–59.

Maclean, G. L. 1972. Problems of display postures in the Charadrii (Aves: Charadriiformes). *African Zoology* 7:57–74.

Mayr, G. 2011. The phylogeny of charadriiform birds (shorebirds and allies)–reassessing the conflict between morphology and molecules. *Zoological Journal of the Linnean Society* 161:916–934.

Maher, K. H., L. J. Eberhart-Phillips, A. Kosztolányi, N. dos Remedios, M. C. Carmona-Isunza, M. Cruz-López, S. Zefania, J. J. H. St. Clair, M. AlRashidi, M. A. Weston, M. A. Serrano-Meneses, O. Krüger, J. I. Hoffman, T. Székely, T. Burke, and C. Küpper. 2017. High fidelity: Extra-pair fertilisations in eight *Charadrius* plover species are not associated with parental relatedness or social mating system. *Journal of Avian Biology* 48:910–920.

Miller, M. P., S. M. Haig, C. L. Gratto-Trevor, and T. D. Mullins. 2010. Subspecies status and population genetic structure in Piping Plover (*Charadrius melodus*). *Auk* 127:57–71.

Mills, L. S., and F. W. Allendorf. 1996. The one-migrant-per-generation rule in conservation and management. *Conservation Biology* 10:1509–1518.

Mitra, S., H. Landel, and S. Pruett-Jones. 1996. Species richness covaries with mating system in birds. *Auk* 113:544–551.

Moritz, C. 1994. Defining 'evolutionarily significant units' for conservation. *Trends in Ecology and Evolution* 9:373–374.

Morrow, E. H., T. E. Pitcher, and G. Arnqvist. 2003. No evidence that sexual selection is an 'engine of speciation' in birds. *Ecology Letters* 6:228–234.

Nabholz, B., S. Glémin, and N. Galtier. 2009. The erratic mitochondrial clock: Variations of mutation rate, not population size, affect mtDNA diversity across birds and mammals. *BMC Evolutionary Biology* 9:54.

Nadachowska-Brzyska, K., C. Li, L. Smeds, G. Zhang, and H. Ellegren. 2015. Temporal dynamics of avian populations during Pleistocene revealed by whole-genome sequences. *Current Biology* 25:1375–1380.

Nixon, K. C., and J. M. Carpenter. 2012. On homology. *Cladistics* 28:160–169.

Nunney, L. 1993. The influence of mating system and overlapping generations on effective population size. *Evolution* 47:1329–1341.

Oring, L. W., R. C. Fleischer, J. M. Reed, and K. E. Marsden. 1992. Cuckoldry through stored sperm in the sequentially polyandrous spotted sandpiper. *Nature* 359:631–633.

Oyler-McCance, S. J., J. St. John, F. L. Knopf, and T. W. Quinn. 2005. Population genetic analysis of Mountain Plover using mitochondrial DNA sequence data. *Condor* 107:353–362.

Oyler-McCance, S. J., J. St. John, R. F. Kysela, and F. L. Knopf. 2008. Population structure of Mountain Plover as determined using nuclear microsatellites. *Condor* 110:493–499.

Paradis, E., S. R. Baillie, W. J. Sutherland, and R. D. Gregory. 1998. Patterns of natal and breeding dispersal in birds. *Journal of Animal Ecology* 67:518–536.

Parham, J. F., P. C. J. Donoghue, C. J. Bell, T. D. Calway, J. J. Head, P. A. Holroyd, J. G. Inoue, R. B. Irmis, W. G. Joyce, D. T. Ksepka, J. S. L. Patané, N. D. Smith, J. E. Tarver, M. van Tuinen, Z. Yang, K. D. Angielczyk, J. M. Greenwood, C. A. Hipsley, L. Jacobs, P. J. Makovicky, J. Müller, K. T. Smith, J. M. Theodor, R. C. M. Warnock, and M. J. Benton. 2012. Best practices for justifying fossil calibrations. *Systematic Biology* 61:346–359.

Pearson, W. J., and M. A. Colwell. 2014. Effects of nest success and mate fidelity on breeding dispersal in a population of Snowy Plovers *Charadrius nivosus*. *Bird Conservation International* 24:342–353.

Prum, R. O., J. S. Berv, A. Dornburg, D. J. Field, J. P. Townsend, E. M. Lemmon, and A. R. Lemmon. 2015. A comprehensive phylogeny of birds (Aves) using targeted next-generation DNA sequencing. *Nature* 526:569–573.

Rheindt, F. E., T. Székely, S. V. Edwards, P. L. M. Lee, T. Burke, P. R. Kennerley, D. N. Bakewell, M. AlRashidi, A. Kosztolányi, M. A. Weston, W. Liu, W. Lei, Y. Shigeta, S. Javed, S. Zefania, and C. Küpper. 2011. Conflict between genetic and phenotypic differentiation: The evolutionary history of a 'lost and rediscovered' shorebird. *PLoS One* 6:e26995.

Rocque, D. A., M. Ben-David, R. P. Barry, and K. Winker. 2006. Assigning birds to wintering and breeding grounds using stable isotopes: Lessons from two feather generations among three intercontinental migrants. *Journal of Ornithology* 147:395–404.

Roure, B., D. Baurain, and H. Philippe. 2013. Impact of missing data on phylogenies inferred from empirical phylogenomic data sets. *Molecular Biology and Evolution* 30:197–214.

Ryder, O. A. 1986. Species conservation and systematics: The dilemma of subspecies. *Trends in Ecology and Evolution* 1:9–10.

Sangster, G. 2014. The application of species criteria in avian taxonomy and its implications for the debate over species concepts. *Biological Reviews* 89:199–214.

Sangster, G., J. M. Collinson, P. A. Crochet, G. M. Kirwan, A. G. Knox, D. T. Parkin, and S. C. Votier. 2016. Taxonomic recommendations for Western Palearctic birds: 11th report. *Ibis* 158:206–212.

Servedio, M. R., and R. Bürger. 2014. The counterintuitive role of sexual selection in species maintenance and speciation. *Proceedings of the National Academy of Sciences* 111:8113–8118.

Sibley, C. G., and J. E. Ahlquist. 1990. *Phylogeny and Classification of Birds: A Study in Molecular Evolution*. Yale University Press, New Haven, CT.

Sibley, C. G., and B. L. Monroe. 1990. *Distribution and Taxonomy of Birds of the World*. Yale University Press, New Haven, CT.

Stenzel, L. E., J. C. Warriner, J. S. Warriner, K. S. Wilson, F. C. Bidstrup, and G. W. Page. 1994. Long-distance breeding dispersal of Snowy Plovers in western North America. *Journal of Animal Ecology* 63:887–902.

Stenzel, L. E., G. W. Page, J. C. Warriner, J. S. Warriner, D. E. George, C. R. Eyster, B. A. Ramer, and K. K. Neuman 2007. Survival and natal dispersal of juvenile snowy plovers in central coastal California. *Auk* 124:1023–1036.

Suh, A., L. Smeds, and H. Ellegren. 2015. The dynamics of incomplete lineage sorting across the ancient adaptive radiation of neoavian birds. *PLoS Biology* 13:e1002224.

Sunnucks, P. 2000. Efficient genetic markers for population biology. *Trends in Ecology and Evolution* 15:199–203.

Thomas, S. M., J. E. Lyons, B. A. Andres, E. Elliot-Smith, E. Palacios, J. F. Cavitt, J. A. Royle, S. D. Fellows, K. Maty, W. H. Howe, E. Mellink, S. Melvin, and T. Zimmerman. 2012. Population size of Snowy Plovers breeding in North America. *Waterbirds* 35:1–14.

Thies, L., P. Tomkovich, N. dos Remedios, T. Lislevand, P. Pinchuk, J. Wallander, J. Dänhardt, B. Þórisson, D. Blomqvist, and C. Küpper. 2018. Population and subspecies differentiation in a high latitude breeding wader, the Common Ringed Plover *Charadrius hiaticula*. *Ardea* 106:163–176.

Tobias, J. A., N. Seddon, C. N. Spottiswoode, J. D. Pilgrim, L. D. Fishpool, and N. J. Collar. 2010. Quantitative criteria for species delimitation. *Ibis* 152:724–746.

van Rhijn, J. G. 1985. A scenario for the evolution of social organization in ruffs *Philomachus pugnax* and other Charadriiform species. *Ardea* 73:25–37.

van Tuinen, M., D. Waterhouse, and G. J. Dyke. 2004. Avian molecular systematics on the rebound: A fresh look at modern shorebird phylogenetic relationships. *Journal of Avian Biology* 35:191–194.

Vaughan, R. 1980. *Plovers*. Terance Dalton Ltd., Lavenham, UK.

Verkuil, Y. I., C. Juillet, D. B. Lank, F. Widemo, and T. Piersma. 2014. Genetic variation in nuclear and mitochondrial markers supports a large sex difference in lifetime reproductive skew in a lekking species. *Ecology and Evolution* 4:3626–3632.

Verkuil, Y. I., T. Piersma, J. Jukema, J. C. E. W. Hooijmeijer, L. Zwarts, and A. J. Baker. 2012. The interplay between habitat availability and population differentiation: A case study on genetic and morphological structure in an inland wader (Charadriiformes). *Biological Journal of the Linnean Society* 106:641–656.

Wake, D. B., M. H. Wake, and C. D. Specht. 2011. Homoplasy: From detecting pattern to determining process and mechanism of evolution. *Science* 331:1032–1035.

Waples, R. S. 1998. Evolutionarily significant units, distinct population segments, and the Endangered Species Act: Reply to Pennock and Dimmick. *Conservation Biology* 12:718–721.

Warren, W. C., L. W. Hillier, C. Tomlinson, P. Minx, M. Kremitzki, T. Graves, C. Markovic, N. Bouk, K. D. Pruitt, F. Thibaud-Nissen, V. Schneider, T. A. Mansour, C. T. Brown, A. Zimin, R. Hawken, M. Abrahamsen, A. B. Pyrkosz, M. Morisson, V. Fillon, A. Vignal, W. Chow, K. Howe, J. E. Fulton, M. M. Miller, P. Lovell, C. V. Mello, M. Wirthlin, A. S. Mason, R. Kuo, D. W. Burt, J. B. Dodgson, and H. H. Cheng. 2017. A new chicken genome assembly provides insight into avian genome structure. *G3: Genes, Genomes, Genetics* 7:109–117.

Warriner, J. S., J. C. Warriner, G. W. Page, and L. E. Stenzel. 1986. Mating system and reproductive success of a small population of polygamous Snowy Plovers. *Wilson Bulletin* 98:15–37.

Wenink, P. W., A. J. Baker, and M. G. Tilanus. 1994. Mitochondrial control-region sequences in two shorebird species, the turnstone and the Dunlin, and their utility in population genetic studies. *Molecular Biology and Evolution* 11:22–31.

West-Eberhard, M. J. 1983. Sexual selection, social competition, and speciation. *The Quarterly Review of Biology* 58:155–183.

Winker, K. 2010. Subspecies represent geographically partitioned variation, a goldmine of evolutionary biology, and a challenge for conservation. *Ornithological Monographs* 67:6–23.

Zayed, A., L. Packer, J. C. Grixti, L. Ruz, R. E. Owen, and H. Toro. 2005. Increased genetic differentiation in a specialist versus a generalist bee: Implications for conservation. *Conservation Genetics* 6:1017–1026.

Zefania S., and T. Székely. 2013. Charadrius spp. pp. 395–403, *The Birds of Africa, Volume VIII: Birds of the Malagasy Region: Madagascar, Seychelles, Comoros, Mascarenes* (R. Safford and F. Hawkins, Eds.). Bloomsbury Publishing, London, UK.

Zink, R. M. 2004. The role of subspecies in obscuring avian biological diversity and misleading conservation policy. *Proceedings of the Royal Society B* 271:561–564.

CHAPTER THREE

Changing Climates and Challenges to *Charadrius* Plover Success throughout the Annual Cycle*

Susan M. Haig

Abstract. The arctic tundra, as well as coastal and inland mudflats and beaches occupied by the 63 *Charadrius* plover species and subspecies around the world, encompass some of the habitats most threatened by current climatic challenges. The migratory habits of most plover species further intensifies these effects as the birds occupy more than one major biome during the annual cycle. And yet, there have only been two plover species where specific issues related to climate change have been addressed. In this chapter, I summarize climate-related issues in areas occupied by the world's *Charadrius* plovers to highlight further research and management to at least slow the negative effects of our changing world on their success. To be most strategic and effective, management and research approaches carried out with full knowledge or investigation of the species' annual cycle and migratory connectivity will be most informative. Given the dearth of climate-related information for this group of birds, future work will likely help not only plovers but other species occupying similar habitats around the world.

Keywords: annual cycle, Arctic tundra, coast, climate change, *Charadrius*, migratory connectivity, phenological mismatch, plovers, sea level rise, water quality.

Owing in large part to increasing greenhouse gas concentrations and population and industrial growth, the temperature of combined land and ocean surfaces increased by 0.85°C between 1880 and 2012 (IPCC 2014a,b). These changes in our world's climates may affect *Charadrius* plovers more than many other groups of avian species because they are so tightly tied to water and water quality (Haig et al. in review). Migrating shorebirds are sentinels of global environmental change as they experience the effects of climate change in several major biomes during their annual cycle (Tables 3.1 and 3.2; Piersma and Lindstrom 2004, Munro 2017). These birds are likely experiencing increases in inland water salinity and decreases of freshwater as well as changing ocean chemistry and rising sea levels (Tables 3.1 and 3.2). If they are sedentary or have high nest site fidelity, they are at the mercy of changing habitat availability and water quality with little chance to escape these factors unless rapid adaptation takes place. In this chapter, I will explore the many changing climatological factors affecting the world's *Charadrius* plovers with the

* Susan M. Haig. 2019. Changing Climates and Challenges to *Charadrius* Plover Success throughout the Annual Cycle. Pp. 45–62 in M. A. Colwell and S. M. Haig (editors). The Population Ecology and Conservation of *Charadrius* Plovers Studies in Avian Biology (no. 52), CRC Press, Boca Raton, FL.

TABLE 3.1

Inland arid, coastal/estuarine, and tundra habitats occupied by Charadrius plovers throughout the annual cycle.

Common name	Latin name	Breeding area	Migration area	Winter area	Non-migrant
Southern Red-breasted Dotterel	*C. obscurus*	I	IC	C	
Northern Red-breasted Dotterel	*C. aquilonius*				C
Lesser Sand-plover	*C. m. mongolus*	I	I	C	
	C. m. pamirensis	I	I	C	
	C. m. atrifrons	I	I	C	
	C. m. schaeferi	I	I	C	
	C. m. stegmanni	I	I	C	
Greater Sand-plover	*C. l. leschenaultii*	I	IC	IC	
	C. l. columbinus	I	IC	IC	
	C. l. scythicus	I	IC	IC	
Caspian Plover	*C. asiaticus*	I	I	I	
Collared Plover	*C. collaris*				IC
Puna Plover	*C. alticola*				I
Two-banded Plover	*C. falklandicus*				C
Double-banded Plover	*C. bicinctus*	IC	IC	IC	
	C. b. exilis	I	IC	C	
Kittlitz's Plover	*C. pecuarius*				IC
Red-capped Plover	*C. ruficapillus*				IC
Malay Plover	*C. peronii*				C
Kentish Plover	*C. a. alexandrinus*	C	C	C	
	C. a. seebohmi	C	C	C	
	C. a. nihonensis				C
	C. a. dealbatus	C	C	C	
Snowy Plover	*C. n. nivosus*	IC	IC	C	
	C. n. occidentalis	C	C	C	
Javan Plover	*C. javanicus*				C
Wilson's Plover	*C. w. wilsonia*	C	C	C	C
	C. w. beldingi				C
	C. w. cinnamominus				C
	C. w. crassirostris				C
Common Ringed Plover	*C. h. hiaticula*	T	IC	C	
	C. h. Arctice	T	IC	C	
Semipalmated Plover	*C. semipalmatus*	T	IC	C	
Long-billed Plover	*C. placidus*	I	IC	IC	
Piping Plover	*C. m. melodus*	C	C	C	
	C. m. circumcinctus	I	IC	C	
Little Ringed Plover	*C. d. dubius*	I	I	I	
	C. d. curonicus	I	I	I	
	C. d. jerdoni				I

(Continued)

TABLE 3.1 (*Continued*)

Inland arid, coastal/estuarine, and tundra habitats occupied by Charadrius plovers throughout the annual cycle.

Common name	Latin name	Breeding area	Migration area	Winter area	Non-migrant
African Three-banded Plover	*C. tricollaris*				IC
Madagascar Plover	*C. thoracicus*				IC
Forbes's Plover	*C. forbesi*				IC
White-fronted Plover	*C. m. marginatus*				IC
	C. m. mechowi				IC
	C. m. arenaceus				IC
	C. m. tenellus				IC
Chestnut-banded Plover	*C. p. pallidus*	I	I	I	
	C. p. venustus				I
Killdeer	*C. v. vociferus*	I	IC	IC	IC
	C. v. ternominatus				IC
	C. v. peruvianus				IC
Mountain Plover	*C. montanus*	I	I	I	
Oriental Plover	*C. veredus*	I	I	C	
Eurasian Dotterel	*C. morinellus*	T	I	I	
St. Helena Plover	*C. sanctaehelenae*				I
Rufous-chested Plover	*C. modestus*	I	I	IC	
Red-kneed Dotterel	*Erythrogonys cinctus*				IC
Hooded Plover	*Thinornis c. cucullatus*				IC
	Thinornis c. tregellasi				IC
Shore Plover	*T. novaeseelandiae*				C
Black-fronted Dotterel	*Elseyornis melanops*				I
Inland Dotterel	*Peltohyas australis*	I	I	I	
Wrybill	*Anarhynchus frontalis*	I	C	C	

SOURCE: Cramp and Simmons 1983, Hayman et al. 1986, and Piersma and Wiersma 1996.
C, coastal; I, inland; T, Tundra.

TABLE 3.2

Habitat use by Charadrius plover species (N) throughout the annual cycle.

Habitat	Breeding	Migration	Winter	Sedentary	Sum	B + S	% Breeding habitat
Tundra	4	0	0	0	4	4	6.15
Coastal and Estuarine	14	9	6	5	34	19	29.23
Arid Inland	2	9	4	16	31	18	27.69

SOURCE: Cramp and Simmons 1983, Hayman et al. 1986, and Piersma and Wiersma 1996.

hope of spurring more climate-related research and informing management of plover species. I begin by exploring the broad aspects of climate change potentially impacting plovers and then delve more specifically into challenges posed to birds using tundra, inland, and coastal systems. I end with research and management suggestions.

Detailed assessments of the effect of climate change on plover species worldwide are limited to two species (Piping Plover [McCauley et al. 2015,

Seavey et al. 2010] and Snowy Plover [Aiello-Lammens et al. 2011]). Even so, there is little information about changes in these species' population estimates, distributions, abundances, or reproductive success that would help assess the effects of climate change (Andres et al. 2012). However, generalized analyses by Partners in Flight (NABCI 2010) and the U.S. Shorebird Conservation Plan (Brown et al. 2001) provide broad insight into the North American plover species. Following up on results from these plans, Galbraith et al. (2014) reported that climate change exacerbates the risk of extinction for nearly 90% of North American shorebirds, including the six breeding Charadrius plover species. They further concluded that the predicted change in extinction risk to Charadrius species based on climate change was at least one rank higher than previously suggested by Partners in Flight (NABCI 2010) for four of the six species (Killdeer and Mountain Plover were exceptions). The following topics, including benefits and alternative hypotheses (Knudsen et al. 2011), are key to examining climate change with respect to Charadrius plovers.

SPATIAL SYNCHRONY VERSUS TEMPORAL MISMATCH

Spatial synchrony refers to the coincident change in abundance across geographically distinct populations (i.e., the Moran Effect; Moran 1953). Extreme weather events can synchronize population fluctuations across an entire metapopulation or community. The Moran Effect has been shown to be an important consideration in metapopulation conservation because as spatial synchrony among populations increases, the chance of demographic rescue of a population in crisis decreases. Thus, there is less of a chance that emigration from another population will occur. Koenig and Liebhold (2016) recently found a strong correlation between temperature changes and increased spatial synchrony among 50 wintering bird species. Their results for Killdeer (860 sites examined) suggested the strongest correlation was among populations that were within 500 km of each other. Further analyses for specific situations, including consideration of temperature and rainfall, may provide greater insight into the effects changing climates are having on metapopulations.

Shifts in the phenology or timing of seasonal activities can cause a temporal or phenological mismatch when the availability of a resource changes temporally with respect to requirements of the consumer. These mismatches are climate-related phenomena that continue to have serious effects on species, food webs, and ecosystems worldwide (e.g., Hughes 2000, Both and Visser 2001, Durant et al. 2007). Two factors—degree of phenological mismatch and migratory distance—influence the effect that changes in phenology will have on migratory populations (Stutzman and Fontaine 2015). For birds, temporal mismatch often results when prey or food items become available sooner (or later) than needed for breeding or migratory activities (Both et al. 2004, Dunn 2004, Stutzman and Fontaine 2015). This phenomenon has not been studied long enough in most avian systems to clearly interpret the effect on the species or the food chain (Visser and Both 2006), although Møller et al. (2008) found that populations of migratory birds that did not show a phenological response to climate change were declining.

Charadrius plovers present a special case in temporal mismatch because their invertebrate prey species need to be available for the chicks as well as for the adults. This extends the date when food must be available. Even so, most studies addressing direct cause and effect of temporal mismatch and chick growth rates or survival are inferential at best, often due to the difficulty in obtaining chick growth rates in the field. For example, studies of arctic-nesting Baird's Sandpiper (*Calidris bairdii*) suggested a possible mismatch between chick hatch dates and peak abundance of their crane fly (Tipulidae) food source (McKinnon et al. 2012). A subsequent paper reported no mismatch as chicks could maintain good growth rates, even when faced with low arthropod availability, due to thermogenic relief (McKinnon et al. 2013). This suggests an important lesson in pursuing multiple hypotheses when addressing cause and effect of climate-induced changes.

Mismatch was predicted in U.K. nesting Golden Plovers (*Pluvialis apricaria*), but the long-term effects have not yet been determined (Pearce-Higgins et al. 2005). More recently, a temporal mismatch between breeding phenology and invertebrate peak was discovered in a U.K. breeding Hudsonian Godwit (*Limosa haemastica*) population, which showed a relationship between lower invertebrate abundance and lower survival in older chicks, but not for younger chicks, highlighting the complex interplay of mismatch effects (Senner et al. 2017). Interestingly, they also show how the same species at different sites exhibits different phenological responses to climate change, with different effects on chick growth and survival.

Perhaps most dramatically, migratory Red Knots (*Calidris canutus*) appear to be demonstrating the negative effects of a temporal mismatch that occurs on the breeding grounds, but negative results are manifested in winter. Van Gils et al. (2016) recently documented reduced body and bill size in Red Knots and hypothesized that phenological mismatch of their invertebrate prey due to early snowmelt in their arctic breeding grounds would lead to smaller body sizes with shorter bills in maturing chicks. They have also observed changes in reducing survival rates in their tropical Southern Hemisphere wintering grounds where shorter-billed knots have lower survival rates. Van Gils et al. (2016) hypothesized that body shrinkage is a genetic microevolutionary response to warming, reasoning that smaller individuals are better able to dissipate body heat because of the larger surface/volume ratio of their bodies (i.e., Bergmann's Rule; Bergmann 1847). Alternatively, climate change may disrupt trophic interactions, potentially leading to malnutrition during an organism's juvenile life stage. Because poor growth may not be compensated for later in life, this would lead to smaller bodies (i.e., shrinkage as a phenotypically plastic response). This work points to the importance of studying birds throughout the annual cycle and determining migratory connectivity among populations. Gauthier et al. (2013) further warns that "long-term monitoring at multiple trophic levels suggests heterogeneity in responses to climate change." Thus, demonstrating the mechanism and effect of a potential mismatch is quite complicated.

MIGRATORY CONNECTIVITY

Determining migratosry connectivity and understanding full life cycle biology is critical in this age of rapidly changing climates. Charadrius plovers present a textbook example of how understanding movement patterns and habitat use throughout the annual cycle is paramount to providing appropriate conservation measures. Tables 3.1 and 3.2 illustrate habitats used by these similar species that have fairly common life histories. For example, most plovers spend a good portion of the annual cycle inland, facing issues related to drought and salinity, as well as part of the year in coastal areas changing as a result of sea level rise and changing sea water chemistry. As in the Red Knot example, addressing issues in more than one phase of the annual cycle may help to understand the entire life cycle and defining connectivity among populations will provide a much better perspective on where management or conservation actions will have the greatest impact. For example, declines in British-breeding populations of Afro-Palearctic migrant birds are linked to a bioclimatic wintering zone in Africa, possibly due to constraints on earlier arrival times (Ockendon et al. 2012). These results would be hard to characterize without knowledge of migratory connectivity and the species' full life cycle biology. Thus, taking migratory connectivity into consideration to tease apart climatic influences on birds is key, although scientific approaches to studying connectivity and appropriate tracking technology are just emerging.

There is agreement among biologists, agencies, and NGOs that understanding migratory connectivity is important as climates change (NABCI 2010). Hedenström et al. (2007) illustrated that, among birds, "departure from the wintering site is advanced in relation to the advancement of spring if the molt is in summer, but not so for species with a winter molt, while arrival at the breeding site is advanced for both molt scenarios." Their modeling also demonstrated that the timing of breeding and the number of successful broods could be affected by the number of weeks spring has advanced. In general, a 6-week spring advancement resulted in a 4-week shift in the breeding season. However, the start of molt was relatively unaffected by climate change. Further, in arctic-breeding plovers (Tables 3.1 and 3.2), spring migration can be very quick as birds are responding to short breeding seasons in the far north (Hedenström et al. 2007). If food availability has been shifted along the migration route, birds may not bring enough reserves to the breeding ground to carry out a successful breeding attempt (Drent et al. 2006). Both et al. (2010) determined that consequences of climate change are most severe for long-distance migrants in seasonal habitats. Iwamura et al. (2013) also found that migratory connectivity magnifies the consequences of habitat loss if a shorebird's migratory route contains a bottleneck such as sea level rise.

An interesting pattern arises when comparing the effects of climate change on migrant Nearctic and Palearctic birds. In general, migrant birds are particularly vulnerable to mismatches in resource availability, as climate change does not act equally across the globe or at each stage of the annual cycle. Jones and Cresswell (2010) examined the potential effect of climate change on 193 subspecies or populations of birds across the Nearctic and Palearctic regions (including Little Ringed Plovers) over a 10- to 20-year period and found that phenology mismatch was correlated with population declines in the Nearctic species, whereas migration distance was more important in predicting migratory schedules of Palearctic species. This suggests that differential global climate change may be responsible for contributing to some migrant species' declines, but its effects may be more important in the Nearctic.

Comparisons of the effects of climate change on species survivorship and reproductive success must be carried out across species annual cycles to identify the most critical bottlenecks. Ockendon et al. (2013) argued that climatic effects on breeding grounds are more important drivers of breeding phenology in migrant birds than carryover effects from wintering grounds. They modeled 45 years of precipitation and temperature data in wintering and breeding areas for 19 migratory species in the UK. They determined that spring temperature accounted for 3.5 times more interannual variation in laying dates than precipitation for 19 species of migratory birds breeding in the United Kingdom. It will be critical to test this hypothesis across species in the future.

ADAPTATION TO GLOBAL CLIMATE CHANGE

There is no doubt that range shifts are often an important means of coping with climate change: evidence for such shifts in response to ongoing climatic changes is overwhelming (e.g., Parmesan and Yohe 2003, Root et al. 2003, Perry et al. 2005, IPCC 2014a,b). Likewise, phenotypic plasticity provides an important mechanism to cope with changing environmental conditions (e.g., Bradshaw 1965, Przybylo et al. 2000). However, we do not yet know how effective microevolution will be in mitigating the consequences of ongoing environmental changes (Gienapp et al. 2007). Results of theoretical treatments suggest that the predicted rate of climate warming, and consequent sea level rise, may be too rapid for many populations to sustain continued response (Lynch and Lande 1993, Bürger and Lynch 1995, Lynch 1996; see also Gomulkiewicz and Holt 1995).

It is important to make the distinction between genetic (evolutionary) and phenotypic (includes a nongenetic, plastic component) responses or adaptations to changing climates (Gienapp et al. 2007). More than 2,300 bird species worldwide are highly vulnerable to climate change (BirdLife International 2013) because they have a combination of high sensitivity to its impacts, for example, through their dependence on other species, low ability to adapt, via dispersal, and a high exposure to changing climates. For migratory plovers in particular, adaptation can be especially challenging given that they face different forms and rates of global climate change during various parts of the annual cycle (Galbraith et al. 2014; Tables 3.1 and 3.2). Conversely, migratory shorebirds may be preadapted to accommodate changes because they exhibit such high degrees of phenotypic flexibility across the annual cycle.

Perhaps the simplest adaptation to climate change that has been recorded in Charadrius plovers is the northward expansion of wintering habitats currently being observed throughout Europe (Austin et al. 2005, MacLean et al. 2008). Godet et al. (2011) examined winter assemblages of *Charadrii* from Europe to Africa including the Common Ringed Plover, Little Ringed Plover, and Kentish Plover, and taken together, found wintering shorebird communities tended to move to warmer areas. Similarly, longer period of prey availability, more moderate breeding conditions, and longer breeding seasons appear to be a positive by-product of warming climates.

MAJOR BIOMES

Most plovers spend a significant portion of the annual cycle in at least two or three generalized areas: Arctic tundra, coastal areas including estuaries, and subarctic arid inland areas (Tables 3.1 and 3.2). The challenges each of these areas faces as a result of climate change are compounded for migratory species having to adapt to each of them.

Tundra

High northern latitudes are experiencing a climate change at an intensity over twice the global average (Smith et al. 2010). The consequences are predicted to have their greatest effect in fragile arctic ecosystems (Parry et al. 2007, Collins et al. 2013). Arctic ecosystems act as a global carbon "sink," buffering carbon emissions from human activities (Keenan et al. 2016). Extreme events like mild winter temperatures with rain and sleet may become more frequent in the Arctic. The resulting changes in snow amount and timing of snowmelt can affect the phenology of plants and insect emergence, as well as phenology and success of shorebirds (Mortensen et al. 2016). However, short-term strong winds, cold, and snowstorms can also drastically reduce arthropod activity (Lancaster and Briers 2008). Meltofte et al. (2007) reported that the amplitude of short-term weather-induced effects could be as large as the seasonal pattern, making it difficult to predict the temporal pattern of arthropod availability during the breeding season. Thus, shorebirds that breed at high latitudes are facing multi-faceted challenges that could result in loss of 66%–83% of their suitable habitat in the next 70 years (Wauchope et al. 2016).

There are some benefits for plovers during these changing conditions (Lindström and Agrell 1999, McKinnon et al. 2013). An earlier start to breeding elongates their usually brief breeding season to potentially incorporate a second nest if the first one fails. Warmer temperatures may also expose more habitats from the snow and result in higher breeding success as a result of more invertebrate emergences (Meltofte et al. 2007). Over the past 30 years, temperatures in the Russian Arctic have warmed enough for North American Semipalmated Plovers to arrive and start breeding in the Chukotsky Peninsula prior to arrival of their Asian congenerics (Tomkovich and Syroechkovski 2005).

Taking a longer perspective, as the Arctic turns greener, the tree line will move north and the tundra will recede. More specifically, the greening tends to be concentrated along riparian corridors so the progression of succession is more like a network of green fingers extending northward (Ju and Masek 2016). Nevertheless, lichens and moss will yield to alders, taller willows, and birches. Combining this view with feedback mechanisms, including changes in herbivory and predator–prey dynamics, and continued sea level rise reducing coastal habitats, shorebirds may not simply move their nests north. For example, in James Bay, Canada, shorebirds may end up having to move south as rising sea levels flood coastal habitat, which is the kind of squeeze that Schlacher et al. (2008) discuss. That is, in the Arctic, shorebirds move landward as rising seas squeeze them into more unsuitable, shrubbier habitats. Places like Hudson and James Bay are experiencing positive relative sea level rise rates because of uplift, but places such as the Mackenzie Delta and coastlines along Alaska are either being increasingly inundated by higher sea level rise-mediated storm surges or are increasingly eroding into the sea as a result of sea level rise (Rühland et al. 2013).

There are only three Charadrius plovers that breed in the Arctic (Tables 3.1 and 3.2: Semipalmated Plovers, Common Ringed Plover, and Eurasian Dotterel), yet their wide distributions across the High North may render them quite susceptible to habitat loss. They might gain more time if they expand their breeding range to the north to accommodate different landmass distances to the sea. This would lengthen their already taxing migration route (although see Conklin et al. 2017) and perhaps even change their traditional migration routes. However, they would ultimately be stopped by the sea (Wauchope et al. 2016). Conversely, their wide distributions may predispose them to tolerate a warming climate better than species with restricted distributions.

Along with changing arctic tundra, habitat composition comes rapidly changing distributions, phenologies, and abundances of plover predators, prey, and competitors (Pearson et al. 2013, Myers-Smith et al. 2015). These aspects are changing faster in the arctic tundra than further south (Smith et al. 2010). Currently, arctic plovers are ground nesters in vast open areas which provide cover from avian and mammalian predators such as jaegers (*Stercorarius* spp.) and Arctic foxes (*Alopex lagopus*). However, variable snowmelt can leave the open patches where birds might nest but are more obvious to predators (Meltofte et al. 2007). These predators also depend on lemmings (*Dicrostonyx* spp., *Lemmus* spp.) which vary widely in their abundance among years. When lemming numbers are down, jaegers and foxes may turn to other food sources such as shorebird eggs. Ruddy Turnstone (*Arenaria interpres*) egg predation was low early in a season at Medusa Bay, Siberia, when lemmings were forced out of their burrows during snowmelt, although egg and chick predation increased after the lemmings occupied their summer burrows (Schekkerman et al. 2004). Conversely, lemming cycles did not appear to have influenced nest success of Semipalmated Plovers over 13 years of monitoring along the southwestern Hudson Bay coast (E. Nol in Meltofte et al. 2007), although the annual variation in nest success was low to begin with (between 52% and 73%, E. Nol, unpublished data). Recent disruptions in lemming cycles have been attributed to warming temperatures and the resulting snow conditions (Kausrud et al. 2008). This collapse puts more pressure on shorebirds as a food source for predators.

Another issue for arctic-nesting plovers and other shorebirds is the phenological mismatch with respect to their arrival times relative to dates when their insect prey emerge (Tulp and Schekkerman 2008, Pearce-Higgins et al. 2005, McKinnon et al. 2012, Liebezeit et al. 2014, Ward et al. 2016). In short, birds are breeding at their usual time, which is not when the insects are emerging. However, McKinnon et al. (2013) recently evaluated this mismatch among Dunlin (*Calidris alpina*) and found that, when food availability was below average, above average chick growth could be maintained in the presence of increasing temperatures as a result of varying parental care. When it was cold and prey availability was low, parents increased their time brooding chicks. Conversely, warmer temperatures provided more time for young chicks (<5 days of age) to forage as they do not need to be brooded as much. Thus, chicks may find physiological relief from the trophic constraints hypothesized by climate change studies.

The fate of arctic shorebirds under projected future climate scenarios is uncertain, but their populations appear to be particularly at risk (Parmesan and Yohe 2003, Lawler et al. 2009). Climatic amelioration may benefit arctic shorebirds in the short term by increasing survival and productivity, whereas in the long term, habitat changes on the breeding grounds and in the temperate and tropical nonbreeding areas may put them under considerable pressure. Their relatively low genetic diversity, which is thought to be a consequence of survival through past climatically driven population bottlenecks, may also put them more at risk to anthropogenic-induced climate variation than other avian taxa (Meltofte et al. 2007).

Coastal and Estuarine

Increasing global temperatures result in thermal expansion of seawater as well as the melting of glaciers that cause sea level rise (Nicholls and Cazenave 2010). Stronger coastal winds emerge from the atmospheric pressure gradient along ocean margins and, together with other factors (e.g., glaciers melting), are predicted to increase global sea levels by more than 1 m by 2100 (DeConto and Pollard 2016). Local relative sea level rise may be much greater or smaller due to the confounding effects of crustal subsidence or uplift. Inundation due to sea level rise could result in the conversion of tidal to subtidal habitat and reductions in the availability of shorebird foraging habitats (Galbraith et al. 2002). Rising seas coupled with increasing number and intensities of storms may have enormous positive and negative impacts on coastal ecosystems and plover species around the world (Galbraith et al. 2002). While strong storms can cause direct mortality at any time of year, rearrangement of sandy coastal beaches by storms can have a more subtle effect: wind scrubbing of beaches that removes successional vegetation can be helpful, movement of beaches along the coast can be positive or negative, but gross loss of beach habitat can be quite negative (Gieder et al. 2014).

Seavey et al. (2010) modeled the effects of climate-induced sea level rise and storms on the Atlantic Coast of New York for Piping Plovers and concluded that the inland movement of beach habitat resulting from sea rise coupled with beach movement and rearrangement produced by storms could provide for more and even better plover habitat, if this movement was not hindered by human development. Of course, this free movement of beaches will be stopped in most places around the world because of human needs (Gittman et al. 2015). This leads these new climatic changes to be of concern for most plover species at some or all phases of their annual cycle. In New York and the Atlantic Coast of North America, this will likely affect breeding, migrating, and nonbreeding Piping Plovers, Wilson's Plovers, and Killdeer, as well as Semipalmated Plovers on migration.

Taking a metapopulation perspective on the effects of sea level rise for Florida's beach-nesting Snowy Plovers, Aiello-Lammens et al. (2011) found a decrease in dry land cover classes and an increase in open ocean areas. Over a 90-year period, they concluded that this would result in decreased population viability for Florida Snowy Plovers based on risk of extinction, risk of decline to a metapopulation size of <20 birds, and the expected minimum metapopulation abundance. The specificity of their model allowed them to pinpoint the most vulnerable points in the life cycle of this population, and they recommended partial mediation to these risks might be attained by enhancing nest success through a variety of means (e.g., predator exclosures on nests). Similarly, Convertino et al. (2012a,b) concluded that climate-induced changes in coastal Florida habitat for the Piping and Snowy plovers would alter large patches of habitat so they would be significantly reduced. They concluded that this could be an important problem, especially for Piping Plovers.

In Britain, Austin and Rehfisch (2003) modeled the effect of increased storms on winter wader habitat and found mixed results. On one hand, changes in estuary morphology tended to significantly decrease the mudflat habitat often used by foraging waders. On the other hand, these changes tended to widen the estuary providing more sandy habitat typically used by Ringed Plovers, Golden Plovers, and Lapwings. Rehfisch et al. (2004) further considered the effects of changing weather on Britain's coastal areas and predicted that by 2020, Ringed Plover would move their wintering areas further east and north and numbers could decline by up to 36% in the Western Isles (which hold almost half of the British population of Ringed Plovers). Similarly, in North America, Galbraith et al. (2002) examined major migratory stopover sites across the continent and found that Bolivar Flats, Texas, a major migration and winter area for all four North American Charadrius plovers, would likely disappear by 2050 because of sea level rise. Lentz et al. (2016) found 70% of the northeastern U.S. coastal habitats had some capacity to respond to rising sea levels, but not enough to curb serious concerns for the area.

An added component to changing sea levels is that ocean acidity is increasing rapidly—more than 10 times faster than at any time in the past 55 million years—and possibly at a rate unprecedented in the past 300 million years (UNEP 2012, IPCC 2014a,b). This change in ocean chemistry may be more important than changes in temperature for the performance and survival of many organisms (Harley et al. 2006). The combination of sea level rise and ocean acidification will take a toll on the crustaceans and other invertebrates that coastal plovers consume for several reasons: (1) As carbon dioxide emissions rise, the pH of seawater drops and becomes less saturated with aragonite, a mineral many coastal invertebrates require for building strong shells. With increasingly acidic seawater, invertebrate shells are smaller, thinner, and weaker. Thus, their larvae grow more slowly, with fewer surviving to settle on the sea bottom (Moy et al. 2009). (2) Ocean circulation, which drives larval transport, will also change with important consequences for invertebrate and subsequently vertebrate population dynamics (Harley et al. 2006). This change, combined with effects of beach nourishment (including offshore deposition), can alter nearshore seawater chemistry and sediment deposition, further altering shorebird food resources.

Finally, one effect of climate change not often discussed is the idea that warmer climates will likely bring more people to beach areas that were otherwise occupied by plovers and other waders. Coombes et al. (2008) examined this possibility for Ringed Plovers in East Anglia, United Kingdom, and found that if beach tourism levels increased, plovers might abandon the area. They recommended beach zoning to keep tourists away from breeding sites. However, zoning is complicated by increasingly changing nesting habitat conditions (and subsequent plover distributions) on beaches during summer seasons when beach attendance is highest. A related situation occurs where sea walls are built to protect buildings and towns from rising sea water. If more sea walls are built, it will remove already declining saltmarshes, sandy beaches, and mudflats used by plovers and other species for foraging. Fujii (2012) recommended setting sea walls further back from the ocean to provide for this habitat. He warned, however, that the subsequent mudflats may not be as productive as in the past, thus monitoring should occur to ensure enough food is available to waders as in the past.

Changing sea levels and increased disturbance can have particularly negative effects on resident and/or island species that have no other options when changes occur. In Tasmania, Hooded and Red-capped Plovers are resident beach species that are susceptible to microhabitat changes predicted to shift under climate-change scenarios (Bock et al. 2016). In Red-capped Plovers, mean beach width and wash gradient were important whereas maximum wave height has more of an effect on Hooded Plovers. This example points to the importance of examining multiple species' needs and not assuming management for one species will fit other members of the community.

Inland Arid Areas

Inland wetlands and open areas are not always recognized as important habitats for plovers, yet 17%–50% of Charadrius plovers spend at least one phase of their annual cycle in these areas (Tables 3.1 and 3.2). In particular, these wetlands provide a critical resource for breeding plovers and their chicks. However, a complex scenario is emerging regarding the needs of chicks. Plovers need access to freshwater for chicks even if adults can take advantage of invertebrates produced by wetlands of varied salinities.

Water quality, defined here as salinity, can have varying effects on many waterbird species including Charadrius plovers (Hannam et al. 2003). On one hand, saline to hypersaline wetlands and the ocean can provide important invertebrate food resources to plovers and other species. Plovers take advantage of these resources, particularly on migration and at postbreeding sites. On the other hand, young plovers cannot metabolize salt as their salt gland is not developed at hatch (Hannam et al. 2003, Rubega and Oring 2004). Thus, not only can they not drink salty water, but they cannot take advantage of the superabundant food resources afforded adults in the salty water. Adults that respond to the need for freshwater by leading their chicks across a dry, hot desert in search of freshwater are not likely to be successful. Thus, a mosaic of fresh, saline, and hypersaline wetlands with freshwater groundwater and inflows provides a stable existence for chicks and adults (Masterson et al. 2014, Haig et al. in press). Understanding movement among these areas during various phases of the annual cycle is key to understanding the system needs for plovers and other waterbirds (Haig et al. 1998).

Water extent and quality, key to viability of inland Charadrius plover populations, are rapidly changing across the globe. Over the past 20 years, 173,000 km^2 of wetlands have been converted to land and 115,000 km^2 have been converted into water (Donchyts et al. 2016). Included in these estimates are results from the Aral Sea, bordering the countries of Kazakhstan, Uzbekistan, and Turkmenistan. Since the 1960s, Russian engineers have diverted rivers away from the Aral Sea, once the fourth largest lake in the world, to irrigate cotton and wheat (Donchyts et al. 2016). Adding in the effects of a warming climate, the lake has almost entirely dried up, losing about 27,650 km^2 of surface water. In the United States, Lake Mead, the largest freshwater reservoir, lost 222 km^2 over the same period (Donchyts et al. 2016). In the Great Basin of North America, not only is water extent decreasing, but also the amount of time water is available has decreased and the salinity of many water sources has increased more over the past 20 years than the previous century (Haig et al. in press).

Water volume often affects water quality (Hamidov et al. 2016). Across the world, inland and coastal wetland salinization is occurring at an ever more rapid pace as a result of climate change (Herbert et al. 2015). Global assessments of increased wetland salinity are difficult to assess due to lack of reporting in many parts of the world (Herbert et al. 2015). However, telling examples exist: government sources predict significantly elevated salt concentrations in 40,000 km of New Zealand and Australia's waterways and associated wetlands by 2050 (Nielsen et al. 2003), and Zhai et al. (2016) have predicted salinity increases in Florida's coastal ecosystems in the York and Chickahominy Rivers. Salinity is also predicted to increase due to projected sea level rise in Chesapeake Bay (Rice et al. 2011).

Unfortunately, much of this dry inland habitat has undergone a drought in recent years resulting in loss or shrinkage of many wetlands as well as salinization of previously fresh sites (Nebel et al. 2008, Shanahan et al. 2009, Buckley et al. 2010, Johnson et al. 2010, Herbert et al. 2015, Ault et al. 2016, Shadkam et al. 2016). In a review of changing temperatures and surface area in 300 wetlands across the world, O'Reilly et al. (2015) found extensive evidence for warming (global mean = 0.34°C per decade) in many areas between 1985 and 2009. Consequences of this extensive warming are predicted to result in a 20% increase in the number of algal blooms and a 5% increase in toxic blooms over the next century as well as declines in lake water level, and in some cases, complete ecosystem loss. None of these changes will benefit the many plovers and their invertebrate food resources occupying these landscapes.

Examples of the effect of these changes on plovers can be found around the globe. In Australia, 80% of the most densely populated shorebird-supporting wetlands are found inland (Nebel et al. 2008). These wetland ecosystems are dependent on a few infrequent heavy rainfalls within or between years. Thus, shorebirds are vulnerable to any change in frequency or magnitude of these events. Climate change that results in extreme wetland evaporation or a reduced frequency of large flood events, exacerbated by extraction of water for agriculture, could be catastrophic for plovers that use a mosaic of wetland habitats at broad spatial scales (Roshier et al. 2001; Haig et al. in press). In North America's Great Basin, significant increases in temperature and decreases in precipitation, especially over the past 20 years, has led to loss of wetlands or loss of their functionality to shorebirds to an extent that threatens the viability of their migratory flyway (Haig et al. in press).

Lake Urmia in northwestern Iran is the second largest hypersaline lake in the world and has undergone substantial desiccation over the past few decades (Shadkam et al. 2016). Much of this has been due to overuse of the lake's resources, but now climate models predict even more serious consequences. Shadkam et al. (2016) modeled predicted future possibilities and recommended stringent water management plans to offset the negative effects of climate change and other human uses of the lake. If implemented, these recommendations could facilitate continued use of the lake by all the plover species that inhabit Iran: Common Ringed Plover, Little Ringed Plover, Lesser Sand-plover, and Caspian Plover.

In the U.S. Northern Great Plains and Prairie provinces of Canada, climate modeling efforts produced a mixture of results with respect to plover habitats. Cook et al. (2015) evaluated 17 general circulation climate models and predicted there will be significant desiccation resulting in severe drought in the region. Peterson (2003) found that northern Great Plains birds, including Mountain Plovers, were more heavily influenced by changing climates (mode of 35% of distributional area lost and up to 400 km shift of range centroid) than nearby montane birds. Steen et al. (2014) concurred with the habitat loss prediction for waterbird species, stating that on average, they would lose 46% of their habitat. Steen et al. (2016) included Killdeer as a species that would likely lose habitat as the climate changed. Similarly, Gratto-Trevor and Abbott (2011) warned that Piping Plovers in the region would be susceptible to negative effects of climate change primarily due to changes in the natural hydrology of the area and from increased human demands on potentially diminished water supplies.

Finally, for most plover species, there are compounding anthropogenic factors that increase the speed and intensity in which climate change will affect them. Examples include the inland island-endemic St. Helena Plover and Madagascar Plover. Here their rarity and lack of dispersal increases their vulnerability to change of any sort. Furthermore, as temperatures increase and human development encroaches, their already alkali habitat will disappear leaving them with few options, especially as nearby ocean beaches are changing in structure and composition.

Perhaps the most compelling example of compounded factors comes from desiccation of the man-made Aral Sea (between Kazakhstan and Uzbekistan). To date, the water level has fallen 23 m and salinity has increased exponentially (Micklin 2007). Bird species diversity decreased from 322 to 168 species, including a decline in shorebird species from 17 to 6 (Kreuzberg-Mukhina 2006, Juger et al. 2012). Among Charadrius plovers, Caspian Plovers, Little Ringed Plovers, and Kentish Plovers still breed on the shores of the Aral Sea, and Ringed Plovers are now the only other seen on migration through the area (Juger et al. 2012).

POTENTIAL SOLUTIONS

An important message from this chapter is that very little specific work has been carried out on Charadrius plovers that address their vulnerability to changing climates. However, Tables 3.1 and 3.2 illustrate that they are occupying some of the areas most affected by climate change and often are using more than one site during the course of their annual cycle. Across the world, *Charadrius* plovers are facing challenging times. Therefore, the following sections outline research and management guidelines that have been recommended for mitigating climate change for specific plovers or closely related taxa in the hopes they will provide guidance for other species. Clearly, plovers are the "canaries in the coal mines" for some of our most treasured natural areas.

Management Considerations

Across all plover habitats, there are several threats resulting from climate change that have potential solutions if considered early in the planning process. For example, predator exclosures may provide enough of a buffer to prolong the productivity of a local population in areas where predation pressure is acute (Aiello-Lammens et al. 2011). Increased energy needs that contribute to global warming are often addressed via development of wind tower fields or solar farms, often at inland or coastal areas favored by plovers. Gratto-Trevor and Abbott (2011) warned that these alternative power structures can have negative effects on Piping Plovers, if constructed in ways that prohibit access to breeding, migration, or winter sites. Alternatives, including smaller structures with much shorter blades, are now available but need to be considered for specific plover needs.

Tundra

With succession advancing so quickly in tundra habitat (Myers-Smith et al. 2015), plovers need to adjust their nest site locations accordingly. Thus, arctic shorebird breeding areas, particularly in Canada, buffers from development will help provide for this adjustment (Wauchope et al. 2016). This might best be achieved by taking a watershed level perspective since a disproportionate amount of change is occurring here and a larger scale may provide more options for mitigation.

Coastal and Estuarine

Coastal areas are losing their sandy beaches, mudflats, and shorebird prey base due to rising sea levels and changing water chemistry (Alrashidi et al. 2012, Fujii 2012). Shorebirds can survive losing beach habitat to some degree if they can take advantage of other prey sources further up into the dune areas (Hunter et al. 2015). And, in some cases, reducing shoreline alterations, such as armoring with beach walls, could prevent loss of beaches and mudflats from erosion (Schlacher et al. 2008; Fujii 2012). Conversely, Seavey et al. (2011) recommended letting dunes and beaches flow freely so the sand and mud can move inland as erosion takes place. This might be the best option if human structures and needs were not issues to contend with. In many places, roads, buildings, and other infrastructure will be at risk as sea levels rise. However, there may be places where the natural movement of beaches can take place particularly if integrated into climate adaptation planning. For example, Chincoteague National Wildlife Refuge and other federal areas along the East Coast of the United States are rezoning parking lots and buildings based on future sea level rise and land response projections, allowing some areas to move naturally (Volpe National Transportation Systems Center 2009).

Inland Arid Areas

Continuation of inland plover habitat revolves around water availability and water quality (Haig et al. in press). Providing access to freshwater is a priority if plovers are to successfully breed. Freshwater is also key for plovers and ranchers following desiccation of a wetland (Forkutsa et al. 2009). While places like the Aral Sea are considered disasters, we can glean important management lessons from those working to restore it (Kreuzberg-Mukhina 2006, Micklin 2007). Certainly, there are always dangers associated with depleting aquifers; yet current water conservation measures can include using more underground water and less surface water. For example, via drip irrigation, water is transferred to the surface resulting in a water system that minimizes water loss (Ibrakhimov et al. 2007). The efficiency of drip irrigation can be increased by 35%–103% compared with that of furrow (traditional) irrigation (Intermountain West Joint Venture 2013). In the U.S. Great Basin, major efforts have been put into working with ranchers to change from mechanical delivery of water to letting spring water remain in fields (flood irrigation). These protocols more closely mimic natural processes and provide water for shorebirds and other waterbirds for longer periods of time in the spring (Intermountain West Joint Venture 2013). However, additional measures in water management could ensure freshwater inflows and water availability to wetlands used by breeding Snowy Plovers and Killdeer as well as migrating Semipalmated Plovers (Moore 2016).

Other climate-related issues that inland birds face include efforts to stop succession as warming temperatures prolong vegetation growth. In the Great Plains in the United States, maintenance of prairie dog (*Cynomys* spp.) towns as well as controlled burns have been effective at keeping habitat open for Mountain Plovers (Augustine and Skagen 2014).

Research Needs

While few efforts address the specific management needs of plovers in changing climates, birds may benefit from efforts implemented in a broad ecosystem approach. Conversely, the dearth of research on the effects of climate change on plovers could lead to implementation of activities that may not benefit their future success. Thus, the ideas listed below provide some general starting points for future research to benefit Charadrius plovers around the world.

With respect to climate change, the most important research that can be carried out on plovers is determining and understanding the phenology of their annual cycle and the migratory connectivity that joins various populations of the same species. This includes understanding where particular habitat or food bottlenecks occur during phases of the annual cycle. When possible, long-term studies will enhance our understanding of how potential mismatch is occurring between plover and invertebrate prey phenologies, evolving habitats, and other limiting factors that are changing with the climate. Species-specific models that examine multiple populations would improve understanding of the idiosyncratic nature of species response to climate change (Peterson 2003). Further, we need to ensure that vulnerability models incorporate each phase of the annual cycle (Small-Lorenz et al. 2013). Better knowledge of migration routes that overlap proposed and existing wind turbine and solar power sites is also necessary to understand potential impacts.

In a broader sense, cross-species multifactorial tools that describe climate patterns across vast geographic areas (Stephens et al. 2016) and include the needs of plovers are critical for effective management and are currently lacking. Given the great similarity among many plover life histories, development and then slight modification of one model could address a number of Charadrius species. Development of system matrices can provide ways of investigating potential actions on a diversity of species in order to lessen the negative effects of climate change and even consider the idea of carefully placed climate change refugia (Morelli et al. 2016). Finally, an expansion of research efforts on sandy beaches that integrate ecological, physical, and socioeconomic disciplines, as well as public outreach (Schlacher et al. 2008) will provide greater perspective on the situation at hand. Worldwide, beaches and associated habitats are some of the most desired habitats by humans and most threatened by changing climates. Thus, future work needs to incorporate the needs and concerns of many to protect the plovers that make these areas their homes.

ACKNOWLEDGMENTS

I thank K.D. Gieder, C. Phillips, B. Ralston, and D. Ruthrauff for helpful comments on this chapter. I am grateful to the scientists in my lab and beyond that have helped me better understand the effects of climate change on shorebirds in the Great Basin, Pacific Coast, and throughout the range of the Piping Plover: Lew Oring, Mark Colwell, Oriane Taft, Peter Sanzenbacher, Nils Warnock, Matthew Johnson, Sean Murphy, Jeff Hollenbeck, Elise Elliott-Smith, and John Matthews. Funding was provided by USGS Forest and Rangeland Ecosystem Science Center and the USGS Climate Change program. Any use of trade, product, or firm names in this publication is for descriptive purposes only and does not imply endorsement by the U.S. Government.

LITERATURE CITED

Aiello-Lammens, M. E., M. Chu-Agor, M. Convertino, R. Muñoz-Carpena, R. Fischer, and I. Linkov. 2011. The impact of sea-level rise on Snowy Plovers in Florida: Integrating geomorphological, habitat, and metapopulation models. *Global Change Biology* 17:3644–3654.

Alrashidi, M., M. Shobrak, M. S. Al-Eissa, and T. Székely. 2012. Integrating spatial data and shorebird nesting locations to predict the potential future impact of global warming on coastal habitats: A case study on Farasan Islands, Saudi Arabia. *Saudi Journal of Biological Science* 19:311–315.

Andres, B. A., P. A. Smith, R. I. G. Morrison, C. L. Gratto-Trevor, S. C. Brown, and C. A. Friis. 2012. Population estimates of North American shorebirds, 2012. *Wader Study Group Bulletin* 119:178–194.

Augustine, D. J., and S. K. Skagen. 2014. Mountain Plover nest survival in relation to prairie dog and fire dynamics in shortgrass steppe. *Journal of Wildlife Management* 78:595–602.

Ault, T. R., J. S. Mankin, B. I. Cook, and J. E. Smerdon. 2016. Relative impacts of mitigation, temperature, and precipitation on 21st-century mega-drought risk in the American Southwest. *Science Advances* 2:e1600873.

Austin, G. E., and M. M. Rehfisch. 2003. The likely impact of sea level rise on waders (*Charadrii*) wintering on estuaries. *Journal of Nature Conservation* 11:43–58.

Austin, G. E., and M. M. Rehfisch. 2005 Shifting distributions of migratory fauna in relation to Climatic change. *Global Change Biology* 11:31–38.

Bergmann, C. 1847. Über die Verhältnisse der Wärmeökonomie der Thiere zu ihrer Grösse. *Göttinger Studien* 3:595–708 (in German).

BirdLife International. 2013. *State of the World's Birds.* BirdLife International, Cambridge, UK.

Bock, A., M. R. Phillips, and E. Woehler. 2016. The role of beach and wave characteristics in determining suitable habitat for three resident shorebird species in Tasmania. In: Vila Concejo, A., E. Bruce, D. M. Kennedy, et al (Eds.), *Proceedings of the 14th International Coastal Symposium (Sydney, Australia). Journal of Coastal Research*, Special Issue, 75: 358–362. Coconut Creek, FL.

Both C., A. V. Artemyev, and B. Blaauw. 2004. Large-scale geographical variation confirms that climate change causes birds to lay earlier. *Proceedings of the Royal Society B* 271:1657–1662.

Both C., C. A. M. Van Turnhout, R. G. Bijlsma, H. Siepel, A. J. Van Strien, and R. P. B. Foppen. 2010. Avian population consequences of climate change are most severe for long-distance migrants in seasonal habitats. *Proceedings of the Royal Society B* 1685:1259–1266.

Both, C., and M. E. Visser. 2001. Adjustment to climate change is constrained by arrival date in a long-distance migrant bird. *Nature* 411:296–298.

Bradshaw, A. D. 1965. Evolutionary significance of phenotypic plasticity in plants. *Advanced Genetics* 13:115–155.

Brown, S., C. Hickey, B. Harrington, and R. Gill (editors). 2001. *The U.S. Shorebird Conservation Plan*, 2nd ed. Manomet Center for Conservation Sciences, Manomet, MA.

Buckley, B. M., K. J. Anchukaitis, D. Penny, R. Fletcher, E. R. Cook, M. Sano, L. C. Nam, A. Wichienkeeo, T. T. Minh, and T. M. Hong. 2010. Climate as a contributing factor in the demise of Angkor, Cambodia. *Proceedings of the National Academy of Science USA* 107:6748–6752.

Bürger, R., and M. R. Lynch. 1995. Evolution and extinction in a changing environment: A quantitative-genetic analysis. *Evolution* 49:151–163.

Collins, M., and R. Knutti, et al. 2013. Long-term climate change: Projections, commitments and irreversibility. Pp. 1029–1136, in Stocker, T. F., D. Qin, G. K. Plattner, et al. (Eds.), *Climate Change 2013: The Physical Science Basis. Contribution of Working Group I to the Fifth Assessment Report of the Intergovernmental Panel on Climate Change.* Cambridge University Press, Cambridge, UK and New York.

Conklin, J. R., N. R. Senner, P. F. Battley, and T. Piersma. 2017. Extreme migration and the individual quality spectrum. *Journal of Avian Biology* 48:10–36.

Convertino, M., A. Bockelie, G. A. Kiker, R. Muñoz-Carpena, and I. Linkov. 2012a. Shorebird patches as fingerprints of fractal coastline fluctuations due to climate change. *Ecological Processes* 1:9

Convertino, M., P. Welle, R. Munoz–Carpena, G. A. Kiker, Ma. L. Chu-Agor, R. A. Fischer, and I. Linkov. 2012b. Epistemic uncertainty in predicting shorebird biogeography affected by sea–level rise. *Ecological Modelling* 240:1–15.

Cook, B. I., T. R. Ault, and J. E. Smerdon. 2015. Unprecedented 21st century drought risk in the American Southwest and Central Plains. *Science Advances* 1:e1400082.

Coombes, E. G., A. P. Jones, and W. J. Sutherland. 2008. The biodiversity implications of changes in coastal tourism due to climate change. *Environmental Conservation* 35:319–330.

Cramp, S., and K. E. L. Simmons (Eds.). 1983. *The Birds of the Western PaleArctic, Vol. III. Waders to Gulls.* Oxford University Press, Oxford, UK. 913 pp.

DeConto, R. M., and D. Pollard. 2016. Contribution of Antarctica to past and future sea level rise. *Nature* 531:591–597.

Donchyts, G., F. Baart, H. Winsemius, N. Gorelick, J. Kwadijk, and N. van de Giesen. 2016. Earth's surface water change over the past 30 years. *Nature Climate Change* 6:810–813.

Drent, R. H., A. D. Fox, and J. Stahl. 2006. Traveling to breed. *Journal of Ornithology* 147:122–134.

Dunn, P. 2004. Breeding dates and reproductive performance. *Advances in Ecological Research* 35:69–87.

Durant, J. M., D. Ø. Hjermann, G. Ottersen, and N. C. Stenseth. 2007. Climate and the match or mismatch between predator requirements and resource availability. *Climate Research* 33:271–283.

Fujii, T. 2012. Climate change, sea-level rise and implications for coastal and estuarine shoreline management with particular reference to the ecology of intertidal benthic macrofauna in NW Europe. *Biology* 1:597–616.

Forkutsa, I., R. Sommer, Y. I. Shirokova, J. P. A. Lamers K. Kienzler B. Tischbein C. Martius and P. L. G. Vlek. 2009. Modeling irrigated cotton with shallow groundwater in the Aral Sea Basin of Uzbekistan: II. Soil salinity dynamics. *Irrigation Science* 27:319–330.

Galbraith, H., D. W. DesRochers, S. Brown, and J. M. Reed. 2014. Predicting vulnerabilities of North American shorebirds to climate change. *PLoS One* 9:e108899.

Galbraith, H., R. Jones, R. Park, J. Clough, S. Herrod-Julius, B. Harrington, and G. Page. 2002. Global climate change and sea level rise: Potential losses of intertidal habitat for shorebirds. *Waterbirds* 25:173–183.

Gauthier, G., J. Bety, M.-C. Cadieux, P. Legagneux, M. Doiron, C. Chevallier, S. Lai, A. Tarroux and D. Berteaux. 2013. Long-term monitoring at multiple trophic levels suggests heterogeneity in responses to climate change in the Canadian Arctic tundra. *Philosophical Transactions of the Royal Society B* 368:20120482.

Gieder, K. D., S. M. Karpantya, J. D. Fraser, D. H. Catlin, B. T. Gutierrez, N. G. Plant, A. M. Turecek, and E. R. Thieler. 2014. A Bayesian network approach to predicting nest presence of the federally-threatened Piping Plover (*Charadrius melodus*) using barrier island features. *Ecological Modelling* 276:38–50.

Gienapp, P., R. Leimu, and J. Merila. 2007. Responses to climate change in avian migration time—microevolution versus phenotypic plasticity. *Climate Research* 35:25–35.

Gittman, R. K., S. B. Scyphers, C. S. Smith, I. P. Neylan and J. H. Grabowski. 2015. Engineering away our natural defenses: An analysis of shoreline hardening in the U.S. *Frontiers in Ecology and the Environment* 13:301–307.

Godet, L., M. Jaffe, and V. Devictor. 2011. Waders in winter: Long-term changes of migratory bird assemblages facing climate change. *Biology Letters* 7:714–717.

Gomulkiewicz, R., and R. D. Holt. 1995. When does evolution by natural selection prevent extinction? *Evolution* 49:201–207.

Gratto-Trevor, C. L., and S. Abbott. 2011. Conservation of Piping Plover (*Charadrius melodus*) in North America: Science, success and challenges. *Canadian Journal of Zoology* 89:401–418.

Haig, S. M., D. W. Mehlman, and L. W. Oring. 1998. Avian movements and wetland connectivity in landscape conservation. *Conservation Biology* 12:749–758.

Haig, S. M., S. P. Murphy, J. D. Matthews, I. Arismendi, and M. Safeeq. Climate-altered wetlands challenge waterbird use and migratory connectivity in arid landscapes. Scientific Reports 9:4666.

Hamidov, A., K. Helming, and D. Bella. 2016. Impact of agricultural land use in Central Asia: A review. *Agronomy and Sustainable Development* 36:6.

Hannam, K., L. Oring, and M. Herzog. 2003. Impacts of salinity on growth and behavior of young American Avocets. *Waterbirds* 26:119–125.

Harley, C. D. G., A. R. Hughes, K. M. Hultgren, B. G. Miner, C. J. B. Sorte, C. S. Thornber, L. F. Rodriguez, L. Tomanek and S. L. Williams. 2006. The impacts of climate change in coastal marine systems. *Ecology Letters* 9:228–2241.

Hayman, P., J. Marchant, and T. Prater. 1986. *Shorebirds: An Identification Guide*. Houghton Mifflin, Boston, MA. 412 pp.

Hedenström A., Z. Barta, B. Helm, A. I. Houston, J. M. McNamara and N. Jonzén. 2007. Migration speed and scheduling of annual events by migrating birds in relation to climate change. *Climate Research* 35:79–91.

Herbert, E. R., P. Boon, A. J. Burgin, S. C. Neubauer, R. B. Franklin, M. Ardón, K. N. Hopfensperger, L. P. M. Lamers and P. Gell. 2015. A global perspective on wetland salinization: Ecological consequences of a growing threat to freshwater wetlands. *Ecosphere* 6:206.

Hughes, L. 2000. Biological consequences of global warming: Is the signal already apparent? *Trends in Ecology and Evolution* 15:56–61.

Hunter, E. A., N. P. Nibbelink, C. R. Alexander, K. Barrett, L. F. Mengak, R. K. Guy, C. T. Moore and R. J. Cooper. 2015. Coastal vertebrate exposure to predicted habitat changes due to sea level rise. *Environmental Management* 56:1528–1537.

Ibrakhimov, M., A. Khamzina, I. Forkutsa, G. Paluasheva, J. P. A. Lamers, B. Tischbein, P. L. G. Vlek and C. Martius. 2007. Groundwater table and salinity: Spatial and temporal distribution and influence on soil salinization in Khorezm region (Uzbekistan, Aral Sea Basin). *Irrigation and Drain Systems* 21:219–236.

Intergovernmental Panel on Climate Change. 2014a. Climate Change 2014: Synthesis Report. Contribution of Working Groups I, II and III to the Fifth Assessment Report of the Intergovernmental Panel on Climate Change. Core Writing Team, R. K. Pachauri, and L. A. Meyer (Eds.). IPCC, Geneva, Switzerland. 151 pp.

Intergovernmental Panel on Climate Change. 2014b. Climate Change 2014: Impacts, Adaptation and Vulnerability. Part B: Regional Aspects. Contribution of Working Group II to the 5th Assessment Report of the IPCC. Barros, V. R., C. B. Field, D. J. Dokken, et al. Cambridge University Press, Cambridge, UK, 688 pp.

Intermountain West Joint Venture. 2013. *Intermountain West Joint Venture Implementation Plan*. Intermountain West Joint Venture, Salt Lake City, UT.

Iwamura, T., H. P. Possingham, I. Chadés, C. Minton, N. J. Murray, D. I. Rogers, E. A. Treml and R. A. Fuller. 2013. Migratory connectivity magnifies the consequences of habitat loss from sea–level rise for shorebird populations. *Proceedings of the Royal Society B* 280:20130325.

Johnson, W. C., B. Werner, G. R. Guntenspergen, R. A. Voldseth, B. Millett, D. E. Naugle, M. Tulbure, R. W. H. Carroll, J. Tracy and C. Olawsky. 2010. Prairie wetland complexes as landscape functional units in a changing climate. *BioScience* 60:128–140.

Jones, T., and W. Cresswell. 2010. The phenology mismatch hypothesis: Are declines of migrant Birds linked to uneven global climate change? *Journal of Animal Ecology* 79:98–108.

Ju, J., and J. G. Masek. 2016. The vegetation greenness trend in Canada and U.S. Alaska from 1984–2012 Landsat data. *Remote Sensing of the Environment* 176:1–16.

Juger, U., T. Dujsebayeva, O. V. Belyalov, et al. 2012. Fauna of the Aralkum. Pp. 199–269, in Breckle, S.-W., W. Wucherer, L. A. Dimeyeva, et al. (Eds.), *Aralkum—A Man-Made Desert: The Desiccated Floor of the Aral Sea (Central Asia)*. Springer, New York.

Kausrud, K. L., A. Mysterud, H. Steen, J. O. Vik, E. Østbye, B. Cazelles, E. Framstad, A. M. Eikeset, I. Mysterud, T. Solhøy and N. C. Stenseth. 2008. Linking climate change to lemming cycles. *Nature* 456:93–97.

Keenan, T., F. I. C. Prentice, J. G. Canadell, C. Williams, H. Wang, M. Raupach, and G. J. Collatz. 2016. Rising human carbon dioxide emissions offset by plants—For now. *Nature Communications*. doi:10.1038/ncomms13428.

Knudsen, E., A. Linden, C. Both, N. Jonzén, F. Pulido, N. Saino, W. J. Sutherland, L. A. Bach, T. Coppack, T. Ergon, P. Gienapp, J. A. Gill, O. Gordo, A. Hedenström, E. Lehikoinen, P. P. Marra, A. P. Møller, A. L. K. Nilsson, G. Péron, E. Ranta, D. Rubolini, T. H. Sparks, F. Spina, C. E. Studds, S. A. Saether, P. Tryjanowski and N. C. Stenseth. 2011. Challenging claims in the study of migratory birds and climate change. *Biological Reviews* 86:928–946.

Koenig, W. D., and A. M. Liebhold. 2016. Temporally increasing spatial synchrony of North American temperature and bird populations. *Nature Climate Change* 6:614–618.

Kreuzberg-Mukhina, E. A. 2006. The Aral Sea Basin: Changes in migratory and breeding waterbird populations due to major human induced changes to the region's hydrology. Pp. 283–284, in Boere, G. C., C. A. Galbraith, and D. A. Stroud (Eds.), *Waterbirds around the World*. The Stationery Office, Edinburgh, UK.

Lancaster, J., and R. Briers (editors). 2008. *Aquatic Insects: Challenges to Populations*. 336 pp., CABI Publishing, Oxfordshire, UK.

Lawler, J. J., S.L. Schafer, D. White, P. Kareiva, E. P. Maurer, A. R. Blaustein and P. J. Bartlein. 2009. Projected climate-induced faunal change in the Western Hemisphere. *Ecology* 90:588–597.

Lentz, E. E., E. R. Thieler, N. P. Plant, S. R. Stippa, R. Horton, and D. B. Gesch. 2016. Evaluation of dynamic coastal response to sea-level rise modifies inundation likelihood. *Nature Climate Change*, doi:10.1038/nclimate295.

Liebezeit, J. R., K. E. B. Gurney, M. Budde, S. Zack and D. Ward. 2014. Phenological advancement in arctic bird species: Relative importance of snow melt and ecological factors. *Polar Biology* 37:1309–1320.

Lindström, Å., and J. Agrell. 1999. Global change and possible effects on the migration and reproduction of Arctic-breeding waders. *Ecological Bulletin* 47:145–159.

Lynch, M. 1996. A quantitative-genetic perspective on conservation issues. Pp. 471–501, in Avise, J., and J. Hamrick (Eds.), *Conservation Genetics: Case Histories from Nature*. Chapman and Hall, New York.

Lynch, M., and R. Lande. 1993. Evolution and extinction in response to environmental change. Pp. 234–250, in Karieva, P., J. Kingsolver, and R. Huey (Eds.), *Biotic Interactions and Global Change*. Sinauer Assoc., Inc, Sunderland, MA.

Maclean, I. M. D., G. E. Austin, M. M. Rehfisch, J. Blew, O. Crowe, S. Delany, K. Devos, B. Deceuninck, K. Günther, K. Laursen, M. Van Roomen, and J. Wahl. 2008 Climate change causes rapid changes in the distribution and site abundance of birds in winter. *Global Change Biology* 14:2489–2500.

Masterson, J. P., M. N. Fienen, E. R. Thieler, D. B. Gesch, B. T. Gutierrez and N.G. Plant. 2014. Effects of sea-level rise on barrier island groundwater system dynamics—Ecohydrological implications. *Ecohydrology* 7:1064–1071.

McCauley, L. A., M. J. Anteau, and M. Post van der Burg. 2015. Consolidation drainage and climate change may reduce Piping Plover habitat in the Great Plains. *Journal of Fish and Wildlife Management* 7:4–12.

McKinnon, L., E. Nol, and C. Juillet. 2013. Arctic–nesting birds find physiological relief in the face of trophic constraints. *Scientific Reports* 3:1816.

McKinnon, L., M. Picotin, E. Bolduc, C. Juillet, and J. Bêty. 2012. Timing of breeding, peak food availability, and effects of mismatch on chick growth in birds nesting in the High Arctic. *Canadian Journal of Zoology* 90:961–971.

Meltofte, H., T. Piersma, H. Boyd, B. McCaffery, B. Ganter, V. V. Golovnyuk, K. Graham, C. L. Gratto-Trevor, R. I. G. Morrison, E. Nol, H-U. Rösner, D. Schamel, H. Schekkerman, M. Y. Soloviev, P. S. Tomkovich, D. M. Tracy, I. Tulp, and L. Wennerberg. 2007. Effects of climate variation on the breeding ecology of Arctic shorebirds. *Meddelelser om Grønland Bioscience* 59:48.

Micklin, P. 2007. The Aral Sea disaster. *Annual Review of Earth and Planetary Sciences* 35:47–72.

Møller A. P., D. Rubolini, and E. Lehikoinen. 2008 Populations of migratory bird species that did not show a phenological response to climate change are declining. *Proceedings of the National Academy of Science USA* 105:195–16.

Moore, J. A. 2016. Recent desiccation of Western Great Basin Saline Lakes: Lessons from Lake Abert, Oregon, U.S.A. *Science of the Total Environment* 554–555:142–154.

Moran, P. A. P. 1953. The statistical analysis of the Canadian lynx cycle. II. Synchronization and meteorology. *Australian Journal of Zoology* 1:291–298.

Morelli, T. L., C. Daly, S. Z. Dobrowski, D. M. Dulen, J. L. Ebersole, S. T. Jackson, J. D. Lundquist, C. I. Millar, S. P. Maher, W. B. Monahan, K. R. Nydick, K. T. Redmond, S. C. Sawyer, S. Stock and S. R. Beissinger. 2016. Managing climate change refugia for climate adaptation. *PLoS One* 11:e0159909.

Mortensen, L. A., N. M. Schmidt, T. T. Høye, C. Damgaard and M. C. Forchhammer. 2016. Analysis of trophic interactions reveals highly plastic response to climate change in a tri trophic High-Arctic ecosystem. *Polar Biology* 39:1467–1478.

Moy, A. D., W. R. Howard, S. G. Bray, and T. W. Trull. 2009. Reduced calcification in modern Southern Ocean planktonic foraminifera. *Nature Geoscience* 2:276–280.

Munro, M. 2017. What is killing the world's shorebirds? *Nature* 541:16–20.

Myers-Smith, I. H., S. C. Elmendorf, P. S. A. Beck, M. Wilmking, M. Hallinger, D. Blok, K. D. Tape, S. A. Rayback, M. Macias-Fauria, B. C. Forbes, J. D. M. Speed, N. Boulanger-Lapointe, C. Rixen, E. Lévesque, N. M. Schmidt, C. Baittinger, A. J. Trant, L. Hermanutz, L. S. Collier, M. A. Dawes, T. C. Lantz, S. Weijers, R. H. Jørgensen, A. Buchwal, A. Buras, A. T. Naito, V. Ravolainen, G. Schaepman-Strub, J. A. Wheeler, S. Wipf, K. C. Guay, D. S. Hik, and M. Vellend. 2015. Climate sensitivity of shrub growth across the Arctic biome. *Nature Climate Change* 5:887–891.

Nebel, S., J. L. Porter, and R. T. Kingsford. 2008. Long–term trends of shorebird populations in eastern Australia and impacts of freshwater extraction. *Biological Conservation* 141:971–980.

Nicholls, R. J., and A. Cazenave. 2010. Sea-level rise and its impact on coastal zones. *Science* 328:1517–1520.

Nielsen, D. L., M. A. Brock, G. N. Rees, and D. S. Baldwin. 2003. Effects of increasing salinity on freshwater ecosystems in Australia. *Australian Journal of Botany* 51:655–665.

Nol, E., M. S. Blanken, and L. Flynn. 1997. Sources of variation in clutch size, egg size and clutch completion dates of Semipalmated Plovers in Churchill, Manitoba. *Condor* 99:389–396.

North American Bird Conservation Initiative, U.S. Committee. 2010. *The State of the Birds 2010 Report on Climate Change, United States of America*. U.S. Department of the Interior, Washington, DC.

Ockendon, N., C. M. Hewson, A. Johnston, and P. W. Atkinson. 2012. Declines in British breeding populations of Afro-Palaearctic migrant birds are linked to bioclimatic wintering zone in Africa, possibly via constraints on arrival time advancement. *Bird Study* 59:111–125.

Ockendon, N., D. Leech, and J. W. Pearce-Higgins. 2013 Climatic effects on breeding grounds are more important drivers of breeding phenology in migrant birds than carry-over effects from wintering grounds. *Biology Letters* 9:20130669.

O'Reilly, C.M., S. Sharma, D. K. Gray, S. E. Hampton, J. S. Read, R. J. Rowley, P. Schneider, J. D. Lenters, P. B. McIntyre, B. M. Kraemer, G. A. Weyhenmeyer, D. Straile, B. Dong, R. Adrian, M. G. Allan, O. Anneville, L. Arvola, J. Austin, J. L. Bailey, J. S. Baron, J. D. Brookes, E. de Eyto, M. T. Dokulil, D. P. Hamilton, K. Havens, A. L. Hetherington, S. N. Higgins, S. Hook, L. R. Izmest'eva, K. D. Joehnk, K. Kangur, P. Kasprzak, M. Kumagai, E. Kuusisto, G. Leshkevich, D. M. Livingstone, S. MacIntyre, L. May, J. M. Melack, D. C. Mueller-Navarra, M. Naumenko, P. Noges, T. Noges, R. P. North, P-D. Plisnier, A. Rigosi, A. Rimmer, M. Rogora, L. G. Rudstam, J. A. Rusak, N. Salmaso, N. R. Samal, D. E. Schindler, S. G. Schladow, M. Schmid, S. R. Schmidt, E. Silow, M. E. Soylu, K. Teubner, P. Verburg, A. Voutilainen, A. Watkinson, C. E. Williamson, and G. Zhang. 2015. Rapid and highly variable warming of lake surface waters around the globe. *Geophysical Research Letters* 42:773–781.

Parry, M. L., O. F. Canziani, J. P. Palutikof, P. J. van der Linden and C.E. Hanson (Eds.) 2007. *Contribution of Working Group II to the Fourth Assessment Report of the Intergovernmental Panel on Climate Change, 2007.* Cambridge University Press, Cambridge, UK and New York.

Parmesan, C., and G. Yohe. 2003. A globally coherent fingerprint of climate change impacts across natural systems. *Nature* 421:37–42.

Pearce-Higgins, J. W., D. W. Yalden, and M. J. Whittingham. 2005. Warmer springs advance the breeding phenology of Golden Plovers *Pluvialis apricaria* and their prey (Tipulidae). *Oecologia* 143:470–476.

Pearson, R. G., S. J. Phillips, M. M. Loranty, P. S. A. Beck, T. Damoulas, S. J. Knight and S. J. Goetz. 2013. Shifts in Arctic vegetation and associated feedbacks under climate change. *Nature Climate Change* 3:673–677.

Perry, A. L., P. J. Low, J. R. Ellis, and J. D. Reynolds. 2005. Climate change and distribution shifts in marine fishes. *Science* 308:1912–1915.

Peterson, A. T. 2003. Projected climate change effects on Rocky Mountain and Great Plains birds: Generalities of biodiversity consequences. *Global Climate Change Biology* 9:647–655.

Piersma, T., and A. Lindstrom. 2004. Migrating shorebirds as integrative sentinels of global environmental change. *Ibis* 146:61–69.

Piersma, T., and P. Wiersma. 1996. Family Charadriidae. Pp. 384–442, in del Hoyo, J., A. Elliott, and J. Sargatal (Eds.), *Handbook of Birds of the World Vol. III: Hoatzin to Auks.* Lynx Edicions, Barcelona, Spain.

Przybylo, R., B. C. Sheldon, and J. Merila. 2000. Climatic effects on breeding and morphology: Evidence for phenotypic plasticity. *Journal of Animal Ecology* 69:395–403.

Rehfisch, M. M., G. A. Austin, S. N. Freeman, M. J. S. Arbitrage and N. H. K. Burton. 2004. The possible impact of climate change on the future distribution and numbers of waders on Britain's non-estuarine coast. *Ibis* 148:70–81.

Rice, K. C., M. R. Bennett, and J. Shen. 2011. Simulated changes in salinity in the York and Chickahominy Rivers from projected sea-level rise in Chesapeake Bay: U.S. Geological Survey Open-File Report 2011–1191, 31 pp.

Root, T. R., J. T. Price, K. R. Hall, S. H. Schneider, C. Rosenzweig and J. A. Pounds. 2003. Fingerprints of global warming on wild animals and plants. *Nature* 421:57–60.

Roshier, D. A., P. H. Whetton, R. J. Allan, and A. I. Robertson. 2001. Distribution and persistence of temporary wetland habitats in arid Australia in relation to climate. *Austral Ecology* 26:371–384.

Rubega, M. A., and L. W. Oring. 2004. Excretory organ growth and implications for salt tolerance in hatchling American Avocets *Recurvirostra americana*. *Journal of Avian Biology* 35:13–15.

Rühland, A. M., W. Paterson, N. Keller, N. Michelutti and J. P. Smol. 2013. Global warming triggers the loss of a key Arctic refugium. *Proceedings of the Royal Society B* 280:20131887

Schekkerman, H., I. Tulp, K.M. Calf, and J. J. de Leeuw. 2004. Studies on breeding shorebirds at Medusa Bay, Taimyr, in summer 2002. Alterra report 922, Wageningen, The Netherlands.

Schlacher, T. A., D. S. Schoeman, J. Dugan, M. Lastra, A. Jones, F. Scapini and A. McLachlan. 2008. Sandy beach ecosystems: Key features, sampling issues, management challenges and climate change impacts. *Marine Ecology* 29:70–90.

Seavey, J. R., B. Gilmer, and K. M. McGarigal. 2010. Effect of sea–level rise on Piping Plover (*Charadrius melodus*) breeding habitat. *Biological Conservation* 144:393–401.

Senner, N. R., M. Stager, and B. K. Sandercock. 2017. Ecological mismatches are moderated by local conditions for two populations of a long-distance migratory bird. *Oikos* 126:61–72.

Shadkam, A., L. Fulco, M. T. H. van Vliet, A. Pastor and P. Kabat. 2016. Preserving the world second largest hypersaline lake under future irrigation and climate change. *Science of the Total Environment* 559:317–325.

Shanahan, T. M., J. T. Overpeck, K. J. Anchukaitis, J. W. Beck, J. E. Cole, D. L. Dettman, J. A. Peck, C. A. Scholz, and J. W. King. 2009. Atlantic forcing of persistent drought in West Africa. *Science* 324:377–380.

Small-Lorenz, S. L., L. A. Culp, T. B. Ryder, T. C. Will and P. P. Marra. 2013. A blind spot in climate change vulnerability assessments. *Nature Climate Change* 3:91–93.

Smith, P. A., G. Gilchrist, M. R. Forbes, J.-L. Martin and K. Allard. 2010. Inter–annual variation in the breeding chronology of Arctic shorebirds: Effects of weather, snow melt and predators. *Journal of Avian Biology* 41:292–304.

Steen, V. A., S. K. Skagen, and C. P. Melcher. 2016. Implications of climate change for wetland dependent birds in the Prairie Pothole Region. *Wetlands* 36 (Suppl 2):445–459.

Steen, V. A., S. K. Skagen, and B. R. Noon. 2014. Vulnerability of breeding waterbirds to climate Change in the Prairie Pothole Region, U.S.A. *PLoS One* 9:e96747.

Stephens, P. A., L. R. Mason, R. E. Green, R. D. Gregory, J. R. Sauer, J. Alison, A. Aunins, L. Brotons, S. H. M. Butchart, T. Campedelli, T. Chodkiewicz, P. Chylarecki, O. Crowe, J. Elts, V. Escandell, R. P. B. Foppen, H. Heldbjerg, S. Herrando, M. Husby, F. Jiguet, A. Lehikoinen, A. Lindstrom, D. G. Noble, J. Y. Paquet, J. Reif, T. Sattler, T. Szep, N. Teufelbauer, S. Trautmann, A. J. van Strien, C. A. M. van Turnhout, P. Vorisek and S. G. Willis. 2016. Consistent response of bird populations to climate change on two continents. *Science* 352:84–87.

Stutzman, R. J., and J. J. Fontaine. 2015. Shorebird migration in the face of climate change. Pp. 145–159, in Wood, E. M., and J. L. Kellermann (Eds.), *Phenological Synchrony and Bird Migration: Changing Climate and Seasonal Resources in North America*. Studies in Avian Biology Series (vol. 47). CRC Press, Boca Raton, FL.

Tomkovich, P. S., and E. E. Syroechkovski, Jr. 2005. Breeding of Semipalmated Plover *Charadrius semipalmatus* in Russia. *Russian Journal of Ornithology* 14(298):795–799 (In Russian with English summary).

Tulp, I., and H. Schekkerman. 2008. Has prey availability for Arctic birds advanced with climate change? Hindcasting the abundance of Arctic arthropods using weather and seasonal variation. *Arctic* 61:48–60.

United Nations Environmental Program GEO5 Global Environment Outlook. 2012. *Environment for the Future We Want*. UNEP Progress Press Ltd., Valletta, Malta.

Van Gils, J. A., S. Lisovski, T. Lok, W. Meissner, A. Ozarowska, J. de Fouw, E. Rakhimberdiev, M. Y. Soloviev, T. Piersma and M. Klaassen. 2016. Body shrinkage due to Arctic warming reduces Red Knot fitness in tropical wintering area. *Science* 352:819–821.

Visser, M. E., and C. Both. 2006. Shifts in phenology due to global climate change: The need for a yardstick. *Proceedings of the Royal Society B* 272:2561–2569.

Volpe National Transportation Systems Center. 2009. *Chincoteague National Wildlife Refuge Alternative Transportation Study*. U.S. Fish and Wildlife Service, Washington, DC.

Ward, D. H., J. Helmericks, J. W. Hupp, L. McManus, M. Budde, D. C. Douglas and K. D. Tape. 2016. Multidecadal trends in spring arrival of avian migrants to the central Arctic coast of Alaska: Effects of environmental and ecological factors. *Journal of Avian Biology* 47:197–207.

Wauchope, H. S., J. D. Shaw, Ø. Varpe, E. G. Lappo, D. Boertmann, R. B. Lanctot and R. A. Fuller. 2016. Rapid climate-driven loss of breeding habitat for Arctic migratory birds. *Global Change Biology* doi:10.1111/gcb.13404.

Zefania, S., R. ffrench-Constant, P. R. Long, and T. Székely. 2008. Breeding distribution and ecology of the threatened Madagascar Plover *Charadrius thoracicus*. *Ostrich* 79:1–9.

Zhai, L., J. Jiang, D. DeAngelis, and L. da Silveira Lobo Sternberg. 2016. Prediction of plant vulnerability to salinity increase in a coastal ecosystem by stable isotope composition (δ18o) of plant stem water: A model study. *Ecosystems* 19:32–49.

CHAPTER FOUR

Plover Breeding Systems*

DIVERSITY AND EVOLUTIONARY ORIGINS

Luke J. Eberhart-Phillips

Abstract. Shorebirds run the gamut of breeding system diversity, with the plover lineage being no exception. Inter- and intraspecific variation in mating and parental strategies has been documented throughout the group, thus providing a model system with which to conduct theoretical and empirical research in the fields of evolutionary biology and behavioral ecology. Here, I review mating strategies and patterns of parental care from plover populations worldwide. In summary, I found that 77% of the 40 plover species were monogamous with biparental care during incubation and brooding, 18% were polygamous or exhibited uniparental care, and the remaining 5% were of unknown status due to insufficient study. Notably, in species with uniparental care, parental desertion (typically by females) occurs during brood rearing, which is likely related to the independent nature of their precocial young. Furthermore, extra-pair mating in plovers is low (0%–9.1% of broods with mixed parentage) compared to other avian taxa, probably owing to the importance of maintaining male involvement during parental care. Local demographic and ecological variation in the environment appears to play a strong role in shaping breeding system expression: adult sex ratio variation is related to parental desertion, presumably via sex-biased mating opportunities, whereas harsh climates impose parental cooperation during care, thus restricting the potential for polygamy. I illustrate these associations with a case study exploring adult sex ratio variation and breeding system expression among six plover populations from around the globe. Lastly, I discuss the importance of integrating breeding system variation into plover conservation and management. Monogamous populations can be vulnerable to inbreeding depression owing to limited gene flow, whereas polygamous populations are characterized by sex-biased effective population sizes vulnerable to density-dependent growth. Effective recovery plans will require routine monitoring coupled with behavioral ecology, demography, and population genetics.

Keywords: adult sex ratio, Charadrius, demography, environment, monogamy, parental care, phylogeny, polygamy, sex roles.

A breeding system describes the general behavioral strategy for obtaining mates. In animals with distinct sexes (i.e., gonochoristic), a breeding system is characterized by the number of mates obtained by each sex, the extent of pair bonding between mates,

* Luke J. Eberhart-Phillips. 2019. Plover Breeding Systems. Pp. 63–88 in M. A. Colwell and S. M. Haig (editors). The Population Ecology and Conservation of *Charadrius* Plovers Studies in Avian Biology (no. 52), CRC Press, Boca Raton, FL.

and the role that each sex has during parental care of the consequential offspring (Emlen and Oring 1977). Shorebirds run the gamut of breeding system diversity, with extreme polygyny in lekking species to sex-role reversal in polyandrous taxa. The plover lineage exhibits several breeding systems (Figure 4.1, Table 4.1), although diversity is noticeably less-pronounced than in their sandpiper relatives (Székely and Reynolds 1995). This lack of diversity is perhaps attributed to the lineage's young evolutionary history (dos Remedios et al. 2015a), with polygamy and uniparental care being recent developments. Phylogenetic analysis of the evolutionary transitions of breeding systems in plovers suggests that the ancestral state consisted of monogamous pair bonding accompanied by biparental care of offspring (Myers 1981, Székely and Reynolds 1995).

The scattering of uniparental species throughout the plover clade (Figure 4.1) implies that this breeding strategy evolved several times independently. Some species have clearly defined sex roles, such as in the polyandrous Eurasian Dotterel, which exhibits unusual courting and parental behaviors not seen elsewhere in the clade (Owens et al. 1994). In others, such as the Kentish Plover, sex roles are more flexible, with observations of either sex deserting or both sexes providing parental care (Székely et al. 2004). European populations of the Kentish Plover display this plasticity in breeding system (Lessells 1984, Székely and Lessells 1993, Amat et al. 1999); however, island populations of the Kentish Plover in the Red Sea and Cape Verde are almost exclusively

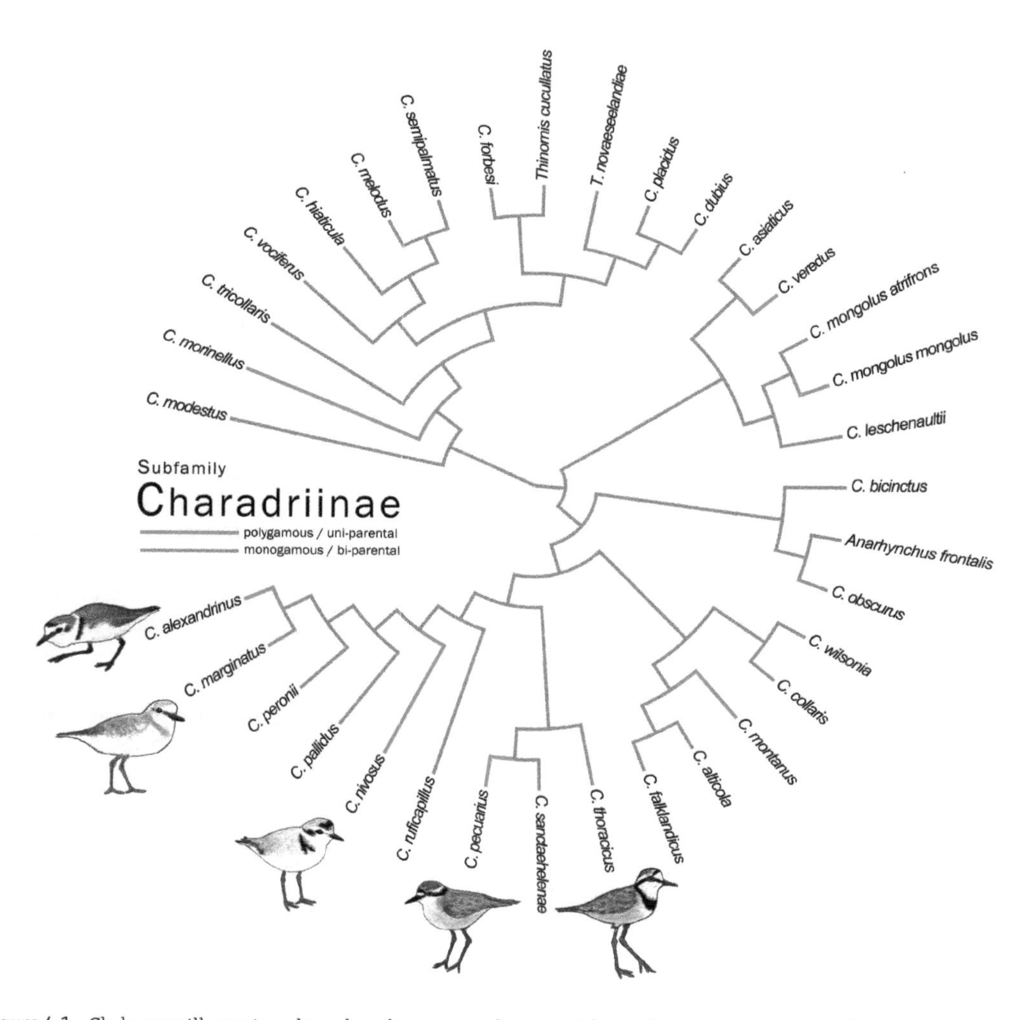

Figure 4.1. Cladogram illustrating plover breeding system diversity. Adapted from dos Remedios et al. 2015a.

TABLE 4.1

Diversity of plover breeding systems, parental care, sex ratios, rates of extra-pair parentage, and breeding climates.

Common name	Social mating strategy	Parental care		ASR	Extra-pair parentage (%)	Main Köppen-Geiger climates in breeding range
		Eggs	Brood			
S. Red-breasted Dotterel	Monogamy	Bi	Bi			Warm
N. Red-breasted Dotterel	Monogamy	Bi	Bi			Warm
Lesser Sand-plover	Serial polygamy	Bi or Uni (M)	Bi or Uni (M)			Arid, Polar
Greater Sand-plover	Monogamy	Bi	Bi			Arid, Warm
Caspian Plover		Bi	Bi			Arid
Collared Plover						Equatorial
Puna Plover						Arid, Equatorial, Warm
Two-banded Plover		Bi	Bi		0	Polar, Warm
Double-banded Plover	Monogamy	Bi	Bi			Warm
Kittlitz's Plover	Serial polygamy	Bi	Uni (M)	M < F	0	Arid, Equatorial, Warm
Red-capped Plover	Monogamy	Bi	Bi		0	Arid, Equatorial, Warm
Malay Plover	Monogamy	Bi	Bi			Equatorial
Kentish Plover	Monogamy or serial polygamy	Bi	Uni (M) or Bi	M = F or M > F	0–3.4	Arid, Equatorial, Polar, Warm
White-faced Plover	Monogamy	Bi	Bi			Equatorial, Warm
Snowy Plover	Serial polygamy	Bi	Uni (M) or Bi	M > F	0	Arid, Equatorial, Warm
Javan Plover		Bi	Bi			Equatorial
Wilson's Plover	Monogamy	Bi	Bi			Arid, Equatorial, Warm
Common Ringed Plover	Monogamy	Bi	Bi		0	Polar
Semipalmated Plover	Monogamy	Bi	Bi		4.2	Polar
Long-billed Plover		Bi	Bi			Arid, Warm
Piping Plover	Monogamy	Bi	Bi			Warm
Madagascar Plover	Monogamy	Bi	Bi	M = F	5	Arid, Equatorial
Little Ringed Plover	Monogamy	Bi	Bi			Arid, Equatorial, Polar, Warm
African Three-banded Plover	Monogamy	Bi	Bi			Arid, Equatorial, Warm
Madagascar Three-banded Plover	Monogamy	Bi	Bi			Arid, Equatorial, Warm
Forbes's Plover	Monogamy	Bi	Bi			Equatorial

(Continued)

TABLE 4.1 (*Continued*)

Diversity of plover breeding systems, parental care, sex ratios, rates of extra-pair parentage, and breeding climates.

Common name	Social mating strategy	Parental care		ASR	Extra-pair parentage (%)	Main Köppen-Geiger climates in breeding range
		Eggs	Brood			
White-fronted Plover	Monogamy	Bi	Bi	M = F	0	Arid, Equatorial, Warm
Chestnut-banded Plover	Monogamy	Bi	Bi			Arid, Equatorial, Warm
Killdeer	Monogamy	Bi	Bi or Uni (M or F)			Arid, Equatorial, Polar, Warm
Mountain Plover	Monogamy or simultaneous polyandry	Uni (M or F)	Uni (M or F)	M > F		Arid
Oriental Plover			Uni (F)			Snow
Eurasian Dotterel	Simultaneous polyandry	Bi or Uni (M)	Uni (M)		9.1	Polar
St. Helena Plover	Monogamy	Bi	Bi			Warm
Rufous-chested Dotterel		Bi	Bi		0	Polar, Warm
Red-kneed Dotterel		Bi	Bi			Arid, Equatorial, Warm
Hooded Plover	Monogamy	Bi	Bi			Warm
Shore Plover	Monogamy	Bi	Bi			Warm
Black-fronted Dotterel	Monogamy	Bi	Bi			Arid, Equatorial, Warm
Inland Dotterel		Bi	Bi			Arid, Warm
Wrybill	Monogamy	Bi	Bi			Warm

Species shown match those shown in Table 1.1. Blank fields indicate insufficient data or unknown information.

monogamous with biparental care (Alrashidi et al. 2013, Carmona-Isunza et al. 2015).

A key life history trait that preadapts plovers as a group to diverse breeding systems is the precocial state of their young (Oring 1986, Thomas and Székely 2005). Within hours of hatching, plover chicks are able to walk, hide, and feed without assistance from a tending parent; chicks are, however, still dependent on parental care for thermoregulation via brooding. But the independent nature of plover chicks creates a situation in which one parent may suffice for brood care. This condition grants one of the parents the opportunity to desert the family and pursue an additional mate, potentially doubling his/her reproductive success. Because of this tight interplay, patterns of parental care and mating strategy generally go hand-in-hand.

Understanding which parent deserts and which parent provides care has puzzled evolutionary biologists since Trivers' (1972) discussions on the role of parental investment in the evolution of sexual strategies. Trivers posited that the sex investing most in parental care up to that point would have more to lose if they deserted, and thus they should continue providing care. However, Trivers' notion was contended by Dawkins and Carlisle (1976), who pointed out that his idea was fallacious because selection will favor a parental strategy that maximizes future fecundity regardless of past decisions. In other words, when deciding to desert or not, a parent must weigh the cost to offspring survival if it deserts with respect to the proportional contribution the offspring represents of their future reproductive potential. Maynard Smith (1977) used mathematical models derived from game theory to illustrate the evolutionarily stable strategies of males and females when deciding to desert or care for their offspring. Deserting has the benefit of possibly increasing fecundity by finding a serial mate, but deserting also has a cost because it could lead to an increased likelihood of chick mortality. These pros and cons to parental decisions generate sex-specific selection pressures on evolutionarily stable strategies of parental care and thus facilitate the evolution of breeding systems.

Plovers were integral to these evolutionary discussions when Lessells (1984) published an empirical study quantifying the social breeding system of a Kentish Plover population inhabiting the saltpans of the Camargue in southern France. Lessells suggested that three potential factors contributed to the diversity of plover breeding systems, including the (1) environment, (2) sex-specific demographic structure of the population, and (3) species' phylogenetic history. Inter- and intraspecific variation in these factors is suggested to have given rise to variation among species and populations in the sex-specific costs and benefits of deserting offspring. Because of the variety of mating and parental strategies documented among plovers and their ease of study in the wild, the group has been used as a model system to apply theoretical and empirical research in the fields of evolutionary biology and behavioral ecology. For example, recent comparative works have shown support for Lessells' original hypotheses, with patterns of matings and parental care matching local variation in the adult sex ratio (ASR) (Eberhart-Phillips et al. 2018) and adapting to pressures imposed by local climate (Vincze et al. 2017).

In this chapter, I examine the diversity of plover breeding systems based on the existing literature. My overview encompasses the facets of both mating and parental behavior documented from natural populations worldwide, and discusses the social and genetic components that contribute to the study and quantification of plover breeding systems. This sets the stage for the following section in which I explore how plover breeding systems are shaped by local variation in the environment and population structure. I then present a cross-population case study that couples behavioral observations of parenting and mating strategies with demographic modeling of over 6,000 individually marked animals from five species distributed worldwide. In conclusion, I consider the conservation implications of demographic and genetic variation in breeding systems.

METHODS

I conducted a comprehensive literature review of the 40 plovers and related species (Chapter One, Table 1.1, this volume) to examine variation in mating strategy and parental care patterns. My literature search was restricted to peer-reviewed publications that were found on Internet search engines such as Google Scholar and Web of Science. To identify relevant literature, my search queries included the common or scientific name(s) of each species in combination with various key words related to the study and quantification of breeding systems (e.g., ["*Charadrius morinellus*" OR "*Eudromias morinellus*" OR "Eurasian Dotterel"] AND [biparental OR breed* OR care* OR copulat* OR court* OR desert* OR divorce OR fertili* OR *gamy OR *gamous OR mat* OR pair* OR parent* OR sex* OR uniparental]). However, for several understudied species, I had to rely on accounts from the Charadriinae chapter of the *Handbook of the Birds of the World* (Piersma and Wiersma 1996) that were based on limited observations from unmarked populations. Unless intraspecific variation in breeding system was well documented, I generalized the observed patterns of mating and parental care over a species.

In my review, the term *breeding system* describes the frequency and timing of mates acquired per individual within an annual reproductive period henceforth called the *breeding season* (Figure 4.2). Breeding season phenology varies according to latitude and local climate, but for plovers the breeding period typically occurs once per year and lasts for at least 2 months. Depending on the length of the breeding season, reproductive individuals can complete one or multiple breeding attempts. I refer to a *breeding attempt* as the period between the initial copulation with a mate and the termination of parental care for the consequential offspring (i.e., fledging or death prior to independence; Figure 4.2). *Monogamy* occurs when a pair restricts copulations to one another within a breeding attempt (Figure 4.2a), whereas *polygamy* describes the case when either sex obtains more than one mating partner within or between breeding attempts (Figure 4.2b, c). Monogamous pair bonds can last for a single breeding attempt, or over multiple breeding attempts that occur in succession during one breeding season (Figure 4.2a), or across many years.

Two principal forms of polygamous breeding systems exist and each is defined by the sex of

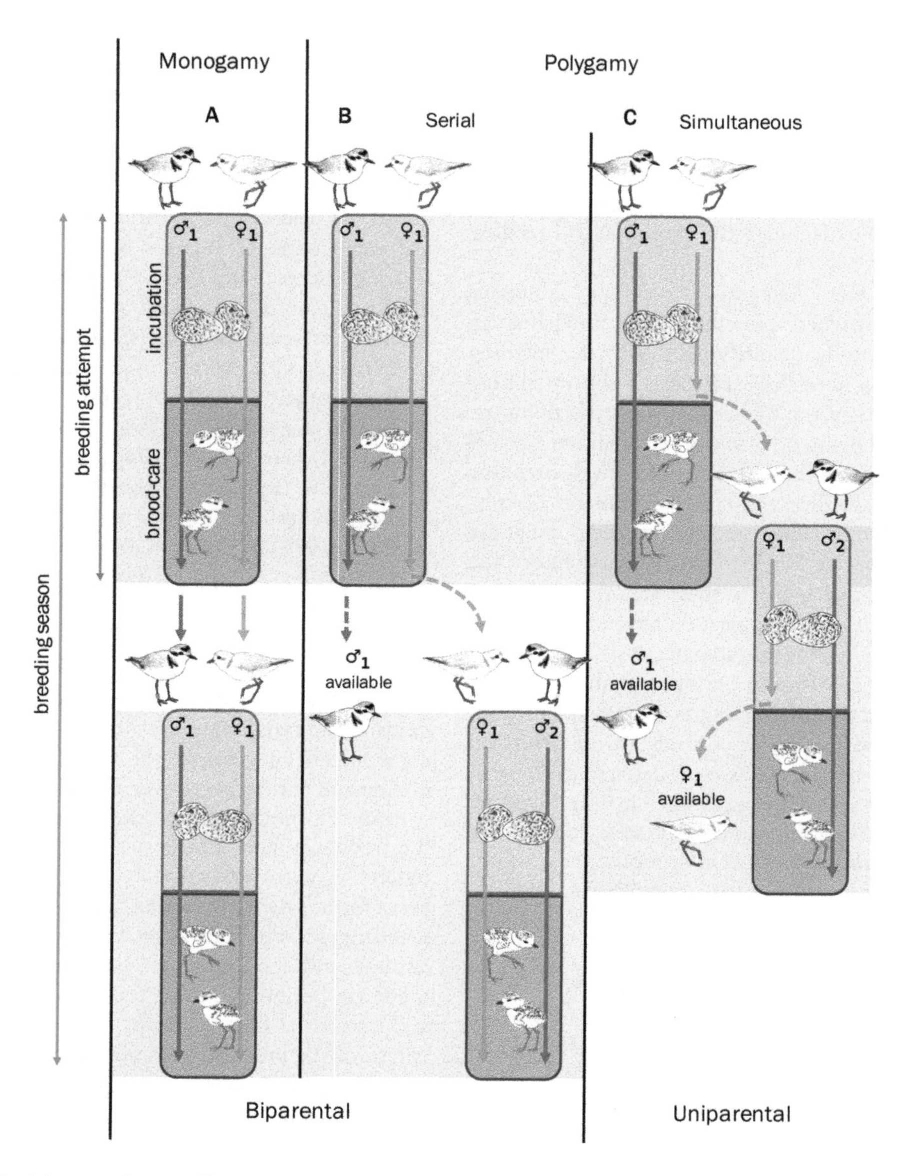

Figure 4.2. Schematic diagram illustrating the terms used to describe (a) monogamy and (b) serial or (c) simultaneous polygamy during breeding attempts throughout the breeding season. Note that monogamy and serial polygamy in plovers usually both have biparental care of eggs and chicks, whereas in simultaneous polygamy, eggs or chicks are tended uniparentally at some stage.

the individual pursuing multiple mates. *Polyandry* describes the mating strategy when a female acquires multiple mates (shown in Figure 4.2b,c), whereas *polygyny* occurs when a male mates with multiple females. In polygamous systems, the acquisition of multiple mates can be done simultaneously within a single breeding attempt (*simultaneous polygamy*, Figure 4.2c) or after a breeding attempt within a single season (*serial polygamy*, Figure 4.2b). Under this definition, simultaneous polygamy will always involve some level of uniparental care in which either the female or male cares solely for the offspring during incubation and/or brood care (Figure 4.2c).

To illustrate the evolutionary origins of contemporary variation in plover breeding systems, I present findings from a recent study that utilized detailed behavioral observations and demographic analysis from six individually marked populations of five species in western Mexico, southern

Turkey, Cape Verde, and southwestern Madagascar (Eberhart-Phillips et al. 2018). I discuss the inter- and intraspecific variation in breeding system among these six populations and interpret these patterns in light of the topics presented in my literature review of other plover species worldwide.

RESULTS AND DISCUSSION

Review of Plover Breeding System Diversity

The most prevalent plover breeding system is monogamy with biparental care of eggs and chicks (77% of species; Figure 4.1, Table 4.1). Of the 18% of plovers that exhibit polygamy and uniparental care, varying degrees of polygamy occur due to variation in the timing with which adults, especially females, abandon their mates. The remaining 5% of species currently have insufficient information about both mating behavior and parental care. I found no documented cases of fully simultaneous polygamy in plovers, as all species appeared to have at least some parental care given by both sexes during incubation or brooding. By contrast, in scolopacid sandpipers, several species show fully simultaneous polygamy in which one sex is exclusively responsible for all parental care. For example, Ruff (*Calidris pugnax*) exhibit lekking polygyny, while Buff-breasted Sandpiper (*Calidris subruficollis*) and Pectoral Sandpiper (*Calidris melanotos*) exhibit simultaneous polygyny and Spotted Sandpiper (*Actitis macularius*) and Wilson's Phalarope (*Phalaropus tricolor*) exhibit simultaneous polyandry. In plovers, the Eurasian Dotterel has the most extreme form of polygamy, with evidence of both serial and simultaneous polyandry. Upon arriving from wintering grounds in North Africa, females aggregate on mating arenas and compete vigorously for access to males (Kålås and Byrkjedal 1984, Owens et al. 1994). Once paired, individuals settle in nearby habitat to initiate a three-egg clutch. Within 2 weeks, the female deserts and returns to the mating arena to pursue additional mates, thus leaving the male to provide all subsequent incubation and brooding (Kålås and Byrkjedal 1984, Owens et al. 1994). Mountain Plovers exhibit a different breeding strategy known as a *rapid multi-clutch* system in which a female lays two consecutive three-egg clutches in separate nests in rapid sequence, which are tended alone by herself and her mate (Graul 1975). Rarely, female Mountain Plovers are polyandrous by simultaneously acquiring a second mate thus laying a total of three clutches each incubated by one parent (Riordan et al. 2015). The only other plover in which exclusive incubation by either sex has been documented is in the Mongolian subspecies of Lesser Sand-plover (*C. m. stegmanni*) in which males tend nests alone (Kruckenberg 2002).

All remaining species exhibit biparental care during incubation. However, during brood care, some species are frequently uniparental. Notably, uniparental brood care is almost always male biased, such as in Kittlitz's Plover (Eberhart-Phillips et al. 2018), Snowy Plover (Warriner et al. 1986), Kentish Plover (Lessells 1984), and Eurasian Dotterel (Owens et al. 1994). The Oriental Plover is reported to have female-biased uniparental care of broods (Piersma and Wiersma 1996) and in some Kittlitz's, Kentish, and Snowy Plover populations, females occasionally care for chicks (Warriner et al. 1986, Székely and Lessells 1993, Székely 1996, Eberhart-Phillips et al. 2018). Although the Kentish and Snowy plovers typically exhibit uniparental care, they sometimes express biparental brood care, especially in late-season breeding attempts (Székely et al. 1999) or in populations where short breeding seasons restrict double brooding (Rittinghaus 1956, Boyd 1972). Furthermore, Killdeers are shown to have flexible parental strategies with either sex deserting broods (Johnson et al. 2006) or both providing care (Lenington 1980, Mundahl 1982). These observations highlight a key component of plover breeding systems—polygamy and parental desertion are facultative traits.

Although biparental care prevails among plover species, the time allocated and type of care varies between sexes. For example, in the Malay Plover, males spend more time defending territories and broods from conspecifics compared to females, which incubate and brood offspring (Yasué and Dearden 2007). In Killdeer, these sex roles are reversed: females provide some parental care but spend more time foraging than males during incubation and brooding phases (Brunton 1988). Brunton (1988) speculated that the sex-biased parental care she observed in Killdeer could be related to the energetic demands of egg production. This hypothesis was tested by Amat et al. (2000) on Kentish Plovers in southern Spain where the authors monitored the metabolic rates and body conditions of males and females over the course of incubation and early chick rearing. In this population, both parents incubated but males typically tended broods alone. No sex difference in energetic costs during parental care was apparent, which led Amat et al. (2000) to conclude that metabolic demands were not prompting females to desert. Rather, female desertion may be prompted by sex differences in the capability of males and females to provide care, as shown by Székely (1996) who noted that broods tended by males had higher survival and were more successful with foraging than broods tended by females (cf. Székely and Cuthill 1999).

In plovers with biparental incubation, males usually incubate at night and females during the day, as seen in the Common Ringed Plover (Wallander 2003), Kentish Plover and Snowy Plover (Vincze et al. 2013), Red-capped Plover (Ekanayake et al. 2015), and Two-banded Plover (St Clair et al. 2010b). Ekanayake et al. (2015) proposed that male Red-capped Plovers incubate at night because they are more ornamented and thus more conspicuous to visual predators during the day. In support of this, the authors demonstrated with an experiment that daytime predation risk of artificial nests "tended" by male decoys exceeded nighttime predation risk. This suggests that the sex-specific circadian rhythm of plover incubation is in part due to intrinsic factors such as sexual dichromatism in plumage. However, in the Rufous-chested Dotterel these sex roles during incubation are reversed (St Clair et al. 2010a), and in the Semipalmated Plover no sex-specific pattern is apparent, with males and females exchanging incubation duties up to 20 times in a single day (Bulla et al. 2016). Plover incubation schedules are thus also shaped by local environmental pressures such as predation risk (Ekanayake et al. 2015, Bulla et al. 2016) or extreme temperature regimes (Alrashidi et al. 2011, Vincze et al. 2013, 2017).

Darwin (1871) originally contemplated the interplay between breeding system and sexual dimorphism by envisioning that "some relation exists between polygamy and the development of secondary sexual characters." However, as not all sexually dimorphic species are polygamous, Darwin was careful not to make a generalization. Although sexual dimorphism does exist in plovers, it is notably less extreme than in other shorebirds such as the Ruff, phalaropes, or Pectoral Sandpiper. Male plovers tend to be more ornamented (Ekanayake et al. 2015) and females tend to be larger (Zefania et al. 2010); in many populations, however, males and females are sexually monomorphic and thus need to be sex-typed genetically for studies of sexual selection (dos Remedios et al. 2015b). In a comparative study across five different populations of the Kentish Plover, Argüelles-Tico et al. (2015) found that the degree of sexual dimorphism in ornament brightness and size varied according to a population's breeding system: males in polygamous populations tended to have darker and smaller ornaments than males of monogamous populations. Conversely, there was no difference in female ornamentation between monogamous and polygamous populations. The results of Argüelles-Tico et al. (2015) illustrate potential evolutionary feedbacks between breeding system and sexually selected traits, and they suggest that the degree of sexual dimorphism may be a useful starting point to evaluate the breeding system of a species or population.

Quantifying Breeding Systems

Lack (1968) concluded that most (~90%) of birds were socially monogamous, with a single male and female pairing and sharing parental care of eggs and chicks (Figure 4.2a). Using molecular markers, ornithologists gained insight into the promiscuity of individuals, which revealed in many birds that social monogamy did not necessarily imply genetic monogamy (Westneat and Stewart 2003, Forstmeier et al. 2014). Determining the breeding system of a species is thus based on two sources of information: the identity of social pair bonds and genetic parentage. Social pair bonds are quantified by detailed observations of parental care interactions between uniquely marked males and females of a population, such as sharing incubation or brooding duties. Alternatively, genetic parentage assignment is determined by DNA profiling of breeding adults and their offspring by, for example, comparing the allele frequencies of offspring and putative parents to find the most likely match. These two sources of information help determine the **social** and **genetic components** of a breeding system, respectively.

Social Components of Breeding Systems

If breeding seasons are long enough for multiple breeding attempts, individuals are faced with the decision of either remaining with the original mate or finding another. Divorcing a mate after a breeding attempt (i.e., both pair members are still alive but choose not to remate with one another;

Figure 4.2b) is generally understood to be an individual's reproductive strategy to maximize fitness (Handel and Gill 2000). Therefore, a fundamental component contributing to divorce or mate fidelity is understanding the relative costs and benefits of either strategy (Choudhury 1995). Maintaining a pair bond could be beneficial in that it increases reproductive success through enhanced parental cooperation and familiarity (Black 1996). Alternatively, divorce may be an advantageous strategy to secure a higher-quality mate or breeding territory (Ens et al. 1993).

In plovers, both between- and within-season divorce rates differ among species, indicating that the costs and benefits vary. Various hypotheses have been proposed to explain interspecific variation in mate fidelity between breeding seasons; however, the two prominent hypotheses are related to site-fidelity and survival. Using published divorce rates and estimates of adult survival across eight Charadrius species, Lloyd (2008) suggested that high survival is related to high mate fidelity. He argued that when adult survival is high, reproductive vacancies are infrequent and few opportunities exist for individuals to switch mates. Notably, site- and mate-fidelity are not necessarily independent phenomena because individuals that return to a breeding site are likely to encounter previous mates, especially in territorial species. For example, in Snowy Plovers, the majority of males (66%) and females (61%) that were site-faithful paired with the same mate as the previous year (Pearson and Colwell 2013). In White-fronted Plovers, 97% of pairs reformed and defended the same territory between seasons (Lloyd 2008).

Reproductive failure is often cited as an important factor provoking divorce. However, no clear evidence exists for this in plovers. In Snowy Plovers of central California, 13% of within-season divorces followed nest failure (Warriner et al. 1986), whereas 34% of Piping Plover pairs divorced during the breeding season following nest failure in southern Manitoba (Haig and Oring 1988a). However, in a White-fronted Plover population in South Africa, only one case of divorce following nest failure was reported among 151 within-season pair bonds, despite extremely high nest predation rates (Lloyd 2008). Between breeding seasons, reproductive failure is also shown to elicit divorce. In Piping Plovers, 81% of the marked population divorced between seasons after low hatching success (Haig and Oring 1988b). However, in Semipalmated Plovers, 40% of previously successful pairs divorced the subsequent year (Flynn et al. 1999). These observations suggest that pair bonding is likely shaped by multiple factors (e.g., sex-specific site fidelity, prior reproductive success), which affect species on a case-by-case basis.

Interspecific variation in pair bond was evaluated by Parra et al. (2014) in an experiment with sympatric populations of the White-fronted Plover and Kittlitz's Plover in south-western Madagascar. In their study, the authors broke pair bonds by removing one of the mates and holding them in captivity. When the captive mate was released, all White-fronted Plovers reestablished pair bonds with their original mate, whereas all Kittlitz's Plovers found new mates and did not re-pair with their original mate. Although this was an experiment and thus not directly comparable to natural patterns of divorce, Parra et al. (2014) interpreted these differences as a consequence of interspecific variation in site fidelity: White-fronted Plovers are highly territorial (Lloyd 2008), whereas Kittlitz's Plovers have more transient territories and exhibit greater dispersal among populations (Eberhart-Phillips et al. 2015). Moreover, captive White-fronted Plovers were quicker to reestablish pair bonds after release (median = 2 days) than Kittlitz's Plovers, which also exhibited sex differences (female median = 6.5 days, male median = 3.3 days, Parra et al. 2014). In an identical mate-removal experiment, male Kentish Plovers took significantly longer than females to find a replacement partner (Székely et al. 1999). These two studies provide experimental evidence that pair bonds are tightly linked to parental care strategies: biparental species exhibit quick pair bond formation, whereas significantly more time is needed in uniparental species and sex biases exist (Figure 4.3). In uniparental species, the reproductive schedules of males and females are likely to be different due to parental desertion, which could create asynchrony in the availability of mates and delay pair-bond formation. In contrast, biparental species rely on cooperation while breeding, which could select for strong pair bonds and high mate-fidelity.

When divorce does occur, female plovers tend to disperse further than males and, curiously, this appears to be unrelated to parental care strategy. In uniparental Snowy Plovers, divorced males moved an average of 0.9 km, whereas females moved 2.2 km (Pearson and Colwell 2013). Similarly, in biparental White-fronted Plovers,

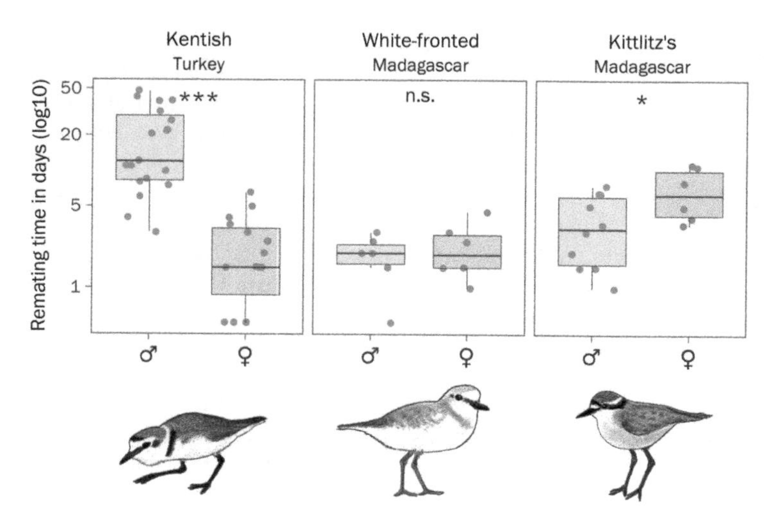

Figure 4.3. Interspecific variation in sex-specific mating opportunities among three plover species. Experimental assessment of sex differences in remating times indicate that Kentish Plover males (N = 19) take longer to find a mate than females (N = 15) after forced divorce. This trend is reversed in Kittlitz's Plovers (N_{male} = 10, N_{female} = 6) and not apparent in White-fronted Plovers (N_{male} = 6, N_{female} = 6). Significance indicated by asterisks (***: P < 0.001, *: P < 0.05, n.s.: P > 0.05). Adapted from Parra et al. 2014.

females dispersed 0.8 km after divorce compared to 0.2 km for males (Lloyd 2008). Biparental Little Ringed Plover males moved on average 4.1 km between breeding sites compared to 5.7 km for females (Pakanen et al. 2015). In contrast, male and female Mountain Plovers dispersed similar distances (average of 2.8 and 3.0 km, respectively) between failed breeding attempts within a breeding season (Skrade and Dinsmore 2010). Inbreeding avoidance is a popular hypothesis explaining sex-biased dispersal, but breeding systems are also likely to play a role (D'Urban Jackson et al. 2017), especially in territorial species that exhibit resource defense, such as the Snowy Plover and White-fronted Plover (Stenzel et al. 1994, Lloyd 2008).

Genetic Components of Breeding Systems

When individuals pursue matings outside their social pair bond, mixed parentage in broods can occur. In plovers, rates of mixed parentage are surprisingly low (0%–9.1%, Table 4.1) compared to passerines (25%–35%, Westneat and Stewart 2003). In the socially polyandrous Eurasian Dotterel and Kentish Plover, rates of mixed parentage in broods were 9.1% (n = 2/22, Owens et al. 1995) and 3.4% (n = 3/89, Küpper et al. 2004), respectively. In the socially monogamous Common Ringed Plover and Semipalmated Plover, mixed parentage rates in broods were 0% (n = 0/21, Wallander et al. 2001) and 4.2% (n = 1/24, Zharikov and Nol 2000), respectively.

Nest parasitism occurs when offspring are not related to both the social mother and father providing parental care. This can happen when a female deposits an egg in the nest of another pair but does not provide parental care. Nest parasitism has been documented in many plovers including: Mountain Plover (Hamas and Graul 1985), Piping Plover (Hussell and Woodford 1965), Kittlitz's Plover (L. Eberhart-Phillips, unpublished data), White-fronted Plover (L. Eberhart-Phillips, unpublished data), Chestnut-banded Plover (Blaker 1966), Semipalmated Plover (Havens 1970), Kentish Plover (Blomqvist et al. 2002), and Killdeer (Mundahl et al. 1981). In most of these cases, the genetic identity of parents was unknown, and researchers inferred parasitism from the presence of anomalously shaped or patterned eggs. An additional source of mixed parentage is quasi nest parasitism, whereby a male is the putative genetic father of all eggs deposited in a common nest by two or more females. In a Turkish population of the Kentish Plover, quasi-parasitism occurred in 3.1% (n = 2/65) of broods (Blomqvist et al. 2002), although quasi-parasitism was not detected in a recent large comparative study that also included this population (Maher et al. 2017).

Extra-pair fertilization, defined as the percentage of fertilizations resulting from copulations outside a social pair bond, is the most common and well-studied mechanism of mixed parentage in birds (Westneat and Stewart 2003). A common hypothesis explaining the adaptive benefit of extra-pair copulations is related to the genetic compatibility of social mates: females may seek extra-pair fertilizations as a strategy to reduce the negative effects of inbreeding or reduced fertility when socially paired to a closely related male. Evidence of this is shown in the Kentish Plover population of southern Turkey, where Blomqvist et al. (2002) documented that extra-pair fertilizations occurred in 4.6% (n = 3/65) of broods and that these three promiscuous pairings had significantly higher genetic relatedness than non-promiscuous pairings. The authors suggest that individuals may be able to assess the genetic similarity of their social mate and adjust their fertilization strategy accordingly to minimize the deleterious consequences of inbreeding. However, the relationship between genetic compatibility and extra-pair matings is likely complicated by an interaction with population size, as individuals may simply have no other option but to breed with closely related kin when populations are small. Such a situation was observed in Snowy Plovers, where four cases of inbreeding occurred after a local population decreased to 19 breeding individuals (Colwell and Pearson 2011).

In a comparative study across 12 plover populations comprising eight species, Maher et al. (2017) tested the relationship between rates of extra-pair fertilization and genetic relatedness of social pairs using a stringent protocol to discriminate between mismatches in parentage introduced by laboratory or field work. The authors found no relationship and, moreover, noted that the degree of extra-pair paternity across all populations was generally low (0%–4.1% of chicks per population) compared to other birds. The authors suggest that this result was most likely explained by the importance of parental cooperation during the period of offspring care. In plovers, male contribution to incubation (Kosztolányi et al. 2003) and brooding (Székely 1996) is particularly important in comparison to other avian groups. Therefore, a female is expected to refrain from extra-pair copulations because her social male could forego caring for unrelated young, thereby reducing the fitness benefits of promiscuity (Thomas and Székely 2005).

The Roles of Ecology and Demography in Plover Breeding Systems

Spatial and temporal variation in the local environment can limit the potential for individuals to adopt polygamous mating strategies (Emlen and Oring 1977). These environmental limitations can be ecological, such as suitability of climate or breeding habitat availability, or demographic, such as conspecific competition or the availability of potential mating partners in the local population (Figure 4.4). Disentangling the relative contribution of ecological and demographic factors on breeding system expression is challenging as they are often not mutually exclusive. For example, Kosztolányi et al. (2006) examined correlates between food availability and brood desertion in two groups of the Kentish Plover occupying different habitats in southern Turkey. The authors observed that females delayed desertion in lakeshore habitat where food availability, as measured by feeding efficiency, was higher than that in saltmarsh habitat. This result contradicted their expectation that desertion would be earlier in high-quality habitat, where food is ample and less parental provision is required. However, the authors also noted that breeding density of the lakeshore population was greater than the saltmarsh, which forced females to delay desertion in favor of defending their broods from conspecific harassment. Plovers exhibit inter- and intraspecific variation in both ecology and demography, which offers the opportunity to investigate how these two components shape local breeding system expression.

Ecological Constraints

The chronology of a breeding season is shaped by temporal fluctuations in a variety of abiotic factors such as water availability and temperature regimes, and biotic factors such as food and habitat availability or predation pressure. Seasonal variation in the local environment constrains the amount of time that breeding individuals have for multiple breeding attempts (Blomqvist et al. 2001), and thus limits breeding system variation. In the Arctic, for example, the breeding window is short and restricted mainly by snow cover and the brief bloom of invertebrate prey. As most plovers require biparental incubation, species dwelling in arctic regions are constrained to monogamy

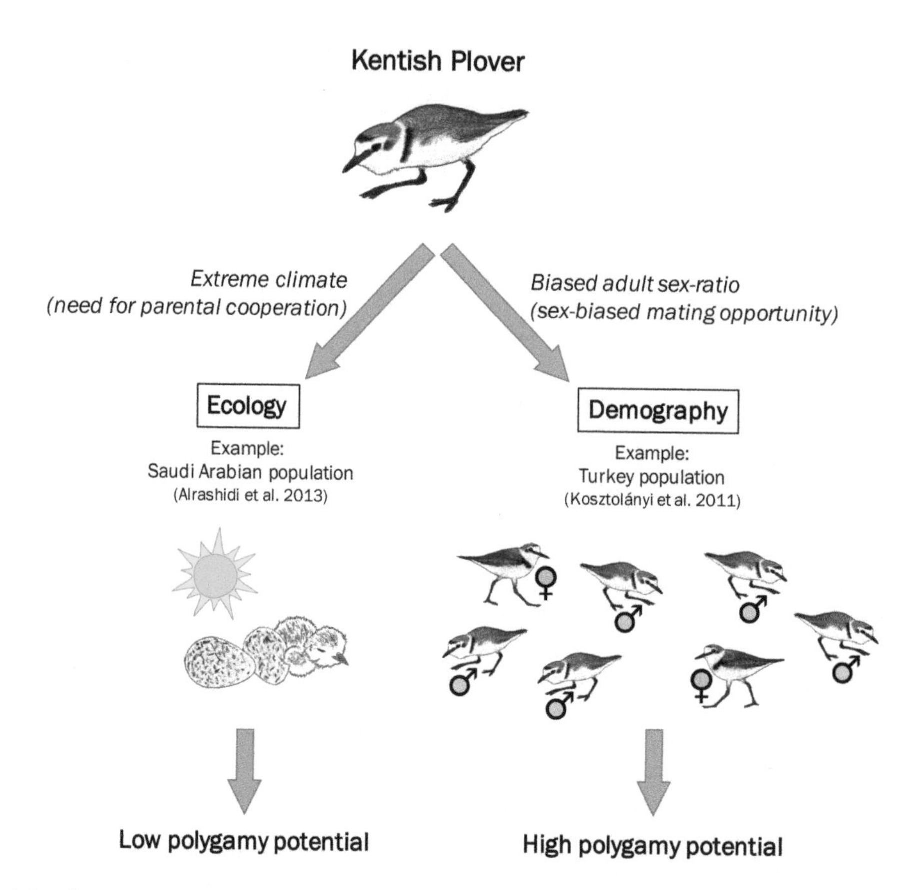

Figure 4.4. Schematic diagram illustrating the ecological and demographic environments that facilitate monogamous or polygamous mating strategies in the Kentish Plover.

because there is simply not enough time to successfully raise a second clutch with a serial mate. This is shown in Common Ringed Plovers breeding in north-eastern Greenland, where pairs initiate nests immediately after the spring snow cover has melted and rarely renest after failure, presumably because late fledglings would not be sufficiently mature to undertake migration (Green et al. 1977). On the other hand, in temperate and tropical latitudes, where breeding seasons are often prolonged, it is possible for females to initiate a second or even third clutch. For example, Malay Plovers breeding in the Gulf of Thailand can fledge up to three broods owing to their 5-month breeding season (Yasué and Dearden 2008), and Snowy Plovers breeding along the North American Pacific coast raise multiple clutches over 6 months (Warriner et al. 1986). These prolonged breeding seasons offer individuals the possibility of acquiring multiple mates either simultaneously or serially (Figure 4.2b,c). In the case of the Snowy Plover, female desertion during brood care translates into sex differences in annual reproductive success because females have the chance to breed with three mates, whereas caring males have only up to two mates. Taken together, the length of the breeding season and the amount of parental cooperation required to care for young seems to dictate polygamy potential for plovers.

Thermoregulation is an important component of parental care during incubation and chick rearing. Therefore, variation in the local climate during the breeding season imposes constraints on polygamous mating opportunities because, under extreme temperature regimes, biparental cooperation is required for eggs and chicks (Figure 4.4), whereas uniparental care could be sufficient under moderate conditions. Lessells (1984) proposed that plover offspring reared in lower latitudes require less thermoregulation than those at higher latitudes, which relieves the demand for biparental care and would thus, increase the potential for polygamy in these regions. Lessells (1984) acknowledged, however, that this notion is somewhat confounded by interspecific variation in the duration of offspring dependence. To address this hypothesis, Reynolds and Székely (1997) conducted a phylogenetic analysis comparing 52 shorebird species (including Eurasian Dotterel, Wilson's Plover, St. Helena Plover, and Kentish Plover) and found no correlation between biparental care and latitude. However, Reynolds and Székely (1997) recognized that latitude is likely a crude proxy of climatic harshness since wet or dry environments, for example, could also select for biparental care.

Aside from affecting offspring thermoregulation, climatic extremes can also induce stress on tending parents, especially during incubation when simultaneous foraging and parental care is not possible. In these severe environs, such as the Arctic or desert, parents must balance both the energetic demands of themselves and their offspring, which requires parental cooperation (Deeming 2002). For example, Kentish Plovers breeding in the Arabian Desert have the highest incidence of biparental care documented in the species, which is attributed to the need for parental cooperation to avoid chicks overheating in an environment which frequently exceeds 45°C (Kosztolányi et al. 2009, Alrashidi et al. 2010). Biparental cooperation in extreme climates also helps alleviate physiological stress to parents. For instance, after prolonged cold spells in the subarctic, biparental Semipalmated Plovers are better able to maintain a consistent body mass during incubation (Graham 2004) than sympatric species of uniparental shorebirds that experience a significant weight reduction (Tulp and Schekkerman 2006). In contrast, parents breeding in temperate and tropical latitudes may forgo incubation to forage during times of the day when ambient temperatures are optimal for embryonic development (15°C–19°C), as seen in Hooded Plovers (Weston and Elgar 2005). Plovers breeding in mild climates are thus relieved from the pressures of parental cooperation and have reduced energetic stress imposed by providing care, which increases polygamy potential by allowing more time and energy for pursuing multiple mates.

The relationship between parental cooperation and local climate was assessed by Vincze et al. (2013) in a comparative analysis of ten geographically distinct populations of the Kentish and the Snowy plovers. The authors found that local climate had a strong effect on incubation patterns among populations: total incubation time allocated by both parents increased in regions of extreme heat, but the relative proportion of time invested by females was less (Figure 4.5), especially at night. In regions of intense heat, males and females frequently switched incubation duty around midday, supposedly to mediate both parental and embryonic thermoregulation. Furthermore, in a follow-up study Vincze et al. (2017) revealed that plovers of these populations adopted optimal parental care in response to stochastic temperature regimes, indicating that plover breeding systems may be flexible to impending climate change. Taken together, the local climate imposes physiological constraints on parents and offspring that intensify or relax the need for biparental care, and hence the opportunity for polygamy.

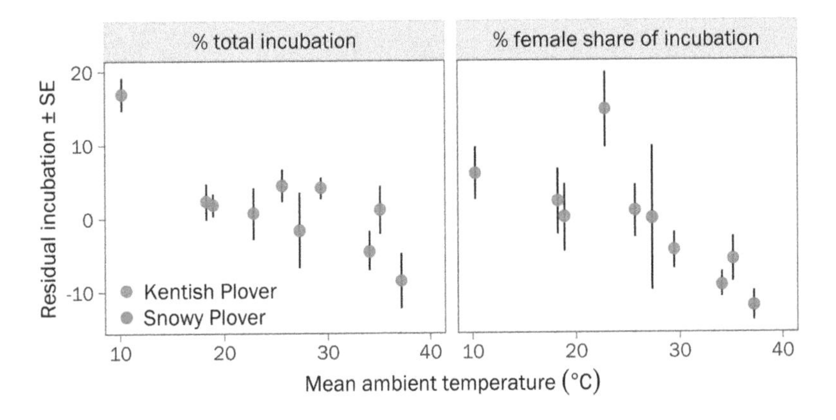

Figure 4.5. Correlations between the mean ambient temperature and (a) total incubation or (b) female share of incubation across ten Kentish Plover and Snowy Plover populations inhabiting North America, Africa, Europe, and the Middle East. Spearman rank correlations, total incubation: $r_s = 20.661$, $P = 0.0440$, female share: $r_s = 20.891$, $P = 0.0014$. Adapted from Vincze et al. 2013.

Demographic Constraints

Sex differences in demographic rates, such as survival or birth rates, can facilitate a sex bias in the population and skew the availability of breeding males and females (Eberhart-Phillips et al. 2017). Under a biased ASR, the abundant sex is forced to compete harder for access to mates because the opposite sex is in limited supply. This situation creates an imbalance in the mating opportunities of males and females and leads to alternative mating strategies (Emlen and Oring 1977). For example, sex differences in the costs and benefits of deserting or providing parental care arise under a biased ASR, which increases the polygamy potential of the limiting sex (McNamara et al. 2000).

The associations between demography and breeding system variation are particularly well studied in shorebirds, where ASR is found to be a strong predictor of mating and parental strategies. Evaluating 18 species of shorebirds (including the Kentish, the Snowy, and the Semipalmated plovers), Liker et al. (2013) found that populations with male-biased ASR had higher incidences of polyandry and male-biased parental care, whereas populations with a female-biased ASR more often exhibited polygyny (Figure 4.6). Causation, however, remains unclear—Is ASR bias driving breeding system evolution or *vice versa*? The relationship between sex ratio and breeding system represents a causality dilemma because, on the one hand, polygamy could impose sex-specific mortality due to the costs of sexual selection, thus facilitating a biased ASR (Székely et al. 2014a). But, on the other hand, a biased ASR creates sex-biased mating opportunities, which facilitates polygamy (Kokko and Jennions 2008). The positive feedback between ASR and breeding behavior reinforces the mating and parenting strategy of individuals because it is the most favorable behavioral response to a biased ASR created by the breeding behavior itself (Figure 4.7).

Although the ASR of the population is an important limitation to mating opportunities, not all breeding individuals may be available to mate, which could dampen the sex bias in mating opportunities or skew it even further. For example, if the breeding density is sparse, individuals may not realize that mating opportunities are in their favor and consequently remain in monogamous pair bonds. Likewise, if the migratory schedules of males and females differ, the local sex ratio will vary over the course of the breeding season and could lead to alternative mating strategies (Carmona-Isunza 2016). The operational sex ratio (OSR) quantifies this spatiotemporal variation in mating availability by describing the local sex ratio of the adults that are available to breed at a given location and point in time (Emlen and Oring 1977).

Uniparental plovers provide a good illustration of how the OSR may vary dramatically across the breeding season because of temporal variation in the number of males and females engaged in or emancipated from parental care. In a Snowy Plover population on the Pacific coast of Mexico, Carmona-Isunza et al. (2017) show that ASR was consistently male biased throughout the breeding season (approximately 60% males), but that OSR varied over short intervals from strongly male biased to female biased (Figure 4.8). The temporal variation in the OSR was attributed to a temporal

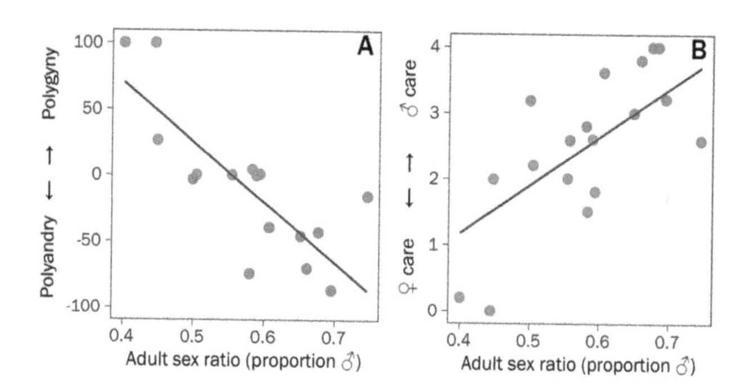

Figure 4.6. Relationships between ASR and mating or parental behavior across 18 species of shorebird. (a) Mating behavior expressed as the percentage of male or female polygamy (phylogenetically corrected regression: $r = 0.79$, $P = 0.001$). (b) Parental care is expressed as the mean of male participation in five parental behaviors: nest building, incubation, nest guarding, chick brooding, and chick guarding (phylogenetically corrected regression: $r = 0.70$, $P = 0.001$). Panels show species values with red and blue dots representing species with reversed and conventional sex roles, respectively. Adapted from Liker et al. 2013.

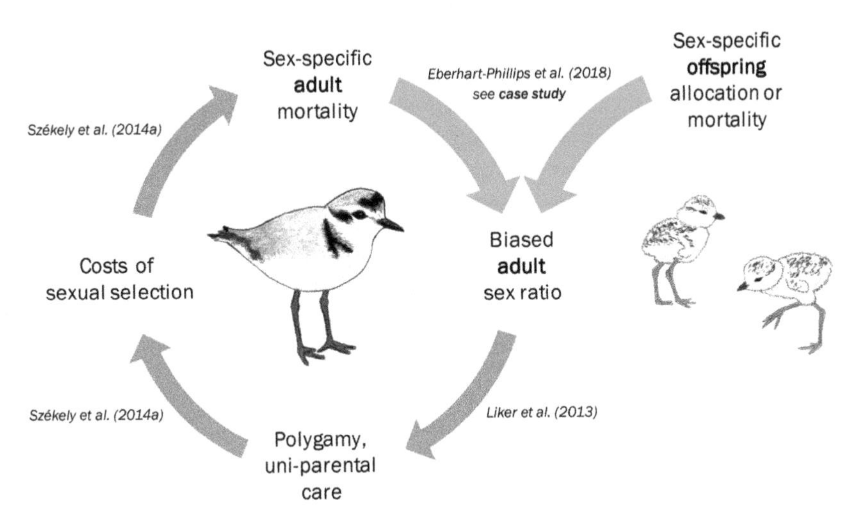

Figure 4.7. Schematic flow diagram illustrating the dynamic evolutionary feedbacks between sex-specific demography and breeding system.

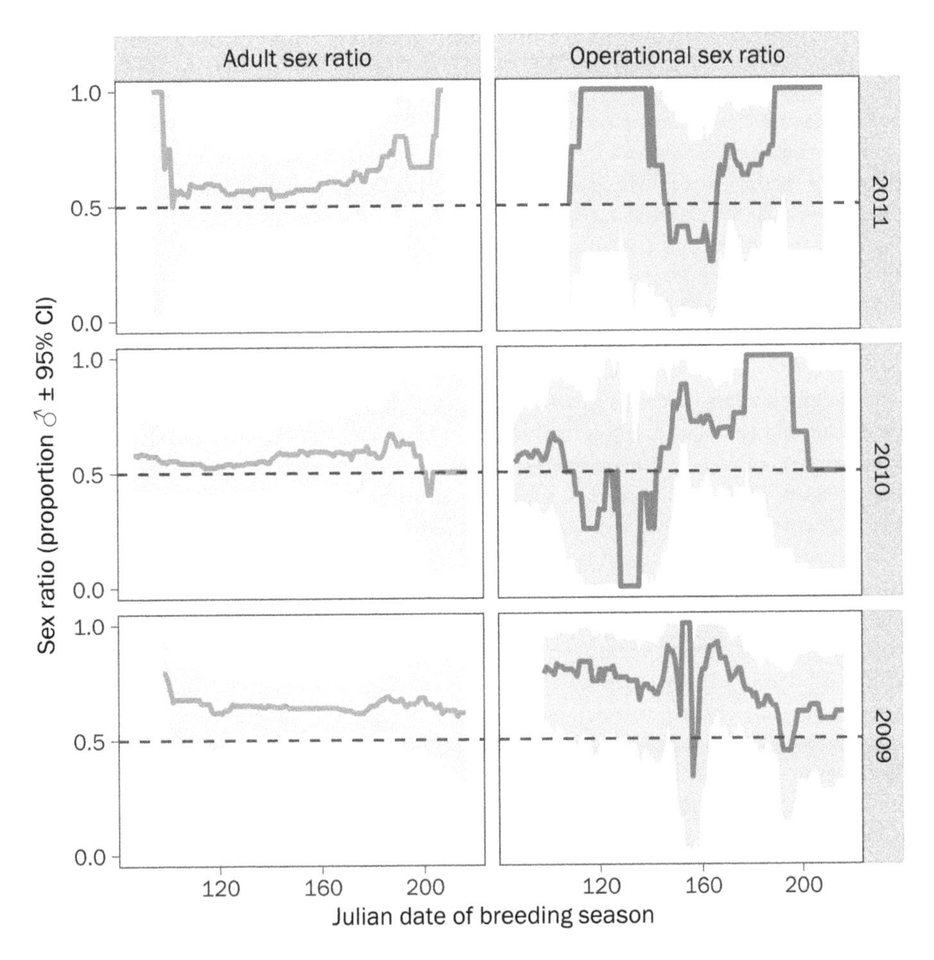

Figure 4.8. Seasonal variation in the ASR and OSR of the polygamous snowy plover population in Ceuta, Mexico. Adapted from Carmona-Isunza et al. 2017.

asynchrony in the availability of either sex to find a mate because of sex roles during parental care. As ASR is male biased in this population, females quickly pair up at the start of the breeding season, thus creating an even stronger male-biased OSR due to the surplus of unmated males. After a month of biparental incubation, these initial nests hatch and females desert their mates, causing a brief female-biased OSR. After another month, the initial males fledge their broods and become available again, swinging the OSR back to being male biased, and so on. The seasonal variation in the OSR of this population is strongly dependent on the nesting success of a given year. If hatching rates are high and nesting habitat is not limited, the breeding chronology of pairs will be more synchronized and create strong cycles in OSR (as seen in 2011, Figure 4.8). However, if conditions create poor nesting success and habitat is restricted, the OSR can imitate the overall ASR due to random and asynchronous pairing and divorce events throughout the season as nests fail (as seen in 2009, Figure 4.8).

CASE STUDY Insights from the Field: ASR Bias and Parental Desertion in Plovers

Due to their remarkable variation in breeding behavior and their ease of study in the wild, plovers offer an ideal system for investigations of breeding system evolution. Their open breeding habitats and cursorial nature makes locating and monitoring nests, broods, and adults relatively straightforward and inexpensive compared to other avian species. A worldwide effort, spearheaded by Tamás Székely at the University of Bath, intensively studies six plover populations using standardized field methods with the aim to understand the causes and consequences of plover breeding system variation. These six populations comprise five sister species: Snowy Plover, Kentish Plover, Kittlitz's Plover, White-fronted Plover, and Madagascar Plover (Figure 4.9a). The latter three species breed sympatrically in southwestern Madagascar, whereas the two populations of Kentish Plover are geographically disparate, inhabiting southern Turkey and the Cape Verde archipelago. Lastly, the Snowy Plover population is located on the Pacific coast of Mexico. All these populations inhabit saltmarsh or seashore habitats characterized by sparsely vegetated substrates.

Earlier works identified that several populations of the Snowy Plover and Kentish Plover with predominantly uniparental care demonstrated significantly male-biased ASR (Stenzel et al. 2011, Kosztolányi et al. 2011), and that sex-biased mating opportunities reflected patterns of parental care in Kittlitz's Plover, White-fronted Plover, and Kentish Plover (Székely et al. 1999, Parra et al. 2014; Figure 4.3). These patterns suggest that ASR is a key component to breeding system variation in plovers. However, it remained unclear which demographic processes generate ASR variation (Székely et al. 2014b) and how biases in the ASR shape mating and parental strategies in plovers. For instance, sex biases may occur at conception or at hatch, or the survival of male and female juveniles may differ to the extent that fewer of one sex survive to adulthood (Figure 4.9b). Furthermore, sex differences in adult survival or maturation rates could create a shortage of the sex that has higher mortality (Figure 4.9b) or slower maturation, and if emigration is not balanced by immigration, sex differences in dispersal behavior could create local biases in ASR.

Individual color banding and resightings of 6,119 juvenile and adult plovers over a combined total of 43 observational years indicate extensive variation in sex-biased demographic rates among these six plover populations. Using mark–recapture analysis and matrix population models, Eberhart-Phillips et al. (2018) estimated the stage- and sex-specific contributions of apparent survival to ASR bias (the term "apparent survival" is used to acknowledge that true mortality cannot be disentangled from permanent emigration within this framework; Lebreton et al. 1992). We found that populations with strong sex biases in juvenile survival produced deviations in ASR, whereas ASR was balanced in populations with little or no juvenile sex bias. Notably, hatching sex ratios did not deviate significantly from parity in any population and sex-biased adult survival had negligible effects on ASR. This demonstrates that ASR variation among these six populations originates from influential sex biases after the period of parental investment but prior to adulthood.

In species where both parents have equal caring capabilities, the desertion of either parent is often influenced by the availability of potential mates (Emlen and Oring 1977)—parental care by the abundant sex is expected to be greater than that of the scarcer sex due to limited future reproductive

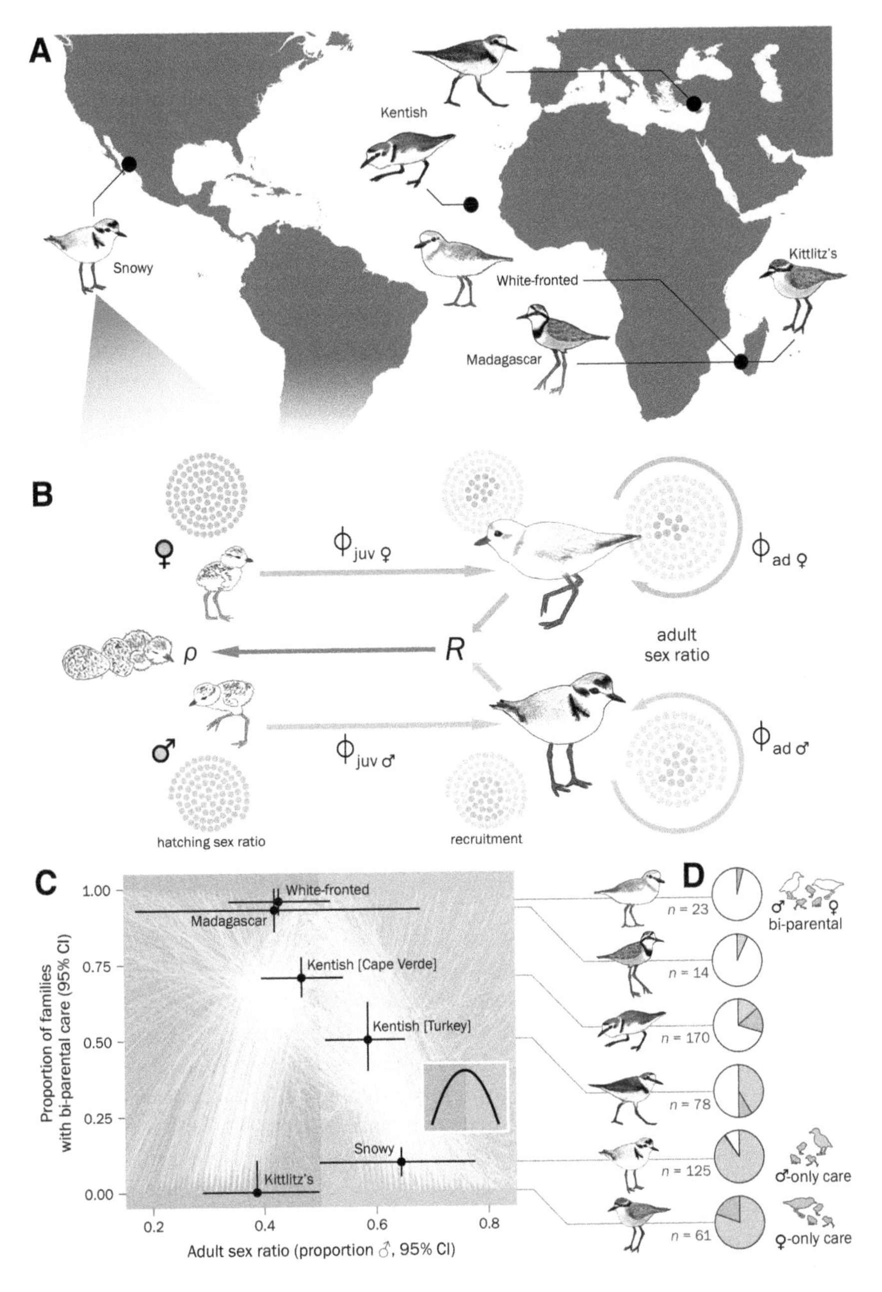

Figure 4.9. Case study exploring the associations between sex-specific survival, ASR bias, and breeding system evolution among plovers. (a) Global distribution of six study populations, comprising five Charadrius species. (b) Schematic diagram of the stage- and sex-specific demographic transitions of individuals from hatching until adulthood and their contributions to the ASR (depicted is the Snowy Plover). The hatching sex ratio (ρ, proportion of male hatchlings) serves as a proxy for the primary sex ratio and allocates progeny to the male or female juvenile stage. During the juvenile ("juv") stage, a subset of this progeny will survive (ϕ) to recruit and remain as adults ("ad"). Dotted clusters illustrate how a cohort is shaped through these sex-specific demographic transitions to derive the ASR (mortality indicated by gray dots). The reproduction function, R, is dependent on mating system and the frequency of available mates. (c) Relationship between biparental care and the ASR. Faint white lines illustrate each iteration of a bootstrap procedure, which randomly sampled an ASR and parental care estimate from each population's uncertainty distribution and fitted them to the *a priori* quadratic model (inset). (d) Proportion of monitored plover families that exhibit parental cooperation (white) or uniparental care by males (green) or females (orange). Sample sizes reflect the number of families monitored per population. Adapted from Eberhart-Phillips et al. 2018.

potential (McNamara et al. 2000). Detailed behavioral observations of 471 plover families revealed high rates of parental desertion in populations with biased ASRs, whereas desertion was rare in unbiased populations (Figure 4.9c). Eberhart-Phillips et al. (2018) evaluated the *a priori* prediction of a quadratic relationship between biparental care and ASR variation using a regression analysis incorporating a bootstrap procedure that acknowledged uncertainty in estimates of ASR and parental care. We found that families in male- or female-biased populations tended to express higher rates of parental desertion, while unbiased populations were more likely to exhibit parental cooperation (Figure 4.9b). This is supported by experimental evidence of sex-biased mating opportunities in three of the populations studied here (Parra et al. 2014; Figure 4.3). Moreover, the relationship between parental cooperation and local ASR bias was apparent in our within-species contrast of the Kentish Plover: the unbiased Cape Verde population exhibited a higher rate of parental cooperation than the male-biased population in Turkey (Figure 4.9c, d). Counterintuitively, we also found a high rate of male-only care in Kittlitz's Plover despite ASR being female biased (Figure 4.9d), although in line with expectations, Kittlitz's Plover also showed the highest proportion of female-only care among our studied populations (Figure 4.9d). This provides partial support for the notion that breeding strategies may respond flexibly to local mating opportunities provided by ASR bias, while also suggesting that other factors may play a role, such as the energetic costs of egg production imposed on females or because of sex differences in parental quality (Amat et al. 2000) or the age at maturation (Fromhage and Jennions 2016).

Several hypotheses may explain why juveniles exhibit sex-biased survival. First, juveniles face multiple novel challenges during the transition from parental independence to sexual maturity, including predation, harsh winter climates, and food shortages, all of which could disproportionately affect either sex. Although this effect is shown in mammals and other birds with sexual size dimorphism (Loison et al. 1999, Martín et al. 2007), such a mechanism is unlikely here given the modest size differences between male and female plovers (about 4%, Küpper et al. 2009). Second, males and females may differ in their premature investment of sexual traits, which could carry survival costs (Clutton-Brock et al. 1985); however, this seems unlikely as sexual ornamentation is negligible in plovers (Küpper et al. 2009). Alternatively, sex-specific ontogeny could favor one sex over the other during development, as seen in the Snowy Plover and Kentish Plover in Turkey, where male hatchlings are larger or grow faster than their sisters (dos Remedios et al. 2015b).

It cannot be discounted that a portion of the sex difference in juvenile apparent survival may be explained by sex-biased dispersal. In fact, I suspect that this is very likely. In birds, natal dispersal is typically female biased, and plovers are no exception (Clarke et al. 1997, Küpper et al. 2012). In the two separate populations of the Kentish Plover, one inhabits a remote island in the Cape Verde archipelago, whereas the other is on the Eurasian mainland (Figure 4.9a). This geographical contrast limits dispersal, and likely contributes to intraspecific variation in sex-biased apparent survival. Yet, two independent studies of the Snowy Plover document higher male than female survival during pre- (Eberhart-Phillips et al. 2017) and post-fledgling (Stenzel et al. 2011) stages, even after accounting for dispersal. This implies that sex differences in mortality are at least partly driven by intrinsic factors, perhaps via genotype-sex interactions (Küpper et al. 2010).

Regardless of whether sex-specific population structure is shaped by mortality, dispersal, or a combination thereof, the intriguing question remains—Why do juvenile sex biases vary? Further study is needed to resolve the underlying mechanisms driving sex-biased juvenile apparent survival; however, the cross-population case study detailed here provides a useful illustration that variation in this early life stage likely has profound consequences on inter- and intraspecific variation in ASR and breeding systems in plovers.

CONCLUSIONS

Conservation Implications

Mating strategies of individuals are driven by short-term fitness gains, such as parental desertion and polygamy; however, these strategies may not be always beneficial for lifetime reproductive success (LRS) or population viability. For example, analysis of LRS in the Snowy Plover revealed that males averaged 2.2 ± 3.0 fledglings and females 1.7 ± 2.3

fledglings (Herman and Colwell 2015), which contradicts the expectation that the deserting sex (females in this case) has greater fecundity. The sex difference in Snowy Plovers could be attributed to the cost of female desertion on chick survival, as was shown in a population in Mexico, where early female desertion decreased brood survival and may contribute to a decline in population productivity (Cruz-López et al. 2017). However, sex differences in LRS might also reflect sex-biased longevity, as female Snowy Plovers are shown to have a higher mortality rate than males (Stenzel et al. 2011, Eberhart-Phillips et al. 2017). Survival and fecundity are the two key demographic components contributing to population growth, and therefore it is important for plover conservation and management to understand and acknowledge the role that breeding system plays in species recovery.

The relationship between site and mate fidelity creates the opportunity for breeding systems to restrict gene flow and facilitate genetic dissimilarity among populations, which could enhance inbreeding depression. For example, the vulnerable Madagascar Plover breeds at low densities along a narrow strip of suitable habitat (Long et al. 2008). High site- and mate-fidelity mean that there is reduced gene flow and low genetic diversity among Madagascar Plover populations (Eberhart-Phillips et al. 2015), a trend also seen in Piping Plover populations of North America (Miller et al. 2013). In contrast, polygamous continental populations of the Snowy Plover, Mountain Plover, and Kittlitz's Plover are genetically panmictic owing to high dispersal and low mate fidelity (Funk et al. 2007, Oyler-McCance et al. 2013, Eberhart-Phillips et al. 2015). In a study assessing genetic population structure across the entire Palearctic range of the Kentish Plover, Küpper et al. (2012) found high levels of gene flow across Eurasia that made polygamous populations on the mainland genetically indiscernible (Chapter Two, this volume). However, monogamous island populations of the species, such as those of the Cape Verde archipelago, were genetically distinct from their mainland counterparts (Küpper et al. 2012). These results demonstrate that breeding system variation plays an important role in shaping phylogeographic patterns of species and subspecies owing to reduced or enhanced dispersal and gene flow (D'Urban Jackson et al. 2017).

Polygamy creates a situation whereby one sex is limiting reproduction, which can compromise population recovery if survival of this sex is reduced (Bessa-Gomes et al. 2004). This makes it especially important to incorporate breeding system dynamics into viability analyses of polygamous populations, because models may underestimate extinction risks if breeding system is ignored, as was shown in Snowy Plovers (Eberhart-Phillips et al. 2017). Taken together, monogamous populations are vulnerable to inbreeding depression owing to limited gene flow, whereas polygamous populations are theoretically characterized by smaller effective population sizes susceptible to density-dependent growth—although this could be offset by higher dispersal. Effective recovery plans should, therefore, require routine monitoring coupled with knowledge of breeding system, sex-specific demography, and population genetics.

Breeding systems are intriguing—as we can personally identify with the social complexities of pair bonds and parental care. This anthropogenic connection is an important attribute that acts as a catalyst for public interest in behavioral ecology, biodiversity, and ultimately the preservation of vulnerable species. However, quantifying and interpreting the breeding system of a species or population is often complex and requires detailed information documenting the social interactions of individually marked animals or intensive molecular analysis to determine sexes or the genetic parentage of offspring. As a group, plovers present a model system for investigating the evolutionary causes and consequences of breeding system variation as they are easy to monitor in the field and exhibit intra- and interspecific variation in mating and parental behavior. These traits enable researchers to collect sufficient sample sizes and conduct comparative analysis across species and populations that reveal powerful inference that can be applied to other lesser tractable organisms.

ACKNOWLEDGMENTS

I thank my colleagues from the Departments of Animal Behaviour and Evolutionary Biology at the University of Bielefeld for engaging discussions about plover ecology, behavior, and demography. In particular, I thank C. Küpper, E. Humble, M. Stoffel, and M. Ottensmann for helpful comments on my chapter and J. Hoffman and O. Krüger for the opportunity to conduct this research under their encouraging supervision. I am especially grateful to T. Székely for his enthusiastic guidance and expertise, M. Colwell for equipping me with his inspirational insight into shorebird ecology and providing steadfast support throughout my academic career, and J. Dierks for her loving companionship.

LITERATURE CITED

Alrashidi, M., A. Kosztolányi, C. Küpper, I. C. Cuthill, S. Javed, and T. Székely. 2010. The influence of a hot environment on parental cooperation of a ground-nesting shorebird, the Kentish plover *Charadrius alexandrinus*. *Frontiers in Zoology* 7:1.

Alrashidi, M., A. Kosztolányi, M. Shobrak, C. Küpper, and T. Székely. 2011. Parental cooperation in an extreme hot environment: Natural behaviour and experimental evidence. *Animal Behaviour* 82:235–243.

Alrashidi, M., A. Kosztolányi, M. Shobrak, and T. Székely. 2013. Breeding ecology of the Kentish Plover, *Charadrius alexandrinus*, in the Farasan Islands, Saudi Arabia. *Zoology in the Middle East* 53:15–24.

Amat, J. A., G. H. Visser, A. P. Hurtado, and G. M. Arroyo. 2000. Brood desertion by female shorebirds: A test of the differential parental capacity hypothesis on Kentish Plovers. *Proceedings of the Royal Society B* 267:2171–2176.

Amat, J. A., R. M. Fraga, and G. M. Arroyo. 1999. Brood desertion and polygamous breeding in the Kentish Plover *Charadrius alexandrinus*. *Ibis* 141:596–607.

Argüelles-Tico, A., C. Küpper, R. N. Kelsh, A. Kosztolányi, T. Székely, and R. E. van Dijk. 2015. Geographic variation in breeding system and environment predicts melanin-based plumage ornamentation of male and female Kentish plovers. *Behavioral Ecology and Sociobiology* 70:49–60.

Bessa-Gomes, C., S. Legendre, and J. Clobert. 2004. Allee effects, mating systems and the extinction risk in populations with two sexes. *Ecology Letters* 7:802–812.

Black, J. M. 1996. *Partnerships in Birds: The Study of Monogamy*. Oxford University Press, Oxford, UK.

Blaker, D. 1966. Notes on the sandplovers *Charadrius* in Southern Africa. *Ostrich* 37:95–102.

Blomqvist, D., J. Wallander, and M. Andersson. 2001. Successive clutches and parental roles in waders: The importance of timing in multiple clutch systems. *Biological Journal of the Linnean Society* 74:549–555.

Blomqvist, D., M. Andersson, C. Küpper, I. C. Cuthill, J. Kis, R. B. Lanctot, B. K. Sandercock, T. Székely, J. Wallander, and B. Kempenaers. 2002. Genetic similarity between mates and extra-pair parentage in three species of shorebirds. *Nature* 419:613–615.

Boyd, R. L. 1972. Breeding biology of the Snowy Plover at Cheyenne Bottoms Waterfowl Management Area, Barton County, Kansas. *MS Thesis*, Kansas State Teachers College.

Brunton, D. H. 1988. Energy expenditure in reproductive effort of male and female killdeer (*Charadrius vociferus*). *Auk* 105:553–564.

Bulla, M., M. Valcu, A. M. Dokter, A. G. Dondua, A. Kosztolányi, A. L. Rutten, B. Helm, B. K. Sandercock, B. Casler, B. J. Ens, C. S. Spiegel, C. J. Hassell, C. Küpper, C. Minton, D. Burgas, D. B. Lank, D. C. Payer, E. Y. Loktionov, E. Nol, E. Kwon, F. Smith, H. R. Gates, H. Vitnerová, H. Prüter, J. A. Johnson, J. J. H. St Clair, J.-F. Lamarre, J. Rausch, J. Reneerkens, J. R. Conklin, J. Burger, J. Liebezeit, J. Bêty, J. T. Coleman, J. Figuerola, J. C. E. W. Hooijmeijer, J. A. Alves, J. A. M. Smith, K. Weidinger, K. Koivula, K. Gosbell, K.-M. Exo, L. Niles, L. Koloski, L. McKinnon, L. Praus, M. Klaassen, M.-A. Giroux, M. Sládeček, M. L. Boldenow, M. I. Goldstein, M. Šálek, N. Senner, N. Rönkä, N. Lecomte, O. Gilg, O. Vincze, O. W. Johnson, P. A. Smith, P. F. Woodard, P. S. Tomkovich, P. F. Battley, R. Bentzen, R. B. Lanctot, R. Porter, S. T. Saalfeld, S. Freeman, S. C. Brown, S. Yezerinac, T. Székely, T. Montalvo, T. Piersma, V. Loverti, V.-M. Pakanen, W. Tijsen, and B. Kempenaers. 2016. Unexpected diversity in socially synchronized rhythms of shorebirds. *Nature* 540:109–113.

Carmona-Isunza, M. C. 2016. Breeding System Evolution in Relation to Adult Sex Ratios. *PhD Dissertation*, University of Bath, Bath, UK.

Carmona-Isunza, M. C., C. Küpper, M. A. Serrano-Meneses, and T. Székely. 2015. Courtship behavior differs between monogamous and polygamous plovers. *Behavioral Ecology and Sociobiology* 69:2035–2042.

Carmona-Isunza, M. C., S. Ancona, T. Székely, A. P. Ramallo-González, M. Cruz-López, M. A. Serrano-Meneses, and C. Küpper. 2017. Adult sex ratio and operational sex ratio exhibit different temporal dynamics in the wild. *Behavioral Ecology* 28:523–532.

Choudhury, S. 1995. Divorce in birds: A review of the hypotheses. *Animal Behaviour* 50:413–429.

Clarke, A. L., B.-E. Sæther, and E. Røskaft. 1997. Sex biases in avian dispersal: A reappraisal. *Oikos* 79:429.

Clutton-Brock, T. H., S. D. Albon, and F. E. Guinness. 1985. Parental investment and sex differences in juvenile mortality in birds and mammals. *Nature* 313:131–133.

Colwell, M. A., and W. J. Pearson. 2011. Four cases of inbreeding in a small population of the snowy plover. *Wader Study Group Bulletin* 118:181–183.

Cruz-López, M., L. J. Eberhart-Phillips, G. Fernández, R. Beamonte-Barrientos, T. Székely, M. A. Serrano-Meneses, and C. Küpper. 2017. The plight of a plover: Viability of an important snowy plover population with flexible brood care in Mexico. *Biological Conservation* 209:440–448.

D'Urban Jackson, J., N. dos Remedios, K. H. Maher, S. Zefania, S. Haig, S. Oyler-McCance, D. Blomqvist, T. Burke, M. W. Bruford, T. Székely, and C. Küpper. 2017. Polygamy slows down population divergence in shorebirds. *Evolution* 115:306.

Darwin, C. 1871. *The Descent of Man and Selection in Relation to Sex*. Murray, London.

Dawkins, R., and T. R. Carlisle. 1976. Parental investment, mate desertion and a fallacy. *Nature* 262:131–133.

Deeming, C. 2002. *Avian Incubation: Behaviour, Environment and Evolution*. Oxford University Press, Oxford, UK.

dos Remedios, N., P. L. M. Lee, T. Burke, T. Székely, and C. Küpper. 2015a. North or south? Phylogenetic and biogeographic origins of a globally distributed avian clade. *Molecular Phylogenetics and Evolution* 89:151–159.

dos Remedios, N., T. Székely, C. Küpper, P. L. M. Lee, and A. Kosztolányi. 2015b. Ontogenic differences in sexual size dimorphism across four plover populations. *Ibis* 157:590–600.

Eberhart-Phillips, L. J., C. Küpper, M. C. Carmona-Isunza, O. Vincze, S. Zefania, M. Cruz-López, A. Kosztolányi, T. E. X. Miller, Z. Barta, T. Burke, T. Székely, J. I. Hoffman, and O. Krüger. 2018. Demographic causes of adult sex ratio and their consequences for parental cooperation. *Nature Communications* 9:1651.

Eberhart-Phillips, L. J., C. Küpper, T. E. X. Miller, M. Cruz-López, K. Maher, N. dos Remedios, M. A. Stoffel, J. I. Hoffman, O. Krüger, and T. Székely. 2017. Sex-specific early survival drives adult sex ratio bias in Snowy Plovers and impacts mating system and population growth. *Proceedings of the National Academy of Sciences USA* 114:E5474–E5481.

Eberhart-Phillips, L. J., J. I. Hoffman, E. G. Brede, S. Zefania, M. J. Kamrad, T. Székely, and M. W. Bruford. 2015. Contrasting genetic diversity and population structure among three sympatric Madagascan shorebirds: Parallels with rarity, endemism, and dispersal. *Ecology and Evolution* 5:997–1010.

Ekanayake, K. B., M. A. Weston, D. G. Nimmo, G. S. Maguire, J. A. Endler, and C. Küpper. 2015. The bright incubate at night: Sexual dichromatism and adaptive incubation division in an open-nesting shorebird. *Proceedings of the Royal Society B* 282:20143026–20143026.

Emlen, S. T., and L. W. Oring. 1977. Ecology, sexual selection, and the evolution of mating systems. *Science* 197:215–223.

Ens, B. J., U. N. Safriel, and M. P. Harris. 1993. Divorce in the long-lived and monogamous oystercatcher, *Haematopus ostralegus*: Incompatibility or choosing the better option? *Animal Behaviour* 45:1199–1217.

Flynn, L., E. Nol, and Y. Zharikov. 1999. Philopatry, nest-site tenacity, and mate fidelity of Semipalmated Plovers. *Journal of Avian Biology* 30:47–55.

Forstmeier, W., S. Nakagawa, S. C. Griffith, and B. Kempenaers. 2014. Female extra-pair mating: Adaptation or genetic constraint? *Trends in Ecology and Evolution* 29:456–464.

Fromhage, L., and M. D. Jennions. 2016. Coevolution of parental investment and sexually selected traits drives sex-role divergence. *Nature Communications* 7:12517.

Funk, W. C., T. D. Mullins, and S. M. Haig. 2007. Conservation genetics of snowy plovers (*Charadrius alexandrinus*) in the Western Hemisphere: Population genetic structure and delineation of subspecies. *Conservation Genetics* 8:1287–1309.

Graham, K. 2004. Semipalmated plover breeding success and adult survival: Effects of weather and body condition. *Thesis*, Trent University, Peterborough, ON, Canada.

Graul, W. D. 1975. Breeding biology of the Mountain Plover. *Wilson Bulletin* 87:6–31.

Green, G. H., J. J. D. Greenwood, and C. S. Lloyd. 1977. The influence of snow conditions on the date of breeding of wading birds in north-east Greenland. *Journal of Zoology* 183:311–328.

Haig, S. M., and L. W. Oring. 1988a. Mate, site, and territory fidelity in Piping Plovers. *Auk* 105:268–277.

Haig, S. M., and L. W. Oring. 1988b. Distribution and dispersal in the Piping Plover. *Auk* 105:630–638.

Hamas, M. J., and W. D. Graul. 1985. A four-egg clutch of the Mountain Plover. *Wilson Bulletin* 97:388–389.

Handel, C. M., and R. E. Gill. 2000. Mate fidelity and breeding site tenacity in a monogamous sandpiper, the Black Turnstone. *Animal Behaviour* 60:471–481.

Havens, P. D. 1970. Aberration in the clutch size of the Semipalmated Plover. *Condor* 72:481.

Herman, D. M., and M. A. Colwell. 2015. Lifetime reproductive success of Snowy Plovers in coastal northern California. *Condor* 117:473–481.

Hussell, D. J. T., and J. K. Woodford. 1965. Piping Plover's nest containing eight eggs. *Wilson Bulletin* 77:294.

Johnson, M., L. W. Oring, and J. R. Walters. 2006. Killdeer parental care when either parent deserts. *Wader Study Group Bulletin* 110:43–47.

Kålås, J. A., and I. Byrkjedal. 1984. Breeding chronology and mating system of the eurasian dotterel (*Charadrius morinellus*). *Auk* 101:838–847.

Kokko, H., and M. Jennions. 2008. Parental investment, sexual selection and sex ratios. *Journal of Evolutionary Biology* 21:919–948.

Kosztolányi, A., S. Javed, C. Küpper, I. C. Cuthill, A. A. Shamsi, and T. Székely. 2009. Breeding ecology of Kentish Plover *Charadrius alexandrinus* in an extremely hot environment. *Bird Study* 56:244–252.

Kosztolányi, A., T. Székely, and I. C. Cuthill. 2003. Why do both parents incubate in the Kentish Plover? *Ethology* 109:645–658.

Kosztolányi, A., T. Székely, I. C. Cuthill, K. T. Yilmaz, and S. Berberoğlu. 2006. Ecological constraints on breeding system evolution: The influence of habitat on brood desertion in Kentish Plover. *Journal of Animal Ecology* 75:257–265.

Kosztolányi, A., Z. Barta, C. Küpper, and T. Székely. 2011. Persistence of an extreme male-biased adult sex ratio in a natural population of polyandrous bird. *Journal of Evolutionary Biology* 24:1842–1846.

Kruckenberg, H. 2002. Only the male breeds: Breeding behaviour of the Mongolian Plover (*Charadrius mongolus*) on the northern coasts of the sea of Okhotsk. Pp. 182–186 in A. V. Andreev and H. H. Bergman (editors), *Biodiversity and Ecological Status Along the Northern Coast of the Sea of Okhotsk*, Russian Academy of Science, Far Eastern Branch, Magaden, Russia.

Küpper, C., A. Kosztolányi, J. Augustin, D. A. Dawson, T. Burke, and T. Székely. 2010. Heterozygosity-fitness correlations of conserved microsatellite markers in Kentish plovers *Charadrius alexandrinus*. *Molecular Ecology* 19:5172–5185.

Küpper, C., J. Augustin, A. Kosztolányi, T. Burke, J. Flguerola, and T. Székely. 2009. Kentish versus Snowy Plover: Phenotypic and genetic analyses of *Charadrius alexandrinus* reveal divergence of Eurasian and American subspecies. *Auk* 126:839–852.

Küpper, C., J. Kis, A. Kosztolányi, T. Székely, I. C. Cuthill, and D. Blomqvist. 2004. Genetic mating system and timing of extra-pair fertilizations in the Kentish plover. *Behavioral Ecology and Sociobiology* 57:32–39.

Küpper, C., S. V. Edwards, A. Kosztolányi, M. Alrashidi, T. Burke, P. Herrmann, A. Argüelles-Tico, J. A. Amat, M. Amezian, A. Rocha, H. Hötker, A. Ivanov, J. Chernicko, and T. Székely. 2012. High gene flow on a continental scale in the polyandrous Kentish Plover *Charadrius alexandrinus*. *Molecular Ecology* 21:5864–5879.

Lack, D. 1968. *Ecological Adaptations for Breeding in Birds*. Methuen Ltd, London, UK.

Lebreton, J. D., K. P. Burnham, J. Clobert, and D. R. Anderson. 1992. Modeling survival and testing biological hypotheses using marked animals: A unified approach with case studies. *Ecological Monographs* 62:67–118.

Lenington, S. 1980. Bi-parental care in killdeer: An adaptive hypothesis. *Wilson Bulletin* 92:8–20.

Lessells, C. M. 1984. The mating system of Kentish Plovers *Charadrius alexandrinus*. *Ibis* 126:474–483.

Liker, A., R. P. Freckleton, and T. Székely. 2013. The evolution of sex roles in birds is related to adult sex ratio. *Nature Communications* 4:1587.

Lloyd, P. 2008. Adult survival, dispersal and mate fidelity in the White-fronted Plover *Charadrius marginatus*. *Ibis* 150:182–187.

Loison, A., R. Langvatn, and E. J. Solberg. 1999. Body mass and winter mortality in red deer calves: Disentangling sex and climate effects. *Ecography* 22:20–30.

Long, P. R., S. Zefania, R. H. ffrench-Constant, and T. Székely. 2008. Estimating the population size of an endangered shorebird, the Madagascar Plover, using a habitat suitability model. *Animal Conservation* 11:118–127.

Maher, K. H., L. J. Eberhart-Phillips, A. Kosztolányi, N. D. Remedios, M. C. Carmona-Isunza, M. Cruz-López, S. Zefania, J. S. Clair, M. Alrashidi, M. A. Weston, M. A. Serrano-Meneses, O. Krüger, J. I. Hoffman, T. Székely, T. Burke, and C. Küpper. 2017. High fidelity: Extra-pair fertilisations in eight *Charadrius* plover species are not associated with parental relatedness or social mating system. *Journal of Avian Biology* 48:910–920.

Martín, C. A., J. C. Alonso, J. A. Alonso, C. Palacín, M. Magaña, and B. Martín. 2007. Sex-biased juvenile survival in a bird with extreme size dimorphism, the great bustard *Otis tarda*. *Journal of Avian Biology* 38:335–346.

Maynard Smith, J. 1977. Parental investment: A prospective analysis. *Animal Behaviour* 25:1–9.

McNamara, J. M., T. Székely, J. N. Webb, and A. I. Houston. 2000. A dynamic game-theoretic model of parental care. *Journal of Theoretical Biology* 205:605–623.

Miller, M. P., S. M. Haig, C. L. Gratto-Trevor, and T. D. Mullins. 2013. Subspecies status and population genetic structure in Piping Plover (*Charadrius melodus*). *Auk* 127:57–71.

Mundahl, J. T. 1982. Role specialization in the parental and territorial behavior of the Killdeer. *Wilson Bulletin* 94:515–530.

Mundahl, J. T., O. L. Johnson, and M. L. Johnson. 1981. Observations at a twenty-egg Killdeer nest. *Condor* 83:180–182.

Myers, J. P. 1981. Cross-seasonal interactions in the evolution of sandpiper social systems. *Behavioral Ecology and Sociobiology* 8:195–202.

Oring, L. W. 1986. Avian polyandry. Pages 309–351 in R. F. Johnston (editors), *Current Ornithology* (vol. 3). Springer US, Boston, MA.

Owens, I. P. F., A. Dixon, T. Burke, and D. B. A. Thompson. 1995. Strategic paternity assurance in the sex-role reversed Eurasian Dotterel (*Charadrius morinellus*): Behavioral and genetic evidence. *Behavioral Ecology* 6:14–21.

Owens, I. P. F., T. Burke, and D. B. A. Thompson. 1994. Extraordinary sex roles in the Eurasian Dotterel: Female mating arenas, female-female competition, and female mate choice. *American Naturalist* 144:76–100.

Oyler-McCance, S. J., J. S. John, R. F. Kysela, and F. L. Knopf. 2013. Population structure of mountain plover as determined using nuclear microsatellites. *Condor* 110:493–499.

Pakanen, V.-M., S. Lampila, H. Arppe, and J. Valkama. 2015. Estimating sex specific apparent survival and dispersal of Little Ringed Plovers (*Charadrius dubius*). *Ornis Fennica* 92:172–186.

Parra, J. E., M. Beltrán, S. Zefania, N. dos Remedios, and T. Székely. 2014. Experimental assessment of mating opportunities in three shorebird species. *Animal Behaviour* 90:83–90.

Pearson, W. J., and M. A. Colwell. 2013. Effects of nest success and mate fidelity on breeding dispersal in a population of Snowy Plovers *Charadrius nivosus*. *Bird Conservation International* 24:342–353.

Piersma, T., and P. Wiersma. 1996. Family Charadriidae (Plovers). Pages 384–443 in J. del Hoyo, A. Elliott, and J. Sargatal (editors), *Handbook of the Birds of the World*. Lynx Edicions, Barcelona, Spain.

Reynolds, J. D., and T. Székely. 1997. The evolution of parental care in shorebirds: Life histories, ecology, and sexual selection. *Behavioral Ecology* 8:126–134.

Riordan, M., P. Lukacs, K. Huyvaert, and V. Dreitz. 2015. Sex ratios of Mountain Plovers from egg production to fledging. *Avian Conservation and Ecology* 10:art3.

Rittinghaus, H. 1956. Untersuchungen am Seeregenpfeifer (*Charadrius alexandrinus* L.) auf der Insel Oldeoog. *Journal of Ornithology* 97:117–155 (in German).

Skrade, P. D. B., and S. J. Dinsmore. 2010. Sex-related dispersal in the Mountain Plover (*Charadrius montanus*). *Auk* 127:671–677.

St Clair, J. J. H., C. Küpper, P. Herrmann, R. W. Woods, and T. Székely. 2010a. Unusual incubation sex-roles in the Rufous-chested Dotterel *Charadrius modestus*. *Ibis* 152:402–404.

St Clair, J. J. H., P. Herrmann, R. W. Woods, and T. Székely. 2010b. Female-biased incubation and strong diel sex-roles in the Two-banded Plover *Charadrius falklandicus*. *Journal of Ornithology* 151:811–816.

Stenzel, L. E., G. W. Page, J. C. Warriner, J. S. Warriner, K. K. Neuman, D. E. George, C. R. Eyster, and F. C. Bidstrup. 2011. Male-skewed adult sex ratio, survival, mating opportunity and annual productivity in the snowy plover *Charadrius alexandrinus*. *Ibis* 153:312–322.

Stenzel, L. E., J. C. Warriner, J. S. Warriner, K. S. Wilson, F. C. Bidstrup, and G. W. Page. 1994. Long-distance breeding dispersal of snowy plovers in western North America. *Journal of Animal Ecology* 63:887–902.

Székely, T. 1996. Brood desertion in Kentish plover *Charadrius alexandrinus*: An experimental test of parental quality and remating opportunities. *Ibis* 138:749–755.

Székely, T., A. Liker, R. P. Freckleton, C. Fichtel, and P. M. Kappeler. 2014a. Sex-biased survival predicts adult sex ratio variation in wild birds. *Proceedings of the Royal Society B* 281:20140342–20140342.

Székely, T., and C. M. Lessells. 1993. Mate change by Kentish Plovers *Charadrius alexandrinus*. *Ornis Scandinavica* 24:317–322.

Székely, T., F. J. Weissing, and J. Komdeur. 2014b. Adult sex ratio variation: Implications for breeding system evolution. *Journal of Evolutionary Biology* 27:1500–1512.

Székely, T., and I. C. Cuthill. 1999. Brood desertion in Kentish plover: The value of parental care. *Behavioral Ecology* 10:191–197.

Székely, T., I. C. Cuthill, and J. Kis. 1999. Brood desertion in Kentish Plover sex differences in remating opportunities. *Behavioral Ecology* 10:185–190.

Székely, T., I. C. Cuthill, S. Yezerinac, R. Griffiths, and J. Kis. 2004. Brood sex ratio in the Kentish Plover. *Behavioral Ecology* 15:58–62.

Székely, T., and J. D. Reynolds. 1995. Evolutionary transitions in parental care in shorebirds. *Proceedings of the Royal Society B* 262:57–64.

Thomas, G. H., and T. Székely. 2005. Evolutionary pathways in shorebird breeding systems: Sexual conflict, parental care, and chick development. *Evolution* 59:2222–2230.

Trivers, R. L. 1972. Parental investment and sexual selection. Pp. 136–179 in B. G. Campbell (editor), *Sexual Selection and the Descent of Man*. Aldine de Gruyter, New York.

Tulp, I., and H. Schekkerman. 2006. Time allocation between feeding and incubation in uniparental arctic-breeding shorebirds: Energy reserves provide leeway in a tight schedule. *Journal of Avian Biology* 37:207–218.

Vincze, O., A. Kosztolányi, Z. Barta, C. Küpper, M. Alrashidi, J. A. Amat, A. Argüelles-Tico, F. Burns, J. Cavitt, W. C. Conway, M. Cruz-López, A. E. DeSucre-Medrano, N. dos Remedios, J. Figuerola, D. Galindo-Espinosa, G. E. García-Peña, S. Gómez del Ángel, C. Gratto Trevor, P. Jönsson, P. Lloyd, T. Montalvo, J. E. Parra, R. Pruner, P. Que, Y. Liu, S. T. Saalfeld, R. Schulz, L. Serra, J. J. H. St Clair, L. E. Stenzel, M. A. Weston, M. Yasué, S. Zefania, and T. Székely. 2017. Parental cooperation in a changing climate: Fluctuating environments predict shifts in care division. *Global Ecology and Biogeography* 26:347–358.

Vincze, O., T. Székely, C. Küpper, M. Alrashidi, J. A. Amat, A. A. Ticó, D. Burgas, T. Burke, J. Cavitt, J. Figuerola, M. Shobrak, T. Montalvo, and A. Kosztolányi. 2013. Local environment but not genetic differentiation influences biparental care in ten plover populations. *PLoS One* 8:e60998.

Wallander, J. 2003. Sex roles during incubation in the Common Ringed Plover. *Condor* 105:378.

Wallander, J., D. Blomqvist, and J. T. Lifjeld. 2001. Genetic and social monogamy—does it occur without mate guarding in the Ringed Plover? *Ethology* 107:561–572.

Warriner, J. S., J. C. Warriner, G. W. Page, and L. E. Stenzel. 1986. Mating system and reproductive success of a small population of polygamous Snowy Plovers. *Wilson Bulletin* 98:15–37.

Westneat, D. F., and I. R. K. Stewart. 2003. Extra-pair paternity in birds: Causes, correlates, and conflict. *Annual Review of Ecology, Evolution, and Systematics* 34:365–396.

Weston, M. A., and M. A. Elgar. 2005. Parental care in Hooded Plovers (*Thinornis rubricollis*). *Emu* 105:283–292.

Yasué, M., and P. Dearden. 2007. Parental sex roles of Malaysian Plovers during territory acquisition, incubation and chick-rearing. *Journal of Ethology* 26:99–112.

Yasué, M., and P. Dearden. 2008. Replacement nesting and double-brooding in Malaysian Plovers *Charadrius peronii*: Effects of season and food availability. *Ardea* 96:59–72.

Zefania, S., R. Emilienne, P. J. Faria, M. W. Bruford, P. R. Long, and T. Székely. 2010. Cryptic sexual size dimorphism in Malagasy Plovers *Charadrius spp. Ostrich* 81:173–178.

Zharikov, Y., and E. Nol. 2000. Copulation behavior, mate guarding, and paternity in the semipalmated plover. *Condor* 102:231–235.

Ram Papish

CHAPTER FIVE

Breeding Biology of Charadrius Plovers*

Lynne E. Stenzel and Gary W. Page

Abstract. Breeding biology figures prominently in the management and conservation of plovers. Accordingly, we reviewed the literature for the 40 species of the Charadrius plover clade to compare the behavioral and physiological investments the plovers make to raise their young in different latitudinal regions. We focus on the incubation and chick care periods and briefly discuss other important aspects of breeding biology. Species laying the most common clutch size, three eggs, occur in all regions whereas, all but one species laying four-egg clutches breed in subarctic and north temperate regions; most species laying two-egg clutches breed wholly or partly in the tropics. Species breeding wholly or partly in tropical regions are smaller than those nesting wholly in temperate or subarctic regions, and their egg sizes are highly positively correlated with their body sizes. Thus, tropical species appear to make a smaller investment in eggs of each clutch than species in other regions. In temperate and tropical regions, breeding seasons are long, most species lay replacement clutches, and many also double-brood. In all regions, biparental egg and chick care are common, but when one sex provides more care than the other it is usually the male. Gaps in our knowledge of plover breeding biology are greatest for tropical and South American temperate breeding species. Understanding variability in elements of plover breeding biology will enhance conservation. We discuss the importance of breeding biology to conservation activities for plovers and identify directions for future research.

Keywords: clutch size, double-brood, egg size, female size, incubation, nest site, nidifugous, parental care, precocial.

Breeding is an energetically expensive and risky endeavor for birds and is closely tied to fitness; consequently, knowledge of breeding biology is of great theoretical and applied importance (Drent and Daan 1980, Magnhagen 1991, Ricklefs 2000). Avian breeding biology comprises the energy expended and behaviors employed to reproduce, particularly the physiological effort to produce eggs and parental activities to care for eggs and young, but also use of space and time, acquisition and retention of mates, and nest building. Among species with precocial young, Charadrius plovers (hereafter, plovers) produce comparatively large eggs, and, like other shorebirds, they have relatively small clutches (Winkler and Walters 1983). As a group, plovers breed on six continents, from arctic to south temperate latitudes, and from sea level to alpine elevations, thereby experiencing a wide range of seasonal, physical, and biotic environments (Chapter Nine, this volume). While phylogeny may constrain some aspects of breeding in this clade, natural and sexual selection (via predation, thermal environment,

* Lynne E. Stenzel and Gary W. Page. 2019. Breeding Biology of Charadrius Plovers. Pp. 91–125 in Colwell and Haig (editors). The Population Ecology and Conservation of Charadrius Plovers (no. 52), CRC Press, Boca Raton, FL.

food availability, mate choice, etc.) have shaped the behaviors, morphologies, and reproductive life history traits of each species (Winkler and Walters 1983, Székely et al. 1994, 2000, Amat and Masero 2004a, b, Kosztolányi et al. 2009).

Despite the variable environments in which plovers breed, many commonalities in breeding traits are present across species, including lengthy breeding seasons within latitudinal limits and strong effects of egg and chick predators on nesting success (Chapters Seven and Nine, this volume). However, some species nest in relatively stable environments, such as the humid Tropics or isolated islands, where species are often at carrying capacity, in contrast to seasonal, less predictable environments at higher latitudes where species must often recover from substantial losses in population size (Skutch 1949, Cody 1966). These same researchers also note that, in stable environments with species at carrying capacity, maximizing reproduction may not be the strongest driver predicting breeding behavior, as it might be in more fluctuating environments. While these principles have been well explored for clutch size, they also can be applied to other breeding parameters, such as egg size, double-brooding, egg-laying intervals, and incubation and fledging periods (Lack 1947, Skutch 1949, Klomp 1970).

Knowledge of breeding biology is essential for conservation, because it identifies each species' capacity to successfully produce offspring under current conditions. Breeding also is the stage of the life cycle currently identified as having the greatest management potential for conserving populations (Chapters Seven and Eleven, this volume). Potential natural challenges and constraints on the breeding grounds range from dependence on sometimes irregular precipitation and limiting nutrients, such as calcium in the Tropics, to short and sometimes inclement breeding seasons in the Arctic and Subarctic (Nol et al. 1997, Patten 2007). Current conservation challenges in many instances center around excessively high levels of egg and chick loss from introduced and subsidized native predators, breeding habitat loss from development, and impacts from numerous other anthropogenic activities (Chapters Seven, Nine, and Eleven, this volume). As current challenges continue or even increase in the future, they will be compounded by global climate change and its consequences, especially accelerated sea level rise, increasing temperatures, disrupted precipitation patterns, and possibly wetland habitat acidification, all of which may affect plovers, their physical habitat, their predators, and their prey (Chapter Three, this volume). Because breeding biology comprises traits closely tied to the fitness of individuals, it is an important determinant of how each species is able to adapt to rapid environmental change.

In this review of plover breeding biology, we examine the inherent geographic variability in clutch and egg size, incubation period, and parental care among species. We briefly review territoriality, mate selection, mate fidelity, site fidelity, and nests acknowledging their importance to breeding biology but recognizing that an in-depth investigation of these parameters is beyond the scope of this review. We anticipate that variable traits among species will suggest promising opportunities to examine intraspecific plasticity, particularly for wide-ranging species. Such comparisons could provide insights into the plovers' potential responses to climate change and other environmental challenges and suggest possible restoration and management actions to promote their conservation. We conclude with a brief discussion of the importance of breeding biology to management and restoration for plovers and suggest promising directions for future research.

METHODS

We reviewed the literature on breeding biology for the 40 species of plover (Table 1.1) to identify the commonalities in behavioral and physiological investments they make to successfully raise young and compared these efforts among species breeding in different geographic regions. We searched online sources (Handbook of the Birds of the World—HBW Alive, and Birds of North America accounts), major avifaunal treatments (Cramp and Simmons 1983, Urban et al. 1986, Marchant and Higgins 1993), primary literature, dissertations, and theses for spatial, temporal, and behavioral features of parental care, and for ovimetric and morphometric details of the breeding biology for each species.

We compared social factors, clutch size, egg size, female mass, and durations of nesting cycle stages, among plovers breeding in six latitudinal extents: Subarctic (spanning Arctic and North Temperate), North Temperate, North Subtropics (spanning North Temperate and the Tropics), Tropics, South Subtropics (spanning South Temperate and the Tropics), and South Temperate to determine how these metrics vary among these large-scale geographic regions.

Egg dimensions have been documented for many species of plover but egg masses much less so. For the 15 species in which both dimensions and masses were available, we obtained the correlation (ρ) between an index of egg volume, length × width^2, and egg mass (g), and used the resulting relationship to obtain an estimate of egg mass for all possible species. We use a single point estimate for each species, preferring midpoints between means (of several years or locations), but also already calculated means or midpoints between pooled minimum and maximum measurements, if those were all that were available.

Data on egg mass frequently were from disparate locations or studies and seldom were reported with information on egg freshness, which adds an unknown source of variation because eggs lose 10%–12% of their mass through the incubation period (Drent 1975). Adult body masses frequently were not separated by sex, season, or location, but, because relatively little size dimorphism exists between sexes, we use adult or, in one case, male body mass when female mass was not available (Piersma and Wiersma 1996). We expected strong patterns to emerge, despite data deficiencies, and hope the relationships revealed will prompt other researchers to collect and publish their ovimetric and morphometric data to allow more complex and refined comparisons.

RESULTS AND DISCUSSION

Breeding Distribution and Habitat

Plovers breed widely from the Arctic throughout south temperate regions, but half of the species occur in the south temperate region (eight species) or in the southern Subtropics (12 species). Five species each breed in either the Tropics or North Subtropics, seven species in north temperate latitudes, and three species in subarctic latitudes (Table 5.1). Only the Snowy Plover's breeding range spans both temperate zones and the Tropics, but we found no published studies of breeding biology for this species from its coastal South American range, so we categorized its range as North Subtropics for comparisons. Of the

TABLE 5.1

Timing of the breeding season of Charadrius clade plovers.

Common name	BDC[a]	Egg-laying period: location, start month[b]	Initiation days	Sources in addition to HBW Alive
S. Red-breasted Dotterel	S3	New Zealand, September	125	
N. Red-breasted Dotterel	S3	New Zealand, August	140	
Lesser Sand-Plover	N3	Central Asia, May	25–30	
Greater Sand-Plover	N3	Central Asia, March	60	
Caspian Plover	N3	Central Asia, April	90	
Collared Plover	S2	Argentina, October	55	
		Brazil, September	120	
		Neotropics (March–December)	NA	
Puna Plover	S2	Argentina, September, rarely January	60	
Two-banded Plover	S3	Argentina, September	140	
Double-banded Plover	S3	Australia high elevation, September	105	
		Australia low elevation, August	155	
Kittlitz's Plover	S2	Malawi, April	245	
		Ethiopia, January	275	
		Madagascar, YR	365	

(Continued)

TABLE 5.1 (*Continued*)
Timing of the breeding season of Charadrius clade plovers.

Common name	BDC[a]	Egg-laying period: location, start month[b]	Initiation days	Sources in addition to HBW Alive
Red-capped Plover	S2R	Australia coast, July	215	
		Australia inland, rain dependent		
Malay Plover	1	Thailand, March	115	Yasué and Dearden (2008a, b)
		Philippines, February	145	
Kentish Plover	N2V[c]	NW Europe, April	~45	
		Socotra Is, November	185	
White-faced Plover	1			
Snowy Plover	N2V[c]	Great Plains/Great Basin, April	105	Page et al. (2009)
		U.S. West Coast, March	140	
		Florida, March	185	
		Coastal Texas (February)		
		Puerto Rico, January	~185	
Javan Plover	1	Java var locs, YR	365	Taufiqurrahman and Subekti (2013)
Wilson's Plover	N2	United States, April	65–75	Corbat and Bergstrom (2000)
		Columbia, March	165	
Common Ringed Plover	4	No. Greenland, June	~30	Pienkowski (1984), Wallander and Andersson (2003)
		North Sea, April	110	
Semipalmated Plover	4	Various loc's, May–June	7–30	Nol and Blanken (2014)
Long-billed Plover	N3	Japan, March	140	Uchida (2007), cited in Katayama et al. (2010)
	R	Russian Far East, April	60	Kolomiytsev and Poddubnaya (2014)
Piping Plover	N3	Manitoba, Canada, April	85	Elliott-Smith and Haig (2004)
Madagascar Plover	S2	Madagascar, August[d]	275	Zefania et al. (2008)
Little Ringed Plover	N2V[c]	Sri Lanka, June	60	
		North Africa, March	90	
		Philippines, February	120	
African Three-banded Plover	S2R	Eritrea/Ethiopia, May	215	Tyler (1978)
		Zambia, February but primarily July	(305) 125	
		Uganda, March	120	
		Zimbabwe, May but primarily July	(215) 90	

(Continued)

TABLE 5.1 (*Continued*)

Timing of the breeding season of Charadrius clade plovers.

Common name	BDC[a]	Egg-laying period: location, start month[b]	Initiation days	Sources in addition to HBW Alive
Madagascar Three-banded Plover	S2R	SW Madagascar (April–August) dry period	155	
		NW (April–May) dry period	60	
		SE (July–August) dry period	60	
		NE (September), dry period	30	
Forbes's Plover	1R	Nigeria, March (wet season)	185	
		Ghana, July (wet season)	60	
		Gabon, June (dry season)	120	
		Zambia, October (dry season)	60	
White-fronted Plover	S2	South Africa, YR	275	Lloyd and Plagányi (2002)
		West Africa, February	210	
		East Africa, May	90	
Chestnut-banded Plover	S2R	East Africa, April	195	
		South Africa, March, September	245	
Killdeer	N2V	Puerto Rico, YR	365	Jackson and Jackson (2000)
		SE United States, March, rare to October	120	
		Washington State, April	NA	
Mountain Plover	N3	Colorado, April	65–75	Knopf and Wunder (2006)
Oriental Plover	N3	Central Asia (May)	~60	Ozerskaya and Zabelin (2006)
Eurasian Dotterel	4	Scotland, May	55	
		Norway, May	25	
St. Helena Plover	1	St. Helena Is, YR	365	McCulloch (1991)
Rufous-chested Dotterel	S3	Falkland Is, September	120	
Red-kneed Dotterel	S2R	Australia inland, July (after rain)	215	Robertson (2013)
Hooded Plover	S3	Australia, August	245	Baird and Dann (2003)
Shore Plover	S3	Chatham Is, October	105	Davis (1994)
Black-fronted Dotterel	S2R	Australia, September	180	Marchant and Higgins (1993)
		New Zealand, August	245	Armitage (2013)

(Continued)

TABLE 5.1 (*Continued*)

Timing of the breeding season of Charadrius clade plovers.

Common name	BDC[a]	Egg-laying period: location, start month[b]	Initiation days	Sources in addition to HBW Alive
Inland Dotterel	S2R	Australia, YR	365	
Wrybill	S3	New Zealand, August	120	Dowding (2013)

[a] Breeding distributional constraints on timing (least to greatest): 1 = Tropical latitudes, 2= Subtropical (combined tropical and temperate latitudes), 3 = Temperate latitudes, 4 = Combined temperate and arctic latitudes; V = Variable over range of species; R = Breeding initiated in response to rainfall; and, for subtropical and temperate latitudes, S = Southern Hemisphere and N = Northern Hemisphere.

[b] Period of the year in which nests may be initiated. YR = year-round; if months are in parentheses, available data are anecdotal for varying latitudes and/or locations, possibly insufficient to determine number of egg laying days; NA is indicated under Initiation days where information is too sparse.

[c] Insignificantly also within tropical latitudes (Kentish and Snowy plovers) or arctic latitudes (Little Ringed Plover). Also, unstudied in these regions.

[d] Egg laying uncommon August–November, with 6.7% of nest and brood observations during surveys.

40 plover species, 34 breed at least in part >50 km from an ocean or sea coast, and for 13 of these, only an interior nesting distribution has been described; breeding ranges of only six species are strictly coastal (Chapter Nine, this volume).

Natural nesting habitats are quite varied for different species but they often share the early successional characteristics of either no, sparse, and/or low-growing vegetation, and overall open "viewscapes" that enable nesting birds to detect distant approaching predators. Most commonly used habitats include sandy coastal beaches, lakeshores, riverbanks, playas, deserts or semideserts near water, short or sparsely vegetated grasslands, and sparsely vegetated tundra. Well over half of the species nest in one or more human-altered landscapes (Chapter Nine, this volume).

Social Organization

Information on spatial patterns of nesting is available for 34 species. Most have been categorized as nesting solitarily, but at least 18 of these species also breed in an aggregated pattern under some conditions (Table 5.2). Although some authors describe these aggregations as semicolonial, there usually is no means to distinguish aggregation due to social attraction versus a patchy distribution of resources (Stamps 1988, Pearson et al. 2014). Plover species known to aggregate are distributed heterogeneously over latitudinal regions, and include all species breeding in subarctic areas, four of seven species breeding in north temperate areas, nine of fourteen species breeding in north or south subtropical areas, but only two of seven species breeding in only south temperate areas and none of the four species breeding only in the Tropics (considering only those species with information on spatial patterns of nests—Tables 5.1 and 5.2). Breeding territories have been described for 33 of the 40 plover species, but details on territory size and function and the duration of territorial defense throughout the pre-laying, laying, incubation, and chick-rearing periods are usually not reported; these types of data are usually painstaking to acquire and boundary disputes may be difficult to document when breeding densities are low (Table 5.2). Eurasian Dotterels, whose polyandrous mating system is discussed in Chapter Four of this volume, may not consistently defend territories throughout their range; they are described as territorial in Scotland but not in Norway (Nethersole-Thompson 1973, Kålås and Byrkjedal 1984, Owens et al. 1994). Nethersole-Thompson (1973) describes already-paired dotterels establishing territories for courtship, copulation, and nesting, and males defending areas of varying distances from nests after incubation commences, but not necessarily a fixed space during brood rearing, except possibly in situations of high-nesting density.

Advertisement

Many plovers signal their presence to potential mates and conspecifics through aerial song displays, in which vocalizing birds circle over nesting areas with exaggerated wing beats or other, sometimes acrobatic, aerial maneuvers (Phillips 1980, Székely et al. 2000, Table 5.2). For example, Greater Sand-Plovers display 30–50 m above the ground, with slow owl-like wing beats, and body tilting side to side while they synchronously call

TABLE 5.2
Aspects of the acquisition and use of space by breeding Charadrius *clade plovers.*

Common name[a]	Aerial display	Breeding concen[b]	Territoriality N FdI FdO[c]	Breeding site fidelity	Sources in addition to HBW Alive
S. Red-breasted Dotterel			Y	high	
N. Red-breasted Dotterel		mixed	Y Y	high	Lord et al. (1997)
Lesser Sand-Plover	Y	mixed	Y[d]		Johnsgard (1981), Kruckenberg et al. (2001)
Greater Sand-Plover	Y	solitary	Y		
Caspian Plover	Y	mixed	Y Y		
Collared Plover	Y			high[e]	
Puna Plover		solitary			
Two-banded Plover		solitary	Y Y		St. Clair (2010)
Double-banded Plover	Y	mixed	Y Y T	high M>F	Phillips (1980), Pierce (1989, 2013)
Kittlitz's Plover		mixed	Y Y Y		
Red-capped Plover		mixed	Y		
Malay Plover		solitary	Y Y	high	Yasué and Dearden (2008a)
Kentish Plover	Y	mixed	Y N	60%–70%	
Snowy Plover	N	mixed	Y Y N	59%–74% M>F	Page et al. (2009) and ref. cited therein, Patrick (2013), Pearson et al. (2014)
Wilson's Plover	N	mixed	Y Y	M: 19%–50%	Bergstrom (1982)
Common Ringed Plover	Y	mixed	Y Y Y	90%	Pienkowski (1984)
Semipalmated Plover	Y	mixed	Y Y N?	yes	Armstrong and Nol (1993)
Long-billed Plover	Y	solitary	Y		Kolomiytsev and Poddubnaya (2014)
Piping Plover	Y	mixed	Y Y	25%–84%	Elliott-Smith and Haig (2004)
Madagascar Plover			Y	high	Zefania et al. (2008)
Little Ringed Plover	Y	mixed	Y Y	high	
African Three-banded Plover		solitary	Y Y T		
Forbes's Plover	Y	solitary			Brown (1948)
White-fronted Plover		mixed	Y Y & T	91%	Lloyd (2008)
Chestnut-banded Plover		solitary	Y		
Killdeer	Y	solitary	Y Y T	high M>F	Lenington (1980), Lenington and Mace (1975), Jackson and Jackson (2000)

(Continued)

TABLE 5.2 (*Continued*)

Aspects of the acquisition and use of space by breeding Charadrius *clade plovers.*

Common name[a]	Aerial display	Breeding concen[b]	Territoriality N FdI FdO[c]	Breeding site fidelity	Sources in addition to HBW Alive
Mountain Plover	Y	mixed	Y Y N	High, M>F	Knopf and Wunder (2006)
Oriental Plover	Y	solitary	Y Y		Ozerskaya and Zabelin (2006)
Eurasian Dotterel	Y	mixed	Y[f] N	no	
St. Helena Plover		solitary	Y Y	high	Burns (2011)
Rufous-chested Dotterel	Y	solitary	Y		St. Clair (2010)
Red-kneed Dotterel	N	mixed	Y		Robertson (2013)
Hooded Plover		solitary	Y	high	
Shore Plover	N	solitary	Y Y Y	high	Davis (1994)
Black-fronted Dotterel	Y	solitary	Y Y	yes	
Inland Dotterel		mixed			
Wrybill		solitary	Y Y	high	Dowding (2013)

[a] No data found for White-faced, Javan, or Madagascar Three-banded plovers. These species are not included here.
[b] Breeding concentration, mixed = both solitary and loosely aggregated or semi-colonial.
[c] Yes or No under N, nest territory (Yes/No), under FdI, feeding in nest territory. Under FdO, feeding outside nest territory, T = feeding territory separated from nest territory, N = feeding on neutral ground.
[d] Male started defending territory after brood hatched in location where female deserted after laying.
[e] Reported from Colombia.
[f] Appears to be defend territories in some areas but not others.

with the rhythm of the wing action (Cramp and Simmons 1983). In central Asia, aerial displays of Oriental Plovers are made by males with wings almost completely stretched and horizontal, with fast beats of small amplitude, and the body vigorously twisting from side to side in a zigzag trajectory at 2–200 m altitude (Ozerskaya and Zabelin 2006). The Oriental Plovers perform the display flight in defense of territories for about 3 months, but the seasonal duration and function of display flights for many species are not reported. Although males typically perform aerial song displays in most species, female Eurasian Dotterels display over large areas, not associated with territories, to attract males throughout the egg-laying period (Kålås and Byrkjedal 1984). Aerial displays are not characteristic of all plovers, as detailed behavioral studies of Kentish Plover detected song flights rarely (Rittinghaus 1956, Dement'ev and Gladkov 1969), and studies of Snowy and Wilson's plovers and Red-kneed Dotterels have failed to detect them at all (Bergstrom 1982, Warriner et al. 1986, Robertson 2013). All ten species breeding in subarctic and north temperate areas perform aerial displays, whereas only six of twenty-two species breeding wholly or partly in the Tropics, and two of eight species breeding only in south temperate areas are reported to do so (Tables 5.1 and 5.2).

Defense of Territories

Before egg laying, territories may be vigorously defended by body postures, chasing, or fighting. For example, Snowy Plovers defend territories against conspecifics with various ground displays and, occasionally, also fight (Warriner et al. 1986). Plumage variation may be associated with the dominance of males in territory and mate defense conflicts; in the case of Kentish Plovers in Turkey, larger breast bands (badges) in males were related to earlier nesting, larger eggs, and possibly higher reproductive success (Lendvai et al. 2004). Among the species, it appears there may be much commonality in the components of these displays, but their comparison is clouded by the use of different descriptive terms by different observers. As a comprehensive analysis of territorial defense in the Charadrius plovers is beyond the scope of this chapter, we refer the reader to other sources for comparative information (Rittinghaus 1956, Graul 1973, Maclean 1977, Phillips 1980, Gochfeld 1984).

Function of Territories

Plovers use territories for courtship and nest placement, and sometimes feeding or brood rearing with functions varying within and among species (Simmons 1956, Pienkowski 1984, Warriner et al. 1986, Armstrong and Nol 1993, Yasué and Dearden 2008a). Many species defend territories for courtship and nest sites, and 16 of the 23 species for which feeding areas are described were noted as foraging in the nest territories. A total of 17 species, including 10 of the 16 feeding within nesting territories, also forage on separate feeding territories or communal feeding areas (Table 5.2). We did not determine a latitudinal relationship in feeding territories because this was reported for so few species but noted that both of the studied species nesting strictly in the Tropics foraged only within their nesting territory. For chick rearing, the adult(s) may defend an area around the chicks either within or outside the nesting territory (Warriner et al. 1986, Yasué and Dearden 2008a). Information on the maintenance and defense of territories after chicks hatch is lacking for most species.

Site and Mate Fidelity

Many plover species exhibit high inter-year fidelity of adults to breeding sites and mates, which may benefit individuals through multi-year familiarity with mates, neighbors, or local landscapes (Table 5.2, Warriner et al. 1986, St Clair 2010, Saunders et al. 2012). Reported site fidelity is complicated by the size and configuration of study sites, which is sometimes arbitrary or dictated by factors unrelated to breeding locations, as perceived by the plovers. For example, some study sites are only part of a larger contiguous breeding area. Nonetheless, reported site fidelity is often greater for males than females. Adult male Semipalmated Plovers returned to their Churchill, Manitoba breeding site at a significantly higher rate (58.9%) than did adult females (41.2%); for males, but especially females, success in hatching chicks the previous year resulted in a higher fidelity rate the following year (Flynn et al. 1999). Similarly, Snowy plovers on the northern California coast moved the shortest distances between years following success in hatching eggs and retaining mates from the prior year (Pearson and Colwell 2014). While fidelity may be correlated to previous nesting success, inter-year site fidelity did not differ between the sexes or depend on nesting success the previous year for Piping Plovers in Manitoba, Canada; however, failed nesters were more likely than successful nesters to change mates the following year (Haig and Oring 1988). Another metric for fidelity is inter-year nest distances. For example, experienced male Double-banded Plovers breeding on South Island, New Zealand, nested an average of 42 m from a previous year's nest site, whereas females averaged 126 m (Pierce 1989). For Piping Plovers, in which the pair share clutch and brood-rearing duties, Saunders et al. (2012) found that female breeding site familiarity was the most important predictor of fledging success, although Haig and Oring (1988) reported male-biased fidelity in this species.

High inter-year mate fidelity also occurs commonly. After pairing one breeding season and surviving to the next, 8 of 18 Double-banded Plover pairs and 19 of 23 Two-banded Plover pairs retained mates the following year (Pierce 1989, St Clair 2010). About 60% of Semipalmated Plover pairs that returned to their arctic breeding area re-paired (Flynn et al. 1999). Pairs of White-fronted Plover were seen together in consecutive years in 117 instances and divorced in 12 instances, suggesting an annual divorce rate of about 10%; some also maintained pair bonds for an entire year (Lloyd 2008). Also, Killdeer in the southern part of their range, Northern and Southern Red-breasted dotterels, and possibly other relatively sedentary plovers retain pair bonds year round (Jackson and Jackson 2000, del Hoyo et al. 2016, Wiersma et al. 2016a).

Courtship and Pair Formation

Many, but not all, species engage in courtship activities in the territory (Warriner et al. 1986, Davis 1994, Yasué and Dearden 2008a). For example, Wilson's Plovers form pairs before territories are established; they scrape at various locations until a territory is chosen and a scrape within it is selected as a nest location (Bergstrom 1988). A few species are reported to be paired when they arrive on their breeding grounds, making it nearly impossible to know the details of pairing (Owens et al. 1994, Kolomiytsev and Poddubnaya 2014). The courtship and pair bonding of the Eurasian Dotterel in Scotland is unique in that females compete for males at communal mating arenas (Owens et al. 1994; Chapter Four, this volume). Currently, the relative importance to pair formation of site and territorial fidelity, advertisement, and prior familiarity with mates is an unexplored question within and among species in the clade.

Scraping is an important behavior for most if not all plovers, because it may lead to pairing, copulation, and nest-site selection. A scraping plover rests its breast on the ground, rotates with body slanting up at about 30° and tail somewhat raised, as it kicks with a backward and forward movement of the legs to form a shallow cup on the ground (Rittinghaus 1956, Phillips 1980). Snowy Plovers typically make several scrapes for courtship, and females lay eggs in one of these (Page et al. 2009). The dual role of scraping for courtship and nest-site selection has received little attention. However, Muir and Colwell (2010) found that plovers scraped in slightly less open habitats than those found around eventual nest sites, suggesting that locations selected for courtship scraping may differ from those eventually selected for the nest sites, a possibility that could be examined for other species.

Nests

The nests of plovers, which are shallow depressions on the ground or in low vegetation, may be placed in open habitat to lower predation risk by providing wide "viewscapes," which enable incubating adults to detect threats and slip off nests at the distant approach of predators or humans. Unvegetated nest sites are typically on clay, mud, sand, or gravel substrates and often are lined with material from the nest vicinity. Substrates for these nests may be so unstable that, in sandy, windy locations, adults must dig them out again after being filled with blowing sand (Buick and Paton 1989, Page et al. 2009).

Nests are sometimes near or next to a distinctive object such as a branch, stone, clamshell, bone, clump of vegetation, strand of kelp, or pad of cattle dung; nests may be located in areas of scattered shell, gravel or small stones on finer substrates (Johnsgard 1981, Knopf and Wunder 2006, Page et al. 2009, Wiersma et al. 2016c). Kentish, Red-capped, and Snowy plovers occasionally use sites partially or completely covered by overhanging vegetation or objects, possibly to provide shading from excessive heat or concealment from aerial predators (Page et al. 1985, Amat and Masero 2004a, b, AlRashidi et al. 2011b, Lomas et al. 2014). Predators of eggs and adults are present in the study areas of all three aforementioned species: primarily a corvid, a larid, and a canine for the Snowy Plover; an accipiter, a harrier, a canine, and mustelids for one, and a corvid, a lark, and introduced and human-subsidized felids for another Kentish Plover site; and a corvid for the Red-capped Plover. Page et al. (1985) reported that hatching success of Snowy Plovers nesting by objects was lower than for plovers nesting in the open or under objects. AlRashidi et al. (2011b) reported that Kentish Plovers breeding in Red Sea islands experienced very high levels of clutch predations but no difference in predation rates between nests in the open and covered nests. Amat and Masero (2004a) and Lomas et al. (2014) found nest temperatures under vegetation to be cooler than those in the open; for Kentish Plover the covered nests were 10°C–15°C cooler than those in the open and plovers nesting under cover were in poorer body condition compared to those nesting in the open, suggesting that heat stress might cause these birds to seek shaded nesting sites (Amat and Masero 2004b). Both Kentish and Red-capped plovers incubating in nests under cover took longer to detect approaching humans and predators and, for the Kentish Plover, the incubating adults were at greater risk of depredation from mammals (Amat and Masero 2004b, Lomas et al. 2014). However, the daily nest survival rate for Red-capped Plovers was slightly greater for nests under cover than those in the open, suggesting that the type of predator may be important in the risk–benefit trade-off of nesting in less open sites (Lomas et al. 2014). Thus, in hot environments it appears that plovers must weigh the relative costs and benefits of using covered rather than open nest sites.

While most species select unvegetated or sparsely vegetated nest sites, there is variation among the species. For example, nests of the Shore Plover are concealed under the cover of deep vegetation, and sometimes under overhanging boulders (Davis 1994). Mountain Plovers nest in vegetation on grazed short grass prairies, frequently in prairie dog (*Cynomys* spp.) colonies, and sometimes in tilled fields (Knopf and Wunder 2006). The nest of the Red-kneed Dotterel may be situated under a plant on mud close to water, and built up with plant stems to form a substantial platform that keeps the eggs dry (MacLean 1977).

Nest-lining materials used by plovers comprise a wide variety of small objects from the vicinity of the nest, including pebbles, shells, small twigs, bits of vegetation, dried invertebrate body parts, small fish bones, bits of dung, and wave-cast debris (Johnsgard 1981, Page et al. 2009, Burns 2011). These materials are placed in the scrape prior to egg laying and/or throughout the incubation period, depending on the species (Jackson and Jackson 2000, Page et al. 2009). Nest linings may reduce the temperature fluctuations of eggs by insulating them from the ambient temperature, provide concealment of eggs from visually searching egg predators, or prevent eggs from blowing out of the scrape (Grant 1982, Szentirmai and Székely 2005). Szentirmai and Székely (2005) documented Kentish Plovers actively regulating the amount of their nest material by covering the eggs with it during the day and uncovering eggs at night. Partial burial of eggs in the nest lining did not appear to have any thermoregulatory benefit for Snowy Plover eggs in the hot environment of the Salton Sea, California (Grant 1982).

A social site selection favored by some plovers is breeding near a colonially nesting, mobbing species, which potentially confers several antipredator benefits (Powell 2001). Most commonly, these associations include Common Tern (*Sterna hirundo*) with Kentish Plover, Least Tern (*S. albifrons*) with Piping and Snowy plovers, Arctic Tern (*S. paradisaea*) with Semipalmated Plover, and Yellow-billed Tern (*Sternula superciliaris*) with Collared Plover; clutches of the aforementioned plovers hatched at higher rates when placed close to the nesting terns compared to farther away (Walters 1957, Burger 1987, Powell 2001, Nguyen et al. 2006, Maugeri 2005). Powell (2001) reported that social factors, particularly proximity to the terns, but not physical nest-site characteristics, predicted successful hatching of Snowy Plover nests.

Nest Attendance during Egg Laying

Sustained incubation by plovers frequently does not begin until after the last egg of the clutch has been laid, although nest attendance and incubation during the laying period are only explicitly described for a few species (Summers and Hockey 1980, Warriner et al. 1986, Weston and Elgar 2005, Ekanayake et al. 2015). Lack of incubation during laying may help to facilitate synchronous hatching of the clutch, but some clutch attendance may be necessary to protect eggs under very cold or hot conditions. In closely watched Snowy Plovers on the California coast, males spent about 17% and females about 10% of daylight hours sitting on or standing over incomplete clutches, compared to 10% and 80%, respectively, after the clutches were complete; females spent 14–87 min on their nest before laying an egg (Warriner et al. 1986). In contrast, male Eurasian Dotterels in Norway started to incubate when the first egg was laid and male daytime attendance averaged 24.3% on the first egg and 73.1% by the second; the female was only seen on the nest when laying an egg (mean 37.7 min, Kålås 1986). Female Lesser Sand-Plovers also may commence incubation with the first of their three eggs (Wiersma et al. 2016c).

Clutch Size

Clutch size is an important attribute of avian reproductive potential, and modal clutch sizes in plovers are limited to two to four eggs (Maclean 1972, Winkler and Walters 1983). However, limited clutch manipulation experiments in one Charadrius and two Vanellinae species of plover suggest that plovers are not determinant layers, as removal of an egg during the laying period may result in females laying more eggs than the modal clutch size, with similar time intervals between all eggs (Klomp 1970, Yogev and Yom-Tov 1996, Wallander and Andersson 2003). Additionally, Claassen et al. (2014) noted continuation of clutches in 28% of 75 cases in which Piping Plovers lost nests during the laying period.

With almost complete information for plovers, we verified that modal clutch size is lowest (two eggs) in species nesting wholly or partly in the Tropics, with the exceptions of the temperate-breeding Wrybill and Rufous-chested Dotterel, which both also have a clutch size of two eggs. Species laying modal four-egg clutches breed primarily in subarctic and north temperate regions, with the exception of the Red-kneed Dotterel, which breeds in south temperate and tropical latitudes, mostly in Australia. Some species in all latitudinal zones lay modal three-egg clutches, the commonest clutch size overall (Figure 5.1, Table 5.3). Published data were nonexistent or

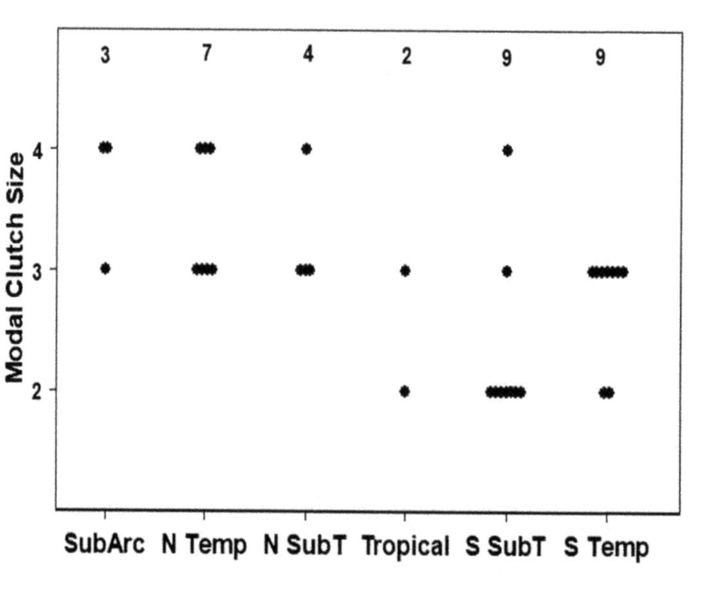

Figure 5.1. Variation in modal clutch sizes of plovers in six latitudinal regions: Subarctic (range extending north of 23.45°N), North Temperate (between 66.55°N and 23.45°N), North Subtropics (between 66.55°N and 23.45°S), Tropics (between 23.45°N and 23.45°S), South Subtropics (between 23.45°N and 66.55°S), and South Temperate (between 66.55°N and 23.45°S), with number of species for each region above. (Data from Table 5.3.)

TABLE 5.3

Egg laying effort of Charadrius clade plovers.

Common name	Modal clutch size[a]	Maximum number of replacements after loss	Maximum number of broods per year[b]	Other sources in addition to HBW alive
S. Red-breasted Dotterel	3			
N. Red-breasted Dotterel	3	3	2	
Lesser Sand-Plover	3	1+	1	
Greater Sand-Plover	3	1+	1	
Caspian Plover	3	1+		
Collared Plover	(2–3)			
Puna Plover	(2–4)			
Two-banded Plover	3	1+	1 (M 2[P] once)	St. Clair (2010), Garcia-Peña (2010)
Double-banded Plover	3	3	1 (2 rarely)	
Kittlitz's Plover	2		2 (w/overlap)	
Red-capped Plover	2	5		
Malay Plover	3	1+	2 (~9 d overlap)	Yasué and Dearden (2008b)
Kentish Plover	3	3	F 3[P], M 2[P]	Fraga and Amat (1996)

(Continued)

TABLE 5.3 (*Continued*)
Egg laying effort of Charadrius clade plovers.

Common name	Modal clutch size[a]	Maximum number of replacements after loss	Maximum number of broods per year[b]	Other sources in addition to HBW alive
White-faced Plover	(3)			
Snowy Plover	3	9	F 3[P], M 2[P]	Colwell et al. (2017)
Javan Plover	(2–3)			
Wilson's Plover	3	2	1 (2 once)	Corbat and Bergstrom (2000)
Common Ringed Plover	4	4	1–3 (N–S of range)	
Semipalmated Plover	4	3	1	Lishman et al. (2010)
Long-billed Plover	4	1+	2[P]	Uchida (2007), cited in Katayama et al. (2010)
Piping Plover	4	4	1, 2 & P rarely	Elliott-Smith and Haig (2004)
Madagascar Plover	2	1+	2	
Little Ringed Plover	4	3	2, 2[P]	
African Three-banded Plover	2	1+?	3 (w/overlap)	Tyler (1978)
Madagascar Three-banded Plover	2			
Forbes's Plover	(2–3)		possibly	Brown (1948)
White-fronted Plover	2	11	2	Summers and Hockey (1980), Lloyd and Plagányi (2002)
Chestnut-banded Plover	2		possibly 2	
Killdeer	4	5	1–2[c]	
Mountain Plover	3	3 (pre-Jun)	F 3[P], M 1	Knopf and Wunder (2006)
Oriental Plover	(3–4)			Ozerskaya and Zabelin (2006)
Eurasian Dotterel	3	2	F 3[P], M 2[P]	Owen et al. (1994)
St. Helena Plover	2	1+	2	Burns (2011)
Rufous-chested Dotterel	2	1		St. Clair (2010)
Red-kneed Dotterel	4			Robertson (2013)
Hooded Plover	3	1+	2	Buick and Paton (1989)
Shore Plover	3	1+	1[d]	Davis (1994)
Black-fronted Dotterel	3	2	2	Marchant and Higgins (1993)
Inland Dotterel	3		2	
Wrybill	2	1+	2	Dowding (2013)

[a] Parentheses indicate modal clutch size is in question due to few clutches found and uncertainty that some of those found were complete or had not previously been reduced. Modal clutch size for White-faced Plover presumed due to its affinity to Kentish Plover, but data on this necessary to validate.

[b] M = male, F = female, [P] or P = polyandry, polygyny, or rapid multiple clutch polygamy. Overlap indicates known cases of subsequent clutch initiated before existing brood fledged.

[c] Three or more broods suspected but not documented in southern part of range.

[d] Single brood believed to be due to limited resources in habitat, species double-brooded in captivity when given sufficient food.

too sparse to determine a modal clutch size for three tropical breeders: White-faced, Javan, and Forbes's plovers; two south subtropical breeders, Collared and Puna plovers; and one north temperate breeder, Oriental Plover. However, limited information on clutch and brood sizes for some of these species is consistent with the overall pattern for known species at the same latitudes (Table 5.3).

Four hypotheses for factors contributing to variation in clutch sizes in precocial birds focus on proximate (physiological and environmental) and ultimate (evolutionary) drivers and extrinsic and intrinsic constraints on the parents (Klomp 1970, Winkler and Walters 1983, Székely et al. 1994, Patten 2007). The egg-formation limitation hypothesis proposes that clutch size is limited by the female's ability to obtain sufficient energy or nutrients, particularly calcium, to produce eggs. The incubation limitation hypothesis postulates that clutch size is limited by the species' ability to adequately incubate the eggs. The predation limitation hypothesis suggests that larger clutches expose the nest and incubating adult to higher rates of predation. The parental care limitation hypothesis postulates that clutch size is limited by the pair's ability to protect clutches and/or hatchlings.

The first two hypotheses, that clutch size is constrained by egg formation and by incubation ability limitations, were examined by Eurasian Dotterel and Kentish Plover. Kålås and Løfaldi (1987) tested the incubation hypothesis and observed greater adult mass loss and possibly increased frequency of absences from the nest by incubating dotterels whose clutches had been experimentally augmented by one egg. Székely et al. (1994) also found supporting evidence for the incubation limitation hypothesis in the prolonged incubation periods and asynchronous hatching of enlarged clutches of Kentish Plovers. They found no support for the predation limitation hypothesis and dismissed the egg-formation limitation because most clutches were laid well before the peak of food availability. If peak food resources occur mid to late breeding season more generally, and resources are a limitation to egg formation, we do not expect lower mean clutch sizes later in the breeding season as has been found for several species—see below. The observation (above) that some plovers are quasi-indeterminate layers also argues against the egg-formation hypothesis. The egg-formation limitation hypothesis suggests that the resources for egg laying are derived from exogenously acquired (income) rather than endogenously stored (capital) energy and nutrients (on an income-capital continuum; Stephens et al. 2009). Shorebirds studied in the Arctic appear to favor the income end of the continuum, but evidence from early season eggs suggests some endogenous resource allocation for breeding (Nol et al. 1997, Klaassen 2001, Morrison and Hobson 2004, Hobson and Jehl 2010). The importance of this distinction for plovers breeding in other habitats deserves further study.

Lengyel et al. (2009) summarized results from several studies of other shorebirds on the effects of augmented clutches and provide further support from their own work on Pied Avocet (*Recurvirostra avocetta*) for the incubation limitation hypothesis; evidence includes physiological stress on the parents or prolonged incubation and hatching periods but not necessarily an inability to hatch an enlarged clutch. For the avocet, they also found that the consequences of augmented clutch size may extend into the chick-rearing period, as parents incubating augmented clutches ultimately fledged fewer young (Lengyel et al. 2009). While the incubation limitation hypothesis appears to be the best supported for Charadrius clade species, limitations on the brood-rearing period have not been examined for this group.

Smaller than modal clutch sizes occur for almost all species, but their cause is usually not determined. Smaller observed clutch sizes may result from partial clutch loss, through weather or flooding, by trampling, or by predation, although predators typically take entire clutches (Fraga and Amat 1996). Alternatively, females may produce smaller-sized clutches, particularly in late-season nests or in more inclement breeding seasons; for many species, these reduced-size clutches are presumed to be due to deteriorated female or environmental conditions or seasonal time constraints (Lack 1947, Winkler and Walters 1983, Nol et al. 1997, Cohen et al. 2009, Claassen et al. 2014). Nol et al. (1997) noted that Semipalmated Plovers reduced clutch, but not egg, sizes in a relatively cold and inclement breeding season. At a hot interior saline lake, Fraga and Amat (1996) also noted a significant increase in small clutch sizes late in the season, but attributed it to partial clutch predation rather than a reduction in clutch sizes laid by females.

Wallander and Andersson (2003) noted that female Common Ringed Plovers laid smaller than modal clutch sizes in late-season nests, whereas egg volume increased by about 5% in late clutches; they hypothesized that females had problems producing a clutch of four full-size eggs late in the season and reduced clutch size but not egg size, presumably not unduly reducing the survival of individual late-season chicks.

Researchers have documented rare supernumerary clutch sizes for several species. The Killdeer, with a modal four-egg clutch rarely hatches five- and six-egg clutches. However, one eight-egg clutch (at a nest for which no adults other than the social pair could be associated) only hatched a single egg and that too with researcher assistance (Jackson and Jackson 2000). Five- or six-egg clutches have been reported for Red-breasted and Black-fronted dotterels, and Double-banded, Kentish, Snowy, and Shore plovers, all typically having three-egg clutches; and four-egg clutches have been reported for the Wrybill and Red-capped Plover, both of whose typical clutch size is two eggs (Johnsgard 1981, Warriner et al. 1986, Marchant and Higgins 1993, Küpper et al. 2004). These atypically large clutches most likely result from facultative intraspecific nest parasitism whereby two females lay eggs in the same scrape, an abnormal behavior that probably also occurs occasionally in other plovers. Amat (1998) and Küpper et al. (2004) examined this behavior for Kentish Plover, using pigmentation patterns on eggs and genetic fingerprinting, respectively, to distinguish different females (or males, for the latter method). Amat (1998) determined that 1% of 883 nests contained eggs from more than one female, including three clutches each of four and six eggs and one clutch of five eggs; none of the five- or six-egg clutches hatched, but two of the four-egg clutches were successful. Küpper et al. (2004) documented extra-pair paternity, extra-pair maternity, or intraspecific brood parasitism in 7 of 89 clutches they examined but none were greater than three eggs. However, they documented one each of four- and six-egg, and two five-egg clutches among 1,291 clutches over multiple years in which egg DNA was not fingerprinted. Amat (1998) suggested that limited nest sites, egg dumping by females that lose a partial clutch, or female–female pairings may be responsible for these larger clutches.

Egg Size

For birds in general, egg size varies little within individual females and appears to have high heritability; however, the causes and fitness consequences of variation among individuals are poorly understood for most species (Christians 2002). Few studies of plovers have examined within-species effects of egg size on chick size at hatch. Several reported a significant positive relationship between egg size and neonate mass (Ricklefs 1984, Amat et al. 2001a, Skrade and Dinsmore 2013). In a few studies of shorebirds, larger neonates were more likely to survive to fledging; in Kentish Plover, the probability of recruiting into the breeding population was positively related to chick mass (Galbraith 1988, Grant 1991, Blomqvist et al. 1997, Amat et al. 2001a). Other factors that have been linked to plover egg size include female body size and condition (despite the general lack of evidence for within-female variability), stage of the breeding season, plumage characteristics of males, male body size, and climatic variations (e.g., drought) between breeding seasons (Amat et al. 2001b, Wallander and Andersson 2003, Lendvai et al. 2004, Lislevand and Thomas 2006, Skrade and Dinsmore 2013). Amat et al. (2001b) further found female size and condition negatively related to variation in egg size within a clutch. However, although clutch sizes of Semipalmated Plover females were smaller in the most inclement year of study at Churchill, Manitoba, egg sizes remained similar to other years (Nol et al. 1997). As noted in the discussion of clutch sizes that were smaller than the mode, increases in egg size have been noted for some species as the breeding season progresses (Fraga and Amat 1996, Wallander and Andersson 2003).

Plover species vary nearly fivefold in body mass, and egg size correlated positively with female size (Table 5.4). A strong relationship exists between the mean egg volume indices and mean egg masses ($\rho = 0.986$, egg mass $= 0.70 + 0.48 \times$ volume index $+ \varepsilon$), which we used to estimate egg masses for other species (Figure 5.2). Estimated egg mass correlated highly with female body mass and represented 14%–25% (median 18.5%) of female body mass; this is at the high end of the <5% to >20% range in all birds (Figure 5.3a, $\rho = 0.946$; Cody 1971, Rahn et al. 1975). The sizeable energy demands that these eggs represent (as gauged by female mass) clearly increases with

TABLE 5.4

Morphometrics and ovimetrics of Charadrius clade plovers.

Common name[a]	Sex[b]	Body mass[c] (g)	Egg mass[c] (g)	Egg width[c] (mm)	Egg length[c] (mm)	Sources in addition to HBW Alive
S. Red-breasted Dotterel	B	(147–179)	(21.4–26.5)	(31.4–34.2)	(42.7–49.7)	
N. Red-breasted Dotterel	B	(128–169)	(18.1–24.3)	(29.2–32.7)	(41.3–46.5)	
Lesser Sand-Plover	B	(39–110)				
Greater Sand-Plover	B	(55–121)		27.7	38.6	
Caspian Plover	B	(60–91)		27	38	
Collared Plover	B	(26–42)	(4–6.3)[d]	20.6–21.6	26.9–30	
Location: Brazil				(20.8–21.8)	(27–28.6)	
Location: Colombia				(20.4–20.8)	(26.6–27.2)	
Location: Uruguay				(20.1–21.6)	(28–32)	
Location: Argentina				(20.8–22.4)	(26.5–30.4)	
Puna Plover	B	(41–49)		(24.2–24.7)[e]	(34.8–34.9)[e]	
Two-banded Plover	F	69.8		(24.9–26.1)	(34.5–39.1)	St. Clair (2010)
Double-banded Plover	B	(47–76)[f]				
Kittlitz's Plover	B	(19–54)	(5.5–10.0)	(21–23)	(29–33.9)	
Red-capped Plover	B	(27–54)				
Malay Plover	B	~42		23.5–23.6	29.2–31.4	Yasué and Dearden (2008b)
Kentish Plover	F*	41.8–42.2	8.4–8.9	23.1–23.3	31.7–32.5	Székely (1992), Fraga and Amat (1996), Szentirmai et al. (2001)
Snowy Plover	F	40.3–42.9	8.5	22.7	31.3	Page et al. (2009)
Wilson's Plover (*wilsonia*)	B	67.5[g]	(12.6–13)	25.9–26.0	35.2–35.4	Corbat and Bergstrom (2000)
Wilson's Plover (*beldingi*)				(25.5–27.2)	(34.9–37.9)	
Common Ringed Plover	B	(42–78)		25.9	35.8	
Semipalmated Plover	B	47.6–48.5	(8.7–9.4)	22.8–24.1	32.4–33.2	Nol and Blanken (2014)
Long-billed Plover	B	(41–70)	11.8–12.07	26.9–28.2	34.8–35.6	g
Piping Plover	B	(43–63)	(9.3–10.7)	23.7–24.8	30.5–32.5	Elliott-Smith and Haig (2004)
Madagascar Plover	F	(31.0–43.5)		(22.4–25.0)	(30.2–35.0)	
Little Ringed Plover	B	(26–53)		22.1	29.8	
African Three-banded Plover	B	(25–49)		(21–24)	(27–33)	

(Continued)

TABLE 5.4 (*Continued*)
Morphometrics and ovimetrics of Charadrius clade plovers.

Common name[a]	Sex[b]	Body mass[c] (g)	Egg mass[c] (g)	Egg width[c] (mm)	Egg length[c] (mm)	Sources in addition to HBW Alive
Madagascar Three-banded Plover	F	(37–43.5)	8	(21.9–23.3)	(29.2–30.9)	
Forbes's Plover	M	(46–49)	(8.3–8.9)	(23–23.8)	(28.7–32)	
White-fronted Plover (*marginatus*)	B	52.9	9.8	(21.3–23.9)[h]	(28.8–33.9)[h]	Summer and Hockey (1980)
White-fronted Plover (*mechowi*)	B	(27–40)				
White-fronted Plover (*tenellus*)	B	(33–45.5)				
Chestnut-banded Plover (*pallidus*)	B	(28–44)		(22–24)[h]	(29–33)[h]	
Chestnut-banded Plover (*venustus*)	B	(20–37)				
Killdeer	B	(72–121)		26.8–27.1	37.9–38.2	Jackson and Jackson (2000)
Mountain Plover	F	102–105[CI]	(13–19)	28.2–28.6	37.0–38.1	Knopf and Wunder (2006)
Oriental Plover	B	~95		(26.5–27.1)[e]	(39.0–40.0)[e]	
Eurasian Dotterel	F	(99–142)		28.9	44.1	
St. Helena Plover				(23.6–25.5)	(30.3–36.5)	
Rufous-chested Dotterel	F	83.14		(28.4–28.8)	(39.5–40.2)	St. Clair (2010)
Red-kneed Dotterel	B	50 (35–77)				Robertson (2013)
Hooded Plover	B	(79–110)	(10.5–16.5)	26.6	36.8	Baird and Dann (2003)
Shore Plover	F	60	12.6	(25.5–25.8)	(35.6–36.3)	Davis (1994)
Black-fronted Dotterel	B	(27–42)				
Inland Dotterel	B	(64–107)				
Wrybill	B	(43–68)				

[a] No data found for White-faced or Javan plovers.
[b] Male, Female, or Both. An asterisk indicated sexual size dimorphism documented.
[c] Values given are mean if single values and are 95% CI if so indicated by superscript. Overall range is indicated by minimums-maximum in parentheses and range of means (from multiple locations and/or years) if not in parentheses.
[d] Combined egg mass for Brazil and Colombia
[e] Sample size = 2 for Puna Plover and = 3 for Oriental Plover.
[f] Body mass for subspecies *bicinctus* given, body mass for subspecies *exilis* 78–89, n = 4.
[g] Minima-maxima for egg mass (8.7–14.2), width (23.1–28.2), length (30.7–38.6) from Uchida (2007); and mass (10.06–13.32), width (24.4–26.9), length (33.1–37.2) from Kolomiytsev and Poddubnaya (2014).
[h] Subspecies for egg dimensions not reported.

clutch size (Figure 5.3b). Large eggs and long incubation period are associated with more advanced stages of development at hatch (Cody 1971, Rahn et al. 1975). Plover neonates are precocial and nidifugous, locating and pursuing prey within hours of hatch; this requires a well-developed optic tectum, a complex brain region associated with visual location of distant prey (Nol 1986). Nol's (1986) review of shorebird neonate foraging methods, egg sizes, and incubation periods

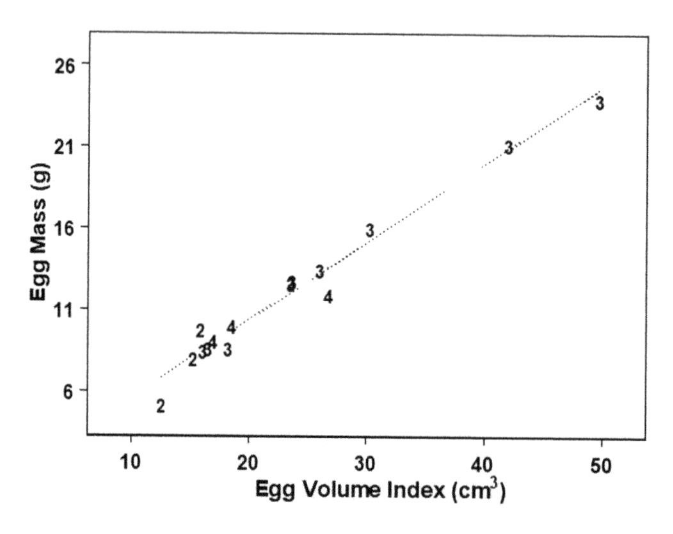

Figure 5.2. Relationship between an index of egg volume (length × breadth²) and egg masses. (Data from Table 5.4.)

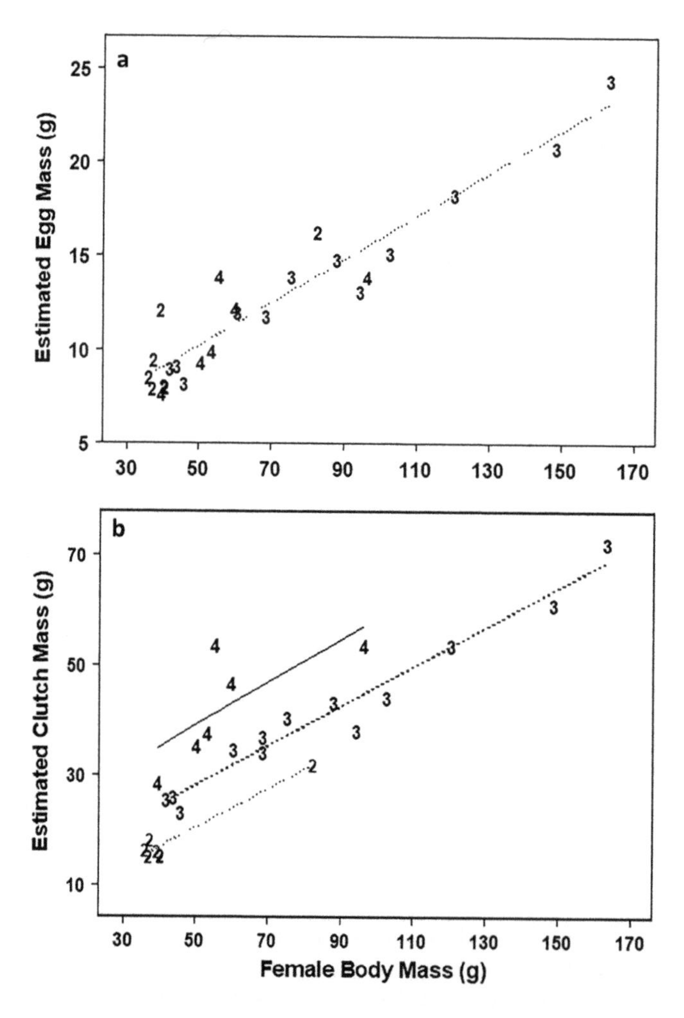

Figure 5.3. Relationship between female body mass and estimated egg mass (a), and estimated clutch mass (b), with symbol indicating clutch size for each species. (Data from Table 5.4 and Figure 5.2.)

suggests that brain development may be a key limitation for plovers, requiring them to produce larger eggs and prolong their incubation periods.

Clutch mass corresponded to a mean of 54% of female plover body mass. Plover females breeding in the Northern Hemisphere invested more in clutch mass relative to their body mass (mean = 63%, SE = 4%) than those breeding in the Southern Hemisphere (mean = 45%, SE = 2%). Further, the four lowest clutch masses as a percent of female mass (37%–43%) were all from species breeding in the Southern Hemisphere, whereas the highest clutch masses (60%–98%) were all from species breeding in the Northern Hemisphere.

Female body mass varies with latitude such that species with all or a portion of their breeding range in the Tropics are smaller than species breeding wholly at higher latitudes. Body masses of only 2 of 18 species with breeding ranges in or extending into tropical latitudes exceeded the median body mass of all the plover species for which we could find estimates (Figure 5.4a). In contrast, body masses of 4 of 20 species with only arctic and/or temperate breeding ranges were ≤ the median

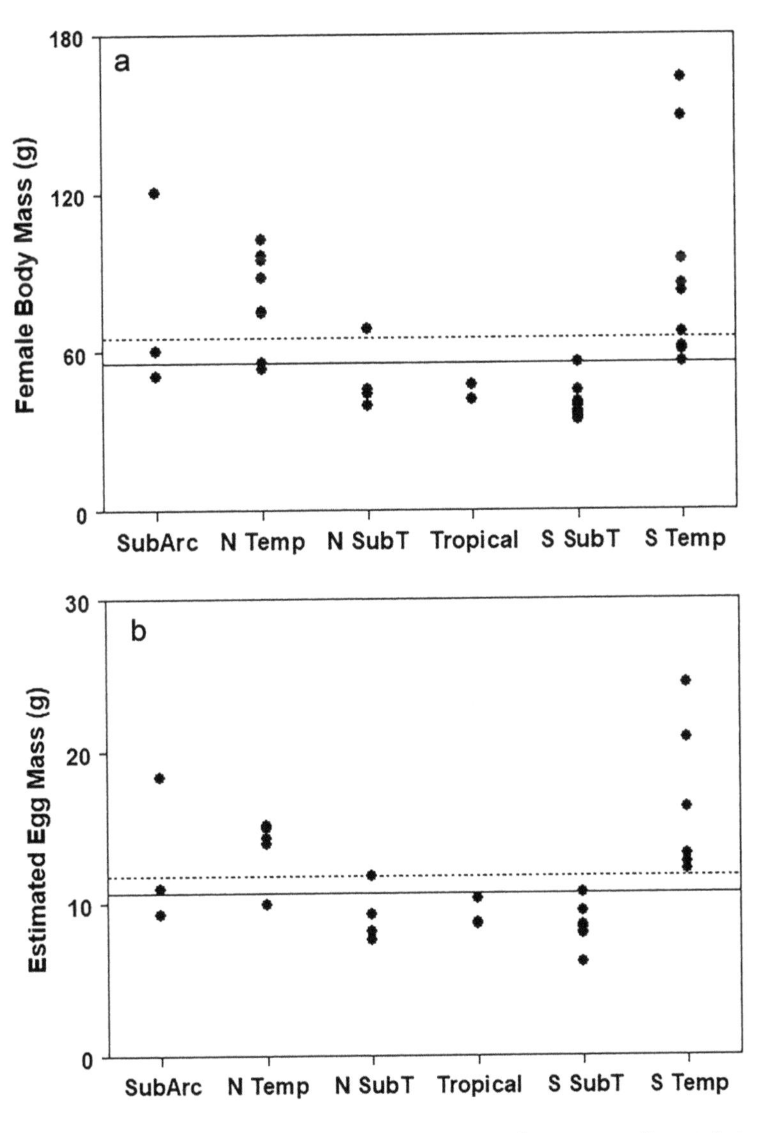

Figure 5.4. Female body masses for 37 species (a), and estimated egg masses for 32 species (b) in six latitudinal regions: Subarctic (north of 23.45°N), North Temperate (between 66.55°N and 23.45°N), North Subtropics (between 66.55°N and 23.45°S), Tropics (between 23.45°N and 23.45°S), South Subtropics (between 23.45°N and 66.55°S), South Temperate (between 66.55°N and 23.45°S), with number of specie for each region above. (Data from Table 5.3 and Figure 5.2.)

body mass (Figure 5.4a). Not surprisingly, eggs of tropical breeding species are smaller than species with arctic and/or temperate breeding ranges (Figure 5.4b). However, because female body mass differs so markedly between species that breed in the Tropics and those that do not, it is not clear how much of this difference in the eggs could be directly due to tropical conditions.

Breeding Season Length and Renesting Opportunities

Females may adjust their investment in eggs by varying egg size, clutch size, or the number of clutches laid. Because modal clutch sizes in plovers are ≤ four eggs, the opportunity to increase the investment in eggs during a breeding season mostly arises through multiple broods or polygamous mating. For most plovers, long breeding seasons allow time for replacement of lost clutches as well as multiple broods.

Breeding season length is, in part, dependent on latitude. At least four species whose breeding ranges extend into the Tropics may nest year round; nests may be initiated for up to a median of 215 days a year for this group (Table 5.1). In comparison, strictly temperate breeders initiate clutches for up to 245 (median 120) days, and subarctic breeders for up to 110 days a year (Table 5.1). These differences, and the precipitation-dependent breeding season of several temperate and tropical nesting species, suggest that climate and food availability constrain breeding season lengths (Piersma and Wiersma 1996; Table 5.1).

Given sufficient remaining time in the breeding season, at least 28, and probably all, plover species lay replacement clutches after loss (Table 5.3). At least 12 species have been documented laying 3–11 replacement clutches in one season; in contrast, individuals from mid to high arctic populations of a few species seldom lay replacement clutches (Pienkowski 1984, Flynn et al. 1999; Table 5.3).

Duration of clutch initiation for the 17 species that regularly produce multiple broods, excluding those that do so through "rapid multiple clutch" polyandry, ranges from 110 to 365 (median 230) days annually (Tables 5.1 and 5.3; Chapter Four, this volume). In contrast, five of the eight species believed to be single brooded (including those for which double brooding is considered rare) initiate clutches over only 30–85 days a year. Only the south temperate-breeding Double-banded, Two-banded, and Shore plovers are thought to be single brooded and initiate clutches over extended periods (105–155 days) (Tables 5.1 and 5.3). The maximum reported durations of nest initiation for the two species employing "rapid multiple clutch" polyandry, Mountain Plover and Eurasian Dotterel, are 75 and 55 days, respectively (Table 5.1). Regularly multiple-brooded species mostly breed in tropical and temperate latitudes, including the Common Ringed Plover in the temperate portion of its breeding range, whereas species regularly producing multiple broods through polygamy are north temperate and subarctic breeders (Tables 5.1 and 5.3).

Durations of Egg Laying, Incubation, and Brood rearing

After pair formation, the periods of egg laying, incubation (last egg laid to the first hatched), and brood care define the minimum time necessary for a successful nesting attempt. The egg-laying period is dependent on the clutch size and the interval between the laying of successive eggs and seldom exceeds one week (Table 5.5). The intervals between eggs for plovers are longer than for other precocial species, such as many waterfowl species, which arrive on the breeding grounds with high lipid stores for egg laying (Lenington 1984). A pattern of decreasing laying intervals with increasing latitude in the Northern Hemisphere, noted for other shorebirds by Colwell (2006), is not mirrored for plovers in the Southern Hemisphere where egg-laying intervals vary greatly (Figure 5.5a). Additionally, the pattern appears to break down completely for species nesting wholly in the Tropics where females typically lay eggs daily (Figure 5.5a, Table 5.5). However, laying intervals are reported in days rather than hours for most species, imparting considerable imprecision in the estimates of this measure, thus potentially disguising geographic or other relationships for this metric (Table 5.5). For example, egg-laying intervals of Snowy Plovers, which were documented in hours, decrease as the nesting season progressed, such that the total laying period for a clutch declined by about a day (Warriner et al. 1986). Studies of other shorebird species have shown that increased food resources may shorten the pre-laying and/or laying periods,

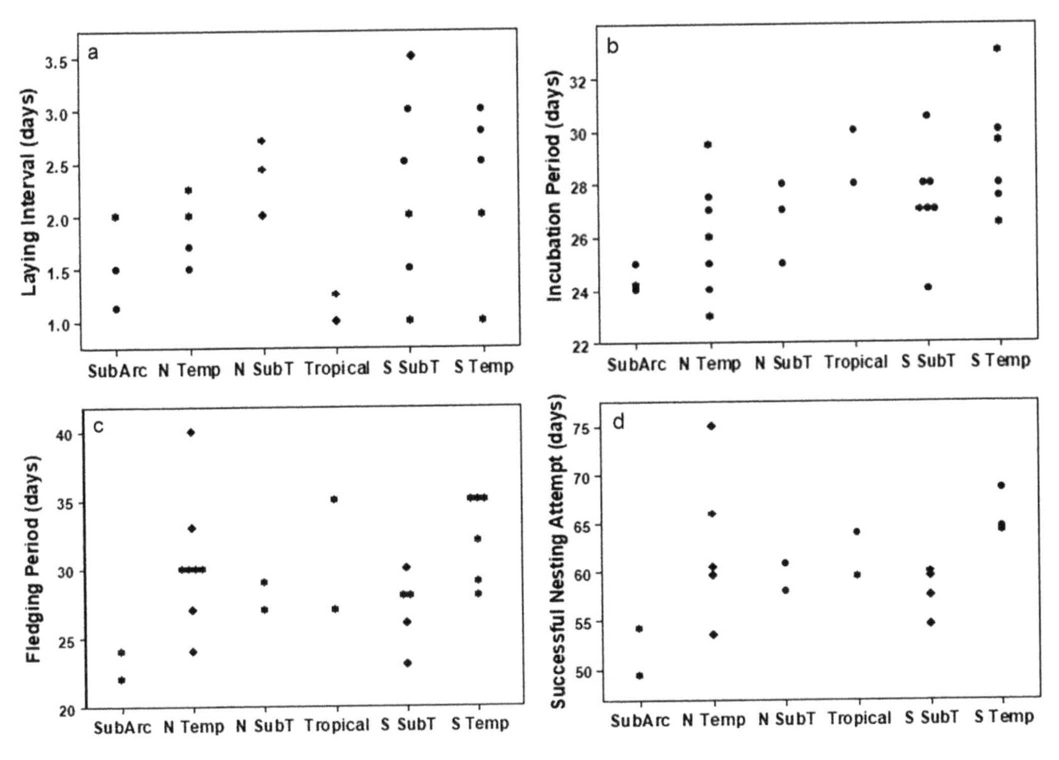

Figure 5.5. A comparison of laying intervals (a), incubation periods (b), fledging periods (c), and duration of a successful nesting attempt (d) for six latitudinal regions: Subarctic (north of 23.45°N), North Temperate (between 66.55°N and 23.45°N), North Subtropics (between 66.55°N and 23.45°S), Tropics (between 23.45°N and 23.45°S), South Subtropics (between 23.45°N and 66.55°S), South Temperate (between 66.55°N and 23.45°S). (Data from Table 5.5.)

TABLE 5.5

Duration of breeding season stages for Charadrius clade plovers.

Common name[a]	Egg laying interval[b]	Incubation period (d)	Min fledging period (days)	DBC[c]	Other sources in addition to HBW alive
N. Red-breasted Dotterel		28–32	28		
Lesser Sand-Plover		22–24	30		
Greater Sand-Plover		24	30		
Caspian Plover			30		
Two-banded Plover	2.8 d	29.6	(29[d])		St. Clair (2010), Garcia-Pena (2010)
Double-banded Plover	<60–120 h	25–28	35		
Kittlitz's Plover	1–2 d	27	26		Clark (1982), Urban et al. (1986)
Red-capped Plover	1 (occ. 2–3) d	30–31			
Malay Plover	1–1.5 d	30	27	9.8	Yasué and Dearden (2006, 2008a, b)

(Continued)

TABLE 5.5 (*Continued*)

Duration of breeding season stages for Charadrius clade plovers.

Common name[a]	Egg laying interval[b]	Incubation period (d)	Min fledging period (days)	DBC[c]	Other sources in addition to HBW alive
Kentish Plover	~2 d	25–29[e]	27	9.3, 7–9	Amat et al. (1999), Székely et al. (2007)
Snowy Plover	58.5 h (n = 25)[e]	23–29[e]	28	6–8	Warriner et al. (1986)
Wilson's Plover	65 h (n = 7)	25 (23–25)	21+?	7.6	Bergstrom (1982), Corbat and Bergstrom (2000)
Common Ringed Plover	1–3 d	21–27	24	~7 (4–9)	Pienkowski (1984)
Semipalmated Plover	24–30 h	24	22	7	
Long-billed Plover	2–2.5 d	26–29	40	7	Uchida (2007), Kolomiytsev and Poddubnaya (2014)
Piping Plover	2 d	26–28[e]	21–30		Elliott-Smith and Haig (2004)
Madagascar Plover	2–3 d	26–28	28		Zefania et al. (2008)
Little Ringed Plover	36 h	22–28	24	4	Johnsgard (1981)
African Three-banded Plover	2–4 d	≤26, 26–28	30, also 21		Tyler (1978)
Forbes's Plover		≥22			
White-fronted Plover	2–5 d	27–29	28	9	Summers and Hockey, (1980), Lloyd and Plagányi (2002)
Killdeer	1–2 d	23–29	30[f]	7	Jackson and Jackson (2000)
Mountain Plover	34–48 h	28–31	33		
Oriental Plover		≤24			Ozerskaya and Zabelin (2006)
Eurasian Dotterel	23–49, 30–36 h	21–29	19?-30	5–11	
St. Helena Plover	1 d	28	35[g]		Burns (2011)
Red-kneed Dotterel	1 d				
Hooded Plover	1–4 d	28	32		
Shore Plover	3.3–4.8 d	27–28	29	17.4	Davis (1994)
Black-fronted Dotterel	2 d	22–26	23	9	Marchant and Higgins (1993)
Inland Dotterel	1 d	26–30			
Wrybill	<48 h	30–36	35		Dowding (2013)

[a] No data found for Southern Red-breasted or Rufous-chested dotterels, or Collared, Puna, White-faced, Javan, or Madagascar Three-banded, Chestnut-banded, or Oriental plovers.

[b] Reported as either days (d) or hours (h).

[c] Days between loss of clutch and first egg of replacement clutch.

[d] Described as period in which both parents attend young, minimum age of flight not reported.

[e] Declining through breeding season.

[f] Not including reported fledging period of 20 days.

[g] Fledging period value use for modeling; mean fledging period reported as 36 days (Burns 2011).

but we are unaware of similar data for plovers (Högstedt 1974, Lank et al. 1985).

The incubation period is generally measured from the time that the last egg is laid until the first egg hatches, but parents of Mountain and Shore plovers may delay sustained incubation sometimes for up to several days after clutch completion (Graul 1973, Davis 1994) and, less commonly, for 1–2 days in the Snowy Plover (G. Page et al. observations). Because of the high variability in incubation periods for plover species in the north temperate and south subtropical regions, we are unable to identify a consistent geographic pattern (Figure 5.5b, Table 5.5). However, incubation periods of subarctic breeders are consistently among the shortest, and those of tropical breeders among the longest within the clade. South temperate breeders average longer incubation periods than north temperate breeders (Figure 5.5b, Table 5.5).

The fledging period, the time of brood care from hatch until the young are capable of flight, is often measured differently among researchers, so variation among species will be at least partly attributable to methodology. For example, after multiple observations of first flight by young, Warriner et al. (1986) defined fledging as reaching the earliest age at which flight ability may be attained. Flight criteria may differ among studies and other considerations may play into setting the threshold age for each species. For example, Graul (1973) determined that chicks fledged when they could fly at least 100 m, a feat usually achieved in 33–34 days, but he did not describe whether chicks were considered fledged when he observed them making this flight or when they reached 33 days of age. Davis (1994) determined that Shore Plover chicks achieved flight when they attained the body mass of 37 g, at 29–63 days of age. Because near fledging-age chicks can become very difficult to locate as they get older, Cairns (1982) established a threshold age quite a bit younger than the known age of fledging, assuming these chicks would eventually fledge because survival increased with age. Some studies evaluated the flight capability of individual chicks to confirm fledging, regardless of age (Warriner et al. 1986, Baird and Dann 2003).

With the above considerations in mind, it is not surprising that considerable variation exists among species in the reported fledging periods at some latitudes (Figure 5.5c). While this makes it more difficult to discern a clear geographic pattern in this metric, the shorter durations of subarctic compared to lower-latitude breeders in fledging period stand out; differences in fledging period between temperate and tropical latitudes are less clear (Figure 5.5c). However, Tjørve et al. (2008) provide evidence suggesting food resources for plover chicks are more limited in subtropical than in temperate or arctic breeding areas, prolonging the fledging period at lower latitudes.

If sustained incubation begins when the last egg is laid and hatching is synchronous, the minimum time required for an entire nesting cycle includes the three periods described above. Because of differences among species breeding at different latitude in length of periods, high-latitude breeders have the shortest duration of a successful nesting cycle (Figure 5.5d). However, a pattern of increasing nesting cycle length with decreasing latitude does not appear to hold up between temperate and tropical regions, with most species breeding wholly or partly in the Tropics appearing to have somewhat shorter nesting cycles than most temperate breeders (Figure 5.5d). This contrast may be an artifact of differences between study methodologies and definitions, and underscores the need for more research using standardized methodologies, which would facilitate comparison among species.

Parental Care

The division of incubation and brood-rearing duties between female and male plovers varies among species, but also within species, by time of day, time in the nesting cycle, time in the nesting season, and, in some cases, location (Table 5.6). Biparental care is most common, followed by male-biased care, and, very uncommonly, female-biased care. Further, while most species employ primarily a single strategy, several species exhibit flexibility in their division of parental duties (Warriner et al. 1986, Fraga and Amat 1996, Vincze et al. 2013, 2016; Chapter Four, this volume).

Females usually perform the majority of diurnal incubation, excluding some populations of Kentish Plover and the polyandrous Lesser Sand-Plover, Mountain Plover, and Eurasian Dotterel (Table 5.6; Chapter Four, this volume). Females and males of three species share incubation approximately equally during the day, whereas

TABLE 5.6
Parental care by Charadrius clade plovers.

Common name[a]	Incubation duties[b] diurnal nocturnal	Brood care duties[c]	Other behaviors[d]	Sources in addition to HBW Alive (2017)
S. Red-breasted Dotterel	F+M M	MF		
N. Red-breasted Dotterel	F+M M	MF	MP(d–m)	
Lesser Sand-Plover	F+M or M[e] M[e]	Mf or M[e]		Kruckenberg et al. (2001)
Greater Sand-Plover	F+M	MF[f]	C	Johnsgard (1981)
Caspian Plover	F+M F	MF	P(months)	Johnsgard (1981)
Collared Plover	F+M	MF[g]		Ruiz-Guerra and Cifuentes-Sarmiento (2013)
Two-banded Plover	F >> M M	MF		St. Clair (2010) St. Clair et al. (2010b), Vincze et al. (2016)
Double-banded Plover	F+M M?	MF	P(2 w)	
Kittlitz's Plover	F≈M M	MF	BCP(2–3 m)	Vincze et al. (2016)
Red-capped Plover	F >> M M	MF	B	Rogers and Eades (1997), Ekanayake et al. (2015), Vincze et al. (2016)
Malay Plover	F>M	MF		Yasué and Dearden (2008a), Vincze et al. (2016)
Kentish Plover	F >> M M >> F	Mf(~6 d) (Fm or MF)	BC	Vincze et al. (2013, 2016)
Snowy Plover	F >> M M >> F	Mf(6 d) (Fm or MF)	ABP(≤18 d)	Page et al. (2009), Vincze et al. (2013, 2016)
Javan Plover		MF[g]		Taufiqurrahman and Subekti (2013)
Wilson's Plover	F>M M>F	MF	AB	Vincze et al. (2016)
Common Ringed Plover	F+M M+F	MF	A	
Semipalmated Plover	F>M M>F	Mf(15 d)		
Long-billed Plover	F+M	MF		Uchida (2007)
Piping Plover	F≈M	MF or Mf (1–17 d)		Elliott-Smith and Haig (2004), Vincze et al. (2016)
Madagascar Plover	F>M	MF		Vincze et al. (2016)
Little Ringed Plover	F+M	Mf	AP(8–25 d)	
African Three-banded Plover	F+M	MF	BP(~10 d)	Tyler (1978)
Madagascar Three-banded Plover	F+M?	MF		
Forbes's Plover	F+M	MF		
White-fronted Plover	F>M M>F	MF	BCP(2–3 m)	Vincze et al. (2016)
Chestnut-banded Plover	F>M M	MF	B	
Killdeer	F≈M M	MF or Mf	BP(~10 d)	Jackson and Jackson (2000)
Mountain Plover	M or M≈F	M or MF[f]		
Oriental Plover	F (n = 1)	FM[h]		Ozerskaya and Zabelin (2006)

(Continued)

TABLE 5.6 (*Continued*)

Parental care by Charadrius clade plovers.

Common name[a]	Incubation duties[b] diurnal nocturnal	Brood care duties[c]	Other behaviors[d]	Sources in addition to HBW Alive (2017)
Eurasian Dotterel	M >> F M >> F	Mf	MP(3 m?)	
St. Helena Plover	F≈M M	MF	C	Pitman (1965), Székely and Reynolds (1995), Burns (2011), Vincze et al. (2016)
Rufous-chested Dotterel	M>F F >> M	MF		Johnsgard (1981), St. Clair et al. (2010a), Vincze et al. (2016)
Red-kneed Dotterel	F+M	MF		Johnsgard (1981)
Hooded Plover	F+M	MF		
Shore Plover	F+M M	Fm	P(4–36 d)	
Black-fronted Dotterel	F+M	MF	BP(≤8 w)	
Inland Dotterel	F>M	MF	C	
Wrybill	F>M	MF		

[a] No data found for Puna or White-faced plovers.

[b] Relative contribution of sexes to incubation during daylight and at night, + indicates both sexes incubate but no information on relative contribution, ≈ indicates approximately equal contribution, lower case indicates greater than, >> indicates much greater than.

[c] Abbreviations for sex(es) providing primary care upper case, for sex providing secondary care (deserting or much less frequently) in lower case. For species in which one sex deserts, the day of brood case by which desertion typically occurs also indicated. Approximately equally shared care indicated by both sexes in upper case. (Second or third strategies, if much less common, are indicated in parentheses.)

[d] Other parental care behaviors recorded include (A) adoption, (B) belly-soaking and/or shading eggs, shading chicks, (C) covering eggs with substrate or dried vegetation, (P) chicks staying with parent past the defined fledging period (for d, days; w, weeks; or m, months).

[e] Females of some populations may desert after laying.

[f] Male and female often split brood.

[g] Single reported observation.

[h] Authors suggest further work needed to understand possibly contradictory evidence observed by other researchers.

females of 13 species and males of only the Rufous-chested Dotterel take the majority of daytime incubation (Table 5.6). For 18 other species, both females and males incubate during the day but the relative contributions of each sex have not been reported. For the sole Oriental Plover clutch studied, only the female incubated during the day (Ozerskaya and Zabelin 2006). Among shorebird species in which females and males share diurnal incubation, the duration of individual incubation sessions tends to be short (median <2 h for five of six Charadrius plovers examined) compared with other species (which are highly variable with median incubation session lengths from 1–19 h; Bulla et al. 2016). As an exception, Two-banded Plover females assume a much greater share of diurnal incubation than males, but this is also true for the Kentish and Snowy plovers, which each exhibit short incubation sessions by members of the pair (Bulla et al. 2016; Table 5.5). For all shorebirds examined, Bulla et al. (2016) posit that the danger of predation rather than energetic needs may determine the duration of individual incubation sessions. The male share of daytime incubation—as well as total daytime incubation by both sexes—has been found to increase under conditions of high ambient temperatures in Kentish and Snowy plovers, two species with female-biased diurnal incubation patterns (Amat and Masero 2004a, Vincze et al. 2013, AlRashidi 2016, AlRashidi et al. 2010, 2011a). For these and an additional ten Charadrius species, the male (female for Rufous-chested Dotterel) share of daytime incubation increased with both higher average and more variable between-year ambient temperatures (Vincze et al. 2016). In this clade, the males are generally more brightly ornamented than the females, which may increase their conspicuousness to visually hunting diurnal predators, and the vulnerability of the clutch when they are incubating, as has been shown for Red-capped Plovers (Ekanayake et al. 2015). Thus, one of the consequences of increasing global temperatures in the future may be increased incubation participation by males and higher rates of clutch loss at some sites.

In contrast to diurnal patterns, males of 17 of 19 species for which there were data perform all or the majority of nighttime incubation (Table 5.6). Of the two exceptions, males of the Caspian Plover perform aerial song flight displays after dark on moonlit nights, which may be the reason for this incubation pattern (Wiersma et al. 2016b). Both males and females of the Rufous-chested Dotterel incubate both day and night, but females take the majority of nocturnal and males the majority of diurnal incubation, in stark contrast to what has been documented for other plovers (Table 5.6).

Incubation behaviors of some species protect eggs from high temperatures and/or predators. High ambient temperatures threaten the viability of eggs and also likely add physiological stress for incubating birds nesting in exposed sites (Purdue 1976, Amat and Masero 2004a, b, AlRashidi et al. 2010). Belly soaking and shading eggs are two common behaviors employed by plovers to regulate their body and egg temperatures in hot environments. Both behaviors have been reported for south subtropical-breeding Kittlitz's, Red-capped, African Three-banded, White-fronted, and Chestnut-banded plovers and Black-fronted Dotterels and north subtropical-breeding Kentish, Snowy, and Wilson's plovers and Killdeer (Table 5.6). In three hot environments, Kentish Plovers place a portion of their nests wholly or partially under vegetation: 9% at Saudi Arabian evaporation lagoons (AlRashidi 2016), 30% at a Spanish saline lake (Amat and Masero 2004b), and 65% on Red Sea islands (AlRashidi et al. 2011a). In Spain, the 70% of pairs that nest in the open use belly soaking and shading to protect their clutches from the heat (Amat and Masero 2004b). Additionally, the male contribution to daytime incubation at exposed, but not covered, nests is greater while the duration of incubation bouts by females decreases with temperature; thus during periods of hot weather, Kentish Plover pairs coordinated their care to provide physiologically stressed females relief from incubation without subjecting the eggs to lethal levels of heat (Amat and Masero 2004b, AlRashidi et al. 2011a). These authors also found that clutch desertion was affected by hot weather, with uncovered nests in close proximity to water (necessary to allow short absences from the nest to belly soak) deserted less frequently than nests at greater distances. Because exposed nests are less vulnerable to ground predators than covered nests, these findings all suggest that behavioral responses to hot environments are important components of plover breeding biology. Some species also cover or bury their eggs, mostly by kicking substrate and debris quickly around the eggs just before leaving the nest, a behavior believed to provide eggs both thermoregulation and camouflage from predators (Summers and Hockey 1981, Amat et al. 2012, Table 5.6).

If fitness consequences result from the manner in which male and female plovers share incubation duties, we would expect observed patterns to be adaptive to important environmental conditions. Predator avoidance, ambient temperature variability, and foraging constraints have been shown or suggested as factors that determine diel and other female–male incubation patterns in a few plover species (Thibault and McNeil 1995, Amat and Masero 2004a, Vicenze et al. 2013, 2016, Ekanayake et al. 2015, Bulla et al. 2016). Sexual selection also may play a role in incubation patterns if, for example, there are asymmetric opportunities for additional pairing as has been reported for the serially polyandrous Kentish and Snowy plovers (Warriner et al. 1986, Székely et al. 1999; Chapter Four, this volume). Nevertheless, the source of the unusual patterns for the Rufous-chested Dotterel is not clear (St. Clair et al. 2010a). Incubation appears to be flexible in the clade, a characteristic demonstrated most clearly under varying temperature regimes, but other potential sources of variability such as habitat type or time within season have not been explored (Amat and Masero 2004a, Vicenze et al. 2013, 2016).

Brood care is most frequently performed by both parents, but in some cases, the male's contribution is considerably greater than the female's (Table 5.6). The greater investment of males in brood care for some species arises primarily from female desertion either after laying eggs or soon after the eggs hatch. Among species with biparental care, the relative attendance of broods by each sex is not usually described. However, Davis (1994) reported female Shore Plovers tending broods in 83%, but males in only 62% of >1,200 brood observations. Thus, Shore Plovers provide biparental care to their broods, but effort is biased toward females, in contrast to most other species with sex-skewed brood care (Table 5.6).

In 28 (74%) species, both parents care for chicks throughout the fledging period, including the Javan Plover, for which biparental care has been documented through only a few opportunistic observations (Table 5.6). In three additional species, Piping and Little Ringed plovers and Killdeer, biparental care of chicks is the rule, although females occasionally desert broods prior to fledging. Regular female desertion is responsible for male-biased brood care for seven species (Table 5.6). For Mountain Plover and Eurasian Dotterel, female desertion occurs mostly after laying and is associated with their "rapid multiple clutch" polyandrous breeding system (Chapter Four, this volume). Similarly, Lesser Sand-Plover females sometimes desert mates immediately after clutch completion but also at hatch or during the chick-rearing period; the relative frequency of various strategies is mostly unstudied for this species (but see Kruckenberg et al. 2001). Kentish and Snowy plover females most commonly desert by the first week of brood care to initiate a new attempt with another mate, but less commonly they raise the brood with the male or, rarely, the female raises the chicks alone. In the Snowy Plover, cases of females staying with the brood most often occur at the end of the breeding season, when there is no opportunity for renesting; in some cases of female-only brood care, we suspected the male's death (G. Page et al. unpublished data). Semipalmated Plover females desert the male when the chicks are ~15 days old and leave the breeding grounds, a pattern common to monogamous arctic nesting sandpipers (Jönsson and Alerstam 1990, Székely and Reynolds 1995).

Of the 38 species for which data are available on both incubation and brood care, 29 (76%) species exhibit biparental care from egg laying to fledging, including 12 species nesting only in arctic or temperate latitudes and 17 species nesting wholly or partly in the Tropics (Table 5.6). In contrast, six arctic to north subtropical nesting species employ male-biased care whereas no species nesting in tropical to south temperate latitudes do so (Table 5.6). Transitional species, for which care is primarily biparental but occasionally male biased, include three north temperate breeders, two of whose ranges extend into the Tropics (Table 5.6).

Plovers exhibit other brood care behaviors, some of which are possible to discern only with marked individuals. Snowy and Wilson's plovers occasionally adopt young from other conspecific broods (Bergstrom 1988, Page et al. 2009). Such adoptions add to the incidences of extra-pair parentage in chicks above that arise from extra-pair eggs laid in the nest (Blomqvist et al. 2002, Küpper et al. 2004). Interactions among Common Ringed and Little Ringed plovers at feeding locations near areas of high-nesting density sometimes lead to interspecific adoptions (Piersma and Wiersma 1996). In at least ten species nesting at all latitudes young remain with parents beyond the time they fledge, but particularly those breeding in the South Subtropics (Table 5.6). The durations of social contact among family members range from up to 10 days for African Three-banded Plover and Killdeer to 14–32 days for Double-banded, Little Ringed, and Shore plovers. However, families of Black-fronted, Northern Red-breasted, and Eurasian dotterels and Kittlitz's and White-fronted plovers sometimes remain together for 2 or more months (Table 5.6).

CONSERVATION IMPLICATIONS AND FUTURE DIRECTIONS

Most conservation efforts for plovers focus on the breeding season and target reproductive success (Chapters Seven and Eleven, this volume). Therefore, management and restoration efforts rely extensively on knowledge of breeding biology, reproductive success, habitat use, anthropogenic disturbance, and predation to produce positive demographic outcomes. The increased pace of climate change, coupled with other anthropogenic stressors, will challenge not only managers to implement efforts at enhancing reproductive success, but also the species themselves and their ability to adapt to novel breeding season conditions. Additionally, any carryover costs of reproduction to adult physiological condition or survival that plovers might currently incur could intensify with environmental change, although, to date, this topic is little studied among Charadriiformes (Reid 1987, Magnhagen 1991, Monaghan and Nager 1997, Hanssen et al. 2005, Santos and Nakagawa 2012). Identifying breeding biology traits that display plasticity currently among locations with varying conditions could provide insights into the plovers' potential responses to climate change and other environmental challenges, a space-for-time approach to prediction (Pickett 1989). These insights may suggest restoration and management actions to promote their conservation. For example, Amat and Masero (2004a,b) showed that Kentish Plovers breeding in a hot environment varied their incubation behaviors according to their body condition, and that some of these behaviors increased their rates of clutch depredation or desertion. These findings suggest that, in similar hot environments where predator management is necessary, restoration that provides water and scattered cover might improve nesting success.

Although some breeding biology parameters, such as clutch size, are fairly fixed for these species, there may be more variation within species in other parameters than is currently recognized. Coordinated studies of breeding biology of species occupying large geographic regions, either broad latitudinal gradients or inland versus coastal portions of their range, could better identify this variation (e.g., Vincze et al. 2013, 2016, Bulla et al. 2016). For example, Semipalmated Plovers rarely renest after failure at 58.77°N in Churchill, Manitoba, but do so commonly at 53°N on Akimiski Island 1,024 km to the southeast, suggesting that warming climate could affect renesting opportunities for some species (Lishman et al. 2010). Although aspects of most species' breeding biology have been studied in at least one location, the large breeding ranges for many species offer further opportunities to explore within-species variability and behavioral plasticity (Chapter Nine, this volume). Two of the most studied species, Kentish and Snowy plovers, are least known in the tropical and subtropical portions of their range, which is, not surprisingly, a region with many other less studied species.

Opportunities to increase our knowledge of plover breeding biology are greatest in the Tropics, but also in North Temperate Asia and South Temperate South America. Because the majority of the 40 plover species breed away from the coast, and breeding ranges of half of them extend into the Tropics (Table 5.1), further studies in these regions would prove valuable.

Some facets of breeding biology pose methodological challenges for study and are not well documented in the literature, even among well-studied species. Territorial use is important in determining the amount of space species require, both for nesting and feeding during incubation and brood rearing, and the carrying capacity of habitat, but it is not documented for many species. Studies often emphasize incubation because the fixed locations of nests facilitate observation, but the success of a breeding effort is ultimately determined by what happens during the chick-rearing period. For species that do not necessarily rear their chicks in territories defended during the incubation period, the increased space necessary for successfully fledging young is an important aspect of their breeding. Radio tracking broods of young or caring parents provides valuable insight into this facet of breeding biology, although broods in areas of low population density often can be tracked by following them directly (Knopf and Rupert 1996, Ruhlen et al. 2003, Wilson and Colwell 2010).

Climate change poses a real threat to plovers worldwide, although the consequences of environmental perturbations are difficult to predict for species with broad geographic distributions (Chapters Three and Nine, this volume). In the Tropics, potential disruption to breeding due to changed precipitation patterns may alter the time or effort that some plovers spend on reproduction. To appreciate the limitations of potential adaptation and to adopt the most effective conservation actions, we believe that detailed studies strongly tied to conservation needs of tropical species will be valuable.

Knowledge of breeding biology is essential to correctly interpret evidence for courting, pairing, laying, incubating, and brood-caring behaviors and forms a basis for research and experimentation into other aspects of life history. Equally important, each species' breeding biology typically tells a natural history story that may inspire the interest and appreciation of the public, thus aiding the conservation and scientific communities in promoting their protection.

ACKNOWLEDGMENTS

We owe an enormous debt to Jane and John Warriner who, starting in 1977, worked tirelessly on Snowy Plovers, shared their experience and knowledge with us, helped grow a small breeding biology study into a major research project, and inspired many people to study and conserve plovers. We thank volume editors Mark Colwell and Susan Haig, reviewers Francesca Cuthbert and Robert Gill, and Douglas George and

R. William Stein, who made many helpful suggestions to early versions of this chapter. We are grateful to our colleagues for their hard work and insights, and for their support during the writing of this review: Frances Bidstrup, Dave Dixon, Jennifer Erbes, Carleton Eyster, Tom Gardali, Doug George, Catherine Hickey, Kriss Neuman, Bernadette Ramer, and R. William Stein. Elizabeth Feucht kindly provided us with hard-to-find literature. This is contribution number 2136 of Point Blue Conservation Science.

LITERATURE CITED

AlRashidi, M. 2016. Breeding biology of the Kentish Plover *Charadrius alexandrinus* in the Sabkhat Al-Fasl Lagoons, Saudi Arabia. *Zoology in the Middle East.*

AlRashidi, M., A. Kosztolanyi, C. Küpper, I. C. Cuthill, S. Javid, and T. Székely. [online]. 2010. The influence of a hot environment on parental cooperation of a ground-nesting shorebird, the Kentish plover *Charadrius alexandrinus. Frontiers in Zoology* 7:1.

AlRashidi, M., A. Kosztolányi, M. Shobrak, C. Küpper, and T. Székely. 2011a. Parental cooperation in an extreme hot environment: Natural behaviour and experimental evidence. *Animal Behaviour* 82:235–243.

AlRashidi, M., A. Kosztolanyi, M. Shobrak, and T. Székely. 2011b. Breeding ecology of the Kentish Plover, Charadrius alexandrinus, in the Farasan Islands, Saudi Arabia. *Zoology in the Middle East* 53:15–24.

Amat, J. A. 1998. Mixed clutches in shorebird nests: Why are they so uncommon? *Wader Study Group Bulletin* 85:55–59.

Amat, J. A., R. M. Fraga, and G. M. Arroyo. 1999. Replacement clutches by Kentish Plovers. *Condor* 101:746–751.

Amat, J. A., R. M. Fraga, and G. M. Arroyo. 2001a. Intraclutch egg-size variation and offspring survival in the Kentish Plover *Charadrius alexandrinus. Ibis* 143:17–23.

Amat, J. A., R. M. Fraga, and G. M. Arroyo. 2001b. Variations in body condition and egg characteristics of female Kentish Plover Charadrius alexandrinus. *Ardea* 89:293–299.

Amat, J. A., and J. A. Masero. 2004a. How Kentish plovers *Charadrius alexandrinus* cope with heat stress during incubation. *Behavioral Ecology and Sociobiology* 56:26–33.

Amat, J. A., and J. A. Masero. 2004b. Predation risk on incubating adults constrains the choice of thermally favourable nest sites in a plover. *Animal Behaviour* 67:293–300.

Amat, J. A., R. Monsa, and J. A. Masero. 2012. Dual function of egg-covering in the Kentish Plover *Charadrius alexandrinus. Behaviour* 149:881–895.

Armitage, I. [online]. 2013. Black-fronted dotterel. In C. M. Miskelly (editor), *New Zealand Birds Online.* <www.nzbirdsonline.org.nz> (10 October 2016).

Armstrong, A. R., and E. Nol. 1993. Spacing behavior and reproductive ecology of the Semipalmated Plover at Churchill, Manitoba. *Wilson Bulletin* 105:455–464.

Baird, B., and P. Dann. 2003. The breeding biology of Hooded Plovers, *Thinornis rubricollis*, on Phillip Island, Victoria. *Emu* 103:323–328.

Bergstrom, P. W. 1982. Ecology of incubation in Wilson's Plover (*Charadrius wilsonius*). *Ph.D. dissertation*, University of Chicago, Chicago, IL.

Bergstrom, P. W. 1988. Breeding biology of Wilson's Plovers. *Wilson Bulletin* 100:25–35.

Blomqvist, D., M. Andersson, C. Küpper, I. C. Cuthill, J. Kis, R. B. Lanctot, B. K. Sandercock, T. Székely, J. Wallander, and B. Kempenaers. 2002. Genetic similarity between mates and extra-pair parentage in three species of shorebirds. *Nature* 419:613–615.

Blomqvist, D., O. C. Johansson, and F. Götmark. 1997. Parental quality and egg size affect chick survival in a precocial bird, the Lapwing *Vanellus vanellus. Oecologia* 110:18–24.

Brown, L. H. 1948. Notes on birds of the Kabba, Ilorin and N. Benin provinces of Nigeria. *Ibis* 90:525–538.

Buick, A. M., and D. C. Paton. 1989. Impact of off-road vehicles on the nesting success of Hooded Plovers *Charadrius rubricollis* in the Coorong Region of South Australia. *Emu* 89:159–172.

Bulla, M., Valcu, M., Dokter, A. M., Dondua, A. G., Kosztolányi, A., Rutten, A. L., Helm, B., Sandercock, B. K., Casler, B., Ens, B. J., Spiegel, C. S., Hassell, C. J., Küpper, C., Minton, C., Burgas, D., Lank, D. B., Payer, D. C., Loktionov, E. Y., Nol, E., Kwon, E., Smith, F., Gates, H. R., Vitnerová, H., Prüter, H., Johnson, J. A., St Clair, J. J. H., Lamarre, J.-F., Rausch, J., Reneerkens, J., Conklin, J. R., Burger, J., Liebezeit, J., Bêty, J., Coleman, J. T., Figuerola, J., Hooijmeijer, J. C. E. W., Alves, J. A., Smith, J. A. M., Weidinger, K., Koivula, K., Gosbell, K., Exo, K.-M., Niles, L., Koloski, L., McKinnon, L., Praus, L., Klaassen, M., Giroux, M.-A., Sládeček, M., Boldenow, M. L., Goldstein, M. I., Šálek, M., Senner, N., Rönkä, N., Lecomte, N., Gilg, O., Vincze, O., Johnson, O. W., Smith, P. A., Woodard, P. F., Tomkovich, P. S., Battley, P. F., Bentzen, R., Lanctot, R. B., Porter, R., Saalfeld, S. T.,

Freeman, S., Brown, S. C., Yezerinac, S., Székely, T., Montalvo, T., Piersma, T., Loverti, V., Pakanen, V.-M., Tijsen, W., and Kempenaers, B. 2016. Unexpected diversity in socially synchronized rhythms of shorebirds. *Nature* 540:109–113.

Burger, J. 1987. Physical and social determinants of nest-site selection in Piping Plover in New Jersey. *Condor* 89:811–818.

Burns, F. E. 2011. Conservation biology of the endangered St. Helena Plover *Charadrius sanctaehelenae*. Ph.D. dissertation, University of Bath, Bath, UK.

Cairns, W. E. 1982. Biology and behavior of breeding Piping Plovers. *Wilson Bulletin* 94:531–545.

Christians, J. K. 2002. Avian egg size: Variation within species and inflexibility within individuals. *Biological Review* 77:1–26.

Clark, A. 1982. Some observations on the breeding behaviour of Kittlitz's Sandplover. *Ostrich* 53:120–122.

Claassen, A. H., Arnold, T. W., Roche, E. A., Saunders, S. P., and Cuthbert, F. J. 2014. Factors influencing nest survival and renesting by Piping Plovers in the Great Lakes region. *Condor* 116:394–407.

Cody, M. 1966. A general theory of clutch size. *Evolution* 20:174–184.

Cody, M. 1971. Ecological aspects of reproduction. pp. 461–512 In D. S. Farner and J. R. King (editors), *Avian Biology* (vol. 1). Academic Press, New York.

Cohen, J. B., Houghton, L. M., and Fraser, J. D. 2009. Nesting density and reproductive success of Piping Plovers in response to storm- and human-created habitat changes. *Wildlife Monographs* 173:1–24.

Colwell, M. A. 2006. Egg-laying intervals in shorebirds. *Wader Study Group Bulletin* 111:50–59.

Colwell, M. A., E. J. Feucht, S. E. McAllister, and A. N. Transou. 2017. Lessons learned from the oldest Snowy Plover. *Wader Study* 124:157–159.

Corbat, C. A., and P. W. Bergstrom. [online]. 2000. Wilson's Plover (*Charadrius wilsonia*). In P. G. Rodewald (editor), *The Birds of North America Online, No. 516*. Cornell Lab of Ornithology, Ithaca, NY. <http://birdsna.org/Species-Account/bna/species/wilplo> (15 July 2016).

Cramp, S., and Simmons K. E. L. 1983. *Handbook of the Birds of Europe, the Middle East and North Africa*. Waders to Gulls (vol. 3). Oxford University Press, Oxford, UK.

Davis, A. M. 1994. Breeding biology of the New Zealand Shore Plover *Thinornis novaeseelandiae*. *Notornis (Supplement)* 41:195–208.

del Hoyo, J., N. Collar, G. M. Kirwan, and P. Boesman. 2016. Northern Red-breasted Plover (*Charadrius aquilonius*). pp. 424–245 In J. del Hoyo, A. Elliott, J. Sargatal, D. A. Christie, and J. de Juana (editors), *Handbook of the Birds of the World Alive*. Lynx Edicions, Barcelona, Spain.

Dement'ev, G. P., and N. A. Gladkov. 1969. *Birds of the Soviet Union* (vol. 5). Israel Program for Scientific Translations, Jerusalem, Israel.

Dowding, J. E. [online]. 2013. Wrybill. In C. M. Miskelly (editor). *New Zealand Birds Online*. <www.nzbirdsonline.org.nz> (10 October 2016).

Drent, R. 1975. Incubation. pp. 333–420 In D. S. Farner and J. R. King (editors), *Avian Biology* (vol. 5). Academic Press, New York, NY.

Drent, R. H., and S. Daan. 1980. The prudent parent: Energetic adjustments in avian breeding. *Ardea* 68:225–252.

Ekanayake, K. B., M. A. Weston, D. G. Nimmo, G. S. Maguire, J. A. Endler, and C. Küpper. 2015. The bright incubate at night: Sexual dichromatism and adaptive incubation division in an open-nesting shorebird. *Proceedings of the Royal Society B* 282:20143028.

Elliott-Smith, E., and S. M. Haig. 2004. Piping Plover (*Charadrius melodus*), in A. Poole (editor), *The Birds of North America Online, No. 2*. Cornell Lab of Ornithology, Ithaca, NY.

Fraga, R. M., and J. A. Amat. 1996. Breeding biology of a Kentish Plover (*Charadrius alexandrinus*) population in an inland saline lake. *Ardeola* 43:69–85.

Flynn, L., E. Nol, and Y. Zharikov. 1999. Philopatry, nest-site tenacity, and mate fidelity of Semipalmated Plovers. *Journal of Avian Biology* 30:47–55.

Galbraith, H. 1988. Effects of egg-size and composition on the size, quality and survival of Lapwing *Vanellus vanellus* chicks. *Journal of Zoology* 214:383–393.

Garcia-Pena, G. E. 2010. Phylogenetic comparative analyses of breeding systems and life-history strategies in shorebirds. Ph.D. dissertation, University of Bath, Bath, UK.

Gochfeld, M. 1984. Antipredator behavior: Aggressive and distraction displays of shorebirds. pp. 289–377 In J. Burger and B. L. Olla (editors), *Shorebirds Breeding Behavior and Populations, Behavior of Marine Animals* (vol. 5). Plenum Press, New York.

Grant, G. S. 1982. Avian incubation: Egg temperature, nest humidity, and behavioral thermoregulation in a hot environment. *Ornithological Monographs* 30:1–75.

Grant, M. C. 1991. Relationship between egg-size, chick size at hatching, and chick survival in the Whimbrel *Numenius phaeopus*. *Ibis* 133:127–133.

Graul, W. D. 1973. Adaptive aspects of the Mountain Plover social system. *Living Bird* 12:69–94.

Haig, S. M., and L. W. Oring. 1988. Mate, site and territory fidelity in Piping Plovers. *Auk* 105:268–277.

Hanssen, S. A., D. Hasselquist, I. Folstad, and K. E. Erikstad. 2005. Cost of reproduction in a long-lived bird: Incubation effort reduces immune function and future reproduction. *Proceedings of the Royal Society B*. 272:1039–1046.

Hobson, K. A., and J. R. Jehl, Jr. 2010. Arctic waders and the capital-income continuum: Further tests using isotopic contrasts of egg components. *Journal of Avian Biology* 41:565–572.

Högstedt, G. 1974. Length of the pre-laying interval in the Lapwing *Vanellus vanellus* L. in relation to its food resources. *Ornis Scandinavica* 5:1–4.

Jackson, B. J., and J. A. Jackson. [online]. 2000. Killdeer (*Charadrius vociferus*). In P. D. Rodewald (editor), *The Birds of North America, No. 517*. Cornell Lab of Ornithology, Ithaca, NY. https://birdsna-org.bnaproxy.birds.cornell.edu/Species-Account/bna/species/killde/intoduction (10 August 2016).

Jönsson, P. E., and T. Alerstam. 1990. The adaptive significance of parental role division and sexual size dimorphism in breeding shorebirds. *Biological Journal of the Linnean Society* 41:301–314.

Johnsgard, P. A. 1981. *The Plovers, Sandpipers and Snipes of the World*. University of Nebraska Press, Lincoln, NE.

Kålås, J. A. 1986. Incubation schedules in different parental care systems in the Dotterel *Charadrius morinellus*. *Ardea* 74:185–190.

Kålås, J. A., and I. Byrkjedal. 1984. Breeding chronology and mating system of the Eurasian Dotterel (*Charadrius morinellus*). *Auk* 101:838–847.

Kålås, J. A., and L. Løfaldli. 1987. Clutch size in the Dotterel (*Charadrius morinellus*): An adaptation to parental incubation behaviour? *Ornis Scandinavica* 18:316–319.

Katayama, N., T. Amano, and S. Ohori. 2010. The effects of gravel bar construction on breeding Long-billed Plovers. *Waterbirds* 33:162–168.

Klaassen, M., Å. Lindström, H. Meltofte, and T. Piersma. 2001. Arctic waders are not capital breeders. *Nature* 413:794.

Klomp, H. 1970. The determination of clutch-size in birds a review. *Ardea* 58:1–124.

Knopf, F. L., and J. R. Rupert. 1996. Reproduction and movements of Mountain Plovers breeding in Colorado. *Wilson Bulletin* 108:28–35.

Knopf, F. L., and M. B. Wunder. [online]. 2006. Mountain Plover (*Charadrius montanus*), in P. D. Rodewald (editor), *The Birds of North America Online, No. 211*. Cornell Lab of Ornithology, Ithaca, NY. <http://bna.birds.cornell.edu.bnaproxy.birds.cornell.edu/bna/species/211> (10 October 2016).

Kolomiytsev, N. P., and N. Y. Poddubnaya. 2014. Breeding biology of the Long-billed Plover *Charadrius placidus* in the northern part of its range in the Russian Far East. *Wader Study Group Bulletin* 121:181–185.

Kosztolányi, A., S. Javed, C. Küpper, I. C. Cuthill, A. Al Shamsi, and T. Székely. 2009. Breeding ecology of Kentish Plover Charadrius alexandrinus in an extremely hot environment. *Bird Study* 56:244–252.

Kruckenberg, H., A. Degen, and A. Hergenhahn. 2001. Only the males work: Breeding behavior of the Mongolian Plover *Charadrius mongolus* on the northern coast of the Sea of Okhotsky, Russia. pp. 182–186 In A. V. Andree and H. Bergmann (editors), *Biodiversity and Ecological Status along the Northern Coast of the Sea of Okhotsky Russia*. Russian Academy of Science, Far Eastern Branch, Magaden, Russia.

Küpper, C., J. Kis, A. Kosztolányi, T. Székely, I. C. Cuthill, and D. Blomqvist. 2004. Genetic mating system and timing of extra-pair fertilizations in the Kentish Plover. *Behavioral Ecology and Sociobiology* 57:32–39.

Lack, D. 1947. The significance of clutch size. *Ibis* 89:302–352.

Lank, D. B., L. W. Oring, and S. J. Maxson. 1985. Mate and nutrient limitation of egg-laying in a polyandrous shorebird. *Ecology* 66:1513–1524.

Lendvai, Á. Z., A. Liker, and J. Kis. 2004. Male badge size is related to clutch volume in the Kentish Plover. *Ornis Hungarica* 14:1–7.

Lengyel, S., B. Kiss, and C. R. Tracy. 2009. Clutch size determination in shorebirds: Revisiting incubation limitation in the pied avocet (*Recurvirostra avosetta*). *Journal of Animal Ecology* 78:396–405.

Lenington, S. 1980. Bi-parental care in Killdeer: An adaptive hypothesis. *Wilson Bulletin* 92:8–20.

Lenington, S. 1984. The evolution of polyandry in shorebirds. pp. 149–167 In J. Burger and B. L. Olla, (editors), *Shorebirds: Breeding Behavior and Populations*. Behavior of Marine Animals (vol. 5). Plenum Press, New York.

Lenington, S., and T. Mace. 1975. Mate fidelity and nesting site tenacity in the Killdeer. *Auk* 92:149–151.

Lishman, C., E. Nol, K. F. Abraham, and L. P. Nguyen. 2010. Behavioral responses to higher predation risk in a subarctic population of the Semipalmated Plover. *Condor* 112:499–506.

Lislevand, T., and G. H. Thomas. 2006. Limited male incubation ability and the evolution of egg size in shorebirds. *Biology Letters* 2:206–208.

Lloyd, P. 2008. Adult survival, dispersal and mate fidelity in the White-fronted Plover *Charadrius marginatus*. *Ibis* 150:182–187.

Lloyd, P., and É. E. Plagányi. 2002. Correcting observer effect bias in estimates of nesting success of a coastal bird, the White-fronted Plover *Charadrius marginatus*. *Bird Study* 49:124–130.

Lomas, S. C., D. A. Whisson, G. S. Maguire, L. X. Tan, P.-J. Guay, and M. A. Weston. 2014. The influence of cover on nesting red-capped plovers: A trade-off between thermoregulation and predation risk? *Victorian Naturalist* 131:115–127.

Lord, A., J. R. Waas, and J. Innes. 1997. Effects of human activity on the behaviour of northern New Zealand Dotterel *Charadrius obscurus aquilonius* chicks. *Biological Conservation* 82:15–20.

Maclean, G. L. 1972. Clutch size and evolution in the Charadrii. *Auk* 89:299–324.

Maclean, G. L. 1977. Comparative notes on black-fronted and red-kneed dotterels. *Emu* 77:199–207.

Magnhagen, C. 1991. Predation risk as a cost of reproduction. *Trends in Ecology and Evolution* 6:183–186.

Marchant, S., and P. J. Higgins. 1993. *Handbook of Australian, New Zealand and Antarctic Birds: Raptors to Lapwings* (vol. 2). Oxford University Press, Melbourne, VIC.

Maugeri, F. G. 2005. Primer registro de nidificación en ambiente fluvial del Gaviotín Chico Común (*Sterna superciliaris*) para la provincial de Buenos Aires y nueva evidencia de su nidificación asociada con el Chorlito de Collar (*Charadrius collaris*). *Ornitología Neotropical* 16:117–121 (in Spanish).

McCulloch, N. M. 1991. Status, habitat and conservation of the St Helena Wirebird *Charadrius sanctaehelenae*. *Bird Conservation International* 1:361–392.

Monaghan, P., and R. G. Nager. 1997. Why don't birds lay more eggs? *Trends in Ecology and Evolution* 12:270–274.

Morrison, R. I. G., and K. A. Hobson. 2004. Use of body stores in shorebirds after arrival on high-arctic breeding grounds. *Auk* 121:333–344.

Muir, J. J., and M. A. Colwell. 2010. Snowy Plovers select open habitats for courtship scrapes and nests. *Condor* 112:507–510.

Nethersole-Thompson, D. 1973. *The Dotterel*. Collins, London.

Nguyen, L. P., K. F. Abraham, and E. Nol. 2006. Influence of Arctic Terns on survival of artificial and natural Semipalmated Plover nests. *Waterbirds* 29:100–104.

Nol, E. 1986. Incubation period and foraging technique in shorebirds. *American Naturalist* 128:115–119.

Nol, E., and M. S. Blanken. [online]. 2014. Semipalmated Plover (*Charadrius semipalmatus*). In P. G. Rodewald (editor), *The Birds of North America, No. 444*. Cornell Lab of Ornithology, Ithaca, NY. <https://birdsna.org/Species-Account/bna/species/semplo> (8 August 2016).

Nol, E., M. S. Blanken, and L. Flynn. 1997. Sources of variation in clutch size, egg size and clutch completion dates of Semipalmated Plovers in Churchill, Manitoba. *Condor* 99:389–396.

Owens, I. P. F., T. Burke, and D. B. A. Thompson. 1994. Extraordinary sex roles in the Eurasian Dotterel: Female mating arenas, female-female competition, and female mate choice. *American Naturalist* 144:76–100.

Ozerskaya, T., and V. Zabelin. 2006. Breeding of the Oriental Plover *Charadrius veredus* in southern Tuva, Russia. *Wader Study Group Bulletin* 110:36–42.

Page, G. W., L. E. Stenzel, and C. A. Ribic. 1985. Nest site selection and clutch predation in the Snowy Plover. *Auk* 102:347–353.

Page, G. W., L. E. Stenzel, J. S. Warriner, J. C. Warriner, and P. W. Paton. 2009. Snowy Plover (*Charadrius alexandrinus*). In P. G. Rodewald (editor), *The Birds of North America Online, No. 154*. Cornell Lab of Ornithology, Ithaca, NY. <https://birdsna.org/Species-Account/bna/species/snoplo5> (31 July 2016).

Patrick, A. M. 2013. Birds of a feather flock (somewhat) together: Semi-colonial nesting in the Snowy Plover. *M.S. Thesis*, Humboldt State University, Arcata, CA.

Patten, M. A. 2007. Geographic variation in calcium and clutch size. *Journal of Avian Biology* 38:637–643.

Pearson, S. F., S. M. Knapp, and C. Sundstrom. [online]. 2014. Evaluating the ecological and behavioural factors influencing Snowy Plover *Charadrius nivosus* egg hatching and the potential benefits of predator exclosures. *Bird Conservation International*, doi:10:1017S0959270914000331.

Pearson, W. J., and M. A. Colwell. 2014. Effects of nest success and mate fidelity on breeding dispersal in a population of Snowy Plovers *Charadrius nivosus*. *Bird Conservation International* 24:342–353.

Phillips, R. E. 1980. Behaviour and systematics of New Zealand plovers. *Emu* 80:177–197.

Pickett, S. T. A. 1989. Space-to-time substitution as an alternative to long-term studies. pp. 110–135 In G. E. Likens (editor), *Long-term Studies in Ecology Approaches and Alternatives*. Springer-Verlag, New York.

Pienkowski, M. W. 1984. Breeding biology and population dynamics of Ringed Plovers *Charadrius hiaticula* in Britain and Greenland: Nest-predation as a possible factor limiting distribution and timing of breeding. *Journal of Zoology* 202:83–114.

Piersma, T., and P. Wiersma. 1996. Family Charadriidae (Plovers). Pp. 384–442 in J. del Hoyo, A. Elliott, and J. Sargatal (editors), *Handbook of the Birds of the World (vol. 3), Hoatzin to Auks*. Lynx Edicions, Barcelona, Spain.

Pierce, R. J. 1989. Breeding and social patterns of Banded Dotterels (*Charadrius bicinctus*) at Cass River. *Notornis* 36:13–23.

Pierce, R. J. [online] 2013. Banded dotterel. In C. M. Miskelly (editor), *New Zealand Birds Online*. <www.nzbirdsonline.org.nz> (10 October 2016).

Pitman, C. R. S. 1965. The eggs and nesting habits of the St. Helena Sand-Plover or Wirebird *Charadrius pecuarius sanctae-helenae* (Harting). *Bulletin of the British Ornithological Club* 85:121–129.

Powell, A. N. 2001. Habitat characteristics and nest success of Snowy Plovers associated with California Least Terns. *Condor* 103:783–792.

Purdue, J. R. 1976. Thermal environment of the nest and related parental behavior in Snowy Plovers, *Charadrius alexandrinus*. *Condor* 78:180–185.

Rahn, H., C. V. Paganelli, and A. Ar. 1975. Relation of avian egg weight to body weight. *Auk* 92:750–765.

Reid, W. V. 1987. The cost of reproduction in the glaucous-winged gull. *Oecologia* 74:458–467.

Ricklefs, R. E. 1984. Egg dimensions and neonatal mass of shorebirds. *Condor* 86:7–11.

Ricklefs, R. E. 2000. Lack, Skutch, and Moreau: The early development of life history thinking. *Condor* 102:3–8.

Rittinghaus, H. 1956. Untersuchungen am Seeregenpfeifer (*Charadrius alexandrinus* L.) auf der Insel Oldeoog. *Journal für Ornithologie* 97:117–155 (in German).

Robertson, H. A. [online]. 2013. Red-kneed dotterel. In C. M. Miskelly (editor), *New Zealand Birds Online*. <www.nzbirdsonline.org.nz/species/red-kneed-dotterel> (16 August 2016)

Rogers, D. I. R., and D. W. Eades. 1997. Belly-soaking and egg-cooling behaviour in a red-capped plover *Charadrius ruficapillus*. *Stilt* 30:53–54.

Ruiz-Guerra, C., and Y. Cifuentes-Sarmiento. 2013. Primeros registros de anidación del Chorlito Collarejo (*Charadrius collaris*) en Colombia. *Ornitología Colombiana* 13:37–43 (in Spanish).

Ruhlen, T. D., S. Abbott, L. E. Stenzel, and G. W. Page. 2003. Evidence that human disturbance reduces Snowy Plover chick survival. *Journal of Field Ornithology* 74:300–304.

St Clair, J. J. H. 2010. Plovers, invertebrates and invasive predators: Aspects of the ecology of some island populations. *Ph.D. dissertation*, University of Bath, Bath, UK.

St Clair, J. J. H., P. Herrmann, R. W. Woods, and T. Székely. 2010b. Female-biased incubation and strong diel sex-roles in the Two-banded Plover. *Journal of Ornithology* 151:811–816.

St Clair, J. J. H., C. Küpper, P. Herrmann, R. W. Woods, and, T. Székely. 2010a. Unusual incubation sex-roles in the Rufous-chested Dotterel *Charadrius modestus*. *Ibis* 152:402–404.

Santos, E. S. A., and S. Nakagawa. 2012. The costs of parental care: A meta-analysis of the trade-off between parental effort and survival in birds. *Journal of Evolutionary Biology* 25:1911–1917.

Saunders, S. P., E. A. Roche, T. W. Arnold, and F. J. Cuthbert. 2012. Female site familiarity increases fledging success in Piping Plovers (*Charadrius melodus*). *Auk* 129:329–337.

Simmons, K. E. L. 1956. Territory in the Little Ringed Plover *Charadrius dubius*. *Ibis* 98:390–397.

Skrade, P. D. B., and S. J. Dinsmore. 2013. Egg-size investment in a bird with uniparental incubation by both sexes. *Condor* 115:508–514.

Skutch, A. F. 1949. Do tropical birds rear as many young as they can nourish? *Ibis* 91:430–455.

Stamps, J. A. 1988. Conspecific attraction and aggregation in territorial species. *American Naturalist* 131:329–347.

Stephens, P. A., I. L. Boyd, J. M. McNamara, and A. I. Houston. 2009. Capital breeding and income breeding: Their meaning, measurement, and worth. *Ecology* 90:2057–2067.

Summers, R. W., and P. A. R. Hockey. 1980. Breeding biology of the White-fronted Plover (*Charadrius marginatus*) in the south-western cape, South Africa. *Journal of Natural History* 14:433–445.

Summers, R. W., and P. A. R. Hockey. 1981. Egg covering behaviour of the White-fronted Plover *Charadrius marginatus*. *Ornis Scandinavica* 12:240–243.

Székely, T. 1992. Reproduction of Kentish Plover *Charadrius alexandrinus* in grasslands and fish ponds: The habitat mal-assessment hypothesis. *Aquila* 99:59–68.

Székely, T., I. C. Cuthill, and J. Kis. 1999. Brood desertion in Kentish plover: Sex differences in remating opportunities. *Behavioral Ecology* 10:185–190.

Székely T., I. Karsai, and T. D. Williams. 1994. Determination of clutch-size in the Kentish Plover *Charadrius alexandrinus*. *Ibis* 136:341–348.

Székely T., A. Kosztolanyi, C. Küpper, G. H. Thomas. 2007. Sexual conflict over parental care: A case study of shorebirds. *Journal of Ornithology* 148:211–217.

Székely, T., and J. D. Reynolds. 1995. Evolutionary transitions in parental care in shorebirds. *Proceedings of the Royal Society B* 262:57–64.

Székely, T., J. D. Reynolds, and J. Figuerola. 2000. Sexual size dimorphism in shorebirds, gulls, and alcids: The influence of sexual and natural selection. *Evolution* 54:1404–1413.

Szentirmai, I., A. Kosztolányi, and T. Székely. 2001. Daily changes in body mass of incubating Kentish Plovers. *Ornis Hungarica* 11:27–32.

Szentirmai, I., and T. Székely. 2005. Diurnal variation in nest material use by the Kentish Plover *Charadrius alexandrinus*. *Ibis* 146:535–537.

Taufiqurrahman, I., and H. Subekti. 2013. Distraction behaviour of breeding Javan Plover *Charadrius javanicus*. *Kukila* 17:17–21.

Thibault, M., and R. McNeil. 1995. Day- and night-time parental investment by incubating Wilson's Plovers in a tropical environment. *Canadian Journal of Zoology* 73:879–886.

Tjørve, K. M. C., L. G. Underhill, and G. H. Visser. 2008. The energetic implications of precocial development for three shorebird species breeding in a warm environment. *Ibis* 150:125–138.

Tyler, S. 1978. Observations on the nesting of the Three-banded Plover (*Charadrius tricollaris*). *Scopus* 2:39–41.

Uchida, H. 2007. Long-billed Plover *Charadrius placidus*. *Bird Research News* 4:1–3.

Urban, E. K., C. H. Fry, and S. Keith (editors). 1986. *The Birds of Africa* (vol. 2). Academic Press, London.

Vincze, O., A. Kosztolányi, Z. Barta, C. Küpper, M. AlRashidi, J. A. Amat, A. A. Ticó, F. Burns, J. Cavitt, W. C. Conway, M. Cruz-López, A. E. Desucre-Medrano, N. dos Remedios, J. Figuerola, D. Galindo-Espinosa, G. E. García-Peña, S. G. Del Angel, C. Gratto-Trevor, P. Jönsson, P. Lloyd, T. Montalvo, J. E. Para, R. Pruner, P. Que, Y. Liu, S. T. Saalfeld, R. Schulz, L. Serra, J. J. H. St Clair, L. E. Stenzel, M. A. Weston, M. Yasué, S. Zefania, and T. Székely. 2016. Parental cooperation in a changing climate: Fluctuating environments predict shifts in care division. *Global Ecology and Biogeography*. doi:10.1111/geb.12540.

Vincze, O., T. Székely, C. Küpper, M. AlRashidi, J. A. Amat, A. A. Ticó, D. Burgas, T. Burke, J. Cavitt, J. Figuerola, M. Shobrak, T. Montalvo, A. Kosztolányi. 2013. Local environment but not genetic differentiation influences biparental care in ten plover populations. *PLoS One* 8:e60998.

Wallander J., and M. Andersson. 2003. Reproductive tactics of the ringed plover *Charadrius hiaticula*. *Journal of Avian Biology* 34:259–266.

Walters, J. 1957. Behaviour during egg-laying period in the Kentish plover (*Charadrius alexandrinus* L.) and the division of labour during the nesting cycle. *Ardea* 45:24–62 (in Dutch).

Warriner, J. S., J. C. Warriner, G. W. Page, and L. E. Stenzel. 1986. Mating system and reproductive success of a small population of polygamous Snowy Plovers. *Wilson Bulletin* 98:15–37.

Weston, M. A., and M. A. Elgar. 2005. Disturbance to brood-rearing Hooded Plover *Thinornis rubricollis*: Responses and consequences. *Bird Conservation International* 15:193–209.

Wiersma, P., G. M. Kirwan, and P. Boesman. [online]. 2016a. Southern Red-breasted Plover (*Charadrius obscurus*) in J. del Hoyo, A. Elliott, J. Sargatal, D. A. Christie, and E. de Juana (editors), *Handbook of the Birds of the World Alive*. Lynx Edicions, Barcelona, Spain.s

Wiersma, P., G. M. Kirwan, and P. Boesman. 2016b. Caspian Plover (*Charadrius asiaticus*). p. 438 In J. del Hoyo, A. Elliott, J. Sargatal, D. A. Christie, and E. de Juana (editors), *Handbook of the Birds of the World Alive*. Lynx Edicions, Barcelona, Spain.

Wiersma, P., G. M. Kirwan, D. A. Christie, and P. Boesman. [online]. 2016c. Lesser Sandplover (*Charadrius mongolus*) in J. del Hoyo, A. Elliott, J. Sargatal, D. A. Christie, and sE. de Juana (editors), *Handbook of the Birds of the World Alive*. Lynx Edicions, Barcelona, Spain. <www.hbw.com/node/53845> (10 August 2016).

Wilson, C. A., and M. A. Colwell. 2010. Movement and fledging success of Snowy Plover (*Charadrius nivosus*) chicks. *Waterbirds* 33:331–340.

Winkler, D. W., and J. R. Walters. 1983. The determination of clutch size in precocial birds. *Current Ornithology* 1:33–68.

Yasué, M., and P. Dearden. 2006. The potential impact of tourism development on habitat availability and productivity of Malaysian plovers *Charadrius peronei*. *Journal of Applied Ecology* 43:978–989.

Yasué, M., and P. Dearden. 2008a. Parental sex roles of Malaysian plovers during territory acquisition, incubation and chick-rearing. *Journal of Ethology* 26:99–112.

Yasué, M., and P. Dearden. 2008b. Replacement nesting and double-brooding in Malaysian Plovers *Charadrius peronii*: Effects of season and food availability. *Ardea* 96:59–72.

Yogev, A., and Y. Yom-Tov. 1996. Indeterminacy in a determinate layer: The Spur-winged Plover. *Condor* 98:858.

Zefania, S., R. ffrench-Constant, P. R. Long, and T. Székely. 2008. Breeding distribution and ecology of the threatened Madagascar Plover Charadrius thoracicus. *Ostrich* 79:43–51.

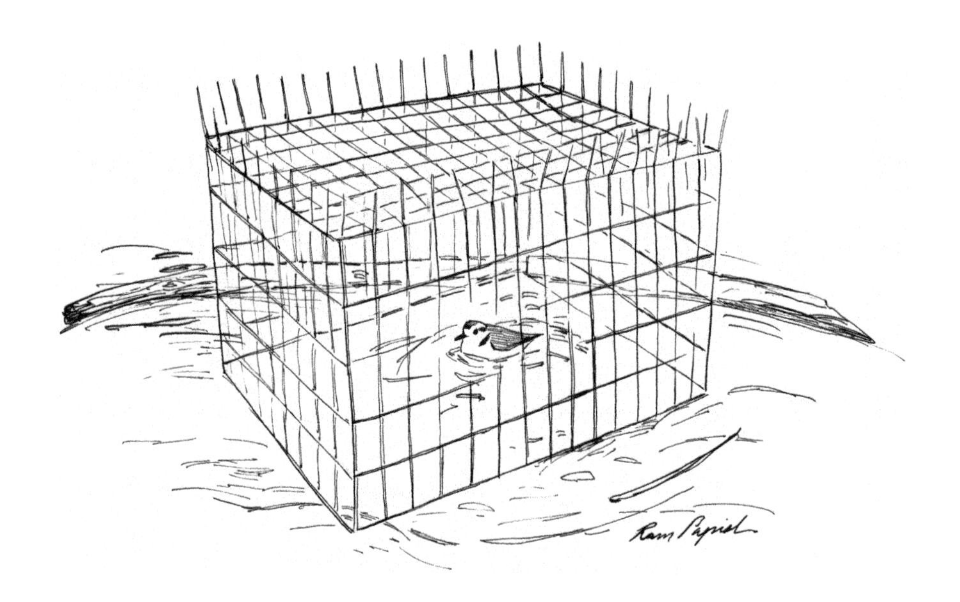

CHAPTER SIX

Predation and Predator Management*

Mark A. Colwell

Abstract. Predation has shaped many facets of the ecology of plovers, including their life history traits, behaviors, and demography. Moreover, conservation often emphasizes managing predators to increase population size. I reviewed literature addressing impacts of predation on survivorship and reproductive success of plovers, as well as four management approaches to ameliorate the negative effects of predation on population growth; most evidence comes from a few species that are threatened or endangered. Three of the four approaches seek to increase productivity by decreasing risk of predation of eggs and chicks by managing: (1) humans, attractants they provide, and disturbance of plovers; (2) habitat to increase availability and quality; (3) predator abundance or behavior using lethal or nonlethal methods; and (4) plover population size directly through captive propagation. The success of predator management is rarely monitored using vital rate data. For the threatened population segment of the Snowy Plover, data suggest that lethal removal has had a strong positive effect on productivity and population growth; even where lethal control is not used, populations are growing owing to immigration from areas where lethal removal of predators is routinely practiced.

Keywords: Charadrius, demography, endangered, plover, population growth, predation, productivity, threatened.

Predation is a strong selective pressure that has shaped the evolution of life history traits and ecologies of all birds (Newton 1994, Martin 1995). For shorebirds, predation has been argued to influence their clutch sizes (Arnold 1999, Summers and Hockey 1980), egg-laying intervals (Colwell 2006), nest-site selection (Amat and Masero 2004), crypsis (Troscianko et al. 2016), parental behavior (Gochfeld 1984, Gómez-Serrano and López-López 2016), sexual dimorphism in plumage (Ekanayake et al. 2015), and strategies of migration and molt (Lank et al. 2003), all of which affect population dynamics and potentially influence plover distributions (Pienkowski 1984, Jackson et al. 2004). Consequently, management of negative effects of excessive or chronic predation has become integral to conservation programs worldwide.

Predators influence the ecologies of their prey in two ways, either via nonlethal effects on individuals or through direct mortality, which affects populations (Luttberg and Kerby 2005, Cresswell 2008). Nonlethal (also referred to as trait-mediated) effects alter the fitness of individuals through subtle influences on behaviors. For example, the presence of a raptor may increase prey vigilance, with a concomitant reduction in foraging efficiency (Cresswell 2008). In this scenario, the "ecology of fear" (*sensu* Brown et al.

* Mark A. Colwell. 2019. Predation and Predator Management. Pp. 127–147 in Colwell and Haig (editors). The Population Ecology and Conservation of Charadrius Plovers (no. 52), CRC Press, Boca Raton, FL

1999) may also cause individuals to leave an area of otherwise suitable habitat or affect sociality by causing them to form flocks to minimize predation risk. McNamara and Houston (1987) indicate that danger can act indirectly to increase mortality. A second way predators influence prey is by killing them, deemed as a density effect (Luttberg and Kerby 2005). Here, I focus on the latter and include mortality of all life stages including eggs, chicks, juveniles, and adults, because these components comprise the vital rates such as survival, productivity, and dispersal that drive population growth. Moreover, direct mortality is relevant to management actions taken to bolster populations, especially of threatened or endangered species. Ironically, adult survival is the vital rate most influential in population growth (Sæther and Bakke 2000), yet it is most difficult to manage.

The impact of predation on plover populations has been assessed directly or indirectly in two ways, which correspond to the vital rates of survival and productivity. Several studies have estimated the percentage of mortality in a local population attributable to predators, especially raptors. During the breeding season, predators killed 23 adult Kentish Plovers at nests, which represented ~2% of a population (Amat and Masero 2004). During the nonbreeding season, Page and Whitacre (1975) identified nine species of raptor, principally the Merlin (*Falco columbarius*) and Short-eared Owl (*Asio flammeus*), as predators of shorebirds wintering (1 October–27 February) at Bolinas Lagoon, California; a simple ratio of prey remains to an estimate of local population size showed that raptors consumed 6.9% of Killdeer. At a small estuary in southeast Scotland, Whitfield (1985) used similar methods over 9 months (August–April) to estimate the impact of raptors, predominantly Peregrine Falcon (*F. peregrinus*), Merlin, and Eurasian Sparrowhawk (*Accipiter nisus*) on waders, which included the Common Ringed Plover. The plover experienced the highest predation rate among waders wintering at the site, with predators taking 19% of the population; juveniles suffered disproportionately. Most predation on plovers occurred from October to March. More recently, van den Hout et al. (2008) estimated that over 7 months, three species of falcon killed 1.9% of Common Ringed Plovers wintering on the Banc d'Arguin, Mauritania. These studies (Page and Whitacre 1975, Whitfield 1985), published several decades ago, likely underestimate contemporary impacts of raptors on plover mortality for two reasons. First, the method of comparing carcasses to population estimates does not account for prey not found. Second, recovery of raptor populations (Ratcliffe 1980, Newton and Wyllie 1992) probably has resulted in an increase in predation rates. Finally, although results suggest that in some winters, predation may strongly impact plover population size, a simple measure of predation-driven mortality does not demonstrate an effect on population dynamics (e.g., as measured in annual variation in censused number). Furthermore, the relationship is complicated by the extent to which other density-dependent sources of mortality affect population size.

Estimates of overwinter survival provide indirect evidence of the role of predation in population dynamics. Survival estimates for adults and juveniles are highly variable. However, it is often difficult to link predation to variation in survival. For example, Stenzel et al. (2007, 2011) showed that estimates of survival for juvenile and adult Snowy Plovers in coastal California varied annually. However, in other papers evaluating the effectiveness of lethal and nonlethal predator management during the breeding season to increase productivity, they concluded that there was no evidence that efforts positively influenced either productivity (i.e., fledging rates; Neuman et al. 2004) or juvenile survival (Stenzel et al. 2007).

Predators also impact population growth via reduction in reproductive success (i.e., per capita fledging success). Predators are widely implicated as the main cause of variation in reproductive success across a range of habitats (Table 6.1). Invasive species have been shown to have especially dramatic negative impacts on island avifaunas, including plovers (e.g., Dowding and Murphy 2001, Baird and Dann 2003, Burns et al. 2013). Removal of invasive predators can have significant positive effects on reproductive success of birds (Lavers et al. 2010). The literature linking predation to variation in reproductive success is richer than that of survival, although often fraught with difficulties of interpretation owing to methodological assumptions.

In this chapter, I review literature that addresses the role of predation in driving plover population dynamics via effects on survivorship and productivity. I assess the various nonlethal and lethal approaches taken to lessen the impact of predation, including (1) managing humans and attractants they provide to predators of plover eggs and chicks, (2) restoring and enhancing habitats to increase productivity, (3) altering the behavior

TABLE 6.1

Examples of management actions taken to ameliorate the negative impacts of predation on reproductive rates, survival, and population growth in plovers.

Method	Description	Objective[a]	Comments	Source(s)
		1. Manage humans		
Reduce food availability	Remove attractant (e.g., road kill, garbage) associated with increased predator densities	MO = reduce predator abundance and predation rate	Artificial nests of the Red-capped Plover depredated faster near fish carcasses	Rees et al. (2015)
Public education	Signage to encourage human behaviors that diminish anthropogenic food sources	MO = reduce the source and amount of food that attracts predators		No evidence available
Restrict access	Limit human access to minimize disturbances and reduce predator attractants	MO = decrease disturbance to plovers and reduced availability of food to predators	3× greater mortality of Snowy Plover chicks on weekends versus weekdays; No predation of Hooded Plover broods occurred in association with disturbance	Ruhlen et al. (2003); Weston and Elgar (2005)
			Snowy Plovers increased at a protected site; nest predation in an unprotected area was double the rate of eggs in the protected area	Lafferty et al. (2006)
Law enforcement	Punish (i.e., issue fines) human behaviors that violate ordinances stipulating no feeding of wildlife	MO = decrease food availability		
		2. Manage habitat		
Habitat restoration	Remove invasive plant species to increase "openness" of habitat, which increases attractiveness as a breeding site and facilitates early detection of predators of incubating adults in detection or avoidance of predators of eggs or chicks	MO = provide high-quality habitat features that facilitate natural behaviors that lead to high reproductive success	Snowy Plovers selected more open habitats in which to court and nest; nest survival doubled restored area	Muir and Colwell (2010); Dinsmore et al. (2014)
			Piping Plovers preferred and had higher nest survival on engineered sandbars than on natural or managed bars	Catlin et al. (2011)
Substrate modification	Enhance substrates to increase camouflage of eggs and chicks in order to increase hatching and fledging success	MO = alter breeding habitats to increase nest and chick survival	Snowy Plover reproductive rates were higher in heterogeneous substrates	Colwell et al. (2007, 2010, 2011); Herman and Colwell (2015)

(Continued)

TABLE 6.1 (*Continued*)

Examples of management actions taken to ameliorate the negative impacts of predation on reproductive rates, survival, and population growth in plovers.

Method	Description	Objective[a]	Comments	Source(s)
		3. Manage predators		
Hazing or scare tactics	Scare predators from an area to decrease their abundance, with concomitant increase in reproductive success	MO = decrease predator abundance in areas occupied by breeding plovers	Corvid effigies resulted in small-scale reduction in Common Raven numbers near Snowy Plovers	Peterson and Colwell (2014)
Cages (i.e., nest exclosures) or fencing	Protect nests with cages or fencing to increase hatching success	MO = increase hatching success by excluding predators from nest sites	Snowy Plover hatching success often positively affected; increased adult mortality; hatching success of Piping Plover clutches increased with exclosure use	Hardy and Colwell (2008); Dinsmore et al. (2014); Ivan and Murphy (2005)
Artificial shelters	Small, human-constructed refuges for chicks	MO = increase fledging success	More Hooded Plover chicks fledged at sites with shelters	Maguire et al. (2011)
Conditioned Taste Aversion	Apply nonlethal doses of poison (e.g., sodium carbonate, methiocarb, carbachol) to mimicked eggs to condition predators to avoid real eggs and increase hatching success	MO = alter foraging behavior of egg predators to increase hatching success	Short-term (i.e., weeks) decrease in predation of untreated (mimicked) Hooded Plovers eggs	Maguire et al. (2009)
			Increased survival of Carbachol-treated Quail (*Coturnix coturnix*) eggs mimicking Snowy Plover eggs	Brinkman et al. (2018)
Contraception	Administer chemicals to predators that reduce breeding propensity or success	MO = reduce predator population size by lowering breeding success	No evidence available	No evidence available
Addled eggs	Reduce reproductive success	MO = reduce predator population size by lowering reproductive success	No evidence available	
Translocation	Capture and release individual predators at another site to reduce the localized negative impacts of the predator	MO = temporarily remove predators to increase reproductive success and/or adult survival	No evidence available	
Temporary removal	Capture and maintain an individual predator in captivity to reduce the localized negative impacts of the predator	MO = temporarily remove predators to increase reproductive success and/or adult survival	No published evidence available	

(Continued)

TABLE 6.1 (*Continued*)

Examples of management actions taken to ameliorate the negative impacts of predation on reproductive rates, survival, and population growth in plovers.

Method	Description	Objective[a]	Comments	Source(s)
Lethal removal	Shoot, poison, trap and euthanize predators to diminish their abundance in an area	MO = permanent reduction in predator abundance to increase reproductive success and/or adult survival	Snowy Plover per capita fledging success was not correlated with lethal removal of foxes and cats	Neuman et al. (2004)
			Piping Plover per capita fledging success increased with number of trapped cats and foxes	Cohen et al. (2009)
		4. Manage plovers		
Alter dispersal behavior	"Do nothing" option; allow reproductive failure, which results in dispersal to sites where predators pose less of a threat	MO = allow natural behaviors of dispersal to facilitate movement to higher quality sites	Snowy Plovers dispersed following nest predation by corvids	Pearson and Colwell (2014)
Translocation	Capture (or captive rear) individuals for release at sites (e.g., predator-free islands)	FO = increase range of wild population and increase population size	Introduced Shore Plovers on islands free of introduced predators did not establish a population; a few captive-reared and released Shore Plovers did stay at the release site	Dowding and Murphy (2001)
Captive rearing	Supplement population size with young reared from eggs	MO = increase (artificially) hatching and fledging success; FO = increase breeding population size	Captive-reared Killdeer had higher hatching and fledging success	Powell et al. (1997)
			Piping Plovers hatched from abandoned eggs and reared in captivity had lower fitness than wild counterparts; captive-reared birds constituted 3% of breeding population	Roche et al. (2008)
Cross-fostering	Use surrogate species to rear young from eggs	MO = increase hatching and fledging success	Killdeer chicks reared in the wild by Spotted Sandpipers fledged in similar numbers to those produced by Killdeer	Powell et al. (1997)

[a] MO = *means objective* (i.e., altering predator abundance or behavior or positively affecting plover reproductive success); FO = *fundamental objective* (i.e., increased plover population size). For details on structured decision-making, see Lyons et al. (2008).

and abundance of predators with the aim of lessening their impacts on plover reproductive success, and (4) bolstering populations via captive rearing. I conclude with a case study of the Snowy Plover to illustrate the challenges of monitoring and evaluating the role of predation and the necessary information required to effect meaningful management decisions.

METHODS

I reviewed the literature for the 40 species of Charadrius plover and related species (see Chapter One, this volume; Table 1.1) to identify how predation influenced population biology and to elucidate the effectiveness of various management practices at increasing population size. In the review of management practices (Table 6.1), I distinguished between fundamental and means objectives (Lyons et al. 2008) in achieving a goal of increasing population size. In the context of conservation, the fundamental objective is to increase plover population size. In the case of species with special conservation status (i.e., endangered or threatened), population growth is linked to recovery objectives that, if achieved, will result in the delisting of a taxon. Means objectives are those that alter facets of the biology of the predator, prey (i.e., plover), habitat, or some presumed causative factor linking predation with plover population biology. A goal of adaptive management (Walters 1986) is to use data derived from monitoring population vital rates, for example, to alter management techniques (means objectives) in order to increase plover populations (fundamental objective); this is seldom achieved.

To illustrate the challenges of monitoring and managing predators in the context of plover conservation, I synthesized information from published and selected unpublished accounts of the threatened population of the Snowy Plover in coastal California. The U.S. government listed the Pacific coast population of the Snowy Plover as threatened in 1993 (USFWS 1993) and issued a recovery plan (USFWS 2007), which identified predation as one of several factors thought to limit population growth. Specifically, predation of eggs and chicks by native vertebrates such as the Common Raven (*Corvus corax*) and introduced vertebrates such as the Red Fox (*Vulpes vulpes*) compromises productivity across the range of the listed population segment (Eberhart-Phillips et al. 2015). While the recovery plan emphasizes the role of predators in lowering reproductive success, it does not explicitly address effects on survivorship of adults or juveniles. Accordingly, much effort has emphasized actions to (1) manage human activities that attract predators and influence their behaviors in areas where plovers breed, (2) restore habitat to make it more attractive to plovers and positively affect their reproductive success by increasing nest and chick survival, (3) directly reduce predation rates on eggs and chicks, and (4) decrease predator populations using nonlethal and lethal methods. In the Snowy Plover population I studied, managers used all but lethal methods to affect plover reproductive success and population recovery.

RESULTS AND DISCUSSION

Review of Predator Management Methods

The strong and varying impact of predation on vital rates has resulted in diverse management approaches (means objectives) to increase plover populations (fundamental objective). Nearly all practices are means objectives, rather than directly increasing population size. In theory, we would like to be able to quantify the effect of any management action on a population's growth rate, but this is often not possible, and we typically use partial measures to assess the effectiveness of our actions. Four main management categories exist to ameliorate negative effects of predation (Table 6.1). Overwhelmingly, management has emphasized techniques to improve reproductive success by increasing nest and chick survival; few methods address predation on adults. Of the four management categories to improve productivity, three involve (1) managing humans, which attract predators or compromise normal reproductive behaviors of plovers; (2) restoring and enhancing habitats such that the negative impacts of predation are lessened; and (3) reducing predator abundance or diminishing the effectiveness of predatory behavior using nonlethal and lethal approaches. A fourth approach directly bolsters plover populations through translocation, captive propagation, or cross-fostering.

Managing Humans

The presence and activity of humans may act in several ways to alter plover distributions and compromise reproductive success (Chapter Eleven, this volume). First, humans may attract egg and chick predators by supplementing food directly (e.g., feeding corvids) or indirectly (e.g., leaving trash). Although it has been argued that increased predator activity may result in greater losses of eggs and chicks, few studies have evaluated this relationship; most evidence is anecdotal. Moreover, assessments of this conjectured relationship involve results following a decrease in human activity. Consequently, it is difficult to determine whether decreased reproductive success stems from humans or predators. For example, in coastal California, Snowy Plovers returned after a long absence to breed at a protected beach after management restricted human recreational access (Lafferty et al. 2006). At this same site, predators consumed simulated plover nests (i.e., with eggs of quail, *Coturnix coturnix*) in an adjacent unprotected area at twice the rate of an area protected by a fence and staffed with docents who hazed corvids and distributed educational materials (Lafferty et al. 2006). In an experimental study, the presence of fish carcasses negatively influenced the survival of artificial nests of Red-capped Plovers; Australian Ravens (*C. coronoides*) caused 80% of artificial nest failures, especially those within 80 m of a carcass (Rees et al. 2015).

A second way in which human disturbance may increase predation stems from changes in behaviors of adults and chicks, which increases the risk of predation and lowers reproductive success. Evidence suggests that nest survival is negatively associated with nest visitation rates of tending adults in ground-nesting passerines (e.g., Lyon and Montgomerie 1987) and shorebirds (e.g., Ashkenazie and Safriel 1979, Safriel 1980). In other words, although eggs, nests, and young may be cryptic, predators may find a clutch (or brood) by viewing an adult depart or return to the nest during normal patterns of incubation. Smith et al. (2007) provided insight into the role that normal (i.e., undisturbed) incubation played in nest survival. They analyzed five arctic-breeding shorebirds (including the Semipalmated Plover), which differed in parental care of eggs: socially monogamous species have biparental incubation whereas care is uniparental in polyandrous species. Nest survival was higher in biparental species that took fewer incubation recesses compared with uniparental species with frequent recesses. In Wilson's Plover, males incubate at night and females during the day; the incubation exchange occurs under low-light conditions of dusk and dawn, which is presumed to decrease nest predation (Thibault and McNeil 1995). Collectively, these results suggest that predation rates should increase when disturbance increases the frequency of incubation recesses above normal, undisturbed levels.

Evidence of a relationship between disturbance and adult tending behavior as a cause of lowered reproductive success is difficult to obtain, especially given the rarity of observing an instantaneous predation event in the context of chronic or episodic human activity. Cameras monitoring nests occasionally provide valuable insights into human-caused predation, including unintended effects of conservationists monitoring nests. For instance, Keedwell and Sanders (2002) used video cameras to examine whether biologists monitoring Double-banded Plover nests on the South Island, New Zealand influenced predation by a suite of invasive mammals. They compared nests that were routinely approached by biologists with those not subjected to this form of disturbance. Similarities in frequency of visits by predators, their angle of approach relative to that of the biologists, and the rate of predation suggested that human disturbance did not influence predation.

In coastal Victoria, Australia, Hooded Plover chicks survived equally well under varying degrees of human disturbance, although chicks foraged and brooded less when humans were present (Weston and Elgar 2005). Piping Plovers (adults and chicks) in artificial wetlands along the U.S. Atlantic coast foraged more and were less vigilant compared with those in areas with higher human recreational activity (Maslo et al. 2012). These behavioral differences were argued to influence differences in reproductive success among sites, although the direct linkage between behavior and reproductive success was unclear. Similarly, human recreational activity may have caused higher mortality of Snowy Plover chicks at Point Reyes National Seashore, California (Ruhlen et al. 2003). Specifically, a threefold increase in chick mortality on weekends, compared with weekdays, was correlated with higher recreational use. In these studies, authors presented no direct linkage between chick mortality and predation, which highlights the difficulty of collecting data on instantaneous predation events and interpreting it in light of evidence of chronic human disturbance. At best, evidence is correlative and suggestive.

Although the link between incubation behavior and disturbance is conjectural, it is supported by evidence from at least one study that used video cameras to characterize the cause of egg predation (Burrell and Colwell 2012). In this study, female Snowy Plovers, which tended to incubate during the day, appeared to depart the nest in response to the presence of a predator's (Common Raven) activity. At one site where clutch loss to predators was especially high, videos revealed that ravens caused most (70%) nest failures. Moreover, ravens typically (63% of predation events) landed within 1 m of the nest usually (50%) within a minute after the adult departed (Burrell and Colwell 2012). Given the camera angle and field of vision, humans did not appear to cause plovers to leave nests; departure stemmed from ravens flying nearby. These results suggest that the synergistic role played by human disturbance in predation events may depend on the species of predator and whether they search for prey visually or using olfactory cues.

Habitat Restoration and Modification

Plovers typically occupy open, sparsely vegetated habitats year round. In several species, evidence links physical features of these habitats with nest-site selection (e.g., Knopf and Miller 1994, Riensche et al. 2015), behaviors that enhance crypsis (Troscianko et al. 2016) or dissuade predators (Gómez-Serrano and López-López 2016), and measures of reproductive success (e.g., Espie et al. 1996, Amat and Masero 2004, Colwell et al. 2011, Nyugen et al. 2013, Que et al. 2015). Accordingly, a common practice is to restore and create habitat that is both attractive to breeding plovers and affects higher reproductive success. Along the Pacific coast, removal of an invasive plant (*Ammophila arenaria*) is used to restore habitat to that attractive to Snowy Plovers. Open, sparsely vegetated habitats comparatively free of this invasive plant are preferred by Snowy Plovers for courting and nesting (Muir and Colwell 2010), and restored areas attract plovers (Leja 2015). Along the Oregon coast, an analysis of a long-term dataset of ~1,500 Snowy Plover nests showed that removal of *Ammophila* correlated with a doubling of nest survival (Dinsmore et al. 2014). Enhancing the cryptic quality of substrates may be employed to increase egg and chick survival. For example, Piping Plovers preferred to breed on engineered riverine sandbars where they had higher nest survival (Catlin et al. 2011).

Social factors play a role in selection of breeding sites, especially for plovers that breed at high densities (i.e., semicolonially; Patrick and Colwell 2018) in suitable habitats. In some cases, the presence of other species breeding in close proximity to waders has been shown to positively influence nest survival via aggressive defense of an area (Dyrcz et al. 1981). For example, Semipalmated Plovers that nested within 100 m of Arctic Tern (*Sterna paradisaea*) colonies had higher nest survival, presumably because aggressive defensive tactics of terns reduced predation on plover nests (Nyugen et al. 2006). No studies have attempted to manipulate this relationship to benefit plovers.

On a finer spatial scale, nest-site selection, and nest and chick survival have been linked to fine scale features of the nest site (e.g., substrates, objects, vegetation), which offers the opportunity to manipulate these features to improve habitat attractiveness and productivity. Killdeer and Mountain Plovers select nest sites on or near cattle feces, potentially an adaptation to reduce predation threat, as incubating adults, eggs, and young may be less obvious to predators when nests are near objects that disrupt the landscape (Graul 1973, Johnson and Oring 2002). Lifetime reproductive success measured as the number of fledged young, was greater for Snowy Plovers breeding in heterogeneous substrates (e.g., a mix of sand, gravel and rock) compared with homogeneous sandy substrates (Colwell et al. 2010, Herman and Colwell 2015). The mechanism underlying this pattern is similar to that for Killdeer and Mountain Plovers: visual predators are the principal cause of low productivity and heterogeneous substrates afford greater camouflage to eggs (Colwell et al. 2011) and chicks (Colwell et al. 2007). Such findings have yielded management recommendations to increase the "clutter" in restored areas to include a mix of natural debris, including shell hash. However, few studies have manipulated habitat features at nests and measured nest survival. Powell and Collier (2000) capitalized on newly created breeding areas to evaluate nest-site selection and nest survival of Snowy Plovers. Their results were mixed. Nest survival varied annually and was initially high in a newly created habitat; subsequent years showed much lower productivity.

During the nonbreeding season, plovers forage and roost in open habitats, a pattern that has been interpreted as a response to predation risk (Chapter Eight, this volume). Snowy Plovers wintering in northern California occurred predictably on wider beaches with less vegetation than random sites; it was argued that these features offered an unobstructed view that favored early detection and response to avian predators (Brindock and Colwell 2011). No study has evaluated the relationships between habitat selection during the nonbreeding period and survival of adults or juveniles. In sandpipers, however, predation risk increases with proximity to habitat edge (Pomeroy et al. 2006), especially for Common Redshanks (*Tringa totanus*) foraging alone (Cresswell and Quinn 2004).

One assumption of restoration to benefit plovers is that suitable habitat is in short supply such that providing more area of higher quality will result in population growth. This assumption hinges on untested relationships among habitat availability, plover breeding density, and predation rates, which are undoubtedly complex. In fact, if predation is density dependent (e.g., Page et al. 1983), then restoration to create attractive habitat (i.e., selection of a breeding site) may foster a scenario in which productivity initially rises when restoration improves the attractiveness of a site, until high breeding densities attract predators, after which productivity declines. In summary, the consequences of habitat management strategies to plover population size and growth remain largely unmeasured. Specifically, research should address whether creation of newly restored areas attracts breeding plovers that experience higher reproductive success (i.e., nest survival, per capita fledging success), with a subsequent increase in population size.

Managing Predator Abundance and Behavior

As ground-nesters, plovers often experience high predation rates on clutches and broods. Worldwide, diverse predators, both native and invasive, have been implicated in the decline of plover populations via their negative effects on productivity and survival. For example, video cameras showed that several invasive mammals in New Zealand were the main cause of egg loss, as well as chick and adult mortality in Double-banded Plovers (Dowding and Murphy 2001, Sanders and Maloney 2002). In coastal California, Common Ravens are widespread and their populations have increased dramatically (Kelly et al. 2002). Direct and indirect evidence indicates that ravens are the most important predator affecting Snowy Plover productivity (Neuman et al. 2004, Burrell and Colwell 2012, Lauten et al. 2015). Based on this evidence, managers have used a variety of nonlethal and lethal methods to mitigate for negative effects on productivity and survival.

Nonlethal methods include altering the predatory behavior or reducing the local abundance of predators by scaring or hazing. Few published accounts exist of a controlled experiment evaluating effects of effigies and scare tactics on corvids. In one study conducted at a location where Common Ravens consistently compromise Snowy Plover reproductive success (Burrell and Colwell 2012), effigies reduced raven abundance in the immediate vicinity (~10 m) of a food attractant (Peterson and Colwell 2014). It was unlikely, however, that this method would deter ravens across large areas of plover breeding habitat. Moreover, the intelligence of corvids suggests that any positive benefits of hazing may be short-lived and negated by learning (e.g., habituation).

Another nonlethal approach to reduce high predation rates of clutches is conditioned taste aversion (CTA). The success of CTA is affected by the species of predator(s), its intelligence, behaviors and local abundance, as well as the diversity of the predator community. With intelligent, synanthropic corvids, for instance, CTA is most likely to be successful when implemented on territorial individuals that learn to avoid eating plover eggs. These conditioned, territorial birds may exclude conspecifics, which reduces predation. Maguire et al. (2009) showed that sodium carbonate injected into quail eggs greatly reduced predation by red foxes on artificial clutches in Hooded Plover breeding areas. Similar results obtained from a study using quail eggs treated with carbachol to increase survival of artificial Snowy Plover nests (Brinkman et al. 2018).

Other nonlethal methods attempt to reduce local abundance of predators by affecting their productivity (e.g., contraception, egg addling) or directly removing them (e.g., capture and translocation or hold captive). No published accounts are available to evaluate the effectiveness of these approaches in the context of plover conservation. Anecdotes, however, occur in unpublished reports for threatened or endangered taxa (e.g., Lauten et al. 2015).

Nest exclosures have been used widely to increase hatching success of shorebirds, especially plovers, with sometimes dramatic results (e.g., Barber et al. 2010, Dinsmore et al. 2014, Tan et al. 2015). When results suggest otherwise, it may stem from the experimental design and analytical approach to comparing success of unprotected and caged nests (Mabee and Estelle 2000). In some instances, the physical design of exclosures (e.g., open versus closed top; size of wire mesh) used to build cages may not effectively exclude some predators (Johnson and Oring 2002). An important and widely recognized shortcoming of exclosures is that adult mortality often increases at specific locations, presumably after a predator learns that incubating adults are especially vulnerable (e.g., Nol and Brooks 1982, Johnson and Oring 2002, Neuman et al. 2004, Hardy and Colwell 2008, Barber et al. 2010, Dinsmore et al. 2014; Figure 6.1).

Lethal methods have been used extensively to remove specific individuals or to reduce local populations of predators. Programs that remove invasive predators can have strong positive impacts on productivity, which translates into population growth, especially for small, ground-nesting species (Lavers et al. 2010). Removal (and translocation) of Great Horned Owls (*Bubo virginianus*) improved survival of Piping Plover chicks on sand bars in the Missouri River (Catlin et al. 2011). In Snowy Plovers, lethal removal has been shown to positively influence nest and chick survival (Dinsmore et al. 2014, 2017). Although Snowy Plover per capita reproductive success did not appear to increase with use of a combination of predator management methods, including lethal removal (Neuman et al. 2004), regionwide population growth correlated positively with an index of lethal predator removal (Eberhart-Phillips et al. 2015). Modeling suggests that the productivity benefit of exclosing nests is outweighed by small increases in adult mortality (Watts et al. 2012).

Captive Rearing, Cross-fostering, and Translocation

For taxa imperiled with extinction, captive breeding for reintroduction is often a last-ditch conservation strategy, which has received critical review (Ebenhard 1995). In a few cases, humans have intervened directly to remove plover eggs and chicks from natural settings in order to implement captive breeding programs associated with research (Malone and Proctor 1966) or reintroduction (Page et al. 1989, Goosen et al. 2011, Neuman et al. 2013). Powell and Cuthbert (1993) used Killdeer as a surrogate for evaluating captive rearing and cross-fostering to bolster Piping Plover populations. They compared the hatching success, behaviors, and growth rates of Killdeer chicks that were: (1) reared by parents in the wild, (2) cross-fostered to wild Spotted Sandpipers (*Actitis macularius*), and (3) captive reared by humans. Their results of normal behaviors and growth suggested that captive rearing was a viable management technique, and was preferred over cross-fostering for a variety of practical reasons.

Figure 6.1. A Great Horned Owl inside an exclosure erected around a Snowy Plover nest on Point Reyes National Seashore (2016), which illustrates the challenges posed by managing predators for productivity versus adult survival. (Photo: Matthew Lau.)

Success of captive rearing programs has been evaluated in several ways including (1) fidelity to a release site, (2) survival, (3) reproductive success, and (4) population viability. Reintroduction to increase a local population may be compromised if individuals disperse from release sites and breed elsewhere. For instance, 53% of Shore Plovers introduced to a predator-free island where no wild conspecifics occurred dispersed to the mainland of New Zealand (Aikman 1999). By contrast, most Snowy Plovers released at a coastal site survived well and bred locally as yearlings with equal propensity to wild conspecifics of the same age (Neuman et al. 2013). In migratory Piping Plovers, however, released juveniles survived less well (Roche et al. 2008, Goosen et al. 2011), and had lower reproductive success than wild conspecifics (Roche et al. 2008). These examples highlight the challenges facing successful reintroduction programs for species that differ in migratory and dispersal behavior. Importantly, lower success of captive-reared birds compared with wild counterparts does not necessarily negate the value of captive propagation, especially if eggs are available (e.g., Neuman et al. 2013).

Ultimately, success of a reintroduction program is measured by population growth, the fundamental objective, which may hinge on resources (e.g., funding), rearing facilities, as well as a species' biology, ecology, and available habitat conditions. Roche et al. (2008) evaluated the contribution of captive rearing to recovery of the Great Lakes population of the Piping Plover. Captive-reared plovers had appreciably lower estimates for all vital rates. Consequently, newly released captive-reared plovers contributed only 30% to population growth while wild-reared individuals contributed 70%. Roche et al. (2008) concluded that a population consisting of captive-reared individuals was not viable. Captive rearing (including cross-fostering and translocation) is often employed to stave off imminent extinction. If investigation identifies predation as the cause of population decline, then management needs to address the cause of the decline for effective conservation. Captive rearing is best used as a site-specific tool for temporarily increasing population growth while conservation plans are developed that removes or neutralizes the agent of decline (Caughley 1994).

CASE STUDY: PACIFIC COAST POPULATION OF THE SNOWY PLOVER

In 1993, the U.S. government listed the Pacific coast population (Washington, Oregon, and California) of the Snowy Plover as threatened under the Endangered Species Act (USFWS 1993). The listing followed evidence that the number of occupied breeding sites and population size had declined (Page and Stenzel 1981, Page et al. 1991). The recovery plan (USFWS 2007) identified predation as one of three factors (including habitat loss and human disturbance) that limit the population. Specifically, predation of eggs and chicks by introduced and native vertebrates compromised reproductive success throughout the plover's range. Consequently, managers have used a variety of nonlethal and lethal methods to affect higher productivity and population recovery (USFWS 2007).

Population Size, Growth, and Vital Rates

The recovery plan established two demographic criteria for delisting: (1) a breeding population of 3,000 adults (maintained for 10 years), apportioned among the six recovery units (RUs), and (2) average per capita productivity of at least 1.0 fledged chick maintained for 5 years. Since 2005, USFWS has coordinated a weeklong, range-wide survey of suitable habitat (i.e., sandy, ocean-fronting beaches; salt ponds; riverine gravel bars) to estimate breeding population size and distribution. Population size (Figure 6.2) has varied from 1537 (2007) to 2260 (2015); most plovers occur in central and southern California (RU 5 and 6; Eberhart-Phillips et al. 2015). As of 2015, the northern region (RU1; Washington and Oregon) has surpassed its population recovery objective (250 adults), but only for three consecutive years. Window survey data represent a minimum population size because some birds are not detected during surveys. For example, in coastal Oregon and Washington, breeding surveys in 2015 indicated 340 adults (64 in Washington; 276 in Oregon). However, detailed records (i.e., of color-marked individuals) indicated a considerably larger population (e.g., 449 in Oregon; Lauten et al. 2015). Similar patterns exist for the population at Monterey Bay, California (469 breeding adults, compared with 305

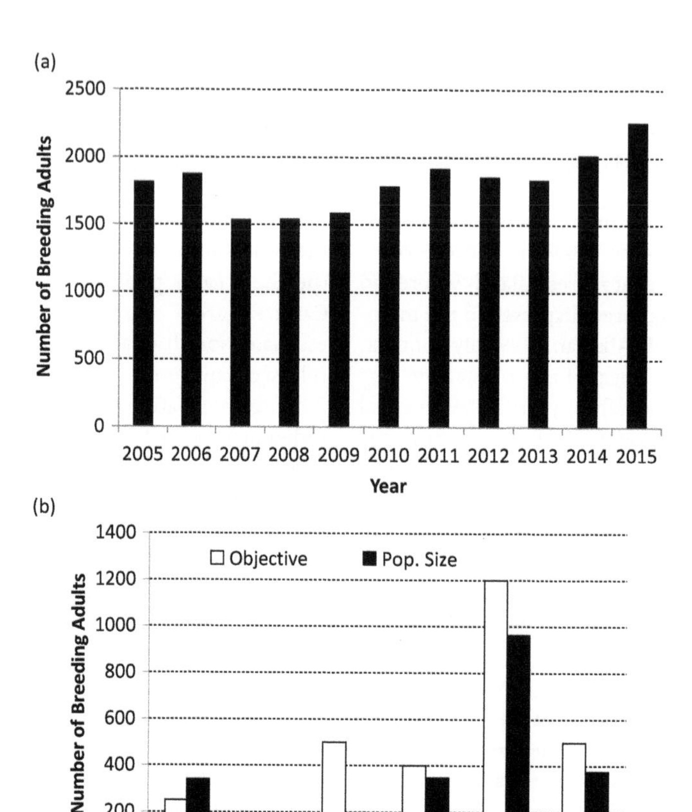

Figure 6.2. Estimates of annual population size of the threatened population of the Snowy Plover (a) along the Pacific coast of the United States based on coordinated surveys conducted during the peak of nesting and (b) broken down by RU with associated population size objectives.

detected during the window survey; Page et al. 2016). Consequently, the USFWS use a correction factor (i.e., multiplier of ~1.3; Nur et al. 1999) to estimate total population size (2938), which is nearing the recovery objective of 3,000 for the listed population.

For management purposes, the USFWS (2007) divided the plover's range into six RUs, each with different population objectives. Within each of the RUs, varying degrees of activity and coordination exist to manage predators, as well as monitor plover vital rates and population growth. Most published accounts of demographic data (e.g., Neuman et al. 2004, Stenzel et al. 2007, Mullin et al. 2010, Eberhart-Phillips and Colwell 2014, Herman and Colwell 2015, Dinsmore et al. 2014, 2017), especially as they relate to predator management, come from areas north of the core of the species' range (Eberhart-Phillips et al. 2015). Overwhelmingly, management has focused on increasing productivity by addressing predation during the breeding season, and predator management has been integral to this approach. Hudgens et al. (2014) provide vital rate data based on review of published and unpublished annual reports.

Productivity

Population viability analyses (Nur et al. 1999, Hudgens et al. 2014) indicate that fledging success of 1.0 chick per male will maintain the population. Where data are available, recent reports indicate that productivity has exceeded this value for multiple years in some areas, especially in Oregon (12-year average = 1.40 ± 0.26; Lauten et al. 2015). This contrasts sharply with chronically low productivity in

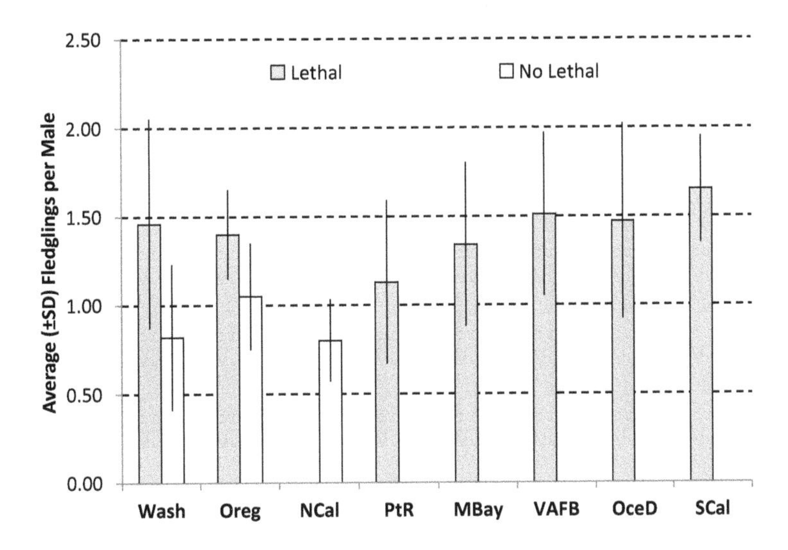

Figure 6.3. Average (±SD) annual per capita fledging success of male Snowy Plovers at locations along the U.S. Pacific coast where lethal removal of predators has and has not been used.

northern California (Colwell et al. 2017). In most southern regions, monitoring efforts do not provide easily interpretable estimates of per capita reproductive success to evaluate the relationship between predator management and productivity.

The effectiveness of predator management at increasing plover reproductive success is best illustrated by contrasts (Figure 6.3) among regions where lethal removal is practiced (Oregon, Lauten et al. 2015, Monterey Bay, Page et al. 2016) and is not (northern California, Colwell et al. 2017). Importantly, these same sites highlight the effectiveness and insights provided by monitoring when it is coupled with detailed accounts of productivity based on individually marked plovers. In Oregon, high productivity is associated with lethal removal of predators at sites where high plover densities occur. In northern California, by contrast, managers used exclosures early on (2001–2006) to enhance hatching success; however, no lethal removal of predators has occurred and per capita fledging success was often below 1.0 chick per male (Figure 6.4a; Colwell et al. 2017). Despite this low productivity, the northern California population has grown steadily for seven consecutive years ending in 2016 where population rate of growth or lambda (λ) is 1.22 based on the ratio of consecutive year population sizes (Figure 6.4b). The reason for this dramatic growth (from a low of 19 adults in 2009 to 61 in 2015) is immigration (Colwell et al. 2017). The principal source population for immigrants is Oregon, where lethal predator removal is routine during the same interval. Specifically, over the 7 years of positive growth, approximately two-thirds of the northern California population consists of Oregon yearlings, many of whom overwinter in northern California and remain to breed there; a small number of immigrants originate from central California coastal regions.

Survivorship

Analyses of mark-resight data indicate that survival of juvenile and adult plovers varies annually (Stenzel et al. 2007, 2011, Mullin et al. 2010). Additionally, population viability analyses suggest that local (Eberhart-Phillips and Colwell 2014) and range-wide (Hudgens et al. 2014) recovery is most strongly influenced by survivorship. Little is known, however, of the cause of annual variation in survival, although there are hints of climate-driven patterns in northern subpopulations. Specifically, decreases in population size (Eberhart-Phillips et al. 2015) and survivorship (Figure 6.4c; Eberhart-Phillips and Colwell 2014) are evident during cold winters. Elsewhere along the Pacific coast (i.e., Monterey Bay), survivorship was especially low in 1997–1998 (Nur et al. 1999, Stenzel et al. 2007), coincident with a prolonged December cold spell and an El Niño weather pattern. In northern California, a low in survivorship coincided with a cold interval in January 2007 (Mullin et al. 2010, Eberhart-Phillips and

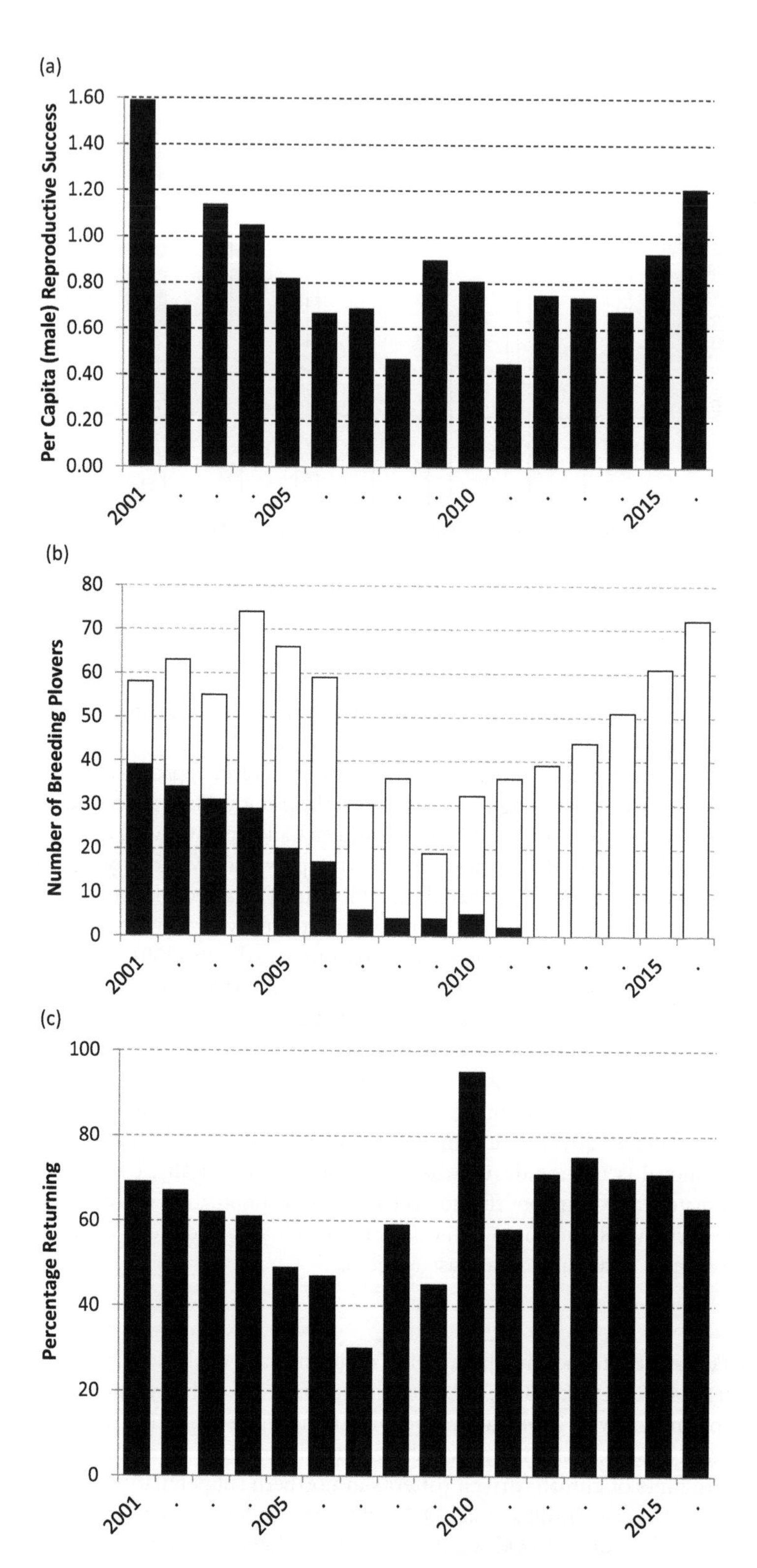

Figure 6.4. A 16-year summary of (a) per capita reproductive success, (b) breeding population size and (c) return rates (i.e., apparent adult survival) in a color-marked population of Snowy Plovers in coastal northern California.

Colwell 2014; Figure 6.4c). Although these observations suggest a "bottom-up" influence involving food availability and increased energetic demands, the proximate cause of high mortality in winter remains unknown, and could include predation.

Predators consume eggs (see above) and kill adult, juvenile, and young plovers year-round. However, the role that predation plays in population growth is poorly understood. Specifically, avian predators (e.g., *Falco peregrinus*, *F. columbianus*) may act in a density-dependent manner to increase mortality rates when local populations produce large numbers of juveniles, which are especially vulnerable in winter flocks. Adult survivorship is the vital rate that most strongly influences population growth in birds (Sæther and Bakke 2000) and Snowy Plovers in particular (Eberhart-Phillips and Colwell 2014). Despite this, few management practices attempt to mitigate for decreased survivorship, especially outside the breeding season. Moreover, the one example of managing predation during the breeding season targets a trade-off between productivity and survivorship. Specifically, use of exclosures to increase nesting success is widespread; however, episodes of increased mortality associated with predation of incubating adults are common (see references above). As a result, managers often stop using exclosures. The only other instances of predator management that seek to increase survivorship are those involving raptors that compromise productivity via egg and chick predation; these same raptors (Northern Harrier, *Circus cyaneus*; Great Horned Owl, Figure 6.1) may also depredate adults (Marcot and Elbert 2015).

Dispersal

Predation may affect dispersal behavior (and, hence, local populations) by influencing an individual's choice to settle at or leave a breeding site. Individuals may emigrate from sites where nest predation is high, especially if they are juveniles breeding for the first time (Oring and Lank 1984). Along the Pacific coast, Snowy Plovers move widely both within and between years (Stenzel et al. 1994, Colwell et al. 2007, Pearson and Colwell 2014). Natal and breeding dispersal distances exhibit a negative exponential pattern (Hudgens et al. 2014) such that most individuals tend to breed within a few 100 km of their natal site (Eberhart-Phillips et al. 2015). Still, some individuals move long distances between breeding (and wintering) sites (Stenzel et. al 1994, 2007), which effectively links the demography of the entire Pacific coast population (Hudgens et al. 2014, Eberhart-Phillips et al. 2015), with likely consequences for genetic structure. Evidence suggests that predation influences individual dispersal behavior, and, therefore, management practices that target predation undoubtedly affect movements. For example, in northern California, most (92%) plovers dispersed <10 km between successive nests, although failed breeders tended to disperse greater distances (median 1.3 km) than those that hatched a clutch (median 0.7 km; Pearson and Colwell 2014). Importantly, the comparatively short distances moved by most individuals did not result in their settling at locations of lower risk of clutch predation; as a result, they experienced poor reproductive success.

A source-sink model of metapopulation dynamics (Levins 1970, Hanski and Gilpin 1991) posits that immigration rates vary among habitat patches that differ in productivity, with source populations providing immigrants that effectively "rescue" sink populations that are otherwise destined for extinction. Consequently, effective predator management in some areas may benefit population growth elsewhere by providing immigrants to sink populations. In northern California, dramatic population growth (average $\lambda = 1.22$) has occurred for six consecutive years (up to 2016), despite poor reproductive success and varying annual survivorship (Colwell et al. 2017). Evidence from color-marked individuals indicates that settlement of yearling immigrants from sites in Oregon (~500 km north), where lethal predator removal has occurred (Dinsmore et al. 2014, 2017), is driving this population growth. Specifically, roughly two-thirds of the northern California breeding populations are immigrants (Colwell et al. 2017).

Monitoring and Adaptive Management

Collectively, results from the Pacific coast population of the Snowy Plover highlight several challenging facets of predator control programs. First, management practices should be coupled with effective monitoring at the appropriate scale (Figure 6.5) to yield information to evaluate and alter practices. Several long-term programs effectively collect vital rate data that are essential to understanding population

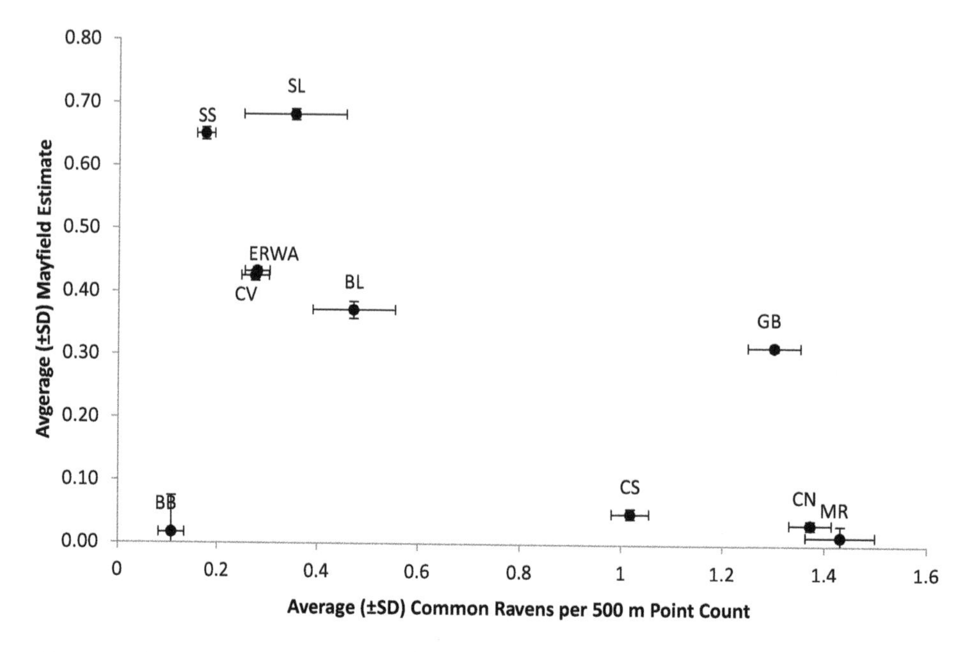

Figure 6.5. Snowy Plover nest survival probability ($\bar{X} \pm$ SE across years) at ten breeding locations in coastal northern California varied inversely with activity ($\bar{X} \pm$ SE number detected per point count) of Common Ravens.

growth and making decisions regarding the utility and effectiveness of predator management techniques (see Dinsmore et al. 2014). However, these programs are concentrated in the northern extent of the species' range, where subpopulations tend to be small. A paucity of coordinated monitoring exists in southern regions, which is the core of the species' range (Eberhart-Phillips et al. 2015). For example, a review of unpublished reports posted to the USFWS website (Arcata Office) revealed that most summaries of vital rate data are limited to simple estimates of nesting success, with few attempts to evaluate the effectiveness of predator management. Similarly, information regarding survival probabilities comes from mark-resight analyses of color-marked populations in the north.

A second challenge associated with Snowy Plover recovery is that much of the predator management effort is directed at increasing productivity. Managers have used two main methods, nest exclosures and predator removal, to decrease nest predation and increase productivity. Detailed studies at several locations show that while nest exclosures increase hatching success, they occasionally cause increased mortality of incubating adults, which has prompted managers to cease using exclosures at specific locations (Neuman et al. 2004; Hardy and Colwell 2008; Dinsmore et al. 2014, 2017). In an analysis of range-wide population growth (derived from breeding window surveys), lethal removal and exclosures had contrasting effects: predator removal correlated with increased growth whereas exclosures were associated with declines (Eberhart-Phillips et al. 2015). This result illuminates the challenges of using a combination of methods to increase per capita reproductive rates. Additionally, although exclosures have been shown to increase nesting success, their use is not often evaluated in relation to per capita reproductive rates. This is especially problematic for affecting population growth if the same predators of plover eggs consume nidifugous chicks after they leave the protective cage around the nest. Neuman et al. (2004) illustrated this paradox by showing that long-term predator management, including widespread use of exclosures, increased hatching success but did not increase per capita fledging success. In the context of structured decision-making (Lyons et al. 2008), Cohen et al. (2016) evaluated the effect of exclosures in affecting population growth (λ) via a trade-off between increased productivity and decreased adult survival in Piping Plovers. Their findings suggested that the benefits of exclosures were limited to sites where nest survival was especially low, but population growth remained negative even when exclosure use at these low-quality sites increased productivity.

Overall, little attention has been directed at increasing survivorship. Given that adult survivorship is recognized as the most influential vital rate driving population growth of birds (Sæther and Bakke 2000), including plovers (Hudgens et al. 2014), efforts should be made to positively affect survivorship of juveniles and adults. Results of a population viability analysis (Eberhart-Phillips and Colwell 2014) showed that increased productivity from exclosure use coupled with decreased (5%) adult survivorship resulted in extirpation of a local population within 50 years. Consequently, the common use of exclosures should be discouraged, as recommended 35 years ago (Nol and Brooks 1982).

CONCLUSIONS

Plovers have contended with predators for an eternity, as evidenced by the multitude of evolved physical traits (e.g., plumages), behaviors (e.g., nest-site selection, distraction displays), and life history characteristics (e.g., clutch sizes, egg-laying intervals). In the Anthropocene (Crutzen 2002), however, we have disrupted natural predator–prey relationships by altering and degrading habitats, subsidizing food sources, and introducing predators where they formerly were absent. As a result, the conservation prospectus for many plovers is dire and management sometimes involves unpopular decisions that run counter to public sentiment and individual ethics. In island ecosystems, introduced predators pose especially challenging threats to plover populations (e.g., New Zealand, St. Helena). In many cases, it may be difficult to reverse these negative population trends imposed by predators. Consequently, increasingly intensive management of humans, habitat, as well as the populations of predators and prey, is required.

ACKNOWLEDGMENTS

I thank the army of students who have worked with me for 17 years on the Snowy Plover project; several deserve special mention: Noah Burrell, Luke Eberhart-Phillips, Elizabeth Feucht, James Hall, Michael Hardy, Dana Herman, Alexandra Hoffmann, Susan Hurley, Teresa King, Matt Lau, Stephanie Leja, Jason Meyer, Cheryl Millett, Jordan Muir, Steve Mullin, Zach Nelson, David Orluck, Allie Patrick, Sara Peterson, Wendy Prestera, and Carol Wilson. Dov Lank and Matt Johnson provided constructive critiques that improved the manuscript.

LITERATURE CITED

Aikman, H. 1999. Attempts to establish shore plover (*Thinornis novaeseelandiae*) on Motuora Island, Hauraki Gulf. *Notornis* 46:195–205.

Amat, J. A., and J. A. Masero. 2004. Predation risk on incubating adults constrains the choice of thermally favourable nest sites in a plover. *Animal Behaviour* 67:293–300.

Arnold, T. W. 1999. What limits clutch size in waders? *Journal of Avian Biology* 30:216–220.

Ashkenazie, S., and U. N. Safriel. 1979. Breeding cycle and behavior of Semipalmated Sandpiper at Barrow, Alaska. *Auk* 96:56–67.

Baird, B., and P. Dann. 2003. The breeding biology of Hooded Plovers, *Thinornis rubricollis*, on Phillip Island, Victoria. *Emu* 103:323–328.

Barber, C., A. Nowak, K. Tulk, and L. Thomas. 2010. Predator exclosures enhance reproductive success but increase adult mortality of Piping Plovers (*Charadrius melodus*). *Avian Conservation and Ecology* 5:6.

Brindock, K. M., and M. A. Colwell. 2011. Habitat selection by Western Snowy Plovers during the nonbreeding season. *Journal of Wildlife Management* 75:786–793.

Brinkman, M. P., D. K. Garcelon, and M. A. Colwell. 2018. Evaluating the efficacy of Carbachol at reducing corvid predation on artificial nests. *Wildlife Society Bulletin* 42:84–93.

Brown, J. S., J. W. Laundré, and M. Gurung. 1999. The ecology of fear: Optimal foraging, game theory and trophic interactions. *Journal of Mammalogy* 80:385–399.

Burns, F., N. McCulloch, T. Székely, and M. Bolton. 2013. The impact of introduced predators on an island endemic, the St Helena Plover, *Charadrius sanctaehelenae*. *Bird Conservation International* 23:125–135.

Burrell, N. S., and M. A. Colwell. 2012. Direct and indirect evidence that productivity of Snowy Plover

Charadrius nivosus varies with occurrence of a nest predator. *Wildfowl* 62:204–223.

Catlin, D. H., J. H. Felio, and J. D. Fraser. 2011. Effect of Great-Horned Owl trapping on chick survival in Piping Plovers. *Journal of Wildlife Management* 75:458–462.

Catlin, D. H., J. D. Fraser, J. H. Felio, and J. B. Cohen. 2011. Piping Plover habitat selection and nest success on natural, managed, and engineered sandbars. *Journal of Wildlife Management* 75:305–310.

Caughley, G. 1994. Directions in conservation biology. *Journal of Animal Ecology* 63:215–244.

Cohen, J. B., A. Hecht, K. F. Robinson, E. E. Osnas, A. J. Tyre, C. Davis, A. Kocek, B. Maslo, and S. M. Melvin. 2016. To exclose nests or not: Structured decision making for the conservation of threatened species. *Ecosphere* 7:e01499.

Cohen, J. B., L. M. Houghton, and J. D. Fraser. 2009. Nesting density and reproductive success of Piping Plovers in response to storm- and human-created habitat changes. *Wildlife Monographs* 173:1–24.

Colwell, M. A. 2006. Egg-laying intervals in shorebirds. *Wader Study Group Bulletin* 111:50–59.

Colwell, M. A., N. S. Burrell, M. A. Hardy, K. Kayano, J. J. Muir, W. J. Pearson, S. A. Peterson, and K. A. Sesser. 2010. Arrival times, laying dates, and reproductive success of Snowy Plovers in two habitats in coastal northern California. *Journal of Field Ornithology* 81:349–360.

Colwell, M. A., E. J. Feucht, M. J. Lau, D. J. Orluck, S. E. McAllister, and A. N. Transou. 2017. Recent Snowy Plover population increase arises from high immigration rate in coastal northern California. *Wader Study* 124:40–48.

Colwell, M. A., S. E. McAllister, C. B. Millett, A. N. Transou, S. M. Mullin, Z. J. Nelson, C. A. Wilson, and R. R. LeValley. 2007. Philopatry and natal dispersal of the Western Snowy Plover. *Wilson Journal of Ornithology* 119:378–385.

Colwell, M. A., J. J. Meyer, M. A. Hardy, S. E. McAllister, A. N. Transou, R. R. LeValley, and S. J. Dinsmore. 2011. Western Snowy Plovers (*Charadrius alexandrinus nivosus*) select nesting substrates that enhance egg crypsis and improve nest survival. *Ibis* 153:303–31.

Cresswell, W. 2008. Non-lethal effects of predation in birds. *Ibis* 150:3–17.

Cresswell, W., and J. L. Quinn. 2004. Faced with a choice, sparrowhawks more often attack the more vulnerable prey group. *Oikos* 104:71–76.

Crutzen, P. J. 2002. Geology of mankind: The anthropocene. *Nature* 415:23.

Dinsmore, S. J., E. P. Gaines, S. F. Pearson, D. J. Lauten, and K. A. Castelein. 2017. Factors affecting Snowy Plover chick survival in a managed population. *Condor* 119:34–43.

Dinsmore, S. J., D. J. Lauten, K. A. Castelein, E. P. Gaines, and M. A. Stern. 2014. Predator exclosures, predator removal, and habitat improvement increase nest success of Snowy Plovers in Oregon, USA. *Condor* 116:619–628.

Dowding, J. E., and E. C. Murphy. 2001. The impact of predation by introduced mammals on endemic shorebirds in New Zealand: A conservation perspective. *Biological Conservation* 99:47–64.

Dyrcz, A., J. Witkowski, and J. Okulewicz. 1981. The nesting of "timid" waders in the vicinity of "bold" ones as an anti-predator adaptation. *Ibis* 123:542–545.

Ebenhard, T. 1995. Conservation breeding as a tool for saving animal species from extinction. *Trends in Ecology and Evolution* 10:438–443.

Eberhart-Phillips, L. J., and M. A. Colwell. 2014. Conservation challenges of a sink: The viability of an isolated population of the Snowy Plover. *Bird Conservation International* 24:327–342.

Eberhart-Phillips, L. J., B. R. Hudgens, and M. A. Colwell. 2015. Spatial synchrony of a threatened shorebird: Regional roles of climate, dispersal and management. *Bird Conservation International* 25:119–135.

Ekanayake, K. B., M. A. Weston, D. G. Nimmo, G. S. Maguire, J. A. Endler, and C. Küpper. 2015. The bright incubate at night: Sexual dichromatism and adaptive incubation division in an open-nesting shorebird. *Proceedings of the Royal Society B* 282:20143026.

Espie, R. H. M., R. M. Brigham, and P. C. James. 1996. Habitat selection and clutch fate of Piping Plovers (*Charadrius melodus*) breeding Lake Diefenbaker, Saskatchewan. *Canadian Journal of Zoology* 74:1069–1075.

Gochfeld, M. 1984. Antipredator behavior: Aggressive and distraction displays of shorebirds. Pp. 289–377 in J. Burger and B. L. Olla (editors), *Shorebirds: Breeding Behavior and Populations*. Plenum Press, New York.

Gómez-Serrano, M. A., and P. López-López. 2016. Deceiving predators: Linking distraction behavior with nest survival in a ground-nesting bird. *Behavioral Ecology* 28:260–269.

Goosen, P. J., R. V. Lee, C. Kruse, C. L. Gratto-Trevor, and S. M. Westworth. 2011. Resightings of captive-reared and wild Piping Plovers from Saskatchewan, Canada. *Wader Study Group Bulletin* 118:1–9.

Graul, W. D. 1973. Possible functions of head and breast markings in Charadriinae. *Wilson Bulletin* 85:60–70.

Hanski, I., and M. Gilpin. 1991. Metapopulation dynamics: Brief history and conceptual domain. *Biological Journal of the Linnaean Society* 42:3–16.

Hardy, M. A., and M. A. Colwell. 2008. The impact of predator exclosures on Snowy Plover nesting success: A seven-year study. *Wader Study Group Bulletin* 115:161–166.

Herman, D. M., and M. A. Colwell. 2015. Lifetime reproductive success of Snowy Plovers in coastal northern California. *Condor* 117:473–481.

Hudgens, B. H., L. Eberhart-Phillips, L. Stenzel, C. Burns, M. Colwell, and G. Page. 2014. Population viability analysis of the Western Snowy Plover. Report prepared for the U.S. Fish and Wildlife Service, Arcata, CA.

Ivan, J. S., and R. K. Murphy. 2005. What preys on piping plover eggs and chicks? *Wildlife Society Bulletin* 33:113–119.

Jackson, D. B., R. J. Fuller, and S. T. Campbell. 2004. Long-term population changes among breeding shorebirds in the Outer Hebrides, Scotland, in relation to introduced hedgehogs (*Erinaceus europaeus*). *Biological Conservation* 117:151–166.

Johnson, M., and L. W. Oring. 2002. Are nest exclosures an effective tool in plover conservation? *Waterbirds* 25:184–190.

Keedwell, R. J., and M. D. Sanders. 2002. Nest monitoring and predator visitation at nests of Banded Dotterels. *Condor* 104:899–902.

Kelly, J. P., K. L. Etienne, and J. E. Roth. 2002. Abundance and distribution of the Common Raven and American Crow in the San Francisco Bay area, California. *Western Birds* 33:202–217.

Knopf, F. L., and B. J. Miller. 1994. *Charadrius montanus*: Montane, grassland, or bare-ground Plover? *Auk* 111:504–506.

Lafferty, K. D., D. Goodman, and C. P. Sandoval. 2006. Restoration of breeding snowy plovers following protection from disturbance. *Biodiversity and Conservation* 15:2217–2230.

Lank, D. B., R. W. Butler, J. Ireland, and R. C. Ydenberg. 2003. Effects of predation danger on migration strategies of sandpipers. *Oikos* 103:303–319.

Lauten, D. J., K. A. Castelein, J. D. Farrar, A. A. Kotaich, and E. P. Gains. 2015. *The Distribution and Reproductive Success of the Western Snowy Plover along the Oregon Coast—*2015. Oregon Biodiversity Information Center, Portland, OR.

Lavers, J. L., C. Wilcox, and C. J. Donlan. 2010. Bird demographic responses to predator removal programs. *Biological Invasions* 12:3839–3859.

Leja, S. D. 2015. Response of Snowy Plovers to human-created and naturally restored habitats. M.Sc. thesis, Humboldt State University, Arcata, CA.

Levins, R. 1970. Extinction. Pp. 77–107 in M. Gertenhaber (editor), *Mathematical Problems in Biology*. American Mathematical Society, Providence, RI.

Luttberg, B., and J. L. Kerby. 2005. Are scared prey as good as dead? *Trends in Ecology and Evolution* 20:416–418.

Lyon, B., and R. Montgomerie. 1987. Ecological correlates of incubation feeding: A comparative study of high Arctic finches. *Ecology* 68:713–722.

Lyons, J. E., M. C. Runge, H. P. Laskowski, and W. L. Kendall. 2008. Monitoring in the context of structured decision-making and adaptive management. *Journal of Wildlife Management* 72:1683–1692.

Mabee, T. J., and V. B. Estelle. 2000. Assessing the effectiveness of predator exclosures for plovers. *Wilson Bulletin* 112:14–20.

Maguire, G. S., A. K. Duivenvoorden, M. A. Weston, and R. Adams. 2011. Provision of artificial shelter on beaches is associated with improved shorebird fledging success. *Bird Conservation International* 21:172–185.

Maguire, G. S., D. Stojanovic, and M. A. Weston. 2009. Conditioned taste aversion reduces fox depredation on model eggs on beaches. *Wildlife Research* 36:702–708.

Malone, C. R., and V. W. Proctor. 1966. Rearing Killdeers for experimental purposes. *Journal Wildlife Management* 30:589–594.

Marcot, B. G., and D. C. Elbert. 2015. Assessing management of raptor predation for Western Snowy Plover recovery. USDA Forest Service General Technical Report PNW-GTR-910. USDA Forest Service Publishers, Portland, OR.

Martin, T. E. 1995. Avian life history evolution in relation to nest sites, nest predation, and food. *Ecological Monographs* 65:101–127.

Maslo, B., J. Burger, and S. N. Handel. 2012. Modeling foraging behavior of Piping Plovers to evaluate habitat restoration success. *Journal of Wildlife Management* 76:181–188.

McNamara, J. M., and A. I. Houston. 1987. Starvation and predation as factors limiting population size. *Ecology* 68:1515–1519.

Muir, J. J., and M. A. Colwell. 2010. Snowy Plovers select open habitats for courtship scrapes and nests. *Condor* 112:507–510.

Mullin, S. M., M. A. Colwell, S. E., McAllister, and S. J. Dinsmore. 2010. Apparent survival and population growth of Snowy Plovers in coastal northern California. *Journal of Wildlife Management* 74:1792–1798.

Neuman, K. K., G. W. Page, L. E. Stenzel, J. C. Warriner, and J. S. Warriner. 2004. Effect of mammalian predator management on Snowy Plover breeding success. *Waterbirds* 27:257–263.

Neuman, K. K., L. E. Stenzel, J. C. Warriner, G. W. Page, J. L. Erbes, C. R. Eyster, E. Miller, and L. A. Henkel. 2013. Success of captive-rearing for a threatened shorebird. *Endangered Species Research* 22:85–94.

Newton, I. 1994. Experiments on the limitation of bird breeding densities: A review. *Ibis* 136:397–411.

Newton, I., and I. Wyllie. 1992. Recovery of a sparrowhawk population in relation to declining pesticide contamination. *Journal of Applied Ecology* 29:476–484.

Nol, E., and R. J. Brooks. 1982. Effects of predator exclosures on nesting success of Killdeer. *Journal of Field Ornithology* 53:263–268.

Nur, N., G. W. Page, and L. E. Stenzel. 1999. *Population Viability Analysis for Pacific Coast Western Snowy Plovers.* Point Reyes Bird Observatory, Stinson Beach, CA.

Nyugen, L. P., K. F. Abraham, and E. Nol. 2006. The influence of Arctic Terns on survival of artificial and natural Semipalmated Plover nests. *Waterbirds* 29:100–104.

Nyugen, L. P., E. Nol, K. F. Abraham, and C. Lishman. 2013. Directional selection and repeatability in nest-site preferences in Semipalmated Plovers (*Charadrius semipalmatus*). *Canadian of Journal Zoology* 91:646–652.

Oring, L. W., and D. B. Lank. 1984. Breeding area fidelity, natal philopatry, and the social system of sandpipers. Pp. 125–147 in J. Burger and B. L. Olla (editors), *Shorebirds: Breeding Behavior and Populations.* Plenum Press, New York.

Page, G. W., K. K. Neuman, J. C. Warriner, C. Eyster, J. Erbes, D. Dixon, A. Palkovic, and L. E. Stenzel. 2016. *Nesting of the Snowy Plover in the Monterey Bay area, California in 2015.* Point Blue Conservation Science, Petaluma, CA.

Page, G. W., P. L. Quinn, and J. C. Warriner. 1989. Comparison of the breeding of hand-reared and wild-reared Snowy Plovers. *Conservation Biology* 3:198–201.

Page, G. W., and L. E. Stenzel (editors). 1981. Breeding population size of the Snowy Plover in California. *Western Birds* 12:1–40.

Page, G. W., L. E. Stenzel, D. W. Winkler and C. W. Swarth. 1983. Spacing out at Mono Lake: Breeding success, nest density, and predation in the Snowy Plover. *Auk* 100:13–24.

Page, G. W., L. E. Stenzel, W. D. Shuford, and C. R. Bruce. 1991. Distribution and abundance of the Snowy Plover on its western North American breeding grounds. *Journal of Field Ornithology* 62:245–255.

Page, G. W., and D. F. Whitacre. 1975. Raptor predation on wintering shorebirds. *Condor* 77:73–83.

Patrick, A. K. M., and M. A. Colwell. 2018. Annual variation in distance to nearest neighbor nest decreases with population size in Snowy Plovers. *Wader Study* 124:215–224.

Pearson, W. J., and M. A. Colwell. 2014. Effects of nest success and mate fidelity on breeding dispersal in a population of Snowy Plovers *Charadrius nivosus*. *Bird Conservation International* 24:342–353.

Peterson, S. A., and M. A. Colwell. 2014. Experimental evidence that scare tactics and effigies reduce corvid occurrence. *Northwestern Naturalist* 95:103–112.

Pienkowski, M. W. 1984. Breeding biology and population dynamics of Ringed plovers *Charadrius hiaticula* in Britain and Greenland: Nest-predation as a possible factor limiting distribution and timing of breeding. *Journal of Zoology* 202:83–114.

Pomeroy, A. C., R. W. Butler, and R. C. Ydenberg. 2006. Experimental evidence that migrants adjust usage at a stopover site to trade off food and danger. *Behavioral Ecology* 17:1041–1045.

Powell, A. N., and C. L. Collier. 2000. Habitat use and reproductive success of Western Snowy Plovers at new nesting areas created for California Least Terns. *Journal of Wildlife Management* 64:24–33.

Powell, A. N., and F. J. Cuthbert. 1993. Augmenting small populations of plovers: An assessment of cross-fostering and captive-rearing. *Conservation Biology* 7:160–168.

Powell, A. N., F. J. Cuthbert, L. C. Wemmer, A. W. Doolittle, and S. T. Feirer. 1997. Captive-rearing Piping Plovers: Developing techniques to augment wild populations. *Zoo Biology* 16:461–477.

Que, P., Y. Chang, L. Eberhart-Phillips, Y. Liu, T. Székely, and Z. Zhang. 2015. Low nest survival of a breeding shorebird in Bohai Bay, China. *Journal of Ornithology* 156:297–307.

Ratcliffe, D. A. 1980. *The Peregrine Falcon.* T. and A. D. Poyser, London.

Rees, J. D., J. K. Webb, M. S. Crowther, and M. Letnic. 2015. Carrion subsidies provided by fishermen increase predation of beach-nesting bird nests by facultative scavengers. *Animal Conservation* 18:44–49.

Riensche, D. L., S. C. Gidre, N. A. Beadle, and S. K. Riensche. 2015. Western Snowy Plover (*Charadrius alexandrinus nivosus*) nest site selection and oyster shell enhancement. *Western Wildlife* 2:38–43.

Robinette, D. P., J. Miller, and J. Howar. 2016. *Monitoring and Management of the Endangered California Least Tern and the Threatened Western Snowy Plover at Vandenberg Air Force Base, 2015*. Point Blue Conservation Science, Petaluma, CA. No. 2065.

Roche, E. A., F. J. Cuthbert, and T. W. Arnold. 2008. Relative fitness of wild and captive-reared piping plovers: Does egg salvage contribute to recovery of the endangered Great Lakes population? *Biological Conservation* 141:3079–3088.

Ruhlen, T.D., S. Abbott, L.E. Stenzel, and G.W. Page. 2003. Evidence that human disturbance reduces Snowy Plover chick survival. *Journal of Field Ornithology* 74:300–304.

Sæther, B.-E., and Ø. Bakke. 2000. Avian life history variation and contribution of demographic traits to the population growth rate. *Ecology* 81:642–653.

Safriel, U. N. 1980. The Semipalmated Sandpiper: Reproductive strategies and tactics. *Ibis* 122:425.

Sanders, M. D., and R. F. Maloney. 2002. Causes of mortality at nests of ground-nesting birds in the Upper Waitaki Basin, South Island, New Zealand: A 5-year video study. *Biological Conservation* 106:225–236.

Smith, P. A., H. G. Gilchrist, and J. N. M. Smith. 2007. Effects of nest habitat, food and parental behavior on shorebird nest success. *Condor* 109:15–31.

Stenzel, L. E., G. W. Page, J. C. Warriner, J. S. Warriner, D. E. George, C. R. Eyster, B. A. Ramer, and K. K. Neuman. 2007. Survival and natal dispersal of juvenile Snowy Plovers (*Charadrius alexandrinus*) in central coastal California. *Auk* 124:1023–1036.

Stenzel, L. E., G. W. Page, J. C. Warriner, J. S. Warriner, K. K. Neuman, D. E. George, C. R. Eyester, and F. C. Bidstrup. 2011. Male-skewed adult sex ratio, survival, mating opportunity and annual productivity in the Snowy Plover *Charadrius alexandrinus*. *Ibis* 153:31–322.

Stenzel, L. E., J. C. Warriner, J. S. Warriner, K. S. Wilson, F. C. Bidstrup, and G. W. Page. 1994. Long-distance breeding dispersal of Snowy Plovers in western North America. *Journal of Animal Ecology* 63:887–902.

Summers, R. W., and P. A. R. Hockey, 1980. Breeding biology of the White-fronted Plover (*Charadrius marginatus*) in the south-western cape, South Africa. *Journal of Natural History* 14:433–445.

Tan, L. X. L., K. L. Buchanan, G. S. Maguire, and M. A. Weston. 2015. Cover, not caging, influences chronic physiological stress in a ground-nesting bird. *Journal of Avian Biology* 46:482–488.

Thibault, M., and R. McNeil. 1995. Day- and night-time parental investment by incubating Wilson's Plovers in a tropical environment. *Canadian Journal of Zoology* 73:879–886.

Troscianko, J., J. Wilson-Aggarwal, C. N. Spottiswoode, and M. Stevens. 2016. Nest covering in plovers: How modifying the visual environment influences egg camouflage. *Ecology and Evolution* 6:7536–7545.

United States Fish and Wildlife Service. 1993. Threatened status for the Pacific coast population of the Western Snowy Plover. *Federal Register* 58:12864–12874.

United States Fish and Wildlife Service. 2007. *Recovery Plan for the Pacific Coast Population of the Western Snowy Plover (Charadrius alexandrinus nivosus)*. United States Fish and Wildlife Service, Sacramento, CA.

Van den Hout, P.J., B. Spaans, and T. Piersma. 2008. Differential mortality of wintering shorebirds on the Banc d'Arguin, Mauritania, due to predation by large falcons. *Ibis* 150 (Suppl. 1):219–230.

Walters, C. 1986. *Adaptive Management of Renewable Resources*. MacMillan Publishing Company, New York.

Watts, C. M., J. Cao, C. Panza, C. Dugaw, M. Colwell, and E. A. Burroughs. 2012. Modeling the effects of predator exclosures on a Western Snowy Plover population. *Natural Resource Modeling* 25:529–547.

Weston, M. A., and M. A. Elgar. 2005. Disturbance to brood-rearing Hooded Plover *Thinornis rubicollis*: Responses and consequences. *Bird Conservation International* 15:193–209.

Whitfield, D. P. 1985. Raptor predation on wintering waders in southeast Scotland. *Ibis* 127:544–558.

Ram Papish

CHAPTER SEVEN

Evolutionary and Ecological Flexibility in Migration of Charadrius Plovers*

Jesse R. Conklin

Abstract. Plovers of the subfamily Charadriinae demonstrate nearly every conceivable annual movement pattern found among birds, including long- and short-distance latitudinal migration, altitudinal migration, irruption, nomadism, and sedentary habits. Within species, and even within some populations, inter-individual differences may range from completely sedentary birds to some of the longest avian migrations yet recorded. This extreme variation makes plovers an interesting group in which to explore the ecological and evolutionary drivers and consequences of movement patterns. In this chapter, I review patterns of movement according to geography, molt, wing morphology, various aspects of annual-cycle strategies, and the known evolutionary history of the group. Of 40 plover species, 26 (65%) are migratory to some degree, and these are found on nearly all global migratory flyways, with the greatest number on the East Asian-Australasian Flyway. The species diversity and the proportion of nonmigratory plovers are highest in the Eastern and Southern Hemispheres. Migration distance increases with breeding latitude; the longest migrations are performed by arctic-breeding plovers, and no Southern Hemisphere breeders migrate further north than the tropics. In general, individual migration strategies are poorly described, but include both short-hop migrations and nonstop flights of at least 5,300 km and perhaps more than 7,000 km in some species. Wing shape varies with migration distance, suggesting that movement patterns and morphology coevolve to some degree, and that some species may be predisposed to greater flexibility in movements. Repeated loss of migration appears to be a significant form of diversification in plovers; recent phylogenetic evidence supports historical radiation from a northern migratory ancestor, and current distributions of sedentary species, particularly on islands, suggest a pattern of isolation from mainland migratory species. Although some species show evidence of evolutionary constraints on migration routes, the present-day diversity of movement patterns implies great flexibility to respond to changing circumstances at multiple time scales. However, the general lack of specific information regarding routes and habitats used during migration is a major obstacle to developing effective global conservation strategies for migratory plovers.

Keywords: biogeography, evolution, flyway, intraspecific variation, molt, non-stop flight, partial migration, sedentary, wing morphology.

* Jesse R. Conklin. 2019. Evolutionary and Ecological Flexibility in Migration of Charadrius Plovers. Pp. 149–182 in Colwell and Haig (editors). The Population Ecology and Conservation of Charadrius Plovers (no. 52), CRC Press, Boca Raton, FL.

Migration defines and organizes many facets of the ecology and life histories of birds, particularly for those that breed in the Arctic and other highly seasonal environments (McNamara et al. 2008, Wingfield 2008). Migration represents two distinct periods of the annual cycle, that is, pre-breeding and postbreeding movements that (1) require preparation and recovery, (2) may entail disproportionate exertion and risk, and (3) must be scheduled strategically with regard to other critical tasks such as reproduction and molt (Alerstam 1990, Alerstam and Lindström 1990). Immense variation exists in the way birds conduct migration and resolve its numerous challenges (Newton 2008). We expect migratory tendencies to be relatively consistent among closely related taxa, due to their shared history and perhaps ecology. When this is not true, it provides a rich context for comparative study and an opportunity to explore the evolutionary and ecological drivers and consequences of migration strategies. The plovers of the subfamily Charadriinae (Table 7.1) represent one of these opportunities.

At one extreme, the Common Ringed Plover features the most spectacular migrations of any plover and remarkable within-species variation. It breeds across an area of ~4.5 million km^2 from central Canada eastward to the Russian Far East, with two fully migratory subspecies (*C. h. psammodromus* and *tundrae*) breeding in the Arctic and sub-Arctic, and a third partially migratory subspecies (*C. h. hiaticula*) breeding in temperate Europe (Figure 7.1). This results in a "leap-frog" migration pattern both within and among subspecies, wherein the lowest-latitude breeders are sedentary, and higher-latitude breeders overfly these birds to spend the nonbreeding season at more southerly latitudes (Salomonsen 1955, Taylor 1980). Despite the vast breeding range (280° of longitude extending from 90°W to 170°E), all individuals spend the nonbreeding season in Europe, Africa, and southwestern Asia (87° of longitude extending from 17°W to 70°E; Figure 7.1). Therefore, the migration of *C. h. psammodromus* necessarily includes nonstop flights of at least 800–2,000 km across the North Atlantic, and the most northeasterly breeders of *C. h. tundrae* travel one-way distances of 9,000–13,000 km (Tomkovich et al. 2017), and perhaps more than 16,000 km to southern Africa (Underhill et al. 1999), which would be the longest migration of any plover.

At the other end of the migratory spectrum, the St. Helena Plover is completely sedentary and inhabits a tropical island of just 121 km^2 year round (Figure 7.1). Bridging these two extremes, the 40 plover species (Table 7.1) include examples of nearly every known type of movement pattern, including short to transhemispheric latitudinal and/or longitudinal migrations, seasonal shifts in altitude, irregular or nomadic movements in response to unpredictable resources, and completely sedentary habits. These patterns may vary dramatically among populations in a species, or even among individuals in a population. Rather than a highly conserved trait, the propensity for seasonal movements in plovers can be best described as flexible: the ecology and evolutionary history of the group have resulted in migratory tendencies that are impressively plastic across time and space.

Before embarking on a more detailed exploration of plover migration, we must identify what distinguishes migration from other types of movement. Migration is surprisingly difficult to define, as it takes many forms and has various meanings in different contexts (Dingle and Drake 2007, Newton 2008). Categorizing types of movement is necessarily arbitrary, as interpretations are scale dependent and such complex behavior is best represented as a continuum without clear divisions. For this chapter, I consider "migration" to include predictable, annually repeated, two-way movements between breeding and nonbreeding sites (see Box 7.1 for definitions of migration-related terminology used in this chapter). This clearly distinguishes it from within-season two-way movements, such as foraging trips away from the breeding site, and from non-repeated one-way movements, such as dispersal. This also means that facultative, unpredictable patterns such as irruptive movements and nomadism are not considered migration here.

By this definition, 14 of the 40 plover species (35%) are nonmigratory (Table 7.1). Four of these are considered entirely sedentary, with adults remaining on or near their breeding territories year round. Eight other species are largely resident year round in their breeding ranges, but show evidence of small local movements outside the breeding season or larger irregular movements according to unpredictable conditions, such as rains or drought. Two Australian species (Inland Dotterel, Red-capped Plover) have poorly understood movements within their large year-round ranges; these species may be essentially nomadic

TABLE 7.1

Breeding range, nonbreeding range, and migratory status of plovers, by subspecies.

Common name	Latin name/ subspecies	Breeding range	Nonbreeding range	Migratory status	Description
Southern Red-breasted Dotterel	*Charadrius obscurus*	Stewart Island, New Zealand	Stewart and South Island, New Zealand	Full	Short-distance altitudinal migrant
Northern Red-breasted Dotterel	*C. aquilonius*	North Island, New Zealand	Same	Non	Resident w/some local movements
Lesser Sand-plover	*C. m. mongolus*	Russian Far East to nw Alaska	Southeast Asia to Australia	Full	Long-distance migrant
	C. m. pamirensis	Pamirs to w China	SW Asia to se Africa	Full	Medium- to long-distance migrant
	C. m. atrifrons	Himalayas, s Tibet	S Asia to Indonesia	Full	Medium- to long-distance migrant
	C. m. schaeferi	E Tibet to s Mongolia	S Asia to Indonesia	Full	Medium- to long-distance migrant
	C. m. stegmanni	Kamchatka to Chukotka	Southeast Asia to Australia	Full	Long-distance migrant
Greater Sand-plover	*C. l. leschenaultii*	W China to s Mongolia, s Siberia	Southeast Asia to Australia	Full	Medium- to long-distance migrant
	C. l. columbinus	Turkey to s Afghanistan	Mediterranean and Red Seas	Full	Short- to medium-distance migrant
	C. l. scythicus	Transcaspia to se Kazakhstan	S Asia to se Africa	Full	Medium- to long-distance migrant
Caspian Plover	*C. asiaticus*	Caspian Sea to w China	S and e Africa	Full	Medium- to long-distance migrant
Collared Plover	*C. collaris*	Mexico to n Argentina, c Chile	Same	Non	Generally resident w/irregular movements
Puna Plover	*C. alticola*	Andes of Peru to nw Argentina	Andes of Peru to nw Argentina, coastal s Peru and n Chile	Partial	Sedentary to short-distance altitudinal migrant
Two-banded Plover	*C. falklandicus*	S Chile, Argentina and Falkland Islands	S South America	Partial	Sedentary to short-distance migrant
Double-banded Plover	*C. bicinctus*	New Zealand, Chatham and Auckland Islands	New Zealand, s and e Australia, New Caledonia, Vanuatu, Fiji	Partial	Sedentary to medium-distance migrant
Kittlitz's Plover	*C. pecuarius*	Sub-Saharan Africa, ne Egypt, Madagascar	Same	Partial	Sedentary to short-distance migrant
Red-capped Plover	*C. ruficapillus*	Australia and Tasmania	Same	Non	Irregular movements
Malay Plover	*C. peronii*	Southeast Asia to Philippines, Indonesia	Same	Non	Generally resident w/dispersive movements

(Continued)

TABLE 7.1 (*Continued*)
Breeding range, nonbreeding range, and migratory status of plovers, by subspecies.

Common name	Latin name/ subspecies	Breeding range	Nonbreeding range	Migratory status	Description
Kentish Plover	*C. a. alexandrinus*	W Palearctic to e China, Japan	SW Europe and Northwest Africa to Southeast Asia	Partial	Sedentary to medium-distance migrant
	C. a. seebohmi	SE India, Sri Lanka	Same	Non	Sedentary
	C. a. nihonensis	Sakhalin and s Kuril Islands	E Asia, Philippines, Indonesia	Full	Medium- to long-distance migrant
White-faced Plover	*C. dealbatus*	SE China to Thailand	Southeast Asia to w Indonesia	Full	Short- to medium-distance migrant
Snowy Plover	*C. n. nivosus*	USA to Mexico, West Indies	USA to Mexico and Caribbean	Partial	Sedentary to medium-distance migrant
	C. n. occidentalis	Coastal Peru to s-c Chile	Same	Non	Sedentary
Javan Plover	*C. javanicus*	Coastal Java, Bali, Kangean Islands	Same	Non	Sedentary
Wilson's Plover	*C. w. wilsonia*	E USA to Belize, West Indies	Gulf of Mexico to e Brazil	Partial	Sedentary to long-distance migrant
	C. w. beldingi	Baja California, Mexico to Ecuador	Baja California, Mexico to s Peru	Partial	Sedentary to medium-distance migrant
	C. w. cinnamominus	Colombia to French Guiana and Caribbean	Same	Non	Sedentary
	C. w. crassirostris	NE Brazil	Same	Non	Sedentary
Common Ringed Plover	*C. h. hiaticula*	W Europe to s Scandinavia	W Europe to Africa	Partial	Sedentary to medium-distance migrant
	C. h. psammodromus	NE Canada, Greenland to Svalbard	SW Europe to West Africa	Full	Medium- to long-distance migrant
	C. h. tundrae	N Scandinavia, n Russia	SW Asia to s Africa	Full	Long-distance migrant
Semipalmated Plover	*C. semipalmatus*	Nearctic, e Chukotka	S USA to South America	Full	Medium- to long-distance migrant
Long-billed Plover	*C. placidus*	E Asia	Japan to Southeast Asia	Partial	Sedentary to long-distance migrant
Piping Plover	*C. m. melodus*	Atlantic coast of North America	S USA and Caribbean	Full	Short- to medium-distance migrant
	C. m. circumcinctus	Great Lakes, prairies of North America	S USA and Caribbean	Full	Medium-distance migrant
Madagascar Plover	*C. thoracicus*	SW Madagascar	Same	Non	Sedentary

(*Continued*)

TABLE 7.1 (*Continued*)

Breeding range, nonbreeding range, and migratory status of plovers, by subspecies.

Common name	Latin name/ subspecies	Breeding range	Nonbreeding range	Migratory status	Description
Little Ringed Plover	*C. d. dubius*	Philippines to New Guinea, Bismarck Archipelago	Same	Non	Resident and locally nomadic
	C. d. curonicus	Palearctic	Africa to Indonesia and New Guinea	Partial	Sedentary to long-distance migrant
	C. d. jerdoni	India, Southeast Asia	Same	Non	Generally resident w/irregular movements
African Three-banded Plover	*C. tricollaris*	Ethiopia to Tanzania, Gabon, Chad and South Africa	Sub-Saharan Africa	Non	Generally resident w/irregular movements
Madagascar Three-banded Plover	*C. bifrontalis*	Madagascar	Same	Non	Sedentary
Forbes's Plover	*C. forbesi*	C Africa	W and c Africa	Full	Short- to medium-distance migrant
White-fronted Plover	*C. m. marginatus*	S Angola to sw Cape Province	Same	Non	Sedentary
	C. m. mechowi	Sub-Saharan Africa to n Angola, Botswana, Mozambique	Same	Partial	Sedentary to short-distance migrant
	C. m. arenaceus	S Mozambique to s Cape Province	Same	Non	Sedentary
	C. m. tenellus	Madagascar	Same	Non	Sedentary
Chestnut-banded Plover	*C. p. pallidus*	S Africa	Same	Partial	Nomadic to short-distance migrant
	C. p. venustus	Rift Valley of Kenya/Tanzania border	Same	Partial	Nomadic to short-distance migrant
Killdeer	*C. v. vociferus*	North America	USA to Central America and Caribbean	Partial	Sedentary to long-distance migrant
	C. v. temominatus	Greater Antilles	Same	Non	Sedentary
	C. v. peruvianus	Peru, nw Chile	Same	Non	Sedentary
Mountain Plover	*C. montanus*	Great Plains of w Canada and USA	SW USA and n Mexico	Full	Short- to medium-distance migrant
Oriental Plover	*C. veredus*	S Siberia, Mongolia, ne China	Indonesia to s Australia	Full	Long-distance migrant

(Continued)

TABLE 7.1 (*Continued*)

Breeding range, nonbreeding range, and migratory status of plovers, by subspecies.

Common name	Latin name/ subspecies	Breeding range	Nonbreeding range	Migratory status	Description
Eurasian Dotterel	*C. morinellus*	N Palearctic to w Alaska	N Africa and Middle East	Full	Medium- to long-distance migrant
St. Helena Plover	*C. sanctaehelanae*	St. Helena Island	Same	Non	Sedentary
Rufous-chested Dotterel	*C. modestus*	S Chile and Argentina, Falkland Islands	S South America	Partial	Sedentary to short-distance migrant
Red-kneed Dotterel	*Erythrogonys cinctus*	Australia, s New Guinea	Same	Non	Generally resident w/irregular movements
Hooded Plover	*Thinornis cucullatus*	S Australia, Tasmania	Same	Non	Resident w/some local movements
Shore Plover	*T. novaeseelandiae*	Rangatira Island, New Zealand	Same	Non	Resident w/some local movements
Black-fronted Dotterel	*Elseyornis melanops*	Australia, Tasmania, New Zealand	Same	Non	Generally resident w/irregular movements
Inland Dotterel	*Peltohyas australis*	S and c Australia	Same	Non	Irregular movements, semi-nomadic
Wrybill	*Anarhynchus frontalis*	South Island, New Zealand	New Zealand	Full	Short-distance migrant

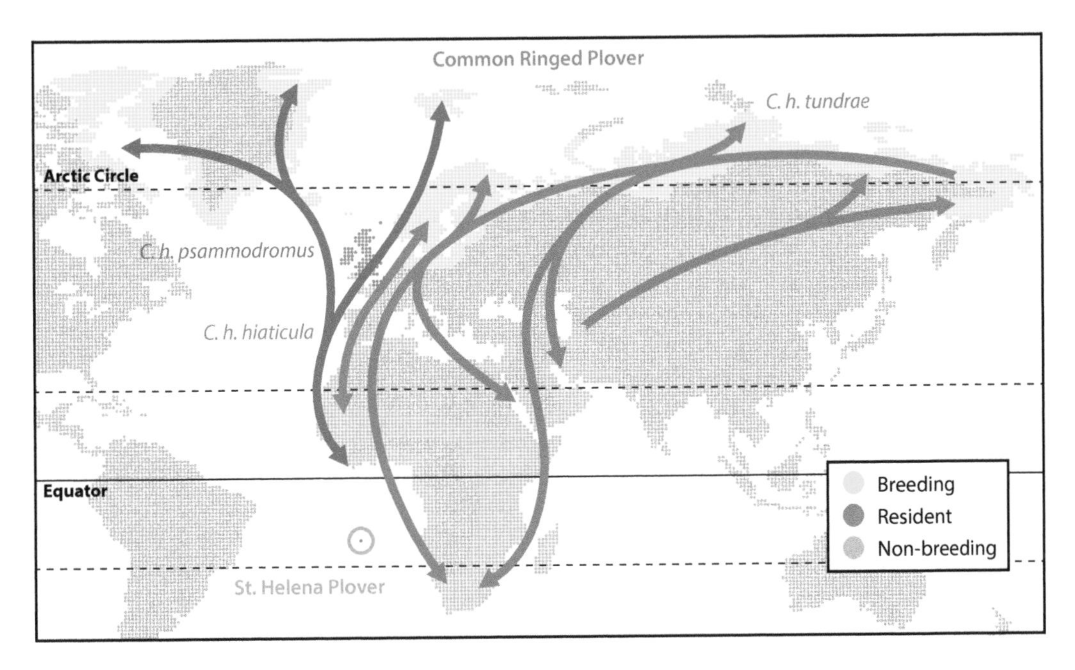

Figure 7.1. Distribution and migration of Common Ringed Plover, which has the greatest within-species variation in migration of any Charadrius plover. For comparison, the year-round range of the sedentary St. Helena Plover is indicated by the orange circle.

BOX 7.1 Glossary of Migration-Related Terminology

Altitudinal migration: migratory movements between higher and lower elevations in the same region.

Contour feathers: feathers that cover the body and contribute to thermoregulation and appearance, but do not power flight.

Differential migration: the situation in which individuals from the same breeding site differ (e.g., by age or sex) in the distance or timing of migration.

Dispersal: one-way movements from a point of origin, such as from an individual's natal site to its adult breeding site, or between breeding sites in successive years.

Flight feathers: feathers that help steer and power bird flight. These include those of the wing, or remiges (primary feathers connected to the bird's "hand," and secondary feathers connected to the "wrist"), and those of the tail, or retrices.

Flyway: a geographical area representing the seasonal connection and flight path (generally along a north–south axis) between breeding and nonbreeding areas used by numerous species of migratory birds. See Figure 7.2.

Fueling: accumulation of nutrient stores to enable migratory flights. This primarily involves storage of energy-rich fat, but also of protein and glycogen.

Fully migratory: a state in which all individuals in a population perform migration.

Irruption: movements from traditional breeding areas that vary substantially among years in direction, spatial scale, and the proportion of a population undertaking them.

Last Glacial Maximum: the period of approximately 18,000–22,000 years before present, when terrestrial habitats, particularly in northern latitudes, were most restricted by the latest Pleistocene glaciation.

(Continued)

BOX 7.1 (*Continued*) Glossary of Migration-Related Terminology

Leap-frog migration: the situation in which individuals breeding at high latitudes migrate beyond the nonbreeding range of lower-latitude breeders, such that the latitudinal distribution is reversed in the nonbreeding season.

Long-jump migration: a strategy in which the distance between breeding and nonbreeding sites is covered in one or a few long flights of several days, each requiring significant staging periods for fueling.

Loop migration: annual movements that involve a different flight path on northward and southward migrations.

Migration: predictable, annually repeated, two-way movements of individuals between traditional (i.e., used seasonally every year) breeding and nonbreeding sites.

Nomadism: irregular (i.e., nonmigratory) movements according to resources that vary unpredictably in time and space, such that neither breeding nor nonbreeding sites are used traditionally by individuals or the population.

Partial migration: the situation in which a population or species includes both migratory and nonmigratory individuals.

Pre-alternate molt: an annual molt that typically takes place prior to the breeding season, representing the transition from basic (nonbreeding or winter) plumage to alternate (breeding or nuptial) plumage. In general, this is a partial molt, involving only some proportion of contour feathers.

Pre-basic molt: an annual molt that typically takes place after the breeding season, representing the transition from alternate plumage to basic plumage. In general, this is a complete molt, involving replacement of all (flight and contour) feathers.

Sedentary: remaining in the breeding range year round, i.e., nonmigratory.

Short-hop migration: a strategy in which the migration distance is covered in numerous short flights of 1 day or less, and only brief stops for fueling.

Stopover and staging sites: places where birds rest and acquire fuel for migration. Generally, locations used for brief fueling (as in short-hop migration) are referred to as stopover sites, whereas longer fueling periods (as in long-jump migration) occur at staging sites.

or irruptive, and are suspected to make large-scale irregular movements according to the availability of water. For many of these nonmigratory species, information on individual movements is scarce, and it is possible that future study will reveal "true" migratory movements. For example, Kittlitz's Plover was previously considered largely sedentary (Piersma and Wiersma 1996), but recent study revealed complicated movement patterns within Africa, in which coastal breeders were sedentary whereas inland birds were nomadic or migratory, depending on breeding latitude (Tree 2001); this presumably reflects differences in seasonality. Given the substantial within-species variation in many plovers, we will surely uncover more unrecognized patterns like this in both purportedly migratory and nonmigratory species.

The remaining 26 species of plover (65%) exhibit true migration to some degree (Table 7.1). This is much higher than the value seen in birds globally, which is roughly 19% of ~10,000 extant species (Kirby et al. 2008, Rolland et al. 2014). The proportion of migratory plovers is very close to the average for the Order Charadriiformes, which includes all shorebirds, gulls, terns, and auks (67%; Rolland et al. 2014), but below that found in the Family Scolopacidae (sandpipers, snipes, and allies), of which >80% are migratory.

Only 12 of 40 plovers (30%) are considered fully migratory, with all adult individuals conducting short- to long-distance movements. Fourteen species (35%) are considered partial migrants, containing both migratory and sedentary individuals. In some cases, this variation manifests in conspicuous intraspecific differences according to geography, with some breeding populations being entirely sedentary or migratory (Table 7.1), but, in many cases, both migratory and nonmigratory individuals exist in the same population or even at the same breeding site (e.g., Stenzel et al. 1994). The magnitude of migration varies from quite short distances (just 10–100 km in New Zealand's Southern Red-breasted Dotterel) to some of the longest migrations known among land birds (Figure 7.2). In eight species, the longest-migrating individuals travel one-way

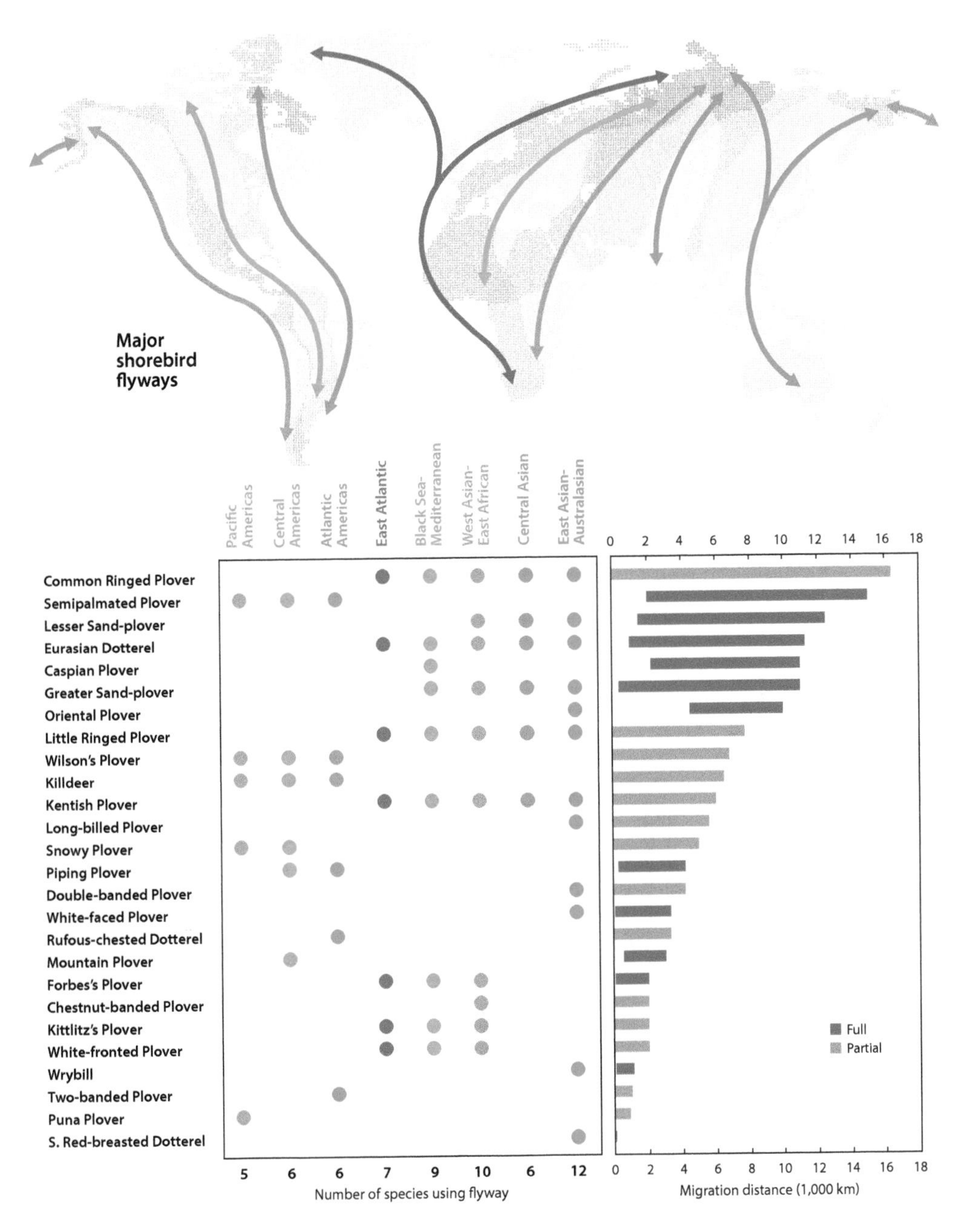

Figure 7.2. Left panel: Species diversity and distribution of the 26 migratory plovers in the subfamily Charadriinae across the eight major shorebird flyways (top). Right panel: Range of within-species variation in migration distance. Bars indicate estimated minimum and maximum one-way migration distances found among individuals in the species. Minimum represents shortest distance between known breeding and nonbreeding ranges (measured in Google Earth, using range maps in Hayman et al. 1986, Piersma and Wiersma 1996). Maximum represents the greatest plausible distance between breeding and nonbreeding sites, given current knowledge of individual movements; these are straight-line (orthodromic, or great-circle) distances, and therefore may underestimate true travel distance if routes are not direct.

distances of <2,000 km; in 11 species, this distance is 3,000–8,000 km; in seven species, it is 10,000–16,000 km or possibly more.

This impressive variation in scale and type of movement makes plovers a profitable and largely unexplored group for investigating relationships among annual-cycle strategies, ecology, and geography. Below, I will describe broad patterns of migration in the group, and discuss what these patterns imply for understanding plover evolution and conservation.

GEOGRAPHIC PATTERNS

Species diversity in the Charadrius group is much higher in the Eastern Hemisphere (Afro-Eurasia and Australasia; 31 species) than in the Western Hemisphere (the Americas; 13 breeding species, but only ten nonbreeding species; Figure 7.3), as might be expected given the greater landmass of the former, particularly in the low and southern latitudes. Three primarily Afro-Eurasian species have breeding ranges that stretch into the Americas (Common Ringed Plover in Greenland and the Canadian Arctic, and Eurasian Dotterel and Lesser Sand-plover occasionally in western Alaska), but all individuals migrate to Afro-Eurasia for the nonbreeding season. Similarly, a small number of Semipalmated Plovers breed in Chukotka (Lappo et al. 2012), but these birds migrate eastward to the Americas. Therefore, no plover species has a distribution spanning both the Eastern and Western hemispheres year round. However, in some cases, morphologically and genetically similar species pairs occupy similar niches in opposite hemispheres (e.g., Snowy and Kentish plovers; Common Ringed and Semipalmated plovers; dos Remedios et al. 2015), suggesting that these species were once globally distributed and migratory in both hemispheres, before isolation of flyway populations led to speciation. By contrast, *Calidris* sandpipers are a similarly diverse, entirely migratory shorebird genus in which several species have Holarctic distributions, such as Dunlin (*C. alpina*) and Red Knot (*C. canutus*). To some extent, this difference between the genera may reflect the somewhat subjective and arbitrary nature of defining species, but could also arise from weaker or more recent ecological isolation in sandpiper populations.

Along with higher species diversity, the occurrence of sedentary plovers is also much higher in the Eastern Hemisphere: only one entirely nonmigratory species occurs in the Americas (Collared Plover), whereas 45% of species (14 of 31) in the Eastern Hemisphere are nonmigratory. Despite

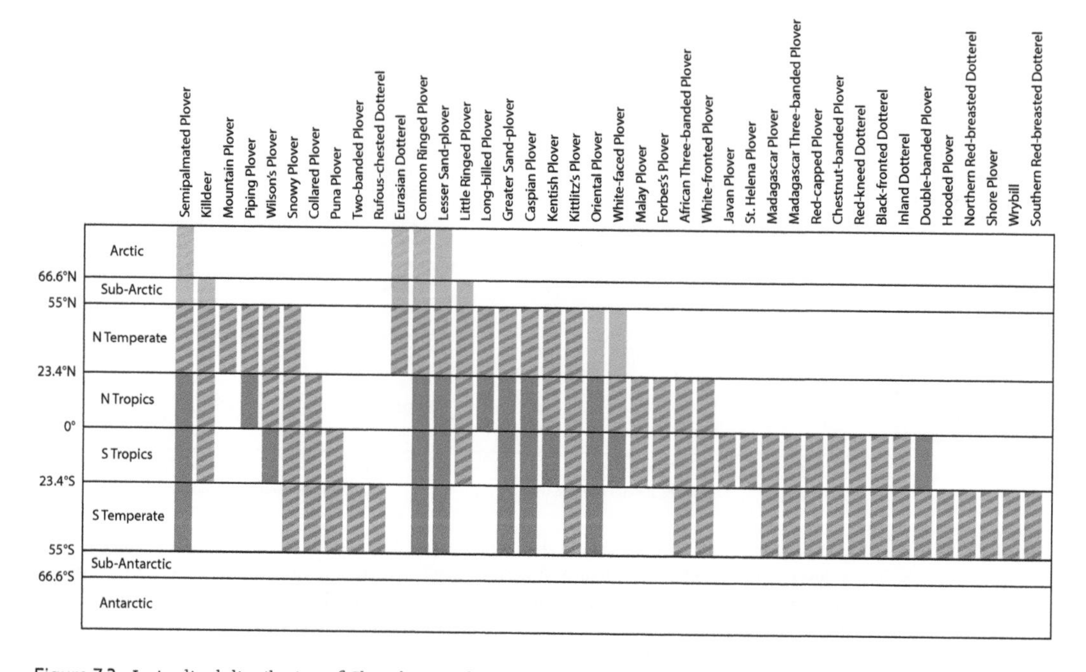

Figure 7.3. Latitudinal distribution of Charadriinae plovers during the breeding and nonbreeding seasons. Green = Western Hemisphere; Blue = Eastern Hemisphere. Light color = breeding; dark color = nonbreeding.

this, species diversity of migrating plovers is still higher within palearctic than nearctic regions (Figure 7.2). The greatest diversity is found on the East Asian-Australasian Flyway, with 12 migratory species; the next highest is the West Asian-East African Flyway, with ten species. Four migratory species (Common Ringed Plover, Little Ringed Plover, Kentish Plover, Eurasian Dotterel) are widely distributed in the Eastern Hemisphere, encompassing all five Palearctic flyways. Three species (Semipalmated Plover, Wilson's Plover, Killdeer) are found on all three Nearctic flyways. No plovers regularly use the Central Pacific Flyway; none of this group breeds on Pacific islands outside of Indonesia and Australasia, or migrates there for the nonbreeding season, as do some other shorebirds that breed in the Beringia region, such as Pacific Golden-plover (*Pluvialis fulva*), Ruddy Turnstone (*Arenaria interpres*), Bristle-thighed Curlew (*Numenius tahitiensis*), and Wandering Tattler (*Tringa incana*). This pattern may reflect a historical lack of trans-Pacific exploration in the plovers, or simply a present-day absence of appropriate foraging conditions in the region.

Patterns by Latitude

As would be expected from increasing seasonality away from the Equator, breeding latitude is highly correlated with migratory tendencies: higher-latitude breeders are more likely to be migratory and also tend to migrate the greatest distances. Four species (Common Ringed Plover, Semipalmated Plover, Lesser Sand-plover, Eurasian Dotterel) have breeding ranges that extend above the Arctic Circle (Figure 7.3); not coincidentally, these species migrate the longest distances (up to 11,000–16,000 km one-way; Figure 7.2). The year-round range of Semipalmated Plover includes the greatest range of latitude, 123° extending from 74°N to 49°S; one would expect the more northerly breeding Common Ringed Plover (spanning 117° from 83°N to 34°S) to have exceeded this, if it were not constrained by the southern limit of Africa (Figure 7.1). No plovers are found above 55°N in the northern winter, demonstrating that all arctic and subarctic breeders migrate to lower latitudes in the nonbreeding season. Among these six species, Eurasian Dotterel is the only species that remains entirely north of the Tropics year round (Figure 7.3).

For species breeding in the temperate zones, the patterns look very different for the Northern and Southern Hemispheres. Among species that breed in the northern temperate zone, all 17 are fully or partially migratory. In two of these species (Oriental Plover, White-faced Plover), all individuals leave the northern temperate zone for more southerly latitudes, and six species include northern temperate breeders that migrate as far as the southern temperate zone (Figure 7.3). Seven of these 17 species (41%) have breeding ranges that extend into the Tropics. In only one species, the fully migratory Mountain Plover, do all individuals remain in the northern temperate zone year round.

By contrast, among 21 species that breed in the southern temperate zone, only ten (48%) are fully or partially migratory; in one of these (Snowy Plover), the particular individuals that breed in the southern temperate zone (subspecies *C. n. occidentalis*) are sedentary. Most of these species (13 of 21; 62%) have breeding ranges that extend into the Tropics (Figure 7.3). Of eight species that breed exclusively in the southern temperate zone, five are at least partially migratory. As expected, these tend to migrate north from breeding areas, but in less impressive movements than found in northern breeders; only two travel distances greater than 1,200 km (some Double-banded Plovers and Rufous-chested Dotterels), and only one contains any individuals that enter the Tropics for the nonbreeding season (the few Double-banded Plovers that reach Fiji from New Zealand). The Wrybill and Southern Red-breasted Dotterel are the only fully migratory species restricted to the Southern Hemisphere, and these never leave the southern temperate zone, migrating only up to ~1,100 and 100 km, respectively, within New Zealand. No southern temperate breeders reach the northern temperate zone on migration.

Of 22 species that breed in the Tropics, 11 (50%) are fully or partially migratory. Again, this includes a number of species that have migratory individuals breeding at higher latitudes, whereas tropical populations are essentially sedentary (Snowy, Wilson's, Little Ringed, and Kentish plovers, Killdeer). Only four species are restricted to the Tropics year round (Figure 7.3), although the two endemic Madagascar species may also be included in this category, as Madagascar just barely extends into the southern temperate zone: of these six species, only Forbes's Plover is migratory to any degree.

Plover populations breeding at low and southern latitudes are more likely to be sedentary.

However, there are examples of sedentary individuals or populations even at latitudes above 50°. In Australasia, the southernmost plovers occur on Auckland Island, more than 450 km south of New Zealand (50.5°S). Double-banded Plovers breeding there, considered by some sources to constitute a separate subspecies (*C. b. exilis*; Piersma and Wiersma 1996), are sedentary. In South America, some Two-banded Plovers and Rufous-chested Dotterels breeding in Tierra del Fuego and the Falkland Islands (52°S–55°S) are also sedentary. In the Northern Hemisphere, a few species include some nonmigratory birds in the temperate zone (Snowy, Wilson's, and Common Ringed plovers, Killdeer): there are sedentary Snowy Plovers north of 41°N in California (Colwell et al. 2007), and sedentary Common Ringed Plovers in northwestern Europe at latitudes of 51°N–53°N (Figure 7.1).

No plovers occur in the antarctic and subantarctic zones (Figure 7.3), which is largely explained by the lack of landmasses, aside from scarce islands, in the Southern Oceans north of Antarctica. Other members of Charadriiformes, such as skuas (*Stercorarius* spp.), terns (*Sterna* spp.), and sheathbills (*Chionis* spp.) do breed and migrate in this region. Overall, only 32% of all landmass is in the Southern Hemisphere, and a large percentage of it consists of islands (including Australia) disconnected from other more northerly landmasses. In addition, climate conditions in the southern temperate zone are milder, and transitions tend to occur more on altitudinal than latitudinal gradients. These factors explain why no Southern Hemisphere breeders migrate north of the Equator, which is true not just in plovers, but across most avian taxa excluding seabirds (Dingle 2008).

Longitudinal Movements

In general, bird migration consists primarily of north–south movements (Newton 2008), as migration is a strategy to exploit seasonal conditions that vary predominantly by latitude, and this is clearly reflected in the orientation of the world's migratory flyways (Figure 7.2). Thus, we expect that individuals in the Northern Hemisphere will winter approximately south of where they breed, and that species with large longitudinal distributions (e.g., Kentish Plover, Common Ringed Plover) will migrate along a number of approximately parallel north–south routes, depending on geographic features such as coastlines and mountain ranges. This is generally true among conspecific populations using adjacent flyways, and parallel migration can even be observed within populations on smaller spatial scales. For example, Piping Plovers breed across the eastern half of North America and winter directly south of this in a similar range of longitude. The wintering population is highly structured, with eastern breeders concentrated on the Atlantic coast of the USA and western breeders along the western Gulf of Mexico, suggesting a predominantly north–south migration for most individuals and strong connectivity between breeding and nonbreeding sites (Gratto-Trevor et al. 2012). However, such patterns are not universally true. For example, geolocator-tracked Little Ringed Plovers from a single breeding site in Sweden migrated in a wide range of directions, arriving at tropical nonbreeding sites spanning >7,500 km from West Africa to India, a longitudinal range of 71° (Hedenström et al. 2013). Similarly, Eurasian Dotterels banded in Finland traveled to sites across the entire nonbreeding range from Morocco to Iran (Pulliainen and Saari 1993), demonstrating weak connectivity between breeding and nonbreeding sites.

More surprisingly, some plover migrations involve an equal or greater change in longitude than in latitude. This is particularly true of short-distance migrations between markedly different breeding and nonbreeding habitats. For example, migratory White-fronted Plovers (*C. m. mechowi*) travel east–west between inland breeding and coastal nonbreeding sites in eastern Africa (Delany et al. 2009); similar movements probably occur in Kittlitz's Plover (Tree 2001). The majority of Rufous-chested Dotterels breeding on the Falkland Islands migrate directly west ~500 km to mainland Argentina. At a larger spatial scale, Mountain Plovers wintering in California have likely traveled further east–west than north–south from breeding areas in central North America (Knopf and Wunder 2006). In Forbes's Plover, migration involves a transition between rocky upland breeding habitat and grassland nonbreeding habitat; although largely overlapping geographically, the winter range extends approximately 1,500 km west of the breeding range in western Africa.

Stunning large-scale longitudinal movements also found in the group may hint at constraints and historical patterns underlying present-day migrations (Sutherland 1998). One example is the Common Ringed Plover, in which migration to the relatively constricted nonbreeding range involves an eastward movement of ≥73° of longitude for the westernmost breeders and ≥100° westward for the easternmost breeders (Figure 7.1). The extreme example is an individual tracked from Chukotka to the Nile River delta, a longitudinal shift of 146° (Tomkovich et al. 2017). Similarly, all Eurasian Dotterels spend the winter in a narrow temperate band from Morocco to the Red Sea, despite a breeding range that spans from western Europe across the Palearctic and into western Alaska. This means that the most easterly breeders travel westward at least 130°, while moving southward by only ~40° of latitude. These examples suggest historical expansions of breeding ranges, while retaining population-level fidelity to nonbreeding areas and perhaps specific migratory routes (e.g., Bairlein et al. 2012). The extension of the breeding range of Semipalmated Plover from the Nearctic into far eastern Russia may be another example, albeit less dramatic and perhaps quite recent (Lappo et al. 2012). Such breeding expansions for northern plovers may have occurred as ice receded from high-latitude regions since the Last Glacial Maximum (Hewitt 2000, Newton 2008).

A unique contrast to the pattern exhibited by northern plovers is found in Double-banded Plovers breeding in mainland New Zealand. While some individuals migrate short distances and remain in New Zealand year round (Pierce 1999), others migrate to southwest Australia (>4,000 km) with very little change in latitude, or as far north as Fiji and New Caledonia. This pattern suggests a historical expansion of nonbreeding range accompanied by little or no colonization of new breeding areas.

Altitudinal Movements

For some plovers that breed in upland habitats, migration primarily involves changes in altitude, and only minor shifts in latitude or longitude. For example, the Puna Plover breeds at altitudes above 3,000 m in the Andes, and the main axis of migration is downslope to the coasts of southern Peru and northern Chile. At a smaller scale, Southern Red-breasted Dotterels breed above 300 m and migrate short distances to coasts of Stewart Island and the southern tip of New Zealand's South Island. Some long-distance migrants also breed in alpine areas up to 3,000–5,500 m (Lesser and Greater sand-plovers, Eurasian Dotterel, Mountain Plover) and so their migrations to distant coastal or lowland areas also include a significant component of altitude.

WITHIN-SPECIES VARIATION

Perhaps even more remarkable than the between-species diversity of movements in this group is the within-species variation demonstrated by many of its members. Even in fully migratory species, the distances traveled by individuals can vary by orders of magnitude. For example, in Greater Sand-plover, one-way migration distance varies from less than 500 km for individuals breeding in Turkey and wintering in the southern Mediterranean (*C. l. columbinus*), to ~11,000 km for Mongolian breeders that travel to southern Australia (*C. l. leschenaultii*). Distances range from 1,000 to 11,000 km in Eurasian Dotterel and 2,000–15,000 km in Semipalmated Plover (Figure 7.2), both of which are considered monotypic species.

In the 15 partially migratory species, individual variation gets more complicated, particularly in those occupying a broad range of breeding latitudes. The Common Ringed Plover exemplifies this, with its leapfrog migration and mix of sedentary and extreme long-distance migrations among temperate and arctic breeders. Killdeer and Little Ringed Plover are the only plover species whose breeding ranges extend from the subarctic into the tropics (Figure 7.3), and these are among the species featuring both sedentary tropical breeders and higher-latitude migratory individuals. Migration distances vary from 0 to >7,500 km in Little Ringed Plover, and from 0 to at least 5,000 km in Killdeer and Kentish, Snowy, Wilson's, and Long-billed plovers (Figure 7.2).

Six partially migratory species are taxonomically monotypic (Table 7.1), and, by definition, comprise one partially migratory population. Among the nine polytypic partially migratory species, seven include at least one completely sedentary subspecies; Common Ringed Plover and Chestnut-banded Plover are the exceptions. Only two partial migrants contain at least one fully migratory subspecies (Common Ringed Plover, Kentish Plover). In at least eight species, migratory individuals meet and mix with sedentary conspecifics during the nonbreeding season (Wilson's, Common, Little Ringed, Snowy, Chestnut-banded, White-fronted, and Kittlitz's plovers, Killdeer).

Age and Sex Differences

Within a population, migration (in terms of distance, routes, or timing) may vary among individuals of different age or sex. Such differential migration is fairly common in birds, and can reflect different strategies or physiological capabilities of respective cohorts (Myers 1981a, Ketterson and Nolan 1983, Cristol et al. 1999). For example, we might expect differential migration by sex in cases of high sexual dimorphism or nonmonogamous mating systems. In plovers, however, sexual dimorphism is generally weak or nonexistent (Piersma and Wiersma 1996). This pattern may reduce both the occurrence of differential migration in plovers, because the sexes are ecologically similar, and the likelihood of it being detected, because the sexes are difficult to distinguish in the field.

Only one example of spatial segregation by sex in the nonbreeding season has been reported for plovers: in Piping Plovers, a greater proportion of males wintered at higher latitudes (Stucker et al. 2010). This general absence of sexual segregation contrasts with other shorebirds, notably sandpipers (Nebel et al. 2002, Nebel 2007). However, some evidence suggests that nonstop migration distance might vary by age: in Eurasian Dotterels, adults appear to fly nonstop from the United Kingdom to nonbreeding areas in northern Africa, whereas some juveniles stop *en route* in mainland Europe (Whitfield et al. 1996). This could result from inexperience or poorer condition of juveniles, or from differences in flight capacity; for example, migrating juvenile Common Ringed Plovers had shorter wings than adults (Meissner 2007). As there is little information on individual migration routes or stopover strategies in plovers, such subtle patterns may be widespread and yet unrecognized.

Reports of differential timing of migration of sex and age classes are more common. Differences between males and females in migration timing are typically interpreted in terms of sexual selection, leading to unequal investment in parental care and acquisition of mates and territories, and therefore unequal time spent on breeding grounds (Myers 1981b, Morbey and Ydenberg 2001). In Semipalmated Plover and Piping Plover, males migrate north before females, but females precede males on southward migration (Elliott-Smith and Haig 2004, Gratto-Trevor et al. 2012, Nol and Blanken 2014). In both species, males defend breeding territories and provide most care for broods. In Eurasian Dotterel, females migrate earlier on both northward and southward migrations (Nankinov 1996, Whitfield et al. 1996); this species is polyandrous, with females competing for breeding opportunities and providing no parental care. In Oriental Plover, males leave earlier on southward migration (Piersma and Wiersma 1996), which reflects exclusive female care of broods.

Age-related differences in migration timing most often involve adults preceding younger birds on southward migration, particularly when one adult sex departs the breeding grounds before the young are independent. Adults migrate south before juveniles by up to four weeks in Semipalmated Plover (Nol and Blanken 2014), Common Ringed Plover (Insley and Young 1981, Wilson 1981), Greater and Lesser sand-plovers (Piersma and Wiersma 1996), Piping Plover (Elliott-Smith and Haig 2004), and Eurasian Dotterel (Nankinov 1996, Whitfield et al. 1996). Interestingly, two New Zealand species show the opposite pattern: juveniles precede adults in departure from the breeding grounds and arrival at nonbreeding sites in Wrybill (Davies 1997) and Double-banded Plover (Pierce 1999). Perhaps this reflects a large benefit or low cost to adults who remain and defend breeding territories after the breeding season. Alternatively, these different patterns may relate to how young birds develop their migratory habits and discover nonbreeding areas: Do they follow experienced adults or migrate naïvely without direct guidance?

Most plovers breed in their second year of life. However, in species with delayed maturation, young birds may remain at nonbreeding sites rather than migrate to breeding sites; this is true in Lesser and Greater sand-plovers (Piersma and Wiersma 1996). In addition, younger breeders may arrive on breeding grounds later than do older birds, as seen in Double-banded Plover (Piersma and Wiersma 1996), which may result from inexperience or poorer condition of first-time breeders.

STRATEGIC ASPECTS OF MIGRATION

The impressive variation in migration across plover taxa suggests enormous differences in selection pressures and a wide range of strategies to address them. Variation in the scale and nature of migration can reveal differences in how various annual-cycle stages (e.g., reproduction, molt) are prioritized, and whether time, energy, or danger impose the most important limitations for populations or individuals (Alerstam and Lindström 1990). Unfortunately, very little information is currently available about how individual plovers conduct their migrations, and how migration might directly affect or result from other aspects of their ecology or demography. Until recently, no individuals in this group had been remotely tracked on migration, largely because lightweight and long-lived tracking devices suitable for migratory birds weighing <80 g have only recently become available (Bridge et al. 2011). Three Charadrius species have now been tracked using miniaturized (~1 g) light-level geolocators, providing the first insights into year-round individual migration strategies (Minton et al. 2011, Hedenström et al. 2013, Lislevand et al. 2017, Tomkovich et al. 2017). Such studies will surely become more common in the near future, as tracking devices continue to decrease in size and cost, and methods for analyzing tracking data continue to improve.

Geolocator-equipped Greater sand-plovers traveled 6,200–7,500 km from a nonbreeding site in northwest Australia to breeding sites in inland China or Mongolia (Minton et al. 2011, Minton et al. 2013). On both northward and southward migrations, most individuals made nonstop flights of 3–4 days and 3,000–3,800 km between Australia and Southeast Asia, and one to three shorter flights between coastal Asia and the breeding grounds. On average, the duration of migration including stopovers was 41–44 days in both directions.

Little Ringed Plovers tracked from a breeding site in southern Sweden traveled 3,000–9,000 km to nonbreeding sites spanning from central Africa to India (Hedenström et al. 2013). Unlike the Greater Sand-plovers, these birds migrated in a series of short flights and often made large detours (generally 20–40%, but up to 157%) from the most direct routes. On average, total duration including stopovers was 72 days for southward migration and 27 days for northward migration.

Common Ringed Plovers tracked from northern Norway traveled 6,800–10,200 km to nonbreeding sites in coastal West Africa (Lislevand et al. 2017); in some cases this included a direct flight of 4,000–5,300 km, the longest nonstop flights thus far demonstrated in Charadrius plovers (Figure 7.4). The total time spent on migration

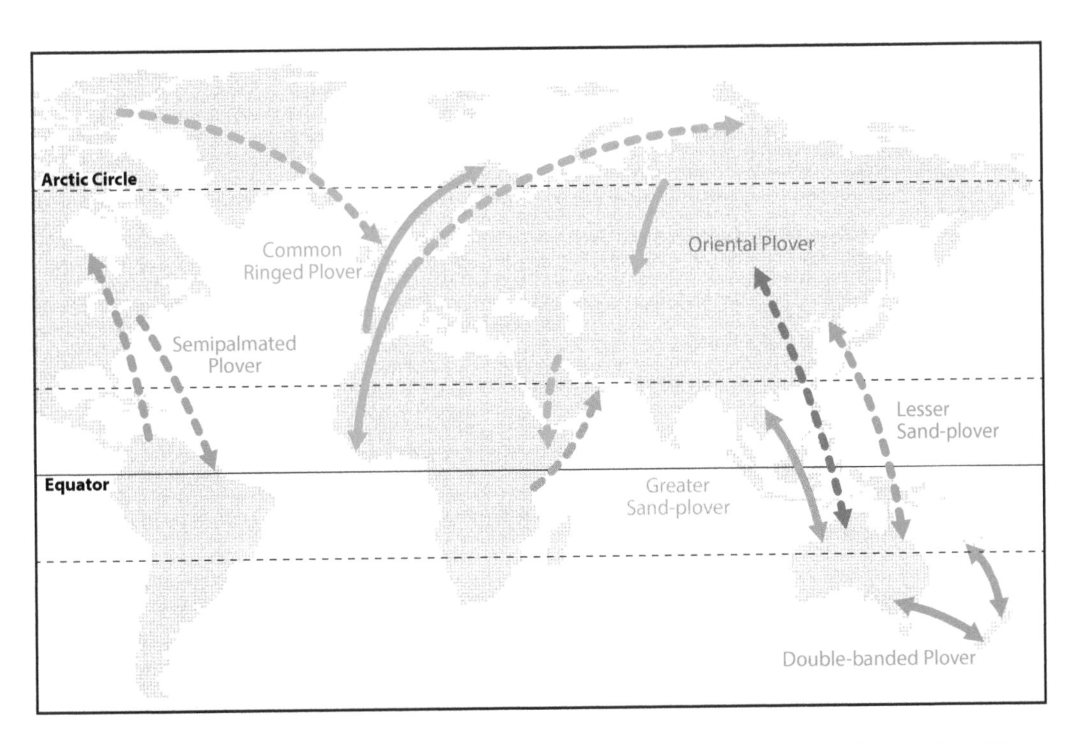

Figure 7.4. Demonstrated (solid lines) and hypothesized (dashed lines) nonstop flights in plovers. Common Ringed Plover, Greater Sand-plover, and Double-banded Plover perform nonstop flights of 1,500–5,300 km. However, the migrations of several other species include plausible direct flights that match or exceed these distances. Due to map projection, flight distances are not to scale.

was variable (28–67 days) but generally similar for the northward and southward journeys and, in both directions, individuals made two to five stopovers of 2–27 days each. Although the specific stopover sites used varied among and within individuals, most birds used the same general route northward and southward. However, one individual performed a "loop" migration, which followed a coastal Atlantic route in autumn, but a more easterly continental route in spring, including northward crossings of the Sahara Desert and Mediterranean Sea.

Common Ringed Plovers breeding in Chukotka, Russia, demonstrated a very different migration strategy, traveling round-trip distances of 20,000–25,000 km with many short flights, very few of which were greater than 1,000 km (Tomkovich et al. 2017). These birds departed their breeding grounds in a northwesterly direction across the Russian Arctic (Figure 7.1), which is counterintuitive given the expected north–south direction of bird migration, but in fact followed the shortest (great-circle) routes toward nonbreeding sites in the Middle East and northeastern Africa, which lie 126°–146° west of Chukotka. Autumn migration included 8,900–12,100 km traveled in 35–56 days, with three to nine major stopovers and many more brief stops of <1 day. In spring, the plovers traveled longer distances (9,500–12,900 km) but in less time (38–40 days, with four to five major stopovers); the longer route through central Asia taken by all individuals is probably necessary because the more direct arctic route is still snow-covered in April and May.

Nonstop Flights and Barrier-Crossing

The different strategies demonstrated in the tracking studies above are similar to the range of patterns found in sandpipers (Scolopacidae), in which variation in migration has been a major focus of research. Sandpiper migrations fall on a continuum from so-called "jump" migrations, in which protracted nonstop flights follow the accumulation of significant stores of migratory fuel, to "hop" migrations, in which birds make numerous short flights interspersed with brief refueling stops (Piersma 1987, Warnock 2010). Where the migrations of other plovers fall on this continuum is at present largely a matter of speculation and circumstantial evidence. In general, long migratory flights occur in the presence of substantial geographic barriers, such as oceans, deserts, or mountain ranges, coupled with the ecological circumstances to allow crossing them, such as assisting wind conditions and rich fueling resources (Alerstam et al. 2003). Of the 26 migratory plovers, seven species have year-round ranges that include no obvious barriers; all of these migrate relatively short distances within one continent including Forbes's, Kittlitz's, Chestnut-banded, and White-fronted plovers in Africa; Puna and Two-banded plovers in South America; and Long-billed Plover in Asia. In the remaining 19 species, individuals must traverse or circumnavigate major geographic barriers; currently, little information exists to evaluate how they manage these tasks.

Water crossings at some scale are an integral part of most migratory flyways (Figure 7.2). For example, a number of plovers routinely cross the Gulf of Mexico (overwater flights of 200–1,000 km; four species) or Mediterranean Sea (15–500 km; five species). For six species that cross the island-rich expanse between Southeast Asia and Australia, the magnitude of flights performed is dependent on migration strategy: some species make the jump in a single flight of 3,000–4,000 km, overflying Indonesia (as in the geolocator-tracked Greater sand-plovers above), but an island-hopping individual might cross this distance in a series of flights of 100–1,000 km each. Other water crossings are evident from the seasonal use of islands; these include Rufous-chested Dotterels flying from the Falkland Islands to mainland Argentina (>500 km), Greater sand-plovers wintering on Madagascar (>400 km from Africa), and Wrybills flying between the South and North Islands of Zealand (50–120 km). Flights by Double-banded Plovers from New Zealand must include 1,500–2,000 km crossings of the Tasman Sea to Australia or 1,500–3,000 km northward to New Caledonia or Fiji (Figure 7.4).

Several species must contend with inland barriers of considerable magnitude during migration. Some Mountain Plovers and Snowy Plovers breeding in central North America must negotiate their way through or around the Rocky Mountains to reach nonbreeding sites on the Pacific and Gulf coasts. The same is true of Little Ringed Plovers for which the European Alps or the Himalayas are situated between their breeding and nonbreeding sites. In Africa, the Sahara Desert presents a barrier for some Common Ringed and Kentish Plovers that winter south of it. Recently, it has been demonstrated that Great Snipes (*Gallinago media*) and a number of passerine species cross directly over this 2,000 km north–south expanse in a single flight, rather than going around it (Klaassen et al. 2011, Adamík et al. 2016, Ouwehand and Both 2016); at least some Common Ringed Plovers do this as well (Lislevand et al. 2017). Perhaps equally impressive is the apparent crossing of the Greenland ice sheet by Common Ringed Plovers migrating between northern Canada and Europe (Taylor 1980, Alerstam et al. 1986). This formidable barrier spans 500–1,200 km east–west, depending on where the crossing is made, and may represent only part of a nonstop flight of more than 3,000 km (Figure 7.4), as has been demonstrated in Red Knots making a similar migration (T. Piersma, pers. comm.).

Aside from geography, observed changes in body mass also indicate that a number of plover species prepare for long nonstop flights. Prior to their long northward flights from northwest Australia, Greater sand-plovers increase from an average nonbreeding mass of 74 g to at least 90–110 g, an increase of 20%–50% in fuel for migration (Barter and Minton 1998). Similar mass gains observed in Greater sand-plovers in Kenya indicate they make a northward flight of >3,000 km directly to the eastern Arabian Peninsula (Jackson 2016). Common Ringed Plovers add at least 30%–60% to their mass prior to northward departure from Mauritania, similar to Bar-tailed Godwits (*Limosa lapponica*) and Red Knots that are expected to fly 4,000–5,000 km nonstop to western Europe (Zwarts et al. 1990). In New Jersey, USA, Semipalmated Plovers staging during northward migration were on average more than 22 g (>50%) heavier than their typical nonbreeding mass of 40 g (Nol and Blanken 2014), suggesting the capability to fly nonstop to breeding sites up to 4,000 km away. In Venezuela, Wilson's Plovers gained average fuel stores of 37% (and up to 70%) of lean mass (Morrier and McNeil 1991), although they apparently migrate no more than 2,000–3,000 km northward along the Atlantic coast of the USA. At the same site, Semipalmated Plovers carried fat stores equivalent to 110% of lean mass (Morrier and McNeil 1991); this approximates the relative fuel stores of Bar-tailed Godwits before the longest nonstop flight known in birds (>11,000 km from Alaska to New Zealand; Piersma and Gill 1998, Gill et al. 2009), and suggests that the plovers could potentially fly 4,000–7,000 km directly to Canadian breeding sites from South America.

In shorebirds (and no other landbird groups to date), recent tracking data have unambiguously demonstrated nonstop migratory flights of 5,000–12,000 km (reviewed in Conklin et al. 2017); these currently include Common Ringed Plover (Lislevand et al. 2017), Pacific Golden-plover (Johnson et al. 2015), and a number of scolopacid sandpipers, in which such extreme flights are fairly common. Several other plovers could join this group, once individuals are tracked on migration (Figure 7.4). In addition to those suggested above, some plausible flights of this magnitude (4,000–7,000 km) include Semipalmated Plovers flying southward from the northeastern USA or southeastern Canada to northeastern South America, Lesser sand-plovers (*C. m. mongolus* and/or *stegmanni*) flying between the Yellow Sea and northern Australia, and Oriental Plovers flying directly from inland breeding areas in Central Asia to Indonesia or Australia. These and other potential extreme nonstop flights await investigation.

Sociality and Consistency on Migration

When a species congregates predictably in large numbers for extended periods at coastal staging sites, we can learn much about the manner and consistency with which individuals conduct their migrations; the fueling rates discussed above illustrate this. However, many plovers migrate much more inconspicuously, trickling across wide expanses in small numbers, and perhaps in short, unpredictable movements. For this reason, migration strategies, in terms of timing, consistency, and flock size, are essentially unknown for most species. However, what little has been described suggests, once again, substantial interspecific variation.

Coastal long-distance migrants tend to form large flocks at stopover sites. For example, more than 30,000 Semipalmated Plovers have been counted in a single day along a 100 km stretch of the upper Bay of Panama during southward migration (Nol and Blanken 2014). Lesser and Greater sand-plovers occur in flocks of 3,000–4,000 individuals during southward migration at various sites in eastern Asia (Conklin et al. 2014). However, actual migratory movements between sites occur in much smaller flocks; for example, among hundreds of staging Common Ringed Plovers in Denmark, flocks observed departing northward included just 6–20 individuals each (Fischer and Meltofte 2015) and northbound flocks from Mauritania averaged 17 individuals, with a maximum of 60 in a flock (Piersma et al. 1990). Observed migrating flocks are similar in some other plovers: up to 20–80 Eurasian Dotterels; up to 30 Caspian Plovers (Piersma and Wiersma 1996); and ≥30 Mountain Plovers (Knopf and Wunder 2006). Other species appear to move in smaller flocks, for example, 3–6 Piping Plovers and 1–10 Little Ringed Plovers (Piersma and Wiersma 1996). For most plovers, however, especially those that migrate inconspicuously through inland areas, no information exists on the size of migrating flocks.

A similar dearth of information exists on whether individuals are consistent from year to year in their migration strategies. In general, long-distance migratory sandpipers appear quite habitual in terms of timing (e.g., Conklin et al. 2013) and use of particular stopover sites during migration (e.g., Harrington et al. 1988, Gudmundsson and Lindström 1992). Some degree of stopover site fidelity has been observed in Kentish and Little Ringed Plovers (Gavrilov et al. 1998), Semipalmated Plover (Smith and Houghton 1984), and Common Ringed Plover (Piersma and Wiersma 1996). Less is known about individual repeatability in migration timing. For the one Little Ringed Plover that was geolocator tracked on southward migration in two consecutive years, departure from the breeding site and arrival on the nonbreeding grounds differed between years by 5 and 6 days, respectively (Hedenström et al. 2013). Short-distance migratory plovers might be expected to show less interannual consistency in migration timing, but this has yet to be investigated.

Differences between migratory routes during northward and southward migration are also unknown for most plover species. At the population level, Caspian Plovers migrating from western Asia to Africa appear to follow the same route through the Middle East in spring and autumn (Nielsen 1971). However, other species show marked differences between the two migrations. Northbound Semipalmated Plovers predominantly take a route through the Gulf of Mexico and central North America, but return to South America by a more coastal/oceanic route via the western Atlantic (Nol and Blanken 2014). Similarly, Eurasian Dotterels appear to take a much more westerly route through Europe on southward than on northward migration (Nankinov 1996). For three individual Greater sand-plovers tracked for an entire year, none used the same route or stopover sites on northward and southward migration (Minton et al. 2013). Among geolocator-tracked Little Ringed Plovers, all four took a direct route northward across the Mediterranean Sea from Africa to Europe, but on northward migration two of these made a major detour through the Middle East (Hedenström et al. 2013). In Common Ringed Plovers, loop migration was performed by all individuals tracked from Chukotka (Tomkovich et al. 2017), whereas four of five individuals tracked from northern Norway used similar routes on northward and southward migrations (Lislevand et al. 2017).

Scheduling of Molt

Migration places time constraints on the annual cycle, not just in terms of achieving optimal timing of the flights themselves, but also in the scheduling of other important annual functions that may be practically or energetically incompatible with travel (Alerstam and Lindström 1990, Wingfield 2005). In birds, molt is a critical annual undertaking, given the importance of feathers for thermoregulation, competition for mates, and flight. Regular replacement of feathers is energetically costly, both directly through feather synthesis, and indirectly through decreased flight and thermoregulatory performance while new feathers are growing (Payne 1972, Murphy and King 1991, Lindström et al. 1993, Swaddle and Witter 1997). In addition, birds might experience higher predation risk during molt or as a consequence of poor-quality feathers (Dawson et al. 2000, Lind et al. 2010). Accordingly, we expect the scheduling of molt to vary with migration strategies (Holmgren and Hedenström 1995, Barta et al. 2008, de la Hera et al. 2009).

The most common pattern of feather replacement in migratory birds includes two separate annually performed molts: (1) the complete "prebasic" molt, in which all contour and flight feathers are replaced and which generally represents the transition from breeding ("alternate") plumage to nonbreeding ("basic") plumage; and (2) a partial "pre-alternate" molt, in which some portion of contour feathers are replaced, in a transition from nonbreeding plumage to breeding plumage (Humphrey and Parkes 1959, Howell et al. 2003). Many variations on this general theme exist, but most Charadrius plovers subscribe to the "Complex Alternate Strategy," which involves these two annual molts in adult birds, and one additional molt (into the "formative plumage," occurring between the juvenile plumage and the first alternate plumage) during a bird's first year of life (Howell et al. 2003). The scheduling of pre-basic molt, typically performed after the breeding season, potentially comes into conflict with post-breeding migration, because flight-feather replacement typically requires 2–4 months to complete and should be scheduled to avoid low flight performance during migration. Therefore, we see two main strategies emerge: pre-basic molt on or near the breeding grounds prior to migration and pre-basic molt on the nonbreeding grounds after postbreeding migration is completed (Holmgren and Hedenström 1995, Barta et al. 2008). Nonmigratory species are relieved of this conflict and consequently appear freer to vary the schedule of molt (Wingfield 2005, Barta et al. 2006).

In many plover species, visible differences between the basic and alternate plumages are quite subtle or nonexistent (although there are also striking counterexamples such as Eurasian Dotterel; Hayman et al. 1986), which makes identification of molts and their timing difficult without detailed study of birds in the hand. There are few such detailed descriptions of molt in plovers, particularly among nonmigratory species. Many nonmigrants have quite long or flexible breeding schedules (Chapter Five, this volume) and timing of molt may be opportunistic and variable among years. I am aware of molt descriptions in just three nonmigratory species. In Collared Plover, primary feather molt partially overlaps with breeding (Torres-Dowdall et al. 2009), despite a discrete breeding season (September–January, varying with latitude) that should allow for temporal separation of these stages. Hooded Plovers have an unusually flexible and protracted primary molt, which may span 5–7 months, completely overlapping their August–November breeding season (Rogers et al. 2014). By contrast, the molt of Shore Plovers follows directly after breeding, with little or no overlap (Dowding and Kennedy 1993).

Migratory species are expected to show less annual variation in molt timing, due to the predictable seasonality of their habitats and annual schedules, and less overlap with their shorter breeding seasons. Among migratory plovers, patterns vary widely. Presumably, most long-distance migrants conduct most or all of pre-basic molt on the nonbreeding grounds (Barta et al. 2008), as occurs in the *Pluvialis* plovers (Johnson and Johnson 1983, Serra and Underhill 2006) and most sandpipers. This strategy avoids both migrating with molting primaries and remaining at high-latitude breeding areas after the end of the short summer. Indeed, we see such a pattern not only in the long-distance Semipalmated Plover (Nol and Blanken 2014) and Common Ringed Plover (subspecies C. *h. tundrae*; Wallander and Andersson 2003), but also in the medium-distance Piping Plover (Elliott-Smith and Haig 2004). In these populations, wing molt occurs on the nonbreeding grounds after southward migration, although some body molt may occur before migration.

In some species, pre-basic molt, including wing molt, may be completed on or near the breeding grounds prior to southward migration. Eurasian Dotterels breeding in Norway start molting their primaries while incubating and finish before departure from breeding sites (Kålås and Byrkjedal 1984); these individuals have a relatively short migration for the species (~3,000 km). Mountain Plover, Snowy Plover, Wilson's Plover, and Killdeer all appear to start pre-basic molt at the end of the breeding season, possibly overlapping with later breeding activity, and complete it prior to southward migration (Corbat and Bergstrom 2000, Jackson and Jackson 2000, Knopf and Wunder 2006, Page et al. 2009).

Other species may adopt the intermediate strategy of initiating wing molt on or near the breeding grounds, suspending molt during southward migration, and then completing it at nonbreeding sites. Some Common Ringed Plovers, Little Ringed Plovers, and Kentish Plovers using the East Atlantic flyway show this pattern (Pienkowski et al. 1976, Pienkowski 1984, Walters 1984).

Due to the high variation in migration distance and phenology within some species, we should also expect some degree of intraspecific variation in molt patterns. In Semipalmated Plover, timing and duration of wing molt varies with wintering latitude: individuals wintering in northern temperate regions complete their wing molt by December, whereas those in the Tropics may protract this into January or February (Nol and Blanken 2014). Common Ringed Plovers demonstrate even more intraspecific flexibility: while northerly breeding birds follow the customary two-molt pattern described earlier, at least some temperate-breeding individuals of *C. h. hiaticula* appear to molt only once a year (subscribing to the "Complex Basic Strategy"; Howell et al. 2003); these birds may have dispensed entirely with the pre-alternate molt to take advantage of a longer breeding season (Wallander and Andersson 2003). It has been suggested that a similar pattern may occur in Killdeer, in which temperate-breeding populations perform a pre-alternate molt, whereas tropical populations do not (Howell 2010). Many more detailed studies are required before we will understand exactly how migration and molt strategies have coevolved within this group.

WING MORPHOLOGY

Beyond behavioral and physiological adaptations for fueling and conducting migration, migratory birds have evolved specific morphological characteristics to enhance flight performance. Among these, wing shape confers the most obvious benefits. Specifically, narrower and more pointed (i.e., high aspect ratio) wings increase both speed and efficiency of flight (Lockwood et al. 1998; Hedenström 2002, 2008; Bowlin and Wikelski 2008). Accordingly, positive correlations between wing pointedness and migration distance have been found among species within a genus (e.g., Marchetti et al. 1995, Mönkkönen 1995), as well as among populations of a single species that differ in migratory tendency (Baldwin et al. 2010). This relationship appears to exist across genera in shorebirds as well (Minias et al. 2015). In addition, contemporary changes in wing shape (within a century) have been linked with changing ecological circumstances favoring increased or decreased mobility (Desrochers 2010, Bitton and Graham 2015). Thus, it seems clear that wing morphology evolves along with migration tendency, although whether one predominantly drives the other is less clear.

Body shape is a relatively conserved feature in plovers, with generally only subtle differences among species. However, wing shape varies substantially in the group, from short, blunt wings (e.g., Red-kneed Dotterel) to long and pointed wings (e.g., Oriental Plover; Figure 7.5a). This variation in wing shape is strongly associated with migratory propensity (the categorical continuum of non-, partial, or fully migratory species); more mobile species have wings with higher aspect ratios (Box 7.2, Figure 7.5b). Furthermore, wing loading (mass per unit area of wing) is positively correlated with aspect ratio; high wing load is associated with higher flight speeds and longer flight distances in birds in general (Rayner 1988, Hedenström and Alerstam 1995) and is a distinctive trait of fast-flying species such as shorebirds and waterfowl.

Two intriguing outliers in Figure 7.5b are the nonmigratory Inland Dotterel and Red-capped Plover, both of which have high wing loadings and aspect ratios, similar to long-distance migrants. The movements of these species are poorly studied, but both are known to make irregular movements within their large ranges in Australia according to unpredictable availability of water and food (Piersma and Wiersma 1996). The wing morphology of these species may enable them to make very long dispersive or nomadic movements when necessary, and perhaps also predisposes them to developing true migratory habits, should conditions favoring regular seasonal movements arise in the future.

Among migratory plovers, the strong correlation between aspect ratio and migration distance (Figure 7.5c) suggests that migratory tendencies are closely tied with evolutionary changes in morphology. The two greatest deviations from the predicted trend are worth mentioning. Well below the trend line, Kittlitz's Plover has the lowest aspect ratio (3.47) of any migratory plover. This species is predominantly sedentary, with a small proportion of birds making regular migratory movements (Tree 2001), which might suggest that these movements have only recently developed in a previously sedentary species. It would be interesting to know whether the migratory portion of the species has a measurably different wing shape than average, but there are currently no data with which to evaluate this. Lying well above the trend line is the fully migratory Wrybill (aspect ratio = 4.30). While having a relatively short migration within New Zealand, most individuals migrate from the South Island to the North Island for the nonbreeding season (Davies 1997); perhaps such overwater crossings invoke selection for wing shape beyond that expected from migration distance alone.

For species demonstrating very high interindividual variation in migration distance, we might expect considerable *intra*-specific variation in wing shape, particularly in partially migratory species in which entire populations are sedentary or fully migratory. The extent to which such populations

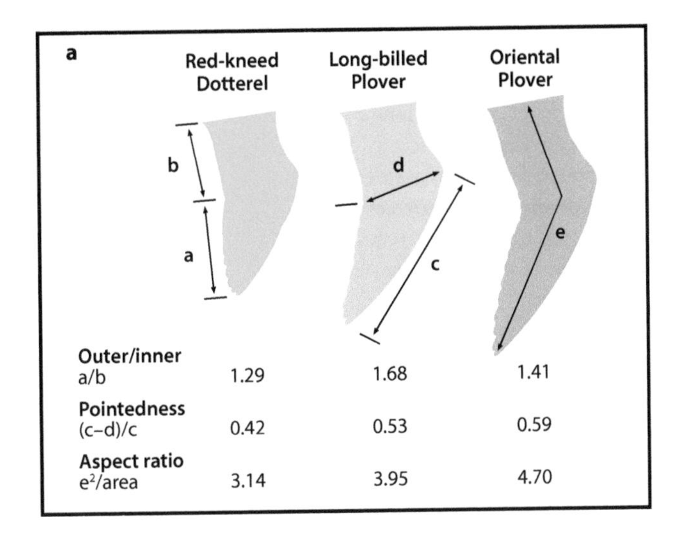

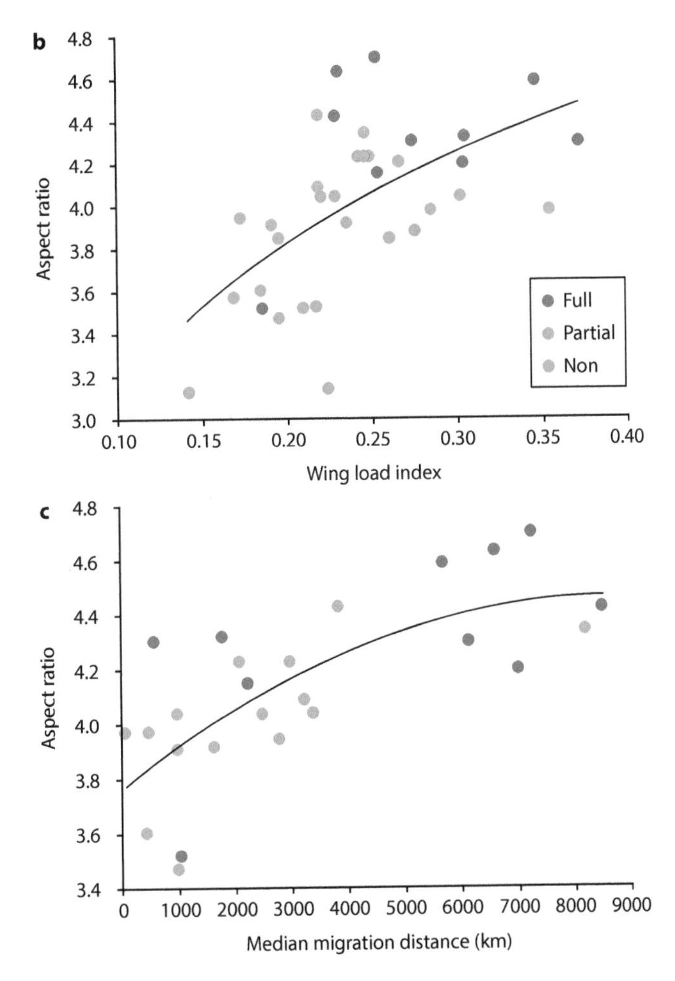

Figure 7.5. (a) Derivation of indices of wing shape from standardized illustrations in Hayman et al. 1986. Example wing silhouettes show the range of variation across non- (orange), partial (blue), and full (purple) migrants. (b) Wing loading increased with wing aspect ratio, which varied with migration tendency. (c) Among migratory plovers, migration distance increased with aspect ratio. See Box 7.2 text and Appendix 7.1 for details.

BOX 7.2 Wing Morphology and Migration

To test the prediction that migration tendency and distance vary with wing shape in this group, I created several indices of wing morphology for interspecies comparisons. The comprehensive guide *Shorebirds: An Identification Guide* (Hayman et al. 1986) includes illustrations that were systematically produced to allow direct interspecies comparison of morphology, in terms of size and shape. Dimensions of museum-specimen wings were quantified with standardized measurements, and then illustrated to scale in identical positions to capture very slight differences in morphology. Their description of plovers included 36 species, according to accepted taxonomy at the time. Javan and White-faced plovers were still considered subpopulations of Kentish Plover (Piersma and Wiersma 1996, Rheindt et al. 2011), and therefore were not illustrated separately; these species are omitted from my analysis. Red-breasted Dotterel was subsequently recognized as two species (del Hoyo et al. 2016), one migratory and one sedentary (Table 7.1); for this analysis, I consider them as one partially migratory species. Three-banded Plover was similarly split into two species (del Hoyo et al. 2016), but as neither is migratory, I consider them here as a single nonmigratory species. Consequently, my analysis includes 36 species.

After scanning and digitizing the illustrations, I used measurement tools in Adobe Illustrator CS6 (Adobe Systems, USA) to extract five standardized straight-line measurements (see Figure 7.5a): (1) length of the trailing edge of the outer wing (i.e., span of the primary flight feathers), (2) length of the trailing edge of the inner wing (i.e., span of the secondary flight feathers), (3) length from the point of the carpal joint to the tip of the wing (i.e., length of the longest primary), (4) length from the point of the carpal joint to the trailing edge where the secondaries meet the primaries (i.e., length of the longest secondary), and (5) length through the center of the wing from the base to the tip (i.e., total wing length). In addition, I traced the outline of the entire wing and calculated the total wing area. From these measurements (which are unitless but to identical scale), I calculated three comparable indices of wing shape: (1) the ratio of outer to inner wing length [a/b], (2) wing pointedness [(c−d)/c], and (3) aspect ratio [e^2/area]. To calculate a wing load index [mass/(2*area)], I summarized mass data presented in Table 5.5 (Chapter Five, this volume); for each species with available mass data ($n = 35$), I calculated a median value from the range of values reported. Lastly, I calculated the median migration distance for each migratory species ($n = 25$), derived from the minimum and maximum distances shown in Figure 7.2. All data are shown in Appendix 7.1.

The outer/inner wing ratio was uncorrelated with wing aspect ratio ($R^2 = 0.035$, $n = 36$, $P = 0.27$) and did not significantly differ among non-, partial, and fully migratory species (Kruskall-Wallis $\chi^2 = 2.42$, $df = 2$, $P = 0.30$). Therefore, it seems the length of the "hand" of the wing as a proportion of total wing length is not strongly related to migratory tendency or performance in plovers. Wing pointedness (the ratio of primary to secondary length) and aspect ratio were sufficiently correlated ($R^2 = 0.90$, $n = 36$, $P < 0.00001$) to be considered essentially redundant indices of wing shape. Wing loading was also correlated with aspect ratio ($R^2 = 0.34$, $n = 35$, $P = 0.0002$; Figure 7.5b), but a trend of increasing wing load with migratory tendency (means: non = 0.22, partial = 0.24, full = 0.28) was not statistically significant ($\chi^2 = 5.30$, $df = 2$, $P = 0.070$). Aspect ratio was strongly related to migration tendency: aspect ratio was higher in migratory (mean = 4.14, $n = 25$) than in nonmigratory species (mean = 3.63, $n = 11$; $\chi^2 = 9.10$, $df = 1$, $P = 0.003$), and higher in fully migratory (mean = 4.31, $n = 10$) than in partially migratory species (mean = 4.02, $n = 15$; $\chi^2 = 6.53$, $df = 1$, $P = 0.011$). Among migratory species, aspect ratio was positively correlated with median migration distance ($R^2 = 0.52$, $n = 25$, $P < 0.0001$; Figure 7.5c).

These analyses are necessarily simple and preliminary, meant only to reveal the broad patterns of migration with wing morphology. Although the illustrations of Hayman et al. (1986) are convenient for comparison, they provided only one wing illustration per species, and therefore each has an effective sample size of one, and cannot represent the range of within- or

(Continued)

BOX 7.2 (Continued) Wing Morphology and Migration

among-population variation that exists. Similarly, the mass data I used were quite coarse and did not consider time of year, and therefore may underestimate important relationships, particularly for migratory species whose mass varies seasonally. Therefore, I hope these preliminary analyses serve as a foundation for more rigid tests that may reveal more subtle patterns in the coevolution of wing morphology and migration in plovers.

vary in wing shape may generally indicate how long these intraspecific differences in migration have existed. It is intriguing to consider how quickly such differences can evolve, and whether specific wing morphology may predispose certain plovers to greater flexibility in the face of changing conditions. Unfortunately, very little information is available on this. In Common Ringed Plover, the longest-migrating birds actually have shorter wings, but this is misleading, as the subspecies *C. h. tundrae* is generally smaller overall (Salomonsen 1955, Hamilton 1961). Collared Plovers in the north of their range are shorter-winged, and are considered by some to be a separate subspecies (Piersma and Wiersma 1996), but this has no apparent relationship with movements, as the entire species is essentially sedentary. It would be very interesting to see more comparative morphology information for partially migratory species.

EVOLUTION OF MIGRATION

It has been long debated whether the plover clade derived from a Northern or Southern Hemisphere ancestor, and each hypothesis has found some degree of empirical and theoretical support (Joseph et al. 1999, Livezey 2010, Barth et al. 2013, dos Remedios et al. 2015; Chapter Two, this volume). Because northern breeding species are much more likely to be migratory (discussed earlier), the consequences of this debate for understanding migration in the group are profound: the southern-origin hypothesis requires that highly migratory species arose from sedentary ancestors through gradual northward expansion of breeding areas, whereas the northern-origin hypothesis suggests that sedentary species arose from migratory ancestors through southward expansion into (and subsequent isolation in) nonseasonal environments. This debate is by no means restricted to plovers, but is rather central to the origin of migration and speciation in many clades, and birds in general (Zink 2002, Rappole et al. 2003, Salewski and Bruderer 2007).

At a glance, the present-day latitudinal distribution of plover diversity appears to make a compelling case for a southern origin of the clade: 18 species are restricted to the Southern Hemisphere year round, whereas only four species are restricted to the Northern Hemisphere (Figure 7.3); 18 additional species inhabit both hemispheres to some extent, so the overall diversity is 22 species in the north and 36 species in the south. However, locally high diversity does not necessarily argue for a point of origin, as it may arise through rates of diversification or extinction that vary geographically (Mittelbach et al. 2007, Weir and Schluter 2007). Alternatively, high diversity may accumulate through immigration rather than *in situ* speciation, as was found in migratory warblers (*Phylloscopus* and *Seicercus* spp.) in the Himalayas (Johansson et al. 2007).

A phylogenetic study using mitochondrial DNA from 13 plovers lent support to a South American origin for the group, suggesting that migratory tendencies arose through the expansion of breeding grounds into more seasonal habitats in the north (Joseph et al. 1999). This pattern has been proposed for birds in general, and has variously been explained as a strategy for increasing reproductive success or annual survival (Rappole et al. 2003, Salewski and Bruderer 2007). A concrete example of how this might occur is provided by within-species variation in Chipping Sparrows (*Spizella passerina*), in which migratory populations in North America were shown to have rapidly evolved from sedentary populations in Central America through northward expansion in the last 18,000 years since the Last Glacial Maximum (Milá et al. 2006). Similarly, long-distance migration in Nearctic thrushes (*Catharus* spp.) has arisen independently multiple times during recent northward expansion in the group (Outlaw et al. 2003, Winker et al. 2006).

In sharp contrast, a recent phylogenetic study argues against northern migratory plovers arising from a southern, sedentary ancestor. Using 33 species and more representative genetic markers (both mitochondrial and nuclear DNA), dos Remedios et al. (2015) showed that the entire group likely arose from an ancestor in the northern Palearctic. From this new phylogeny (Chapter Two, this volume), we can see that migration exists to some extent in every descendant clade: from 33% in the southern African group, including Kittlitz's, St. Helena, and Madagascar plovers, to 100% in the northern Asian group of the sand-plovers, Oriental Plover, and Caspian Plover. Migration in every descendant clade is an unlikely result if the ancestral state were sedentary, particularly as <20% of birds in general are migratory (Kirby et al. 2008, Rolland et al. 2014). In addition, there is no clear evidence that a migratory plover species ever arose from a nonmigratory lineage (dos Remedios et al. 2015), although the partially migratory Kittlitz's Plover is one possible case of migration arising in a previously (and still largely) sedentary species. These findings strongly suggest that migratory is the ancestral state of the clade and that subsequent diversification occurred primarily through the loss rather than the gain of migratory tendencies.

Accordingly, a recent meta-analysis of bird diversity, including nearly 10,000 extant species, concluded that, although sedentary behavior is ancestral, migratory species have lower extinction rates and are more likely to diversify through the generation of subpopulations that then become sedentary (Rolland et al. 2014). Interestingly, the study also suggested that migration first arose *ca.* 80 million years ago, which is close to the time that the common ancestor of Charadrius is thought to have diverged from the fully migratory *Pluvialis* group (C. Küpper, pers. comm.; Chapter Two, this volume). The current distribution of sedentary plovers appears to fit well with a scenario in which migratory species generate new populations and eventual species by expanding into nonseasonal habitats where migration is no longer advantageous. At least eight partially migratory species include sedentary birds on islands (Table 7.1): Little Ringed Plover (Philippines, New Guinea), White-fronted and Kittlitz's plovers (Madagascar), Killdeer (Greater Antilles), Kentish Plover (Sri Lanka), Double-banded Plover (Chatham and Auckland Islands), Two-banded Plover (Falkland Islands), and Long-billed Plover (Japan); some of these represent entirely sedentary subspecies, whereas others are sedentary portions of partially migratory populations. Six other nonmigratory species also include, or are completely represented by, isolated island populations: Javan Plover (Indonesia), Malay Plover (Philippines, Indonesia), Madagascar and Madagascar Three-banded plovers (Madagascar), St. Helena Plover (St. Helena Island), and Shore Plover (Rangitara Island). These examples suggest a tendency for birds dispersing from mainland populations to become isolated through the loss of migratory habits. Morphological and phylogenetic similarities between some sedentary island species and more migratory sister species on the adjacent mainland may attest to such speciation occurring in recent evolutionary history. For example, St. Helena Plover likely diverged from Kittlitz's Plover, and Javan and Malay plovers from Kentish Plover.

Regardless of whether the adaptive loss or gain of migration is more common, the diversity of patterns in plovers demonstrates a surprising flexibility with regard to migratory habits and implies that a clear division between migratory and sedentary birds does not really exist. It was once expected that migratory birds, due to their specific requirements and capabilities (e.g., fueling, orientation, and timing), could be characterized by a "syndrome," essentially a suite of co-occurring traits not found in sedentary birds (Piersma et al. 2005, Dingle 2006). This somewhat rigid and deterministic view of migration is ostensibly supported by the finding that many aspects of migration have a heritable component (Berthold 1991), and the identification of specific genomic regions associated with multiple migration-related traits (e.g., Delmore et al. 2016). However, it has gradually become clear that the essential traits for migration are present in all or most birds, in the form of "hidden" variation that can be suppressed or activated as necessary, even on very short-time scales (Pulido 2007). Although existing genetic variation may allow rapid microevolution of migratory behavior (e.g., Berthold and Pulido 1994), it is also not clear that genetic change need occur for birds to adapt their migration tendencies to changing circumstances (Rappole et al. 2003). A population of sedentary House Finches (*Haemorhous mexicanus*) introduced in the eastern USA became migratory just since the 1960s (Able and Belthoff 1998), suggesting significant behavioral plasticity in migration. On the other hand, migration is not infinitely flexible, as demonstrated by apparent evolutionary constraints on the adoption of optimal migration routes (Sutherland 1998, Alerstam et al. 2003).

Plovers, with their impressive within-species variation, are clearly a group to watch for examples of adaptation and diversification of migration strategies within ecological time scales. For example, migratory birds are expected to have breeding, migratory, and nonbreeding locations that are very clearly defined, both temporally and geographically. At least two migratory plovers elude even this simple generalization. Eurasian Dotterels have been reported to exhibit migratory double-breeding, in which nesting is first attempted after migration to Scotland, and then again after further migration to Scandinavia (Rohwer et al. 2012). In California, Snowy Plovers have been observed making nesting attempts on winter grounds, either before or after migration to the traditional breeding area (Stenzel et al. 1994). These apparent behavioral oddities indicate ways in which plovers might discover novel migration strategies and may represent precursors to behavioral isolation and strategic diversifications in populations.

CONSERVATION IMPLICATIONS

Migratory bird populations have shown declines worldwide and are generally considered to be highly vulnerable to climate change and human-induced habitat degradation, due to their demanding annual routines and a reliance on precisely timed resources at a network of habitats along their migration routes (Myers et al. 1987, Kirby et al. 2008). This is particularly true of migratory shorebirds, which are declining at alarming rates on most of the world's flyways. For example, on the East Asian-Australasian Flyway, 24 of 25 migratory shorebird populations with known trends are declining (Conklin et al. 2014), including populations of Greater and Lesser Sand-plover and Long-billed Plover; a number of other plover species have unknown population trends on the flyway.

Despite these general and justifiable concerns about migratory birds, among plovers it appears that nonmigratory species actually face the greatest threats of extinction. According to the International Union for the Conservation of Nature (IUCN) Red List criteria (IUCN 2012), 13 of 40 species in the group are listed at Near Threatened status or higher (Table 1.1), and more than half of these are nonmigratory (Table 7.1). Of the three Endangered or Critically Endangered species, two are nonmigratory (Shore Plover and St. Helena Plover) and one is the shortest-distance migrant in the group (Southern Red-breasted Dotterel). Of 14 nonmigratory species, seven (50%) are Near Threatened or worse, compared to 2 of 14 (14%) partial migrants and 4 of 12 (33%) full migrants. It is clear that sedentary island species are generally the most threatened plovers: 8 of 13 listed species are restricted to one or a few islands. These tend to have small populations, and perhaps fewer options than mainland species to deal with natural or human-induced alterations to their habitats, including the introduction of predators.

At a glance, it appears that the longest-migrating plovers are actually faring the best: none of the species on the IUCN Red list migrates farther than ~4,000 km. However, this may stem from current methods for prioritizing species for conservation, rather than reflecting their true relative vulnerability. The Red List criteria underemphasize species with large ranges and large populations, two factors common to the longest-migrating plovers; all plover populations larger than 150,000 individuals are highly migratory (Table 1.1) and most of these are widely distributed. Their large ranges also make identifying specieswide population trends quite difficult. For example, both Greater and Lesser sand-plovers have unknown trends at the species level; however, dramatic declines of nonbreeding populations in Australia (*C. l. leschenaultii*, *C. m. mongolus*, and *C. m. stegmanni*) warrant Vulnerable or Endangered status when the Red List criteria are applied at the regional level of the East Asian-Australasian Flyway (Garnett et al. 2011, Conklin et al. 2014). These declines are linked to degradation and direct loss (e.g., through reclamation) of intertidal staging areas, particularly in eastern Asia (MacKinnon et al. 2012), which would be expected to have the greatest effect on arctic-breeding species needing to accumulate large fuel stores for their long flights.

In general, the inter- and intraspecific variation in migration tendencies demonstrated by plovers implies a historical ability to adapt to natural climate change and other environmental perturbations, and perhaps the flexibility to do so in the future. Although there have been contemporary changes in breeding phenology or distribution in response to climate change in migratory plovers of the genus *Pluvialis* (Pearce-Higgins et al. 2005, Maclean et al. 2008), there are few such reports in Charadriinae. I am aware of only one documented case: at migratory stopover sites in central Europe, Little Ringed Plovers have advanced their passage timing during northward migration, but delayed passage on southward migration (Adamík and Pietruszková 2008). However, it is likely that the migrations of many plovers will be altered or threatened by anthropogenic climate change (Chapter Three, this volume); the proximate causes may include sea level rise altering coastal sites, drought-induced drying of inland wetlands, and disruption of prevailing winds supporting migration. In an assessment of climate threats to waterbirds on Afro-Eurasian flyways, four migratory species (Caspian Plover, White-fronted Plover, Greater Sand-plover, Eurasian Dotterel) and one nonmigratory species (Three-banded Plover) were identified among those likely to face some recognizable threats, and the Chestnut-banded Plover was considered critically threatened (Maclean et al. 2007). In the East Asian-Australasian Flyway, projected sea level rise is expected to cause loss of stopover habitat and an associated disruption of migratory routes for a number of shorebirds, including Lesser Sand-plover (Iwamura et al. 2013); however, the projected effect on Lesser sand-plovers was less severe than for many sandpiper species, indicating that their migration was less "bottlenecked" by the reliance of a large proportion of the population on specific stopover sites. Sea level rise will surely affect migratory plovers in other regions as well, particularly those species that rely on specific, patchily distributed intertidal areas for fueling.

The variation in movement patterns found in plovers means that conservation strategies cannot be uniform across species or even populations within a species. For long-jump migrants that concentrate predictably at specific sites (e.g., Semipalmated Plover and some Common Ringed Plovers), an obvious solution is to identify these sites and work toward their preservation. This approach has proven effective in the case of Red Knots migrating through Delaware Bay, USA (Baker et al. 2004; McGowan et al. 2011, 2015). Perhaps the greatest challenge in such cases is to foster the international collaboration required to conserve a network of such sites for populations that cross multiple continents in their journeys (Myers et al. 1987); despite the conceptual simplicity of this flyway approach, success globally has fallen short of expectations (Runge et al. 2015).

The challenge is very different for short-hop inland migrants; in many of these species, movements are poorly described, largely because they may occur irregularly, inconspicuously in small groups, or in a broad front across wide expanses of habitat. Devising management plans for such plovers is difficult, because we have a poor understanding of their year-round requirements, and their opportunistic or low-density use of sites complicates the identification of specific locations or habitats for meaningful conservation action. This makes many plovers a poor fit for conservation approaches that focus on discrete sites and view species distributions as static (Runge et al. 2014).

For species such as the Common Ringed Plover, management strategies must account for ecology and demography that may differ as dramatically among populations with different migration strategies as between very different species. Migration entails a trade-off between annual survival, which is increased by exploitation of seasonal resources in multiple locations, and reproductive output, which is potentially reduced by a shorter breeding season (Greenberg and Marra 2005, Newton 2008). Therefore, we may expect sedentary Common Ringed Plovers in western Europe to adopt a more r-selected strategy prioritizing reproduction in a given year, whereas those breeding in Greenland might adopt a more k-selected strategy of prioritizing long-term survival over current reproduction (MacArthur and Wilson 1967, Pianka 1970). These strategies predict different fecundity, survival, age structure, and perhaps population trends, even in partially sympatric segments of the same species. The central challenge is identifying the potential effects of migration strategy on key demographic parameters, and thereby devising effective population-specific conservation measures.

APPENDIX 7.1

Wing morphology and migration status of plovers.

Species	Migratory status	Median migration distance (km)[a]	Median mass (g)[b]	Outer wing [a][c]	Inner wing [b]	Primary length [c]	Secondary length [d]	Wing length [e]	Wing area	Outer/inner [a/b]	Pointedness [(c – d)/c]	Aspect ratio [e^2/area]	Wing load index [mass/(2*area)]
Semipalmated Plover	Full	8,500	48.1	12.42	8.01	16.89	7.15	21.57	105.2	1.55	0.58	4.42	0.23
Common Ringed Plover	Partial	8,200	60.0	13.79	8.04	18.81	8.04	23.01	121.9	1.72	0.57	4.34	0.25
Oriental Plover	Full	7,250	95.0	16.27	11.57	22.98	9.31	29.70	187.7	1.41	0.59	4.70	0.25
Lesser Sand-plover	Full	7,000	74.5	12.94	7.99	18.12	8.08	22.69	122.6	1.62	0.55	4.20	0.30
Caspian Plover	Full	6,600	75.5	15.18	11.25	20.83	8.60	27.55	163.9	1.35	0.59	4.63	0.23
Eurasian Dotterel	Full	6,150	120.5	13.87	11.37	20.02	8.94	26.39	161.9	1.22	0.55	4.30	0.37
Greater Sand-plover	Full	5,700	88.0	12.79	10.47	18.41	7.95	24.15	127.1	1.22	0.57	4.59	0.35
Little Ringed Plover	Partial	3,850	39.5	10.72	7.66	15.97	7.25	20.00	90.3	1.40	0.55	4.43	0.22
Wilson's Plover	Partial	3,400	67.5	11.74	8.34	16.14	7.44	21.24	111.6	1.41	0.54	4.04	0.30
Kentish Plover	Partial	3,250	42.0	11.24	7.27	15.82	6.96	19.83	96.1	1.55	0.56	4.09	0.22
Killdeer	Partial	3,000	96.5	17.13	10.30	23.10	10.00	28.66	194.2	1.66	0.57	4.23	0.25
Long-billed Plover	Partial	2,800	55.5	14.63	8.72	20.17	9.44	25.18	160.6	1.68	0.53	3.95	0.17
Snowy Plover	Partial	2,500	41.6	10.47	7.60	14.78	6.98	19.17	91.0	1.38	0.53	4.04	0.23
Piping Plover	Full	2,250	53.0	12.16	7.59	16.60	7.45	20.83	104.5	1.60	0.55	4.15	0.25
Double-banded Plover	Partial	2,100	61.5	13.50	8.71	17.89	7.57	23.13	126.5	1.55	0.58	4.23	0.24
Mountain Plover	Full	1,800	103.5	15.19	10.25	20.57	9.11	27.08	169.7	1.48	0.56	4.32	0.30
Rufous-chested Dotterel	Partial	1,650	83.1	14.09	11.08	19.98	9.97	26.32	176.8	1.27	0.50	3.92	0.24
Forbes's Plover	Full	1,050	47.5	12.37	7.67	16.72	8.25	21.26	128.4	1.61	0.51	3.52	0.18
Chestnut-banded Plover	Partial	1,000	32.0	9.50	7.82	13.40	6.53	18.11	83.9	1.21	0.51	3.91	0.19
Kittlitz's Plover	Partial	1,000	36.5	9.71	7.46	13.67	7.07	18.02	93.5	1.30	0.48	3.47	0.20
White-fronted Plover	Partial	1,000	43.0	10.17	8.70	14.74	7.13	19.84	97.4	1.17	0.52	4.04	0.22
Wrybill	Full	600	55.5	11.51	8.28	16.42	7.38	20.88	101.3	1.39	0.55	4.30	0.27
Two-banded Plover	Partial	500	69.8	11.98	8.89	17.12	8.06	22.07	122.6	1.35	0.53	3.97	0.28

(Continued)

APPENDIX 7.1 (*Continued*)

Wing morphology and migration status of plovers.

Species	Migratory status	Median migration distance (km)[a]	Median mass (g)[b]	Outer wing [a][c]	Inner wing [b]	Primary length [c]	Secondary length [d]	Wing length [e]	Wing area	Outer/inner [a/b]	Pointedness [(c – d)/c]	Aspect ratio [e²/area]	Wing load index [mass/(2*area)]
Puna Plover	Partial	450	45.0	11.59	8.33	16.20	8.19	20.97	121.9	1.39	0.49	3.61	0.18
Red-breasted Dotterel[d]	Partial	50	153.5	15.91	11.91	22.45	10.52	29.35	216.8	1.34	0.53	3.97	0.35
Black-fronted Dotterel	Non	NA	34.5	11.10	8.02	15.07	8.40	19.52	121.9	1.38	0.44	3.12	0.14
Collared Plover	Non	NA	34.0	9.84	7.41	13.95	6.61	18.30	87.1	1.33	0.53	3.85	0.20
Hooded Plover	Non	NA	94.5	14.30	10.50	19.01	9.02	25.79	171.6	1.36	0.53	3.88	0.28
Inland Dotterel	Non	NA	85.5	15.68	9.69	21.47	9.34	27.15	174.2	1.62	0.56	4.23	0.25
Madagascar Plover	Non	NA	37.3	9.29	7.65	13.95	7.22	17.70	89.0	1.21	0.48	3.52	0.21
Malay Plover	Non	NA	42.0	9.13	7.35	13.74	6.70	17.61	80.6	1.24	0.51	3.85	0.26
Red-capped Plover	Non	NA	40.5	9.25	7.61	13.33	6.12	17.89	76.1	1.22	0.54	4.20	0.27
Red-kneed Dotterel	Non	NA	56.0	10.80	8.38	14.79	8.65	19.81	125.2	1.29	0.42	3.14	0.22
Shore Plover	Non	NA	60.0	11.77	9.53	16.33	8.57	22.05	138.1	1.24	0.48	3.52	0.22
St. Helena Plover	Non	NA	–[e]	11.08	8.68	15.93	9.13	20.44	136.8	1.28	0.43	3.05	–[e]
Three-banded Plover[f]	Non	NA	37.0	12.16	6.95	15.45	7.44	19.78	109.7	1.75	0.52	3.57	0.17
Minimum		50	32.0	9.13	6.95	13.33	6.12	17.61	76.1	1.17	0.42	3.05	0.14
Mean		3,266	63.1	12.37	8.81	17.25	8.07	22.31	127.4	1.41	0.53	3.98	0.24
Maximum		8,500	153.5	17.13	11.91	23.10	10.52	29.70	216.8	1.75	0.59	4.70	0.37

See Box 7.2 and Figure 7.5a for details. Except for migration distance (km) and mass (g), values are unitless but to scale for comparability.

[a] Median of minimum and maximum migration distances shown in Figure 7.2.

[b] Source data and references for body mass can be found in Table 5.5, Chapter Five, this volume.

[c] Letters in this and subsequent columns refer to measurements illustrated in Figure 7.5a.

[d] Includes both Northern and Southern Red-breasted Dotterels; see Box 7.2 for explanation.

[e] Mass data unavailable.

[f] Includes both African and Madagascar Three-banded Plovers; see Box 7.2 for explanation.

LITERATURE CITED

Able, K. P. and J. R. Belthoff. 1998. Rapid 'evolution' of migratory behaviour in the introduced house finch of eastern North America. *Proceedings of the Royal Society B* 265:2063–2071.

Adamík, P., T. Emmenegger, M. Briedis, L. Gustafsson, I. Henshaw, M. Krist, T. Laaksonen, F. Liechti, P. Procházka, V. Salewski, and S. Hahn. 2016. Barrier crossing in small avian migrants: Individual tracking reveals prolonged nocturnal flights into the day as a common migratory strategy. *Scientific Reports* 6:21560.

Adamík, P. and J. Pietruszková. 2008. Advances in spring but variable autumnal trends in timing of inland wader migration. *Acta Ornithologica* 43:119–128.

Alerstam, T. 1990. *Bird Migration*. Cambridge University Press: Cambridge, UK.

Alerstam, T., A. Hedenström, and S. Åkesson. 2003. Long-distance migration: Evolution and determinants. *Oikos* 103:247–260.

Alerstam, T., C. Hjort, G. Högstedt, P. E. Jönsson, J. Karlsson, and B. Larsson. 1986. Spring migration of birds across the Greenland Inlandice. *Meddelelser om Grønland, Bioscience* 21:1–38.

Alerstam, T. and Å. Lindström. 1990. Optimal bird migration: The relative importance of time, energy and safety. Pp 331–351 in E. Gwinner (ed.), *Bird Migration: Physiology and Ecophysiology*. Springer-Verlag: Berlin, Germany.

Bairlein, F., D. R. Norris, R. Nagel, M. Bulte, C. C. Voigt, J. W. Fox, D. J. Hussell, and H. Schmaljohann. 2012. Cross-hemisphere migration of a 25 g songbird. *Biology Letters* 8:505–507.

Baker, A. J., P. M. González, T. Piersma, L. J. Niles, I. L. S. do Nascimento, P. W. Atkinson, N. A. Clark, C. D. T. Minton, M. K. Peck, and G. Aarts. 2004. Rapid population decline in red knots: Fitness consequences of decreased refuelling rates and late arrival in Delaware Bay. *Proceedings of the Royal Society B* 271:875–882.

Baldwin, M. W., H. Winkler, C. Organ, and B. Helm. 2010. Wing pointedness associated with migratory distance in common-garden and comparative studies of stonechats (*Saxicola torquata*). *Journal of Evolutionary Biology* 23:1050–1063.

Barta, Z., A. I. Houston, J. M. McNamara, R. K. Welham, A. Hedenström, T. P. Weber, and O. Feró. 2006. Annual routines of non-migratory birds: Optimal moult strategies. *Oikos* 112:580–593.

Barta, Z., J. M. McNamara, A. I. Houston, T. P. Weber, A. Hedenström, and O. Feró. 2008. Optimal moult strategies in migratory birds. *Philosophical Transactions of the Royal Society of London, Series B* 363:211–229.

Barter, M. and C. Minton. 1998. Can pre-migratory weight gain rates be used to predict departure weights of individual waders from north-western Australia? *Stilt* 32:3–13.

Barth, J. M. I., M. Matschiner, and B. C. Robertson. 2013. Phylogenetic position and subspecies divergence of the endangered New Zealand Dotterel (*Charadrius obscurus*). *PLoS One* 8:e78068.

Berthold, P. 1991. Genetic control of migratory behaviour in birds. *Trends in Ecology & Evolution* 6:254–257.

Berthold, P. and F. Pulido. 1994. Heritability of migratory activity in a natural bird population. *Proceedings of the Royal Society B* 257:311–315.

Bitton, P.-P. and B. A. Graham. 2015. Change in wing morphology of the European starling during and after colonization of North America. *Journal of Zoology* 295:254–260.

Bowlin, M. S. and M. Wikelski. 2008. Pointed wings, low wingloading and calm air reduce migratory flight costs in songbirds. *PLoS One* 3:e2154.

Bridge, E. S., K. Thorup, M. S. Bowlin, P. B. Chilson, R. H. Diehl, R. W. Fléron, P. Hartl, R. Kays, J. F. Kelly, W. D. Robinson, and M. Wikelski. 2011. Technology on the move: Recent and forthcoming innovations for tracking migratory birds. *Bioscience* 61:689–698.

Colwell, M. A., S. E. McAllister, C. B. Millett, A. N. Transou, S. M. Mullin, Z. J. Nelson, C. A. Wilson, and R. R. Levalley. 2007. Philopatry and natal dispersal of the Western Snowy Plover. *Wilson Journal of Ornithology* 119:378–385.

Conklin, J. R., P. F. Battley, and M. A. Potter. 2013. Absolute consistency: Individual versus population variation in annual-cycle schedules of a long-distance migrant bird. *PLoS One* 8:e54535.

Conklin, J. R., N. R. Senner, P. F. Battley, and T. Piersma. 2017. Extreme migration and the individual quality spectrum. *Journal of Avian Biology* 48:19–36.

Conklin, J. R., Y. I. Verkuil, and B. R. Smith. 2014. *Prioritizing Migratory Shorebirds for Conservation Action on the East Asian-Australasian Flyway*. WWF-Hong Kong: Hong Kong.

Corbat, C. A. and P. W. Bergstrom. 2000. Wilson's Plover (*Charadrius wilsonia*). In P. G. Rodewald (ed.), *The Birds of North America Online*. Cornell Lab of Ornithology: Ithaca, NY.

Cristol, D. A., M. B. Baker, and C. Carbone. 1999. Differential migration revisited: Latitudinal segregation by age and sex class. *Current Ornithology* 15:33–88.

Davies, S. 1997. Population structure, morphometrics, moult, migration, and wintering of the Wrybill (*Anarhynchus frontalis*). *Notornis* 44:1–14.

Dawson, A., S. A. Hinsley, P. N. Ferns, R. H. C. Bonser, and L. Eccleston. 2000. Rate of moult affects feather quality: A mechanism linking current reproductive effort to future survival. *Proceedings of the Royal Society B* 267:2093–2098.

de la Hera, I., J. Pérez-Tris, and J. L. Telleria. 2009. Migratory behaviour affects the trade-off between feather growth rate and feather quality in a passerine bird. *Biological Journal of the Linnean Society* 97:98–105.

del Hoyo, J., A. Elliott, J. Sargatal, D. A. Christie, and E. de Juana (eds). 2016. *Handbook of the Birds of the World Alive*. Lynx Edicions: Barcelona, Spain.

Delany, S., D. Scott, A. Helmink, T. Dodman, S. Flink, D. Stroud, and L. Haanstra. 2009. *An Atlas of Wader Populations in Africa and Western Eurasia*. International Wader Study Group and Wetlands International: Wageningen, The Netherlands.

Delmore, K. E., D. P. Toews, R. R. Germain, G. L. Owens, and D. E. Irwin. 2016. The genetics of seasonal migration and plumage color. *Current Biology* 26:1–7.

Desrochers, A. 2010. Morphological response of songbirds to 100 years of landscape change in North America. *Ecology* 91:1577–1582.

Dingle, H. 2006. Animal migration: Is there a common migratory syndrome? *Journal of Ornithology* 147:212–220.

Dingle, H. 2008. Bird migration in the southern hemisphere: A review comparing continents. *Emu* 108:341–359.

Dingle, H. and V. A. Drake. 2007. What is migration? *Bioscience* 57:113–121.

dos Remedios, N., P. L. Lee, T. Burke, T. Székely, and C. Küpper. 2015. North or south? Phylogenetic and biogeographic origins of a globally distributed avian clade. *Molecular Phylogenetics and Evolution* 89:151–159.

Dowding, J. E. and E. S. Kennedy. 1993. Size, age structure and morphometrics of the shore plover population on South East Island. *Notornis* 40:213–222.

Elliott-Smith, E. and S. M. Haig. 2004. Piping Plover (*Charadrius melodus*). in P. G. Rodewald (ed.), *The Birds of North America Online*. Cornell Lab of Ornithology: Ithaca, NY.

Fischer, K. and H. Meltofte. 2015. Departure directions of Sanderlings and 'tundra' Common Ringed Plovers from the northernmost Danish Wadden Sea in spring. *Wader Study* 122:25–30.

Garnett, S. T., J. K. Szabo, and G. Dutson. 2011. *Action Plan for Australian Birds 2010*. CSIRO: Collingwood, Australia.

Gavrilov, E., S. Erokhov, and A. Gavrilov. 1998. Between-year recapture rates of waders ringed on migration in south-eastern Kazakhstan: Constancy in timing and location of flyway routes. *International Wader Studies* 10:414–416.

Gill, Jr. R. E., T. L. Tibbitts, D. C. Douglas, C. M. Handel, D. M. Mulcahy, J. C. Gottschalck, N. Warnock, B. J. McCaffery, P. F. Battley, and T. Piersma. 2009. Extreme endurance flights by landbirds crossing the Pacific Ocean: Ecological corridor rather than barrier? *Proceedings of the Royal Society B* 276:447–457.

Gratto-Trevor, C., D. Amirault-Langlais, D. Catlin, F. Cuthbert, J. Fraser, S. Maddock, E. Roche, and F. Shaffer. 2012. Connectivity in piping plovers: Do breeding populations have distinct winter distributions? *Journal of Wildlife Management* 76:348–355.

Greenberg, R. and P. P. Marra. 2005. *Birds of Two Worlds: The Ecology and Evolution of Migration*. Johns Hopkins University Press: Baltimore, MD.

Gudmundsson, G. A. and Å. Lindström. 1992. Spring migration of Sanderlings *Calidris alba* through SW Iceland: Wherefrom and whereto? *Ardea* 80:315–326.

Hamilton, T. H. 1961. The adaptive significances of intraspecific trends of variation in wing length and body size among bird species. *Evolution* 15:180–195.

Harrington, B. A., J. M. Hagan, and L. E. Leddy. 1988. Site fidelity and survival differences between two groups of New World Red Knots (*Calidris canutus*). *Auk* 105:439–445.

Hayman, P., J. Marchant, and T. Prater. 1986. *Shorebirds: An Identification Guide*. A&C Black: London, UK.

Hedenström, A. 2002. Aerodynamics, evolution and ecology of avian flight. *Trends in Ecology & Evolution* 17:415–422.

Hedenström, A. 2008. Adaptations to migration in birds: Behavioural strategies, morphology and scaling effects. *Philosophical Transactions of the Royal Society of London, Series B* 363:287–299.

Hedenström, A. and T. Alerstam. 1995. Optimal flight speed of birds. *Philosophical Transactions of the Royal Society of London, Series B* 348:471–487.

Hedenström, A., R. H. G. Klaassen, and S. Åkesson. 2013. Migration of the Little Ringed Plover *Charadrius dubius* breeding in South Sweden tracked by geolocators. *Bird Study* 60:466–474.

Hewitt, G. 2000. The genetic legacy of the Quaternary ice ages. *Nature* 405:907–913.

Holmgren, N. and A. Hedenström. 1995. The scheduling of molt in migratory birds. *Evolutionary Ecology* 9:354–368.

Howell, S. N. G. 2010. *Molt in North American Birds*. Houghton Mifflin Harcourt: New York.

Howell, S. N. G., C. Corben, P. Pyle, and D. I. Rogers. 2003. The first basic problem: A review of molt and plumage homologies. *Condor* 105:635–653.

Humphrey, P. S. and K. C. Parkes. 1959. An approach to the study of molts and plumages. *Auk* 76:1–31.

Insley, H. and L. Young. 1981. Autumn passage of ringed plovers through Southampton water. *Ringing & Migration* 3:157–164.

IUCN. 2012. *IUCN Red List Categories and Criteria: Version 3.1*, 2nd ed., International Union for Conservation of Nature: Gland, Switzerland and Cambridge, UK.

Iwamura, T., H. P. Possingham, I. Chadès, C. Minton, N. J. Murray, D. I. Rogers, E. A. Treml, and R. A. Fuller. 2013. Migratory connectivity magnifies the consequences of habitat loss from sea-level rise for shorebird populations. *Proceedings of the Royal Society B* 280:20130325.

Jackson, C. H. W. 2016. Clues towards the migration route for Greater and Lesser Sand Plovers spending the non-breeding season in Kenya. *Biodiversity Observations* 7: Article 36.

Jackson, B. J. and J. A. Jackson. 2000. Killdeer (*Charadrius vociferus*). in P. G. Rodewald (ed.), *The Birds of North America Online*. Cornell Lab of Ornithology: Ithaca, NY.

Johansson, U., P. Alström, U. Olsson, P. Ericson, P. Sundberg, and T. Price. 2007. Build-up of the Himalayan avifauna through immigration: A biogeographical analysis of the *Phylloscopus* and *Seicercus* warblers. *Evolution* 61:324–333.

Johnson, O. W. and P. M. Johnson. 1983. Plumage-molt-age relationships in "over-summering" and migratory Lesser Golden-Plovers. *Condor* 85:406–419.

Johnson, O. W., R. R. Porter, L. Fielding, M. F. Weber, R. S. Gold, R. H. Goodwill, P. M. Johnson, A. E. Bruner, P. A. Brusseau, N. H. Brusseau, K. Hurwitz, and J. W. Fox. 2015. Tracking Pacific golden-plovers *Pluvialis fulva*: Transoceanic migrations between non-breeding grounds in Kwajalein, Japan and Hawaii and breeding grounds in Alaska and Chukotka. *Wader Study* 122:13–20.

Joseph, L., E. P. Lessa, and L. Christidis. 1999. Phylogeny and biogeography in the evolution of migration: Shorebirds of the *Charadrius* complex. *Journal of Biogeography* 26:329–342.

Kålås, J. A. and I. Byrkjedal. 1984. Breeding chronology and mating system of the Eurasian Dotterel (*Charadrius morinellus*). *Auk* 101:838–847.

Ketterson, E. D. and V. Nolan Jr. 1983. The evolution of differential bird migration. pp. 357–402 in R. F. Johnston (ed.), *Current Ornithology*, vol. 1. Plenum Press: New York.

Kirby, J. S., A. J. Stattersfield, S. H. Butchart, M. I. Evans, R. F. Grimmett, V. R. Jones, J. O'Sullivan, G. M. Tucker, and I. Newton. 2008. Key conservation issues for migratory land and waterbird species on the world's major flyways. *Bird Conservation International* 18:S49–S73.

Klaassen, R. H. G., T. Alerstam, P. Carlsson, J. W. Fox, and Å. Lindström. 2011. Great flights by great snipes: Long and fast non-stop migration over benign habitats. *Biology Letters* 7:833–835.

Knopf, F. L. and M. B. Wunder. 2006. Mountain Plover (*Charadrius montanus*). In P. G. Rodewald (ed.), *The Birds of North America Online*. Cornell Lab of Ornithology: Ithaca, NY.

Lappo, E. G., P. S. Tomkovich, and E. E. Syroechkovski. 2012. *Atlas of Breeding Waders in the Russian Arctic*. UF Ofsetnaya Pechat: Moscow, Russia.

Lind, J., S. Jakobsson, and C. Kullberg. 2010. Impaired predator evasion in the life history of birds: Behavioral and physiological adaptations to reduced flight ability. pp. 1–30 in C. F. Thompson (ed.), *Current Ornithology*, vol. 17. Springer: New York.

Lindström, Å., G. H. Visser, and S. Daan. 1993. The energetic cost of feather synthesis is proportional to basal metabolic rate. *Physiological Zoology* 66:490–510.

Lislevand, T., M. Briedis, O. Heggøy, and S. Hahn. 2017. Seasonal migration strategies of common ringed plovers *Charadrius hiaticula*. *Ibis* 159:225–229.

Livezey, B. C. 2010. Phylogenetics of modern shorebirds (Charadriiformes) based on phenotypic evidence: Analysis and discussion. *Zoological Journal of the Linnean Society* 160:567–618.

Lockwood, R., J. P. Swaddle, and J. M. V. Rayner. 1998. Avian wingtip shape reconsidered: Wingtip shape indices and morphological adaptations to migration. *Journal of Avian Biology* 29:273–292.

MacArthur, R. and E. Wilson. 1967. *The Theory of Island Biogeography*. Princeton University Press: Princeton, NJ.

MacKinnon, J., Y. I. Verkuil, and N. Murray. 2012. IUCN situation analysis on East and Southeast Asian intertidal habitats, with particular reference to the Yellow Sea (including the Bohai Sea). Occasional Paper of the IUCN Species Survival Commission No. 47. IUCN, Gland, Switzerland and Cambridge, UK.

Maclean, I., G. E. Austin, M. M. Rehfisch, J. Blew, O. Crowe, S. Delany, K. Devos, B. Deceuninck, K. Guenther, and K. Laursen. 2008. Climate change causes rapid changes in the distribution and site abundance of birds in winter. *Global Change Biology* 14:2489–2500.

Maclean, I., M. M. Rehfisch, S. Delany, and R. A. Robinson. 2007. *The Effects of Climate Change on Migratory Waterbirds Within the African-Eurasian Flyway*. British Trust for Ornithology: Norfolk, UK.

Marchetti, K., T. Price, and A. Richman. 1995. Correlates of wing morphology with foraging behaviour and migration distance in the genus *Phylloscopus*. *Journal of Avian Biology* 26:177–181.

McGowan, C. P., J. E. Hines, J. D. Nichols, J. E. Lyons, D. R. Smith, K. S. Kalasz, L. J. Niles, A. D. Dey, N. A. Clark, and P. W. Atkinson. 2011. Demographic consequences of migratory stopover: Linking red knot survival to horseshoe crab spawning abundance. *Ecosphere* 2:1–22.

McGowan, C. P., J. E. Lyons, and D. R. Smith. 2015. Developing objectives with multiple stakeholders: Adaptive management of horseshoe crabs and red knots in the Delaware Bay. *Environmental Management* 55:972–982.

McNamara, J. M., Z. Barta, M. Wikelski, and A. I. Houston. 2008. A theoretical investigation of the effect of latitude on avian life histories. *American Naturalist* 172:331–345.

Meissner, W. 2007. Different timing of autumn migration of two Ringed Plover *Charadrius hiaticula* subspecies through the southern Baltic revealed by biometric analysis. *Ringing & Migration* 23:129–133.

Milá, B., T. B. Smith, and R. K. Wayne. 2006. Postglacial population expansion drives the evolution of long-distance migration in a songbird. *Evolution* 60:2403–2409.

Minias, P., W. Meissner, R. Włodarczyk, A. Ożarowska, A. Piasecka, K. Kaczmarek, and T. Janiszewski. 2015. Wing shape and migration in shorebirds: A comparative study. *Ibis* 157:528–535.

Minton, C., K. Gosbell, P. Johns, M. Christie, M. Klaassen, C. Hassell, A. Boyle, R. Jessop, and J. Fox. 2011. Geolocator studies on Ruddy Turnstones *Arenaria interpres* and Greater Sandplovers *Charadrius leschenaultii* in the East Asian-Australasia Flyway reveal widely different migration strategies. *Wader Study Group Bulletin* 118:87–96.

Minton, C., K. Gosbell, P. Johns, M. Christie, M. Klaassen, C. Hassell, A. Boyle, R. Jessop, and J. Fox. 2013. New insights from geolocators deployed on waders in Australia. *Wader Study Group Bulletin* 120:37–46.

Mittelbach, G. G., D. W. Schemske, H. V. Cornell, A. P. Allen, J. M. Brown, M. B. Bush, S. P. Harrison, A. H. Hurlbert, N. Knowlton, and H. A. Lessios. 2007. Evolution and the latitudinal diversity gradient: Speciation, extinction and biogeography. *Ecology Letters* 10:315–331.

Mönkkönen, M. 1995. Do migrant birds have more pointed wings? A comparative study. *Evolutionary Ecology* 9:520–528.

Morbey, Y. E. and R. C. Ydenberg. 2001. Protandrous arrival timing to breeding areas: A review. *Ecology Letters* 4:663–673.

Morrier, A. and R. McNeil. 1991. Time-activity budget of Wilson's and Semipalmated Plovers in a tropical environment. *Wilson Bulletin* 103:598–620.

Murphy, M. E. and J. R. King. 1991. Nutritional aspects of avian molt. pp. 2186–2193 in B. D. Bell (ed.), *Acta XX Congressus Internationalis Ornithologici*. New Zealand Ornithological Congress Trust Board: Wellington, New Zealand.

Myers, J. 1981a. A test of three hypotheses for latitudinal segregation of the sexes in wintering birds. *Canadian Journal of Zoology* 59:1527–1534.

Myers, J. P. 1981b. Cross-seasonal interactions in the evolution of sandpiper social systems. *Behavioral Ecology and Sociobiology* 8:195–202.

Myers, J. P., R. I. G. Morrison, P. Z. Antas, B. A. Harrington, T. E. Lovejoy, M. Sallaberry, S. E. Senner, and A. Tarak. 1987. Conservation strategy for migratory species. *American Scientist* 75:18–26.

Nankinov, D. 1996. Dotterel in Bulgaria and routes of its migration in Eurasia. *Berkut* 5: 141–146.

Nebel, S. 2007. Differential migration of shorebirds in the East Asian-Australasian flyway. *Emu* 107:14–18.

Nebel, S., D. B. Lank, P. D. O'Hara, G. Fernandez, B. Haase, F. Delgado, F. A. Estela, L. J. E. Ogden, B. Harrington, B. E. Kus, J. E. Lyons, F. Mercier, B. Ortego, J. Y. Takekawa, N. Warnock, and S. E. Warnock. 2002. Western sandpipers (*Calidris mauri*) during the nonbreeding season: Spatial segregation on a hemispheric scale. *Auk* 119:922–928.

Newton, I. 2008. *The Migration Ecology of Birds*. Academic Press: London, UK.

Nielsen, B. P. 1971. Migration and relationships of four Asiatic plovers Charadriinae. *Ornis Scandinavica* 2:137–142.

Nol, E. and M. S. Blanken. 2014. Semipalmated Plover (*Charadrius semipalmatus*). In P. G. Rodewald (ed.), *The Birds of North America Online*. Cornell Lab of Ornithology: Ithaca, NY.

Outlaw, D. C., G. Voelker, B. Mila, D. J. Girman, and R. Fleischer. 2003. Evolution of long-distance migration in and historical biogeography of Catharus thrushes: A molecular phylogenetic approach. *Auk* 120:299–310.

Ouwehand, J. and C. Both. 2016. Alternate non-stop migration strategies of pied flycatchers to cross the Sahara desert. *Biology Letters* 12:20151060.

Page, G. W., L. E. Stenzel, J. S. Warriner, J. C. Warriner, and P. W. Paton. 2009. Snowy Plover (*Charadrius nivosus*). in P. G. Rodewald (ed.), *The Birds of North America Online*. Cornell Lab of Ornithology: Ithaca, NY.

Payne, R. B. 1972. Mechanisms and control of molt. pp. 103–155 in D. S. Farner and J. R. King (eds), *Avian Biology*. Academic Press: New York.

Pearce-Higgins, J., D. Yalden, and M. Whittingham. 2005. Warmer springs advance the breeding phenology of golden plovers *Pluvialis apricaria* and their prey (Tipulidae). *Oecologia* 143:470–476.

Pianka, E. R. 1970. On r- and K-selection. *American Naturalist* 104:592–597.

Pienkowski, M. W. 1984. Breeding biology and population dynamics of Ringed plovers *Charadrius hiaticula* in Britain and Greenland: Nest-predation as a possible factor limiting distribution and timing of breeding. *Journal of Zoology* 202:83–114.

Pienkowski, M. W., P. J. Knight, D. J. Stanyard, and F. B. Argyle. 1976. The primary moult of waders on the Atlantic coast of Morocco. *Ibis* 118:347–365.

Pierce, R. 1999. Regional patterns of migration in the Banded Dotterel (*Charadrius bicinctus bicinctus*). *Notornis* 46:101–122.

Piersma, T. 1987. Hop, skip or jump? Constraints on migration of arctic waders by feeding, fattening, and flight speed. *Limosa* 60:185–194.

Piersma, T. and R. E. Gill Jr. 1998. Guts don't fly: Small digestive organs in obese bar-tailed godwits. *Auk* 115:196–203.

Piersma, T., J. Pérez-Tris, H. Mouritsen, U. Bauchinger, and F. Bairlein. 2005. Is there a "migratory syndrome" common to all migrant birds? *Annals of the New York Academy of Sciences* 1046:282–293.

Piersma, T. and P. Wiersma. 1996. Family Charadriidae (Plovers). pp. 384–443 in J. del Hoyo, A. Elliot, and J. Sargatal (eds), *Handbook of the Birds of the World*, vol. 3. Hoatzin to Auks. Lynx Edicions: Barcelona, Spain.

Piersma, T., L. Zwarts, and J. H. Bruggemann. 1990. Behavioural aspects of the departure of waders before long-distance flights – flocking, vocalizations, flight paths and diurnal timing. *Ardea* 78:157–184.

Pulido, F. 2007. The genetics and evolution of avian migration. *Bioscience* 57:165–174.

Pulliainen, E. and L. Saari. 1993. Ring recoveries of Finnish Dotterels *Charadrius morinellus*. *Wader Study Group Bulletin* 67:54–56.

Rappole, J. H., B. Helm, and M. A. Ramos. 2003. An integrative framework for understanding the origin and evolution of avian migration. *Journal of Avian Biology* 34:124–128.

Rayner, J. M. V. 1988. Form and function in avian flight. pp. 1–66 in R. F. Johnston (ed.), *Current Ornithology*, vol. 5. Springer: New York.

Rheindt, F. E., T. Székely, S. V. Edwards, P. L. Lee, T. Burke, P. R. Kennerley, D. N. Bakewell, M. Alrashidi, A. Kosztolányi, M. A. Weston, W.-T. Liu, W.-P. Lei, Y. Shigeta, S. Javed, S. Zefania, and C. Küpper. 2011. Conflict between genetic and phenotypic differentiation: The evolutionary history of a 'lost and rediscovered' shorebird. *PLoS One* 6:e26995.

Rogers, K. G., D. I. Rogers, and M. A. Weston. 2014. Prolonged and flexible primary moult overlaps extensively with breeding in beach-nesting Hooded Plovers *Thinornis rubricollis*. *Ibis* 156:840–849.

Rohwer, S., V. G. Rohwer, A. T. Peterson, A. G. Navarro-Sigüenza, and P. English. 2012. Assessing migratory double breeding through complementary specimen densities and breeding records. *Condor* 114:1–14.

Rolland, J., F. Jiguet, K. A. Jønsson, F. L. Condamine, and H. Morlon. 2014. Settling down of seasonal migrants promotes bird diversification. *Proceedings of the Royal Society B* 281:20140473.

Runge, C. A., T. G. Martin, H. P. Possingham, S. G. Willis, and R. A. Fuller. 2014. Conserving mobile species. *Frontiers in Ecology and the Environment* 12:395–402.

Runge, C. A., J. E. M. Watson, S. H. M. Butchart, J. O. Hanson, H. P. Possingham, and R. A. Fuller. 2015. Protected areas and global conservation of migratory birds. *Science* 350:1255–1258.

Salewski, V. and B. Bruderer. 2007. The evolution of bird migration—A synthesis. *Naturwissenschaften* 94:268–279.

Salomonsen, F. 1955. The evolutionary significance of bird-migration. *Danske Biologiske Meddelelser* 22:1–62.

Serra, L. and L. G. Underhill. 2006. The regulation of primary molt speed in the grey plover, *Pluvialis squatarola*. *Acta Zoologica Sinica* 52:451–455.

Smith, P. W. and N. T. Houghton. 1984. Fidelity of semipalmated plovers to a migration stopover area. *Journal of Field Ornithology* 55:247–249.

Stenzel, L. E., J. C. Warriner, J. S. Warriner, K. S. Wilson, F. C. Bidstrup, and G. W. Page. 1994. Long-distance breeding dispersal of snowy plovers in western North America. *Journal of Animal Ecology* 63:887–902.

Stucker, J. H., F. J. Cuthbert, B. Winn, B. L. Noel, S. B. Maddock, P. R. Leary, J. Cordes, and L. C. Wemmer. 2010. Distribution of non-breeding Great Lakes piping plovers (*Charadrius melodus*) along Atlantic and Gulf of Mexico coastlines: Ten years of band sightings. *Waterbirds* 33:22–32.

Sutherland, W. J. 1998. Evidence for flexibility and constraint in migration systems. *Journal of Avian Biology* 29:441–446.

Swaddle, J. P., and M. S. Witter. 1997. The effects of molt on the flight performance, body mass, and behavior of European starlings (*Sturnus vulgaris*): An experimental approach. *Canadian Journal of Zoology* 75:1135–1146.

Taylor, R. C. 1980. Migration of the Ringed Plover *Charadrius hiaticula*. *Ornis Scandinavica* 11:30–42.

Tomkovich, P. S., R. Porter, E. Y. Loktionov, and E. E. Syroechkovskiy. 2017. Transcontinental pathways and seasonal movements of an Asian migrant, the Common Ringed Plover *Charadrius hiaticula*. *Wader Study* 124:175–184.

Torres-Dowdall, J., A. H. Farmer, E. H. Bucher, R. O. Rye, and G. Landis. 2009. Population variation in isotopic composition of shorebird feathers: Implications for determining molting grounds. *Waterbirds* 32:300–310.

Tree, A. 2001. Kittlitz's Plover as an intra-African migrant. *Honeyguide* 47:10–16.

Underhill, L. G., A. Tree, H. Oschadleus, and V. Parker. 1999. *Review of Ring Recoveries of Waterbirds in Southern Africa*. Avian Demography Unit, University of Cape Town: Cape Town, South Africa.

Wallander, J. and M. Andersson. 2003. Reproductive tactics of the Ringed Plover *Charadrius hiaticula*. *Journal of Avian Biology* 34:259–266.

Walters, J. 1984. The onset of primary moult in breeding *Charadrius* plovers. *Bird Study* 31:43–48.

Warnock, N. 2010. Stopping vs. staging: The difference between a hop and a jump. *Journal of Avian Biology* 41:621–626.

Weir, J. T. and D. Schluter. 2007. The latitudinal gradient in recent speciation and extinction rates of birds and mammals. *Science* 315:1574–1576.

Whitfield, D. P., K. Duncan, D. Pullan, and R. D. Smith. 1996. Recoveries of Scottish-ringed Dotterel *Charadrius morinellus* in the non-breeding season: Evidence for seasonal shifts in wintering distribution. *Ringing & Migration* 17:105–110.

Wilson, J. 1981. The migration of High Arctic shorebirds through Iceland. *Bird Study* 28:21–32.

Wingfield, J. C. 2005. Flexibility in annual cycles of birds: Implications for endocrine control mechanisms. *Journal of Ornithology* 146:291–304.

Wingfield, J. C. 2008. Organization of vertebrate annual cycles: Implications for control mechanisms. *Philosophical Transactions of the Royal Society of London, Series B* 363:425–441.

Winker, K., C. L. Pruett, and J. Klicka. 2006. Seasonal migration, speciation, and morphological convergence in the genus *Catharus* (Turdidae). *Auk* 123:1052–1068.

Zink, R. M. 2002. Towards a framework for understanding the evolution of avian migration. *Journal of Avian Biology* 33:433–436.

Zwarts, L., B. J. Ens, M. Kersten, and T. Piersma. 1990. Molt, mass and flight range of waders ready to take off for long-distance migrations. *Ardea* 78:339–364.

CHAPTER EIGHT

Nonbreeding Ecology*

Erica Nol

Abstract. The nonbreeding season refers to the stationary period when birds are neither breeding nor migrating, and is generally the longest period of the avian annual cycle. Among plovers, some move only short distances or not at all from their breeding grounds, whereas others migrate thousands of kilometers to spend 4–6 months on their nonbreeding grounds. The temporal limits of the nonbreeding period are difficult to assess for both sedentary species and migrants because birds can and do move during this period, but not in the regular patterns that constitute formal migration, as birds often move to follow food resources. In this chapter, I review the evidence that food, habitat (either foraging or roosting), or predation limit plover populations during the nonbreeding season. Plovers are rarely territorial, observations of aggressive interactions and potential competition for food are rare, and few data exist to suggest that food during the nonbreeding season can be depleted with negative consequences for individuals and populations. However, spacing of plovers while foraging might limit the total number of birds that can use a site and the positive relationship between habitat patch size and number of birds suggests that habitat limitation is important. At medium and large geographic scales, plovers track habitats with the most food resources. Habitat quality (either through size of the patch or the degree of disturbance on the patch) appears to also have carryover effects to annual survival. Plover behavior and social organization during the nonbreeding season is also altered by the presence of predators, which affect their habitat selection, and may translate into population-level effects. I conclude that, while the exact mechanism of population limitation is not always clear, nor will it necessarily be the same for all plover species, sufficient evidence suggests that events on the nonbreeding grounds are critical for the determining the health of plover populations.

Keywords: bottom-up, food resources, habitat limitation, limiting factors, non-breeding ecology, plover, predation, roosting habitat, top-down.

The nonbreeding season of plovers is the period in the annual cycle when plovers are stationary (or resident) and neither breeding nor migrating. This period is usually longer than other stages of the annual cycle (Table 8.1), but, for many years, was poorly studied. In part, the case for overlooking the nonbreeding season was practical: tropical nonbreeding (and breeding) areas were in less accessible locations, or birds were widely dispersed, moved more frequently and were, therefore, less easy to study. Primarily, however, studies during the breeding season took precedence for researchers because breeding was assumed to be the key period underlying

* Erica Nol. 2019. Nonbreeding Ecology. Pp. 185–215 in Colwell and Haig (editors). The Population Ecology and Conservation of Charadrius Plovers (no. 52), CRC Press, Boca Raton, FL.

TABLE 8.1
A summary of the nonbreeding distribution, typical flock sizes, months spent on nonbreeding grounds, and evidence for nonbreeding territoriality of Charadrius plovers.

Common name	Distribution[a]	Flock sizes	Months spent on nonbreeding area[a]	Evidence for nonbreeding territoriality	Sources and comments
S. Red-breasted Dotterel	Stewart I., Southern tip of South Island, NZ	<10	January–August	Yes	Myers et al. (1979a), Dowding and Moore (2006)
N. Red-breasted Dotterel	Coastal regions, North Island, NZ	Up to 10	January–July	NR[b]	Dowding and Moore (2006)
Lesser Sand-Plover	Coastal regions of eastern Africa, Arabian Peninsula, India, S. Pacific Islands, Thailand Peninsula, Australia	Up to 13,000	August–February	No	Balachandran and Hussain (1998), Panov 1963 in Colwell (2000), Round (2006), Milton et al. (2014)
Greater Sand-Plover	Coastal regions of eastern Africa, India, S. Pacific Islands, Thailand Peninsula, Australia	10–10,000	July–February	NR	Pandiyan and Asokan (2016), Sripanomyom et al. (2011)
Caspian Plover	Eastern Africa throughout upland areas far from coast	Small flocks up to 200	August–mid-April	NR	Sinclair (1978)
Collared Plover	Coastal areas of s. Mexico, Central America, near water throughout South America but not in s. Argentina	Solitary up to 10	March–August	NR	Scherer and Petry (2012)
Puna Plover	Coastal areas of w. South America, pampas areas of Peru, Bolivia and Chile	Up to 30	February–August	NR	Laredo (1996), Van Den Hout and Martin (2011)
Two-banded Plover	Coastal areas of n. Argentina, Chile, Uruguay and s. Brazil	Up to 1,500	March–August	NR	Alfaro et al. (2008), Botto et al. (1998), Gonzalez (1996), Bala et al. (2002)
Double-banded Plover	Coastal regions of s. and e. Australia, Tasmania, Norfolk I. and Lord Howe I.; uncommon visitor in New Caledonia, Vanuatu and Fiji. New Zealand (resident population)	Up to 3,700	February–August	NR	Ferrari et al. (2002), McConkey and Bell (2005), Veitch (1978)

(Continued)

TABLE 8.1 (*Continued*)

A summary of the nonbreeding distribution, typical flock sizes, months spent on nonbreeding grounds, and evidence for nonbreeding territoriality of Charadrius plovers.

Common name	Distribution[a]	Flock sizes	Months spent on nonbreeding area[a]	Evidence for nonbreeding territoriality	Sources and comments
Kittlitz's Plover	NE Egypt; sub-Saharan Africa from Senegal to Sudan and South Africa; Madagascar	2–5 birds, rarely 50 up to 500	April–August (but up to 10 months)	NR	Owino (2002), Lipshutz et al. (2011)
Red-capped Plover	E Nepal and n.e. India to n. Indochina, s. China, South Korea and Japan	Up to 5,000		NR	Chapman and Lane (1997), Evans et al. (2010), Gosbell and Grear (2005)
Malay Plover	S Thailand, Malay Peninsula, s. Cambodia and s. Vietnam to Sumatra, Borneo, Philippines, Sulawesi, and Bali to Timor	In pairs but roosts up to 27 birds	July–February	Yes	Yasué and Dearden (2009)
Kentish Plover	Coastal areas of Mediterranean, s. to sub-Saharan Africa, s. Asia and w. Indonesia	Up to 30 common, but up to 600	October–March	NR	Masero et al. (2000)
White-faced Plover	Coastal areas of Vietnam, Gulf of Thailand, Malay Peninsula and Sumatra	Unknown	August–March	No information	Kennerley et al. (2008), Iqbal et al. (2010)
Snowy Plover	Pacific coastal areas US to s. Chile, Gulf of Mexico, Caribbean	Loose flocks up to 300	August–January	NR	Myers et al. (1979b), Page et al. (2009), Elliott-Smith et al. (2004)
Javan Plover	Coastal Vietnam, Gulf of Thailand, Malay Peninsula and Sumatra	Up to 100	September–April	No Information	Trainor (2011), Iqbal et al. (2013)
Wilson's Plover	Coastal areas Baja California, Central America, Caribbean, n. S. America	Up to 30	October–April	No (although forage solitarily so spacing)	Wunderle et al. (1989), Morrier and McNeil (1991)
Common Ringed Plover	Caspian Sea and SW Asia s. to South Africa, British Isles s. to Africa	1–60	October–March	No	Masero et al. (2000), Piersma et al. (1990)

(Continued)

TABLE 8.1 (*Continued*)

A summary of the nonbreeding distribution, typical flock sizes, months spent on nonbreeding grounds, and evidence for nonbreeding territoriality of Charadrius plovers.

Common name	Distribution[a]	Flock sizes	Months spent on nonbreeding area[a]	Evidence for nonbreeding territoriality	Sources and comments
Semipalmated Plover	Coasts of North and South America, from California to n. Chile and from s. Virginia to Patagonia; also Bermuda, West Indies, Galapagos Islands	Up to 4,000	October–April	Yes, Myers et al. (1979b), No, Morrier and McNeil (1991), Rose and Nol (2010), Nol et al. (2014)	Almeida and Ferrari (2010), Myers et al. (1979b) some habitats, others not. Nol et al. (2014), Fedrizzi (2003)
Long-billed Plover	E Nepal and n.e. India to n. Indochina, s. China, South Korea and Japan	Up to 10	Unknown	No information	Murose (1998)
Piping Plover	Atlantic coast of s. United States, Gulf of Mexico, Bahamas, locally in Greater Antilles; also w. Mexico (Sonora)	Up to 60	August–March	<5% of time	Johnson and Baldassarre (1988), Gratto-Trevor et al. (2016), Noel and Chandler (2008)
Madagascar Plover	Coastal areas of w. and s. Madagascar	2–50	June–July	No information	Long et al. (2008)
Little Ringed Plover	Africa s. of Sahara, Arabia, India, e. China and Indonesia	Up to 12	August–February[b]	No	Simmons (1956)
African Three-banded Plover	Eritrea to Tanzania, S Democratic Republic of Congo and Gabon, and S to South Africa; nonbreeding also around l Chad	Solitary, small groups or rarely up to 40	October–March	No information	del Hoyo et al. (2016)
Madagascar Three-banded Plover	Madagascar	Solitary or up to 15	October–June	No information	del Hoyo et al. (2016)
Forbes's Plover	Inland Guinea to s.w. South Sudan, s. through Democratic Republic of Congo, w. Uganda locally to c. Angola, n. Zambia	Solitary but sometimes 15–20	September–February (Variable)	No	del Hoyo et al. (2016)
White-fronted Plover	West Africa, s. Africa, w. Madagascar	Small, Occurrence up to 375	May–July	Some are territorial all year, others in flocks (probably in poorer territories)	Lloyd (2008)

(Continued)

TABLE 8.1 (*Continued*)

A summary of the nonbreeding distribution, typical flock sizes, months spent on nonbreeding grounds, and evidence for nonbreeding territoriality of Charadrius plovers.

Common name	Distribution[a]	Flock sizes	Months spent on nonbreeding area[a]	Evidence for nonbreeding territoriality	Sources and comments
Chestnut-banded Plover	Rift Valley soda-lakes on Kenya–Tanzania border. SW coast Angola to Mozambique; inland n. Namibia, n. Botswana and n.c. South Africa	Up to 50–60	Variable and not consistent across range	No information	BirdLife International (2016)
Killdeer	C. and S US, Central America, n. South America, Pacific coast to s. Peru	Up to 30	October–March (N) Resident elsewhere	No	Myers et al. (1979b) some habitats only. Sanzenbacher and Haig (2002), Nol et al. (2014)
Mountain Plover	C. California to Baja California, east to s. Texas, n.e. Mexico, inland locations	Up to 600	October–mid-March	No	Knopf and Rupert (1995). Loose moving flocks
Oriental Plover	NW & NC Australia, sparsely in S, inland areas	Up to 100, exception 1,440,000	September–March	No	Large numbers feeding on grasshoppers (Piersma and Hassell 2010)
Eurasian Dotterel	N Africa and Middle East E to W Iran, inland areas	30–80	August–February	Gregarious so NR	del Hoyo et al. (2016), Hayman et al. (1986)
St. Helena Plover	St Helena, in S Atlantic Ocean, upland and coastal	Up to 15	February–September	No information	del Hoyo et al. (2016)
Rufous-chested Dotterel	N to N Chile, N Argentina, Uruguay and S Brazil, occasionally farther N to Peru and SE Brazil, Falkand Islands	15–200 birds, exception 350 (Falkland Is	May–August	Yes	Myers et al. (1979b), Alfaro and Clara (2007), Alfaro et al. (2008), Isacch and Martínez (2003)
Red-kneed Dotterel	Inland wetlands of extreme S New Guinea and Australia	10–100	February–July	No information	del Hoyo et al. (2016)
Hooded Plover	Coastal areas of S Australia, Tasmania	10–40, occurrence 200	May–July	No	Weston et al. (2009)
Shore Plover	Coastal areas of Chatham Is: n Rangatira I (South East I), Mangere I, New Zealand	Up to 15	February–September	No	Davis and Aikman (1997), Dowding and O'Connor (2013)

(Continued)

TABLE 8.1 (*Continued*)

A summary of the nonbreeding distribution, typical flock sizes, months spent on nonbreeding grounds, and evidence for nonbreeding territoriality of Charadrius plovers.

Common name	Distribution[a]	Flock sizes	Months spent on nonbreeding area[a]	Evidence for nonbreeding territoriality	Sources and comments
Black-fronted Dotterel	Inland areas of Australia, Tasmania and New Zealand	Up to 175	April–July	Not recorded	Taylor (2004)
Inland Dotterel	Scattered locations, W, SC and EC Australia	Up to 20	Variable, breeds any time	Not recorded	MacLean (1976)
Wrybill	Northern South Is, North Island, New Zealand	Up to 1,000s	February–August	Not reported	Veitch (1978)

[a] Most nonbreeding distributions and periods come from del Hoyo et al. (2016) and Birdlife International (2016).
[b] Not reported.

the population dynamics of most birds. This view changed with the excellent body of work on the nonbreeding ecology of Eurasian Oystercatcher (*Haematopus ostralegus*) from western Europe (summarized in Goss-Custard 1996) and the Sanderling (*Calidris alba*) from North America (Myers et al. 1979a, b; Myers 1984). Insights from studies of these two species drew attention to the importance of the nonbreeding season to population dynamics and the potential for important carry-over effects, where population regulation can be influenced by limiting factors operating in one or more of an animal's life stages (Norris and Taylor 2006). More recently, several authors, including some studying Piping Plovers, have demonstrated that the quality of migratory stopover or wintering habitats can directly impact apparent annual survival (Gibson et al. 2018, Piersma et al. 2016). These findings are critical for informing our conservation actions because they draw attention to critical ecological effects that drive population dynamics.

Two primary limiting factors, food and predation, have been proposed to regulate shorebird and other bird populations, and the relative importance of each factor during the nonbreeding season has been a focus of study. While examples from plovers are, in general, less well developed, studies of Eurasian Oystercatchers show that access to food during the nonbreeding season can affect the quality of individuals during the subsequent breeding season, which influences reproductive success and longevity (Goss-Custard 1996). For nonbreeding Black-bellied Plovers (*Pluvialis squatarola*), defense of feeding territories during the nonbreeding season occurs in greater than half of individuals in South Africa (Turpie 1995). The enhanced intake rate of territorial individuals allows them to leave earlier in the spring and arrive sooner in northern breeding grounds (Turpie 1995). Taken together, these observations of the importance of food to nonbreeding shorebirds suggest that bottom-up processes during the nonbreeding period may regulate populations (Aarif et al. 2014). For other shorebirds, and again, without any studies directed to the biology of plovers, predation may strongly influence their ecology during the nonbreeding season. Studies of Western Sandpipers (*Calidris mauri*) during the migratory period (Pomeroy 2006) and Common Redshank (*Tringa totanus*) during the nonbreeding period (Cresswell 1994) have emphasized the role of predation risk in structuring the behavior of individuals in flocks, body mass dynamics, migratory timing and, to some extent, population dynamics. These observations support a top-down mechanism of control of populations.

While food and predation have received considerable attention in understanding the distribution of nonbreeding birds, differences in the locations of the nonbreeding grounds have also been attributed to different patterns in the timing of molt (Meltofte 1996). Molt is an energetically challenging process that is integral to the nonbreeding ecology of most plovers. Many plover species molt shortly after arrival on their nonbreeding grounds, resulting in additional energy needs (e.g., Wrybill, Davies 1997; Double-banded Plover, Thomas 1972; Lesser Sand-plover, Balachandran and Hussain 1998; Common Ringed Plover, Meissner et al. 2010). When plovers molt into their basic plumage, their often strongly contrasting, black or brown and white alternate plumage is muted, which may provide some protection from detection by predators. Wrybill is one of the few plover species for which molt has been studied in some detail in the context of where this species spends the nonbreeding season. The time to complete a molt in Wrybill can last 100 days (Davies 1997), and Wrybill may choose northerly sites where it is easier to molt, prior to migrating further south for the remainder of winter. While most species molt outside of the breeding season, Hooded Plover's primary molt overlaps entirely with its breeding season (Rogers et al. 2014) as does that of the Double-banded Plover (Barter and Minton 1987). Molt can vary among populations and migration patterns within and between species (Meltofte 1996, Pyle 2008, Howell 2010), but there are few published data on which to base any understanding of the adaptive significance of timing or speed of molt in plover species. Other adaptive explanations for variation in the spatial distribution of nonbreeding ranges, including differential survival probabilities, competitive exclusion of particular ages or sexes, or the distribution of resources have also not been well supported nor tested in studies of plovers nor other shorebirds (Myers et al. 1985, Meltofte 1996, but see Hockey et al. 1992). In three plover species, the sex ratios of nonbreeding flocks have been confirmed using molecular methods, and in two (Semipalmated Plover, Storm-Suke 2012; Common Ringed Plover, Summers et al. 2013) there is no disparity from a 1:1 ratio of males to females at nonbreeding sites. In Hooded Plover, however, 69% of captured birds at a salt lake in western Australia were male, indicating possible habitat segregation by sex (Weston et al. 2004). No evidence exists of latitudinal segregation between the sexes for Snowy Plover (Page et al. 1995). Lower resighting probabilities of male Piping Plover on the nonbreeding grounds have led to the suggestion that males may winter in less accessible or infrequently surveyed locations (LeDee et al. 2010). Plovers have less well-developed sexual size dimorphism (Zefania et al. 2010, Nol et al. 2013) than what exists in some sandpipers (e.g., *Calidris mauri*, Nebel et al. 2002). This suggests that selective pressures for the two sexes to diversify their prey or their geographic distributions are less pronounced. The general lack of well-developed sexual size or plumage dimorphism also means that birds need to be sexed molecularly to definitively test hypotheses about habitat or geographic range differentiation.

While many, but not all, plovers spend the nonbreeding season in coastal environments, they may do so because of the lower prevalence of parasites in these habitats (Mendes et al. 2005). This hypothesis has also been invoked as an explanation for the differences within species in the distance flown from breeding to non-wintering areas across a latitudinal gradient, with higher latitude, and colder non-wintering areas resulting in lower parasite prevalence for birds (Aharon-Rotman et al. 2016). Finally, Hockey et al. (1992) proposed that more Palearctic migratory shorebirds spend the nonbreeding period in the southern than in the northern hemispheres because the seasonal pulse of reproduction of estuarine invertebrate prey coincides with the period when shorebirds acquire nutrients for northbound migration. Thus, there are many potential explanations, probably not mutually exclusive, that can help future researchers to understand why plovers spend their nonbreeding periods where they do and for how long.

In this chapter, I review the ecology of plovers during the nonbreeding season, including their distributions, fluctuations in numbers, whether these nonbreeding distributions are the same (i.e., year-long resident species) or separate (i.e., migratory species) from breeding distributions, the length of time spent on nonbreeding grounds, and the general patterns of habitat use and diet of plovers during the nonbreeding season. I examine the literature for evidence that factors acting during the nonbreeding season may indeed limit populations of plovers. Through a more detailed examination of patterns of foraging, diet and intake rates, the presence and requirement for nocturnal foraging and evidence of territorial behavior, I explore the importance of food as a limiting factor for plovers in the nonbreeding season. Demonstration of food limitation would support the existence of bottom-up control. I also review the evidence that habitat size or quality can limit plover populations. I then turn to the evidence for the importance of predators in affecting patterns of plover habitat use and flock sizes, which would support a top-down view of population limitation. In the context of predation risk, I also assess whether roosting habitats can limit nonbreeding plover populations. As roosts consist of congregations of birds and are widely acknowledged to have an antipredator function, evidence about roosting habitat limitation would indirectly support the importance of top-down control. Exciting new (albeit limited) evidence on carryover effects from the nonbreeding period and the importance of the nonbreeding season in limiting populations will be reviewed. Finally, I discuss research needs that will further our understanding of plover ecology and, therefore, enhance their conservation. This chapter does not cover the ecology of nonbreeding migratory plovers who "over summer" (*sensu* McNeil 1970), that is, first year or older birds who may not move from their nonbreeding grounds to their breeding grounds either because they forego breeding because of poor conditions on the breeding grounds (Hughey 1985) or because they have delayed maturity (McNeil 1970).

NONBREEDING DISTRIBUTION

Plovers spend their nonbreeding season on all continents except Antarctica, although they are scarce in North America above 37°N and east of 81°W and above 48°N and west of 106°W, and in Europe and Asia above 58°N (Table 8.1). The geographic extent of nonbreeding areas appears proportional to population size, possibly reflecting an ideal-free distribution (Fretwell and Lucas 1970, Ould Aveloitt et al. 2013). Nonbreeding distributions of long-distant Nearctic migrants that breed across vast expanses of arctic and subarctic habitats also tend to be extensive. For example, the nonbreeding range of the Semipalmated Plover on the east coasts of North and South America extends from South Carolina to northeastern Brazil (Nol and Blanken 2014), a straight-line distance of over 8,000 km. The long-distant migrant, Common Ringed Plover, has a similarly large wintering range from southern Spain to South Africa (9,600 km, del Hoyo et al. 2016; Chapter Seven, this volume). The extent of geographic ranges for austral migrants during the nonbreeding season is generally less than Nearctic migrants (e.g., migratory Two-banded Plover winters from Tierra del Fuego to Uruguay, 2,600 km, del Hoyo et al. 2016). Year-long resident (sedentary) species have nonbreeding distributions that broadly overlap their breeding distribution (Figure 7.3, Chapter Seven, this volume) but often are accompanied by different patterns of habitat use (e.g., Red-capped Plover, Hobbs 1972; Hooded Plover, Weston et al. 2009).

Timing of the Nonbreeding Season

For most migrant plovers that spend the nonbreeding season away from their breeding grounds, the usual length of stay is 4–6 months annually. This pattern persists for both long- and short-distant migrants, and species in northern and southern hemispheres. For long-distance migrants, additional time outside the usual 2-month breeding season is spent in transit, while in the case of short-distance migrants, this time may be spent on the breeding grounds where the species typically have more than one clutch or an extended breeding season (see Chapter Five, this volume). Sedentary, nonmigratory species sometimes move short distances from their breeding grounds to form small flocks for a few months (Table 8.2) while in other years, there is little movement away from the breeding grounds after breeding is complete (e.g., Hooded Plover, Weston et al. 2009; Malay Plover, Crossland 2000, 2002; Southern Red-breasted Dotterel, Dowding and Chamberlain 1991). Within species there is evidence for a latitudinal gradient in migratory or resident status as northerly populations are migratory while those in southern temperate environments are not (e.g., Snowy Plover, Page et al. 2009; Killdeer, Jackson and Jackson 2000).

Length of time spent on the nonbreeding grounds can also vary within a species. For example, most Double-banded Plovers are sedentary, although small flocks migrate 1,500 km to Australia, and some individuals are nomadic (Dowding and Moore 2006). In other cases, age-related differences in distribution and movement are apparent. Wrybill juveniles spend up to 7.5 months on nonbreeding grounds while adults spend 6.5 months (Hay 1984). Variability within species in the time spent on the nonbreeding areas can also extend to variation in space use. For example, individual sedentary Killdeer breeding in agricultural settings of the Willamette Valley, Oregon, had smaller home ranges during the nonbreeding season than the migrant Killdeer who spent the nonbreeding season at the same location (Sanzenbacher and Haig 2002).

Sizes of Nonbreeding Flocks

The numbers of plovers at a wintering site also varies, and, at least for some plovers, may be proportional to the distance migrated, with those

TABLE 8.2

Observations of the ecology of nonbreeding plovers and strength of evidence in the literature (see text) to support bottom-up or top-down regulation of populations.

Observation	Bottom-up	Top-down	Explanation
Lower mass variation between breeding and nonbreeding grounds	Moderate	Strong	Lower food requirements to avoid winter starvation and greater escape performance
Flocking during roosting	NA	Strong	Selfish herd hypothesis: safety in numbers
Small flock sizes	Moderate	Weak	Small flock sizes are more vulnerable to predation, but also reduce competition for food
Food depletion	Weak	NA	Food supply determines upper limit of birds
Food specialization	Weak (many opportunistic species)	NA	Suggests sufficient food to allow them to be selective
Competition (intra- or interspecific)	Weak	NA	Suggests food is limited
Nocturnal foraging	Moderate	Moderate	Depending on study, birds have lower or higher foraging success at night
Mortality greater in winter than other seasons	Weak	Weak	Heightened risk of starvation or predation
Molt during winter	Weak	Weak to moderate	Adequate food to molt and not vulnerable to predation, often exhibit cryptic (less contrasting) plumage

migrating longer distances relatively more concentrated in their distribution than short-distant migrants (Table 8.1). However, unlike sandpipers (Myers et al. 1987), plovers do not typically occur in large (e.g., >10,000) numbers during the nonbreeding season. For many species, small (<50) and dispersed flocks of plovers exist throughout the nonbreeding range (Table 8.1), and the smallest of flocks are more likely to occur at the edges of the nonbreeding distribution (e.g., a single Semipalmated Plover in Buenos Aires Province, Argentina, during December to March at the southern extent of their range, Blanco et al. 2006).

Substantial temporal variability in numbers can reflect movement away from nonbreeding areas even during the "stationary" nonbreeding months (Table 8.1). Thousands of Semipalmated Plovers occur regularly on wide mudflats in northeastern Brazil, but only through the month of December; by January, these flocks have almost completely dissipated (Rodrigues 2000, Fedrizzi 2003). This species then only shows up at sites in southern Brazil in January (Scherer and Petry 2012). By contrast, smaller numbers of Semipalmated Plover are relatively stable throughout the nonbreeding season (e.g., Semipalmated Plover wintering in Cayo Guillermo, Cuba, Nol et al. 2014). At other nearby sites (e.g., coastal Venezuela, coastal Brazil), increases in winter numbers of the Semipalmated Plover in December and January are suggested to consist of staging populations that undergo regional movements away from these sites (Morrier and McNeil 1991, Rodrigues 2000). The Two-banded Plover is numerous (~100 individuals) on Buenos Aires mudflats in March and April, but then declines to smaller numbers (~50 individuals) in May to July (Martínez-Curci et al. 2015). During winter the numbers of Red-capped Plover increase from 200 to 400 in two New Zealand estuaries, suggesting that they come from elsewhere (Veitch 1978). A two- to threefold increase in numbers of Lesser and Greater Sand-Plovers in coastal Malaysia occurs from November to January (Norazlimi and Ramli 2014). In other cases, nonbreeding areas remain unoccupied in some years, usually due to the complete absence of prey, or factors that affect the availability of that prey. For example, rains in East Africa trigger the appearance of nonbreeding Caspian Plover only in some years (Sinclair 1978). Similarly, the largest concentration of plovers ever recorded was more than 144,000 nonbreeding Oriental Plovers foraging on grasshoppers in agricultural fields in northwestern Australia during a drought; the flock stretched for 75 km of beach and numbers declined greatly 2 weeks later (Piersma and Hassell 2010). Annual variation in numbers can also be caused by variation in reproductive success, as gauged by the number of juveniles (e.g., Kentish Plover, Lesser and Greater sand-plovers, Eiamampai et al. 2014). Thus, birds can move among sites and there can be considerable annual variation in numbers. While information is often only available from a few sites, evidence suggests that numbers of birds vary substantially both temporally and spatially within the nonbreeding season.

HABITAT USE

At first glance, there is a relative uniformity in the habitats used by nonbreeding plovers. Many species forage on wet, firm sandy or muddy substrates in large and exposed coastal areas or small coastal bays and estuaries (e.g., Withers and Chapman 1993, Young et al. 2006). However, substantial variation in both habitat use and diet exists among congeners (Skagen and Oman 1996). About half of all species spend the nonbreeding season at inland habitats (15 of 32 species, Kraaijeveld 2008). For example, Red-kneed Dotterel occur near freshwater ponds (del Hoyo et al. 2016) and Eurasian Dotterel forages in upland, drier habitats, including stony and shrubby steppe, semidesert, ploughed farmland and the margins of cultivation (del Hoyo et al. 2016). Caspian Plover forages in upland areas after rain (Sinclair 1978), Oriental Plover forages in grasslands (Piersma and Hassell 2010), and Mountain Plover feed in grazed alfalfa fields (Wunder and Knopf 2003) or fields burnt following harvest, as well as in freshly tilled fields (Knopf and Rupert 1995).

Some inland breeding plovers descend to coastal habitats during winter (e.g., Puna Plover, Hughes 1984) while others move between inland and coastal habitats. For example, Double-banded Plover prefers foraging on mudflats in one harbor in Auckland, New Zealand, but also forages in fields during the nonbreeding season (Veitch 1978). Whether the species winters at inland or coastal habitats does not appear to be related to whether they are migrants, residents, or partial migrants (Kraaijeveld 2008) or whether they are inland or coastal breeders. Wrybills breed on braided rivers but winter in mudflats of harbors and estuaries (Hay 1984). By contrast, some inland nesting species (e.g., Killdeer) rarely forage in tidal flats (F. Sanders, pers. comm.; Engilis et al. 1998).

Coastal species prefer mud over sand flats for foraging (Ntiamoa-Baidu et al. 1998, Pandiyan and Asokan 2016), presumably because of greater organic matter in the mud that supports larger populations of invertebrates (Rose and Nol 2010, Aarif et al. 2014). Piping Plover used primarily sand and algal flats (Drake et al. 2001); the algae presumably have greater invertebrate communities.

Regardless of habitats used, all plover species forage predominantly on invertebrates (Skagen and Oman 1996, Isacch et al. 2005, del Hoyo et al. 2016), with some species occasionally including fish in their diets (e.g., Northern Red-breasted Plover, del Hoyo et al. 2016). Whether species forage on estuarine mudflats on small or large invertebrates, and/or on drier habitats may relate in part, to bill lengths (Strauch and Abele 1979) as many short-billed plovers winter in coastal estuaries feeding on benthic invertebrates, but those with longer bills forage on crustacean or macroinvertebrate diets (Strauch and Abele 1979). While this morphology–ecology link applies to the three species (Wilson's, Semipalmated and Collared plovers) studied by Strauch and Abele (1979) in Costa Rica, the general pattern may or may not extend to other plover species. For example, the Lesser Sand-plover, with a short, stubby bill, forages regularly on both polychaetes and crustaceans (Aarif 2009). In general, bill lengths of plovers vary little, especially in comparison to the greater variation in bill lengths seen among species in the other major shorebird family, the Scolopacidae (Colwell 2010). In African wetlands, elements of the foraging behavior of plovers, such as size of the prey, and distance at which prey are attacked, do not appear to vary along a body mass gradient (Hockey et al. 1999). Instead, diet in the suite of African plovers seems to vary primarily along a gradient of food items captured with strong (e.g., crabs) versus weak visual signals (e.g., polychaetes; Hockey et al. 1999). There also exists a relationship between the lengths of the legs and the bills, which together will determine how far forward an individual can bend during prey capture (Strauch and Abele 1979). Other features of their life history including age of maturity and adult survival may be associated with plovers whose diets include crustaceans or larger macroinvertebrates compared to those who forage primarily on polychaetes and small invertebrates, but currently these features are known.

POTENTIAL LIMITING FACTORS

Evidence for Bottom-up Control of Populations

Multiple lines of evidence suggest that variation in food availability during the nonbreeding season affects population dynamics of plovers (Table 8.2). Territoriality, for example, may indicate defense of food resources. Evidence for direct competition would also support this view. Site fidelity might indicate that nonbreeding locations promote overwinter survival not only through access to food, but also possibly through access to areas with low predation risk. Evidence for habitat or food limitation directly is also reviewed, as is diet, the breadth of which can indicate the degree to which plovers can respond to changing food resources under environmental degradation. Finally, I review the literature on time budgets to determine whether these data suggest that food acquisition monopolizes a large portion of the daily (including nocturnal) activity patterns of plovers.

Evidence for Territoriality

Myers et al. (1979b), in their review of territoriality during the nonbreeding period, defined territoriality as "nonbreeding shorebirds obtaining control of resources within a defended area through aggressive spatial defense." Territoriality suggests that food, and the habitat that supports that food, limits the number of birds that can occupy that habitat patch. Territorial behavior on breeding grounds is the general rule in plovers although the scale at which this occurs may vary (i.e., at a nest site or a territory) among species or populations. However, few studies specifically mention territoriality during the nonbreeding season, nor explicit mention of conspecific nor interspecific aggression (e.g., chases, fights) during foraging or roosting (Table 8.1). A few authors specifically make note of the absence or rarity of this behavior. For example, time devoted to agonistic behavior in Semipalmated Plovers in Venezuela was always extremely low (0.1% ± 0.1%) and did not change over the nonbreeding season (e.g., in March and April, Morrier and McNeil 1991). Agonistic behavior was also very low for Wilson's Plovers (0.6%) but increased significantly to 1.2–1.6% in March and April in comparison to other months (Morrier and McNeil 1991). Competitive, intraspecific interactions in Semipalmated Plover and Killdeer were rare in Cuba, where <5% of interactions in focal individuals were competitive (Nol et al. 2014). Myers and McCaffery (1984) report evidence of territoriality in Semipalmated Plover during 3 days of observation in March in Paracas, Peru, as well as non-territorial aggression, observations that contrast with those of Duffy et al. (1981) from the same area. Duffy et al. (1981) suggested that there was neither food limitation nor defense of food resources at this nonbreeding site. In a flock of 35 Semipalmated Plovers in the same site at Paracas, Peru in November 2017, some individuals repeatedly chased conspecifics while most individuals foraged among conspecifics without noticeable interaction (E. Nol, pers. obs.).

There was no evidence of territoriality in a small flock of released Shore Plover until the onset of the breeding season (December), although that flock was intensively monitored (Dowding and O'Connor 2013). Several reviews of territoriality in shorebirds included plovers (Myers et al. 1979b, Colwell 2000). Of the eight species mentioned by Colwell (2000), none have reported territoriality or if it occurred, it was very uncommon (e.g., <5% of time in Piping Plover, Johnson and Baldassarre 1988). Myers et al. (1979b) studied four of these same species (Snowy Plover, Semipalmated Plover, Killdeer, Two-banded Plover). While territoriality in some individuals was recorded on some habitats, it was absent in other habitats (Myers et al. 1979b). While infrequent territoriality is noted for Snowy Plovers on beaches (www.westernsnowyplover.org/pdfs/plover_natural_history.pdf), this source did not provide any supporting data. Aggressive displays between nonbreeding Rufous-chested Dotterel in coastal Argentina were noted in Myers et al. (1979b), but the dates of this observation relative to the onset of breeding and the sexes of the antagonists were not recorded.

Spacing can be an indicator of territoriality, if it is accompanied by intraspecific aggression (Myers et al. 1979b). Plovers maintain larger intraspecific distances during foraging than during roosting (e.g., Nol et al. 2014), likely due to their method of run-and-stop visual foraging (Strauch and Abele 1979). Many species forage solitarily (e.g., Puna Plover, Van den Hout and Martin 2011). Others, like Wilson's Plover that feed predominantly on fiddler crabs (*Uca* spp.), forage alone far from conspecifics. Solitary foraging is thought to occur so that the foraging bird can be still enough to allow a sufficient density of crabs to gather around it to increase its intake rate (Thibault and McNeil 1995). This density requirement likely produces some degree of spacing and possibly an upper limit to the number of birds that a habitat can support.

Myers et al. (1979b) suggested, and subsequently rejected, the hypothesis that some chases and signs of territoriality occurred in early spring, prior to migration, not for defense of food, but simply because of the increase in circulating sex hormones (e.g., "residual," Hamilton 1959). In Semipalmated Plover in coastal Georgia and South Carolina, observations of chasing during the nonbreeding season occurred in late March and throughout April, and were usually of males who had completed their pre-alternate molt (E. Nol, pers. obs.), and likely had increased levels of circulating testosterone. During roosting, many plovers occur in loose or tight aggregations (MacLean 1977, Rose and Nol 2010), which suggests sociality and a lack of territorial behavior. Given the rarity with which agonistic behavior has been recorded (Table 8.1), limited aggression observed during the nonbreeding season in plovers does not generally appear to be a consequence of defending feeding areas but may be indicative of early signs of breeding season territorial behavior (*sensu* agonistic behavior in Black-bellied Plover, during migration, Buchanan 2011). Alternatively, within plovers, as has been observed in Sanderling (Myers et al 1979a), there may be two strategies, with individuals adjusting their territorial behavior both on the basis of prey density and on the number of competitors. Further studies using marked individuals and conducted across the nonbreeding period are needed to distinguish these two alternatives.

Evidence and Opportunity for Competition

Food and foraging are principal components of the nonbreeding ecology of plovers. The small, dispersed flocks while foraging suggest that competition may be prevalent. Competition is inferred when defense of feeding areas during nonbreeding is observed. Migrant plovers often winter at the same sites as resident heterospecific plovers, and sometimes also with resident conspecifics. Semipalmated Plover and Collared Plover form mixed flocks in multiple sites in South America (Withers and Chapman 1993, Almeida and Ferrari 2010, Scherer and Petry 2012). Migrants will also winter with year-long residents of their own species (Sanzenbacher and Haig 2002). While intraspecific and interspecific competition are thought to expand and constrain diet breadth for a species, respectively (Svanbäck and Bolnick 2007, Bolnick et al. 2010), there appears to be enough niche differentiation among co-occurring species to reduce niche overlap and hence interspecific agonistic behavior (Strauch and Abele 1979). These authors indicated "no obvious aggressive interactions were seen," for co-occurring Semipalmated Plover, Collared Plover, and Wilson's Plover. Niche differentiation can also be indicative of competition during past interactions between species (i.e., the "ghost of competition past," Connell 1980), but without conducting removal experiments, and in the absence of observations of interspecific interactions, the presence of contemporary competition is difficult to assess. Interspecific interactions have been observed in coastal Alabama, among Piping Plover, Snowy Plover, and Semipalmated Plover (Johnson and Baldassarre 1988). The lack of temporal overlap of wintering migratory flocks from the northern hemisphere with wintering resident plovers in Patagonia was suggested as support for the evolution of annual cycles that avoided competition between the two groups (Isacch and Martínez 2003, see also Meltofte 1996). These few examples, coupled with the above information on spacing, suggest that neither inter- nor intraspecific competition is common or widespread in contemporary co-existing populations of plovers (Boland 1990). That said, it only takes one occasional "event" of food scarcity to influence population size via competition. Disentangling these two drivers of agonistic behavior, territoriality and competition, is challenging, but no strong evidence seems to support either.

Nonbreeding Site Fidelity as an Indicator of Defense over Food Resources

Site fidelity, the return of an individual to an area usually in successive years, suggests that the site is valuable for attaining food resources (or avoiding predation) and therefore enhancing overwinter survival. While large movements within a nonbreeding season can occur, some studies report moderate or strong interannual site fidelity, although the scale at which individuals display fidelity varies substantially. The site faithfulness to nonbreeding locations has rarely been reported with the precision recorded for breeding plovers (e.g., <100 m, particularly males, Flynn et al. 1999).

Site fidelity to a large barrier island occurs in Piping Plover in the Gulf of Mexico (Johnson and Baldassarre 1988) and coastal Georgia (Noel and Chandler 2008) while most resighted Piping Plovers in the Bahamas were seen at the same beach or within 6 km of the original capture location (Gratto-Trevor et al. 2016). Site fidelity was strong for adult Wrybills but juveniles wandered more freely (Hay 1984). Recaptures of Wrybills between two wintering sites 100 km apart implies that site fidelity is weak for some individuals, although most birds tended to be site faithful (Davies 1997). Between 4% and 8% of banded Lesser Sand-Plovers were recaptured in subsequent years at the same wintering sites in India. Similar totals of birds occurred at the Gulf of Mannar over three consecutive winters (Balachandran and Hussain 1998), suggesting site fidelity to location for the population although not necessarily for individual birds. Site fidelity has been inferred for Rufous-chested Dotterel and Tawny-throated Dotterel (*Oreopholus ruficollis*) in pampas grasslands of Argentina because previous use correlated with current use (Isacch and Martínez 2003). By contrast, limited or no site tenacity occurs in Caspian Plover. Individuals performed local movements, appearing in some areas for a week or two only to move on, and then return several weeks later (Sinclair 1978).

In Mountain Plover, the scale and variability in movement patterns have been well documented using radio telemetry (Knopf and Rupert 1995). Prior to migration, plovers foraged and roosted in loose flocks of 2 to >1,100 birds, with average flock size increasing as the season progressed, but the Mountain Plovers with radios were, with one exception, not relocated together. Mountain Plovers sometimes moved over 1 km a day, but this distance was variable (0.2–1.9 km/day) as birds remained at one location for a few days then moved to a new area. The distance between weekly resighting locations of Mountain Plover exceeded 55 km on seven occasions with one bird moving 127 km over a week period.

Variation in the patterns of site fidelity within species may depend on whether the individuals are permanent or nonbreeding residents. For radio-tracked Hooded Plovers, whose breeding and nonbreeding ranges overlap, movements were greater during the nonbreeding season than during breeding and there was no difference between the sexes in the distances moved (Weston et al. 2009). Hooded Plover in small flocks (4–6 individuals) moved an average of 225 m/day during the nonbreeding season. Resident Killdeer exhibited some site fidelity throughout the year and movements averaged less than 2 km, whereas nonbreeding migrant Killdeer in the same general study area exhibited no site fidelity with movements averaging about 5 km (Sanzenbacher and Haig 2002). In Piping Plovers, a large average home range size and considerable linear movement (average of 3.3 linear km) between successive locations were reported for 48 radio-tracked Piping Plovers (average of 12.6 km^2, Drake et al. 2001), during their 6-month nonbreeding season.

High site fidelity during the nonbreeding season occurred in a marked population of Double-banded Plovers, whereas other populations appeared to be nomadic (Barter and Minton 1987). More than 95% of adult Double-banded Plovers were re-trapped at the original banding site (Barter and Minton 1987, Minton 1987). Southern Red-breasted Dotterels are highly faithful to flocks and locations during the nonbreeding period (Dowding and Moore 2006). Similarly, resighting probabilities for individually banded Semipalmated Plovers from a nonbreeding flock at the south end of Cumberland Island, Georgia, varied between 0.29 and 0.93, over three consecutive 6-month nonbreeding periods (S. Williams, unpublished data). Piping Plover detection probabilities during the nonbreeding season ranged from 0.25 to 0.50, increasing over the study period as the skill of volunteer observers improved (LeDee et al. 2010), and were about 10% higher for female than for male plovers. The factors that affect the substantial variation in the degree to which plovers are site faithful to their nonbreeding areas are currently unknown. Combining data on the quality of nonbreeding locations, based on either food resources or predation risk, with data on site fidelity would greatly facilitate our understanding of the causes of movement both within and among species. The degree of site fidelity within and among species may have evolved in response to the predictability with which their foraging habitats provide adequate nutrition during the nonbreeding season.

Habitat Limitation

Nonbreeding habitats provide food resources, areas with low predation risk, or both. The best evidence for nonbreeding habitat limitation comes from natural experiments where disturbance or loss of habitat results in a reduction in period-specific survival (Fernández and Lank 2008, Piersma et al. 2016). The latter would not demonstrate that habitat *per se* was the cause of lower survival as lower survival could be directly related to a decline in per capita food availability caused by the lack of habitat. These two factors are notoriously difficult to disentangle (Myers et al. 1979a). Declines in population sizes owing to habitat loss or degradation (e.g., from human disturbance) in foraging or roosting habitats might also indicate that habitat is limited, although declines in local nonbreeding populations can also simply be indicative of movements away from these to other better wintering sites (see above; Owino 2002). Evidence to evaluate the relationship between habitat limitation and plovers largely comes from changes in population size, rather than changes in survival.

In West Africa, Common Ringed Plover, Kittlitz's Plover, and White-fronted Plover abundance increased 300% after a wetland was protected, whereas numbers decreased at other sites along the Ghana coast that had become degraded or lost (Ntiamoa-Baidu et al. 2000). Similarly, Escapa et al. (2004) document higher densities of Two-banded Plover at one site, presumably in response to an increase in food on introduced oyster beds. A decline in Wrybill numbers after construction of sewage treatment plant on a coastal mudflat suggested that habitat was limited (Veitch 1978). In coastal China, shorebird diversity and abundance of Kentish Plovers and Greater Sand-plover correlated positively with mudflat area (Zou et al. 2008). Similarly, Lesser Sand-Plover decreased in abundance in southern India, presumably due to large-scale development of shrimp farms (Sandilyan et al. 2010). Finally, numbers of Collared Plover in a riverine system in Paraguay were negatively correlated with water levels on exposed freshwater mudflats. There was much less foraging habitat during periods in the annual cycle when water from precipitation upstream overwhelmed the river's drainage system (Hayes and Fox 1991). Thus, more mudflat habitat supported more birds and suggests that habitat is limiting. These same authors also reported a negative correlation between the density of shorebirds and polychaetes and suggested that food was being depleted (see also Sánchez et al. 2006 but few plovers in that sample from southern Spain). In tidal flats of Cumberland Island, Georgia, the density of Semipalmated Plovers foraging on tidal mudflats appears to asymptote at about 14–16 per ha which was, for that site, about 800 individuals (Rose and Nol 2010). This density and total number may be limited by maximum spacing of birds, which suggests habitat limitation. Many additional studies have shown that plovers are more abundant in habitats with higher densities of prey (Long-billed Plover, Zou et al. 2008; Snowy Plover, Brindock and Colwell 2010; Semipalmated Plover, Rose and Nol 2010).

Common Ringed Plover shifted their nonbreeding distribution from the milder southeast to the eastern estuaries of Britain, as winter temperatures increased in eastern Britain in the 1990s, relative to the 1970s, where they then capitalized on richer food resources (Austin and Rehfisch 2005). Common Ringed Plover in Europe also preferred to feed in coastal habitats close to towns where nutrient inputs were higher (Burton et al. 2002), which presumably created better conditions for their invertebrate prey. Caspian Plover is adapted to dry conditions and can tolerate dry periods before rains that make their foraging habitats suitable but they can experience food shortages if storms are rare or rains fail (Sinclair 1978). All the above observations suggest that adequate amount or quality of foraging habitat could act as a limiting factor for plovers (Sinclair 1978), although the evidence is correlative, and therefore somewhat weak.

The use of human-modified habitats suggests that natural shoreline or wetland habitats are limited. Alternatively, use of these habitats can also simply indicate that human activities have concentrated and augmented food resources and the birds are benefiting from the use of these food subsidies (Okes et al. 2008, Dias 2009). Numerous reports of plovers using human-modified habitats during the nonbreeding season exist. Collared Plover used rice fields in Uruguay (Alfaro et al. 2008); Rufous-chested Dotterel occurred in pasture (Alfaro et al. 2008); Common Ringed Plover,

Kittlitz's Plover, and Kentish Plover occupied salt ponds (Gbogbo 2007, Dias 2009); and African Three-banded Plover used wastewater treatment wetlands (Harebottle et al. 2008). In coastal British Columbia, Canada, Killdeer preferred fertilized fields over non-fertilized fields for foraging (Evans Ogden et al. 2008). Alkali flats, historically extensive but virtually nonexistent today, were the most favored habitat for Mountain Plover, where available (Knopf and Rupert 1995). These authors concluded that the lack of natural alkali flats forced plovers to use cultivated lands during winter in the San Joaquin Valley, California, while birds in relict populations were dependent upon remaining, small, core areas of native habitat. Other plovers use salt works or aquaculture sites, suggesting flexibility in habitat choices (e.g., Kentish Plover in China, Ma et al. 2002; Javan Plover in Timor-Leste, Trainor 2011). Common Ringed Plover and Kentish Plover feed on salt pans in southern Spain (Masero et al. 2000). In using human modified habitats, it is not clear whether the natural and remaining mudflats can no longer provide sufficient food to meet energetic demands of these species (*sensu* Johnson et al. 2004) or whether these species use human-modified habitats because food is simply easier to obtain.

Food Limitation

While many published works document the diets of plovers, the objectives of this section are to review these studies in the contexts of the degree of specialization in diet, which might reflect vulnerability to a change and hence vulnerability to possible food limitations, and, similarly, the degree to which specialization might reflect, in the absence of other more direct evidence, either intra- or interspecific competition for food and therefore, food limitation.

Diets and Degree of Specialization

If food is limiting during the nonbreeding season, then one would predict that the diets would be broad, and foraging birds would be opportunistic in their prey selection (Estabrook and Dunham 1976). Direct observation of predation in many plover species is infrequently recorded because of the small size of the prey and the generally short handling times (Rose and Nol 2010). However, visual identification of the prey has occurred for Common Ringed Plover, Greater and Lesser Sand-plovers. Hockey et al. (1999) observed these species foraging on crabs, shrimp, and fish, and similarly Aarif (2009) observed Lesser Sand-plover foraging on both polychaetes and crabs. Diets during the nonbreeding season have often been established through analyses of fecal samples and stomach contents (e.g., Dann 1991). These methods have been applied to a number of species (e.g., Mountain Plover, Knopf 1998; Wilson's Plover, Collared Plover, Semipalmated Plover, Strauch and Abele 1979; Common Ringed Plover and Kentish Plover, Perez-Hurtado et al. 1997; Two-banded Plover, D'Amico and Bala 2004b, Musmeci et al. 2013; Black-fronted Dotterel, Taylor 2004). Many species of small-billed plover with coastal nonbreeding distributions have broad diets with a wide range of families of aquatic invertebrates, but with an underlying preference for polychaetes (Strauch and Abele 1979, D'Amico and Bala 2004b, Rose and Nol 2010). Diets of the similarly sized Kentish Plover and Common Ringed Plover in coastal Spain during the nonbreeding season differ substantially in the proportions of dipterans, molluscs, and polychaetes, with polychaetes only dominating the diet of Common Ringed Plover (Perez-Hurtado et al. 1997). Polychaetes are also the most important prey type for Two-banded Plover in Peninsula Valdes, Argentina (D'Amico and Bala 2004b, Ribeiro et al. 2004). Diet of Black-fronted Dotterel includes chironomid larvae and ostracods (Taylor 2004). Eurasian Dotterel eats more terrestrial insects, snails, and sometimes plant material (del Hoyo et al. 2016). Oriental Plover specialize on grasshoppers in winter in grasslands of Australia (Piersma and Hassell 2010). Snowy Plover feed on kelp flies and other beach wrack-associated invertebrates (Page et al. 1995). The diet of Wilson's Plover consists of 98.6% fiddler crabs (Morrier and McNeil 1991).

The diets of the remaining plover species are less well known. Greater and Lesser Sand-Plovers and Red-capped Plover feed on small bubble crabs in addition to polychaetes and fly larvae (Yasué and Dearden 2009, Evans et al. 2010) and Double-banded Plover feeds primarily on crabs (Dann 1991). The nocturnal Inland Dotterel, in western New South Wales, Australia, feeds on succulent leaves during the day, presumably for water, and insects at night (MacLean 1976).

The degree to which plovers select prey species or size classes of prey varies as much as the variation in diet. Visually foraging plovers, including Kittlitz's Plover, were highly selective, feeding on the largest polychaetes regardless of their abundance (Kalejta 1993). This species also exhibits some degree of microhabitat differentiation between resident and itinerant (nonbreeding) flocks in South Africa (Lipshutz et al. 2011), which suggests differences in patterns of diet selectivity. Although diets of Semipalmated Plover were broad in the estuaries of Cumberland Island, Georgia, birds feeding in salt marshes and on mudflats selected intermediate-sized polychaetes and avoided the smallest (Rose et al. 2016). Kentish Plover and Common Ringed Plover took the most abundant prey items and sizes in a Portuguese estuary, which suggests no diet selectivity (Pedro and Ramos 2009). Two-banded Plover consumed the most abundant prey items suggesting an opportunistic diet (D'Amico and Bala 2004a). Prey items identified from stomachs of wintering Mountain Plover included 2,092 different food items representing 13 Orders and at least 16 Families of invertebrates. Diets at each collection locale differed greatly, with coleopterans and hymenopterans dominating one sample, lepidopterans a second, and coleopterans and orthopterans in samples from the Salton Sea, California. Diets of males and females were similar. These findings counter the perception that Mountain Plover diets are specialized on coleopterans and orthopterans and suggest that Mountain Plover is a dietary generalist/opportunist (Knopf 1998). Except for a few specialized species, this conclusion may apply generally to many plovers (Isacch et al. 2005) and provide resiliency to substantive changes to prey species abundance. Specialization, such as that seen in Wilson's Plover, suggests heightened vulnerability to any environmental change that might impact the health of the major prey population, the fiddler crab.

Time Budgets

If food is limiting, then one would predict that plovers should forage a large percentage of available hours, both during the day and night. Analyses of time budgets of some plover species suggest that time spent foraging often exceeds 50%. Piping Plover in coastal Alabama foraged 90% of daylight hours (Johnson and Baldassarre 1988). In coastal Cuba both Semipalmated Plover and Killdeer forage more than 60% of daylight hours (Nol et al. 2014). Seventy-five percent of a flock of 25 adult and juvenile Double-banded Plovers were observed feeding during every visit to a New Zealand estuary (McConkey and Bell 2005). Common Ringed Plover and Kentish Plover both spend more than 85% of their time foraging during winter in southern Spain (Masero et al. 2000). Both species also forage at night, and in January, the highest proportion (30%) also forage during the full moon (Masero et al. 2000). Feeding in salt pans occurs only at high tide when other sites are unavailable, and, at low tide by small numbers of individuals, suggesting that food at salt pans supplements energy acquired from intertidal mudflats (Masero et al. 2000). During falling tides in coastal Georgia, only about 30% of the local population of Semipalmated Plover fed in a grazed salt marsh which had relatively low prey densities, whereas nearly 100% of birds foraged at a mudflat at low tide. This suggests that some individuals could not obtain sufficient energy from mudflats alone (Rose and Nol 2010). Similarly, Greater and Lesser sand-plovers and Kentish Plover in Thailand foraged at higher densities at low tide in supratidal than in intertidal areas, which the authors suggested supplemented their prey from intertidal areas alone (Yasué and Dearden 2009).

While few other papers report on time budgets during the nonbreeding season, there was no evidence that any of the three species, Lesser Sand-Plover, Greater Sand-Plover, and Common Ringed Plover, foraging on mudflats in Kenya had difficulty obtaining enough food to meet their daily energy requirements, given the time that the tide allowed for foraging, prey abundance, and intake rates (Hockey et al. 1999). Two-banded Plover forages at a relatively high rate in an Argentinian mudflat (1.8 polychaetes/min, D'Amico and Bala 2004b). These authors concluded that there was no evidence of food limitation even though this species feeds primarily on intermediate tidal heights during tidal cycle (D'Amico and Bala 2004b). By contrast, Collared Plover fed throughout the tidal cycle (Barbieri and Sato 2000). Both the abundance of food sampled in the mudflats and the observations of Semipalmated Plover and Wilson's Plover moving to roost while foraging areas were still available, were presented as evidence that food was not limited for those species in a Panama wintering site (Strauch and Abele 1979). In a second study in Panama, low densities of Wilson's Plover and Semipalmated Plover and no decline in prey densities suggest an absence of food limitation (Schneider 1985). In Argentina, a mixed-species group of shorebirds (including Two-banded, American Golden-Plover, Black-bellied Plover, and Semipalmated Plover) reduced densities of polychaetes (Botto et al. 1998). However, at another site, there was no reduction in polychaetes outside an experimental exclosure. The authors suggested that this was likely due to lower predator density and concluded that there was no evidence that Two-banded Plover experienced food limitation (Botto et al. 1998).

Several authors have described nocturnal foraging in plovers, and for Wilson's Plover, Kentish Plover, and Common Ringed Plover, the dual functions of foraging at night to either supplement diurnal energy intake or to avoid predation from diurnal predators have been evaluated (Robert et al. 1989, Morrier and McNeil 1991, Thibault and McNeil 1994, 1995, Lourenço et al. 2008). In the Common Ringed Plover and the Kentish Plover, night foraging was less efficient and was thought to supplement daytime foraging (Lourenço et al. 2008), whereas for Wilson's Plover, the primary function was thought to be avoidance of predation (see below and Chapter Six, this volume). In Australia, nocturnal habitat use has been documented for Lesser Sand-Plover, Double-banded Plover, and Red-capped Dotterel (Rohweder and Baverstock 1996, Rohweder and Lewis 2004) and, at least for Double-banded Plover, accompanied by a change in habitat. Night foraging has also been documented on salt pans in South Africa by Kittlitz's Plover, White-fronted Plover, and the Chestnut-banded Plover (Velasquez 1992), but whether the function was to supplement food or avoid predation was not established. For Semipalmated Plover, nocturnal foraging was only recorded under moonlit conditions and when the tide was sufficiently low (Dodd and Colwell 1998). In a study of Kentish Plover in Japan, night feeding was more efficient for gathering food than foraging during daylight (Kuwae 2007). These observations, taken together, suggest that plovers can and do supplement their daytime foraging, reflecting that foraging during the daytime hours, particularly in tidal environments, is probably not sufficient for these species to meet their energetic requirements.

Unusual foraging methods may allow plovers to supplement their usual intake. Common Ringed Plover foraged using "sandpiper-type" techniques of moving through shallow water and capturing brine shrimp (*Artemia* spp.) in hypersaline pans in coastal Spain, suggesting opportunism (Masero et al. 2007).

Thus, plovers appear to exhibit a pattern of generalized diets and opportunistic prey capture (with some exceptions, e.g., Wilson's Plover who specialize on a diet of fiddler crabs). Within the numerous descriptions of diet, there can also occur a degree of specialization, either in the invertebrate groups taken or preferred (e.g., polychaetes), or in the sizes of individuals within those groups (e.g., largest individuals). There is only one example where prey reduction is suggested in the case of Two-banded Plover foraging alongside other, more numerous shorebird species (Botto et al. 1998). While the evidence is far from conclusive, it appears that there is sufficient flexibility in foraging behavior of most plovers that food limitation during the nonbreeding season is uncommon or rare. A caveat of this conclusion, however, is that there are many species of plover, particularly in tropical regions, where changes in food abundance or quality could have substantial changes to populations (e.g., Caspian Plover), but for which, there is currently too little information.

Do Studies of Plover Energetics Inform Ideas about Food Limitation?

Energy budgets calculate necessary requirements, given the basal metabolic rates (BMRs) of species and the availability of food. While plovers wintering in the tropics have relatively low BMRs (and hence lower foraging requirements) compared to those wintering in temperate zones (Kersten et al. 1998), the most detailed analysis of an energy budget for a plover comes from Stillman et al. (2005) on Common Ringed Plover, a species for which part of the population spends its nonbreeding season at the relatively high latitude of the United Kingdom. The Stillman et al. (2005) model, which incorporated both rates of prey depletion and the energetic requirements of birds, predicted that Common Ringed Plover would stay in patches of prey as long as their intake rate exceeded their energy-expenditure rate and mass did not decline (the "satisficing" model). The model predicted that birds would occupy a wide range of suitable patches, suggesting that as long as there was enough food to meet energetic minimums food was not likely to be limited. The model predicted

100% winter survival of Common Ringed Plover given adherence to a satisficing model and no effect of survival rates on Common Ringed Plover even under a simulation of 50% habitat loss (Stillman et al. 2005). The "satisficing" model predicted that birds used a wider variety of patches than predicted using a model that assumed birds always chose the patch that maximized their rate of intake ("the rate maximizing model"). The use of a wider range of food patches presumably reduced intraspecific competition over what occurred if all birds chose the best habitats.

A second study, using calculations of total energy requirements, suggested that plovers migrating to the southern hemisphere timed their arrival and duration on the nonbreeding areas to coincide with the growth of benthic prey populations (Hockey et al. 1992). The study also compared the energy balance of Common Ringed Plovers who stayed north near their breeding grounds to those who undertook the long migrations to the equatorial region. The higher abundance of food near the equator was calculated to more than counteract any costs from long-distance migration. The authors concluded that shorebirds track the carrying capacity of coastal wetlands across a wide latitudinal range, thus strongly supportive of a bottom-up model.

Evidence for Top-down Regulation of Plover Populations

Many of the patterns attributed to food limitation have alternative explanations that derive from a view that predation limits populations. As with evidence for bottom-up regulation of plover populations, the importance of predation comes from a variety of observations of nonbreeding plovers (Table 8.2). While predation risk has been proposed as a primary determinant of site use in some nonbreeding Calidridine sandpipers (Nebel and Ydenberg 2005), specific mention of predation on adult plovers during the nonbreeding season is rare. Single instances include New Zealand Falcon (*Falco novaeseelandiae*) predation on Double-banded Plover (Hyde and Worthy 2010), Peregrine Falcon (*Falco peregrinus*) predation on Javan Plover (Kholil 2014), and raptor predation (probably *Falco* spp.) on Common Ringed Plover (Whitfield 1985). High rates of predation on Shore Plover released from captivity suggest that, for small populations of resident species, owl predation could be important and limiting, such that it affects choices of habitat, dispersal, and behavior (Davis and Aikman 1997). While evidence of predation *per se* does not necessarily reflect predation risk, it provides some perspective on the importance of this factor.

Foraging at night by Wilson's Plover is thought to be a tactic to avoid predation in Venezuela, rather than a way for this species to supplement its diet (see above, Thibault and McNeil 1994, 1995). By contrast, the author of a study that directly compared diurnal and nocturnal foraging by Kentish Plover in Japan explicitly rejected the hypothesis that nocturnal foraging served to avoid predation. Kentish Plover captured 3.7×more prey at night than during the day, and nocturnal foraging was deemed a major activity to meet energy requirements (Kuwae 2007). As many plovers forage in small flocks, they may be more vulnerable to falcon predation than large flocks of sandpipers (Van Den Hout et al. 2008). For example, Wilson's Plovers forage alone and have long prey handling times, so they may be less attentive to potential predators. Thus, the function of nocturnal foraging varies among species and may also depend on their mode of foraging.

Owens and Goss-Custard (1976) proposed that the relatively loud alarm calls of some shorebirds might promote flocking which then would reduce an individual's chance of being eaten (Owens and Goss-Custard 1976). Many plovers have relatively soft alarm calls, but do often join other species to form larger flocks. For example, Semipalmated Plovers join evading flocks of Dunlin (*Calidris alpina*) during nonbreeding in Georgia and South Carolina (E. Nol, pers. obs.) and Venezuela (Morrier and McNeil 1991) when a predator appears. The cryptic coloration of plovers in their basic plumage might help reduce predation during the nonbreeding season and some species (e.g., Snowy Plover, Mountain Plover) sit in depressions to decrease their visibility to predators (Knopf and Rupert 1995). An alternative view, although not mutually exclusive to the predation hypothesis, is that the depressions provide a thermal environment that is favorable during high winds and low temperatures during the nonbreeding season.

Roosting Habitat: Does It Serve an Antipredator Function and Is It Limiting?

Roosting in plovers usually occurs in flocks, although specific information on the size, location, and function of roosts for many species is lacking. Flocking suggests that this behavior serves an antipredator function because birds are not foraging while on roosts. Recher (1966) suggested that population size of plovers feeding at coastal locations was limited more by the scarcity of roosting sites than the availability of food, a view shared by others (Strauch and Abele 1979, Senner and Howe 1984). If roost habitat is indeed limiting, it would support a top-down view of population regulation, as selection appears to strongly favor flock formation during roosting. In coastal regions, plovers roost regularly during high tides, when foraging areas are unavailable (Rohweder 2001, Mönke and Seelig 2009, Rose and Nol 2010). Roosting Lesser Sand-Plover flocks did not seem to react to flying raptors but flushed and exhibited erratic flight when disturbed by humans or dogs (Mönke and Seelig 2009). This observation suggests that roosting in this species has evolved specifically in response to mammalian predators. Semipalmated Plover in coastal Brazil roost on rooftops 2–2.5 km from the nearest feeding areas during spring high tides coinciding with onshore winds (Cardoso and Zeppelin 2013) or on the lower branches of mangroves in coastal Brazil (K. MacCulloch, pers. obs.). These observations suggest that safe roost sites might be limited along water and may become more so as sea level rise proceeds. After a hurricane, the loss of a large sand spit where Semipalmated Plovers roosted led to a reduction in their local population (Rose and Nol 2010). After the loss of several roosts due to residential development in Queensland, Australia, citizens concerned about impacts of the loss called for construction of artificial roosts (Dening 2005). Four plovers (Red-capped Plover, Double-banded Plover, Lesser and Greater sand-plovers) readily used the newly created roosts in the same relative composition of plovers that had been present prior to the destruction of roosts (Dening 2005). These examples suggest that safe roosting habitat could be an important determinant of local population sizes and essential for maintaining local populations of nonbreeding plovers (Sprandel et al. 2000).

Whether there are predictable daily patterns of roosting by those species that forage and spend their nonbreeding season at inland sites is not well documented, although Mountain Plovers primarily use burned fields for night roosting (Knopf and Rupert 1995). The absence of tidal effects on habitat availability suggests that populations that spend the nonbreeding period at inland locations may be less vulnerable to loss of adequate roosting sites that potentially reduce an individual's risk of predation. Additionally, the potential for using cryptic coloration to avoid predation may be greater in inland habitats that exhibit greater structural heterogeneity than coastal habitats. Finally, the absence of tides also implies that foraging habitat should always be available (unless prey availability is affected by hot or dry conditions), thus potentially lessening the need for roosting.

Mortality During the Nonbreeding Season

Only a handful of studies on nonbreeding plovers have quantified mortality rates as a proxy for the risk of predation. The use of radio transmitters can indicate with some clarity whether individuals are alive or dead. Forty-nine Piping Plovers were outfitted with transmitters on the nonbreeding grounds in coastal Texas. There was no mortality from 1 December to 15 February (75 days) or from 16 February to 25 April (37 days, Drake et al. 2001). A second study on a much smaller sample of this species (7), captured, color-marked, and radio-tagged between 5 December and 10 February in Oregon Inlet, North Carolina, United States, also reported 100% survival until 4 March (Cohen et al. 2008). Thirty-five Piping Plovers were outfitted with transmitters in San Padre Island, Texas (Newstead and Vale 2014). Daily survival probabilities over 2,695 telemetry days were calculated as 0.9993, and for a 120-day period, survival was 0.9139.

In Mountain Plovers, the daily survival rate for the 44 birds outfitted with transmitters was 0.9996 based upon 2,395 telemetry days of information (Knopf and Rupert 1995). The calculated survival probability for the 1 November to 15 March winter period was 0.9474. Mortality of the two birds that died was attributed to kit fox (*Vulpes macrotis mutica*) predation. Extending this rate of mortality to the entire 12-month period would equate to an annual survival of 0.866. Published records of annual survival (0.68, Dinsmore et al. 2003; Chapter Ten, this volume) suggest that mortality must be higher during the combined breeding and migratory periods than during winter unless the estimates come from different years or include subsets of the population from different regions. Using repeated resightings of color-banded Semipalmated Plover in coastal Georgia, USA, S. Williams (unpublished data) has calculated an apparent survival rate of 0.78 for the 6-month period from October to March, which, once extended to annual mortality (0.61), is similar to estimates of apparent annual survival derived from mark-recapture data from both the nonbreeding grounds (range: 0.67–0.72; S. Williams, unpublished data) and the breeding grounds (range: 0.46–0.91; Badzinski 2001).

Annual survival of White-fronted Plover and other resident species is higher than migratory species (Lloyd 2008), but where or in which season the mortality occurs has not been studied. It is likely that data on season-specific mortality may be available, but currently unpublished, in studies where birds were intensively color marked and resighted throughout the annual cycle (e.g., Double-banded Plover, Barter 1991; Shore Plover, Davis and Aikman 1997; White-fronted Plover, Lloyd 2008; Hooded Plover, Weston et al. 2009). For example, annual mortality of Wrybill was calculated as 17% from an intensive banding study (Davies 1997) but not partitioned by season (see Evans and Pienkoswki 1984 for a variety of wader nonbreeding survival rates, but not Charadrius spp.). Of the seven species of wintering shorebird in southeast Scotland, Common Ringed Plover experienced the highest predation rate (by raptors: 19%, Whitfield 1985). Shore Plover in New Zealand suffered high rates of mortality after gradual release from a captive breeding facility to supplementing food and then to full independence (>50%, Davis and Aikman 1997). Predation by introduced mammals has also been recorded on Wrybills during the nonbreeding season (Battley and Moore 2004). These latter examples of New Zealand species exposed to evolutionarily recent forms of predation are probably not representative of the majority of plover species where nonbreeding survival (with some exceptions, Whitfield 1985) appears to be quite high. In addition, while observations of direct predation on nonbreeding shorebirds could be rare, predators can still influence habitat use, flock sizes, flock formation (e.g., mixed-species flocks), roosting habitat choice, and time budgets (Lank and Ydenberg 2003).

Seasonal Variation in Body Mass: Contrasting Views of Adaptive Significance

While lower body masses of plovers during the nonbreeding season suggest that food is limiting, showing support for bottom-up processes, low masses could also be an adaptive antipredator response (Nebel and Ydenberg 2005, Zimmer et al. 2011). Alternatively, higher weights recorded during the breeding season could serve as a buffer against the impacts of low temperatures on energy requirements (Davies 1997), which would suggest that selection pressures on food intake during the nonbreeding season are relaxed relative to those during the breeding season.

Weights are lower and less variable for wintering Semipalmated Plover than for birds in other stages of the annual cycle (Nol and Blanken 2014, Table 8.3) but not for Snowy Plover, a short-distance migrant (Page et al. 2009). Morrier and McNeil (1991) reported that mass increased slightly in Wilson's Plover at the beginning of the breeding season. For most other species for which there are data, breeding season masses tend to be higher than those during the nonbreeding (Table 8.3). Very few studies have published fat levels in nonbreeding plovers, although similar to the results of body mass, Wilson's Plover maintained a low fat level throughout winter (18%–24%) until just before breeding (Morrier and McNeil 1991). The most parsimonious explanation for consistently lower masses (possibly due to low fat) during the nonbreeding season across this relatively small sample of species (seven) is that this is an adaptive response. Rather than reflecting an inability to acquire food, the lower mass is an adaptation to either increase escape performance (Burns and Ydenberg 2002) or has evolved because energetic requirements are less than during the breeding or migratory seasons.

Linking the Quality of Nonbreeding Habitat with Survival or Reproductive Success

Regardless of the mechanism that limits plover populations during the nonbreeding season, a link between a reduction in nonbreeding habitat quality and either reproductive success or annual survival would strengthen the message that nonbreeding habitats are critically important for conservation. Several studies are now emerging that do just that. Not surprisingly, both studies are on the United States' federally threatened Piping Plovers where there exist marked populations with high migratory connectivity. In the first, annual survival was estimated for seven breeding populations and the lowest survival

TABLE 8.3

Published information of breeding and/or nonbreeding masses (g) of nine plover species.

Species	Breeding mass	Nonbreeding mass	Significance?	Sources
Semipalmated Plover	47.6 ± 0.16 (272)	41.4 ± 0.24 (112)	Yes, P<0.05	Nol and Blanken (2014)
Common Ringed Plover	63.7 (140)	53.5 (22)	Not tested	Cramp (1983)[B], Summers and Waltner (1979)[W]
Killdeer	85.0 (3)	80.5 (2)	Not tested	Cramp (1983)
White-fronted Plover	52.5 (65)	47.3 (186)	Yes, P<0.05	Summers and Waltner (1979)
Chestnut-banded Plover		36.0 ± 2.92 (10)	NA	Summers and Waltner (1979)
Kittlitz's Plover	41.9 (28)	41.6 (9)	No	Summers and Waltner (1979)
Double-banded Plover	58.6	59.4 ± 3.9	Not tested	Marchant and Higgins (1993)[B], Barter and Minton (1987)[W]
Three-banded Plover		34.4 ± 3.3 (55)	NA	Summers and Waltner (1979)
Greater Sand-Plover	106.8 (4)	77 (2)	NA	Summers and Waltner (1979)
Wrybill		54.9 (111)	NA	Davies (1997)

Means plus standard errors (when available) and sample sizes in parentheses are provided. NA = No masses available for one season, Not tested = insufficient sample size or raw data not available, Need B = breeding, W = non-breeding.

was associated with those wintering in the same location (Roche et al. 2010). A second study (Gibson et al. 2018) links human disturbance at nonbreeding locations with low annual survival. Given the high site fidelity observed in wintering Piping Plovers, this result is probably not due to movement away from the poor quality nonbreeding site. These two very important studies are augmented by similar works on other shorebird species (Piersma et al. 2016), which suggest that habitat loss or degradation of either migratory stopover or wintering sites could have strong negative effects on the survival of individuals, and by extension, of population sizes.

RESEARCH NEEDS

The literature on plovers during the nonbreeding season is dominated by papers that provide seasonal patterns of numbers of plovers at sites or even single-day counts (e.g., Caziani et al. 2001 and many others), rather than detailed ecological studies of habitat relationships, behavior, or limiting factors. Fortunately, a few excellent studies do exist. Unlike breeding plovers (see Chapter Five, this volume), papers on nonbreeding plovers are not dominated by common species breeding in the northern hemisphere. However, detailed understanding of survival still comes from a few very limited studies of birds for whom special legislative attention has been given (e.g., endangered species legislation in Australia, Canada, and the United States), and therefore where monetary support for this research is very high. Below, I provide a list of research priorities for further understanding whether the nonbreeding season plays an important role in regulating plover populations. I also remark on the practicality of furthering research in these priority areas.

1. More studies are required that use tracking devices throughout the nonbreeding season to document survival and sources of mortality (e.g., Drake et al. 2001), including the impact of habitat loss (both foraging and roosting) through rising sea levels. Small, low-cost transmitters and a large network of towers with receiving units (e.g., Motus network, Woodworth et al. 2014) and the continuing miniaturization of GPS tags and lower costs of these devices will make this possible for more species in the future.

2. More studies of nonbreeding behavior are required. For many of the species for which I concluded that there was no evidence of territoriality (Table 8.1), researchers had spent a substantial amount of time in the field with the species, so I assumed obvious aggression either between conspecifics or heterospecifics, would have been documented. Only two papers explicitly measured aggression while collecting data on time budgets. Studies on individually marked birds would be most helpful, given that there may be individual, condition- or sex-dependent strategies among nonbreeding birds (e.g., Myers et al. 1979a).
3. Explicit documentation of antipredator behavior (e.g., Morrier and McNeil 1991) would shed light on how plovers are impacted by predation (although some of this would come more directly from 1 above).
4. Plover researchers are encouraged to always report the stage of the annual cycle of their study period, especially in studies reporting distributions of partial migrants or sedentary species.
5. Studies on the energetics of foraging plovers, and, in particular, the use of nocturnal foraging and how that might impact the energy budgets of shorebirds (*sensu* Morrier and McNeil 1991) would be valuable. Modeling exercises which explicitly measure the costs of food limitation and predation risk (e.g., Quaintenne et al. 2011 for food) can also be extended to other plover species. Models that explicitly test the satisficing model (*sensu* Stillman et al. 2005) can also help to guide conservation actions. In terms of food availability, a modeling approach could identify whether many small, uneven patches or only the largest patches with the richest food supply should be targeted for land acquisition to aid in shorebird conservation.
6. Laboratory-based studies on plovers, especially if done in conjunction with zoo populations, could move energetic studies forward. The recent use of stress indicators to indirectly assess habitat quality is also promising (Aharon-Rotman et al. 2016).
7. We need more published papers on mass variation throughout the annual cycle (e.g. Summers and Waltner 1979). While I conclude that lower and less variable masses during the nonbreeding period are evidence of adaptive mass variation due to lower energetic needs during winter and/or as an antipredator response, the number of studies comparing seasonal patterns of mass or body condition could readily be augmented, perhaps from unpublished sources. As with studies of distributions, researchers are urged, especially when contributing to larger review works like the Birds of North America or del Hoyo online accounts, to make explicit the season (and dates) in which birds were measured and weighed (e.g., Cramp 1983). The International Wader Study Group could make online repositories of these data available.
8. Further understanding of whether plover nonbreeding distributions are segregated by sex and/or age may be facilitated through the study of museum specimens.
9. Molt in the plovers is poorly understood. Many, but not all species appear to molt into their basic plumage in the first few months of the nonbreeding season, but data on how long this takes, whether there are behavioral modifications associated with it, how much energy is required, or whether there are different strategies among or within species (e.g., partial molt during migration, *sensu* Barshep et al. 2013) during this period are unknown.
10. Finally, studies of carryover effects that truly link different stages of the life cycle, and especially events during plover nonbreeding seasons with reproductive success on breeding grounds are urgently required. While Hughey (1985) showed that numbers of Wrybill that oversummered on the nonbreeding grounds were correlated with habitat quality during the previous breeding season, true carryover effects, where individual condition on the nonbreeding grounds as a result of variation in habitat quality, has consequences for subsequent breeding performance (*sensu* Sillett and Holmes 2002, Norris et al. 2004) are only beginning (Gibson

et al. 2018). Species with high (e.g., Piping Plover, Gratto-Trevor et al. 2012) rather than low migratory connectivity (e.g., Little Ringed Plover, Hedenström et al. 2013) are the best candidates for this work.

The species chosen for any of the above suggested research areas, but most importantly for work linking habitat quality with nonbreeding survival, should be living near human settlements where loss or degradation of the nonbreeding habitat will probably have the greatest impact on the conservation of that species. These species are not necessarily rare or common because, for example, habitat degradation could impact a large proportion of a relatively common species. Sites most severely and imminently impacted by rising sea levels should be prioritized. Plovers, with their mixture of resident and migratory populations and species, and their use of sites that vary substantially in food resources, predators, and human impacts, are excellent model organisms for elucidating mechanisms of population limitations in birds.

LITERATURE CITED

Aarif, K. M. 2009. Some aspects of feeding ecology of the Lesser Sand Plover *Charadrius mongolus* in three different zones in the Kadalundy Estuary, Kerala, South India. *Podoces* 4:100–107.

Aarif, K. M., S. B. Muzaffar, S. Babu, and P. K. Prasadan. 2014. Shorebird assemblages respond to anthropogenic stress by altering habitat use in a wetland in India. *Biodiversity and Conservation* 23:727–740.

Aharon-Rotman, Y., K. L. Buchanan, N. J. Clark, M. Klaassen, and W. A. Buttemer, W. A. 2016. Why fly the extra mile? Using stress biomarkers to assess wintering habitat quality in migratory shorebirds. *Oecologia* 182:385–395.

Alfaro, M., A. Azpiroz, T. Rabau, and M. Abreu. 2008. Distribution, relative abundance, and habitat use of four species of neotropical shorebirds in Uruguay. *Ornitología Neotropical* 19:461–472.

Alfaro, M. and M. Clara. 2007. Assemblage of shorebirds and seabirds on Rocha Lagoon sandbar, Uruguay. *Ornitología Neotropical* 18:421–432.

Almeida, B. and S. F. Ferrari. 2010. Seasonal and longitudinal variation in the abundance and diversity of shorebirds (*Aves, Charadriiformes*) on Atalaia beach in northeastern Brazil. *Ornitología Neotropical* 21:56–580.

Austin, G. E. and M. M. Rehfisch. 2005. Shifting nonbreeding distributions of migratory fauna in relation to climatic change. *Global Change Biology* 11:31–38.

Badzinski, D. 2001. Population Dynamics of Semipalmated Plovers Breeding at Churchill, MB. *M.Sc. Thesis*, Trent University, Peterborough, Ontario, Canada.

Bala, L. O., V. L. D'Amico, and P. Stoyanoff. 2002. Migrant shorebirds at Peninsula Valdes, Argentina: Report for the year 2000. *Wader Study Group Bulletin* 98:14–19.

Balachandran, S., and S. A. Hussain. 1998. Moult, age structure, biometrics and subspecies of Lesser Sand Plover *Charadrius mongolus* wintering along the southeast coast of India. *Stilt* 33:3–9.

Barbieri, E. and T. Sato. 2000. Information analysis of foraging behavior sequences of the Collared Plover (*Charadrius collaris*). *Ciencia e Cultura (Sao Paulo)* 52:178–184.

Barshep, Y., C. D. Minton, L. G. Underhill, B. Erni, and P. Tomkovich. 2013. Flexibility and constraints in the molt schedule of long-distance migratory shorebirds: Causes and consequences. *Ecology and Evolution* 3:1967–1976.

Barter, M. and C. Minton. 1987. Biometrics, moult and migration of double-banded plovers *Charadrius bicinctus bicinctus* spending the non-breeding season in Victoria. *Stilt* 10:9–14.

Barter, M. 1991. Addendum to survival rate of double-banded plovers *Charadrius bicinctus bicinctus* spending the non-breeding season in Victoria. *Stilt* 19:5.

Battley, P. F. and S. J. Moore. 2004. Predation on non-breeding wrybills in the Firth of Thames. *Notornis* 51:233–234.

BirdLife International. [online]. 2016. Species factsheet: *Charadrius pallidus*. <www.birdlife.org> (18 April 2016).

Blanco, D. E., P. Yorio, P. F. Petracci, and G. Pugnali. 2006. Distribution and abundance of non-breeding shorebirds along the coasts of the Buenos Aires Province, Argentina. *Waterbirds* 29:381–390.

Boland, J. M. 1990. Leapfrog migration in North American shorebirds: Intra-and interspecific examples. *Condor* 92:284–290.

Bolnick, D. I., T. Ingram, W. E. Stutz, L. K. Snowberg, O. L. Lau, and J. S. Paull. 2010. Ecological release from interspecific competition leads to decoupled changes in population and individual niche width. *Proceedings of the Royal Society B* 277:1789–1797.

Botto, F., O. O. Iribarne, M. M. Martínez, K. Delhey, and M. Carrete. 1998. The effect of migratory shorebirds on the benthic species of three southwestern Atlantic Argentinean estuaries. *Estuaries* 21:700–709.

Brindock, K. M. and M. A. Colwell. 2010. Habitat selection by Western Snowy Plovers during the nonbreeding season. *Journal of Wildlife Management* 75:786–779.

Buchanan, J. B. 2011. Behavior of spring migrant Black-bellied Plovers at Totten Inlet Washington: Agonistic or courtship function. *Washington Birds* 11:8–17.

Burns, J. G. and R. C. Ydenberg. 2002. The effects of wing loading and gender on the escape flights of least sandpipers (*Calidris minutilla*) and western sandpipers (*Calidris mauri*). *Behavioral Ecology and Sociobiology* 52:128–136.

Burton, N. H., M. J. Armitage, A. J. Musgrove, and M. M. Rehfisch. 2002. Impacts of man-made landscape features on numbers of estuarine waterbirds at low tide. *Environmental Management* 30:857–864.

Cardoso, T. A. L. and D. Zeppelin. 2013. Migratory shorebirds roosting on a roof in Paraíba, Brazil: Response to a new habitat or loss of the natural ones? *Ornitología Neotropical* 24:225–229.

Caziani, S. M., E. J. Derlindati, A. Tálamo, A. L. Sureda, C. E. Trucco, and G. Nicolossi. 2001. Waterbird richness in Altiplano wetlands of northwestern Argentina. *Waterbirds* 24:103–117.

Chapman, A. and, J. A. K. Lane. 1997. Waterfowl usage of wetlands in the south-east arid interior of Western Australia 1992–93. *Emu* 97:51–59.

Cohen, J. B., S. M. Karpanty, D. H. Catlin, J. D. Fraser, and R. A. Fischer. 2008. Winter ecology of piping plovers at Oregon Inlet, North Carolina. *Waterbirds* 31:472–479.

Colwell, M. A. 2000. A review of territoriality in nonbreeding shorebirds (*Charadrii*). *Wader Study Group Bulletin* 93:58–66.

Colwell, M. A. 2010. *Shorebird Ecology, Conservation, and Management*. University of California Press: Berkeley, CA.

Connell, J. H. 1980. Diversity and the coevolution of competitors, or the ghost of competition past. *Oikos* 1980:131–138.

Cramp, S. 1983. *The Birds of the Western Palearctic: Waders to Gulls. V.3*. Oxford University Press: Oxford, UK.

Cresswell, W. 1994. Flocking is an effective antipredation strategy in redshanks, *Tringa totanus*. *Animal Behaviour* 47:433–442.

Crossland, A. C. 2000. Notes on the waders wintering at three sites at the north-western tip of Sumatra, Indonesia. *Stilt* 36:4–6.

Crossland, A. C. 2002. Seasonal abundance of waterbirds at Tanah Merah Beach, a newly formed inter-tidal habitat on the edge of reclaimed land in Singapore. *Stilt* 42:4–9.

D'Amico, V. L., and L. O. Bala. 2004a. Prey selection and feeding behavior of the two-banded plover in Patagonia, Argentina. *Waterbirds* 27:264–269.

D'Amico, V. L. and L. O. Bala. 2004b. Diet of the two-banded plover at Caleta Valdes Peninsula Valdes, Argentina. *Wader Study Group Bulletin* 104:85–87.

Dann, P. 1991. Feeding behavior and diet of Double-banded Plovers *Charadrius bicinctus* in Western Port, Victoria. *Emu* 91:179–184.

Davies, S. 1997. Population structure, morphometrics, moult, migration, and wintering of the Wrybill (*Anarhynchus frontalis*). *Notornis* 44:1–14.

Davis, A. and H. Aikman. 1997. *Establishment of shore plover (Thinornis novaeseelandiae) on Motuora Island. Part 1 and 2*. Science for Conservation 46. Department of Natural Resources: Wellington, New Zealand.

del Hoyo, J., A. Elliott, J. Sargatal, D. A. Christie, and E. de Juana (eds). [online]. 2016 *Handbook of the Birds of the World Alive*. Lynx Edicions: Barcelona, Spain. <www.hbw.com/node/53833> (7 June 2016).

Dening, J. 2005. Roost management in south-East Queensland: Building partnerships to replace lost habitat. pp. 94–96 in P. Straw (ed.), *Status and Conservation of Shorebirds in the East Asian-Australasian Flyway. Proceedings of the Australasian Shorebirds Conference 13–15 December* 2003. Sydney, NSW. Wetlands International Global Series 18, International Wader Studies 17. Sydney, Australia.

Dias, M. P. 2009. Use of salt ponds by wintering shorebirds throughout the tidal cycle. *Waterbirds* 32:531–537.

Dinsmore, S. J., G. C. White, and F. L. Knopf. 2003. Annual survival and population estimates of Mountain Plovers in southern Phillips County, Montana. *Ecological Applications* 13:1013–1026.

Dodd, S. L. and M. A. Colwell. 1998. Environmental correlates of diurnal and nocturnal foraging patterns of nonbreeding shorebirds. *Wilson Bulletin* 110:182–189.

Dowding, J. E. and S. P. Chamberlain. 1991. Annual movement patterns and breeding-site fidelity of the New Zealand Dotterel (*Charadrius obscurus*). *Notornis* 38:89–102.

Dowding, J. E. and S. J. Moore. 2006. *Habitat Networks of Indigenous Shorebirds in New Zealand*. Science for Conservation 261. Science and Technical Publishing: Wellington, New Zealand.

Dowding, J. E. and S. M. O'Connor. 2013. Reducing the risk of extinction of a globally threatened shorebird: Translocations of the Shore Plover (*Thinornis novaeseelandiae*), 1990–2012. *Notornis* 60:70–84.

Drake, K. R., J. E. Thompson, K. L. Drake, and C. Zonick. 2001. Movements, habitat use, and survival of nonbreeding Piping Plovers. *Condor* 103:259–267.

Duffy, D. C., N. Atkins, and D. C. Schneider. 1981. Do shorebirds compete on their wintering grounds? *Auk* 98:215–229.

Eiamampai, K., S. Nimnuan, T. Sornsa, D. Phothieng, S. Thong-Aree, K. Ittiponr, K. G. Rogers and P. D. Round. 2014. Proportions of first-year individuals in cannon-net catches of waders in Thailand with a comparison to Australia. *Stilt* 65:17–24.

Elliott-Smith, E., S. M. Haig, C. L. Ferland, and L. Gorman. 2004. Winter distribution and abundance of Snowy Plovers in eastern North America and the West Indies. *Wader Study Group Bulletin* 104:28–33.

Engilis Jr., A., L. W. Oring, E. Carrera, J. W. Nelson, and A. M. Lopez. 1998. Shorebird surveys in Ensenada Pabellones and Bahia Santa Maria, Sinaloa, Mexico: Critical winter habitats for Pacific flyway shorebirds. *Wilson Bulletin* 10:332–341.

Escapa, M., J. P. Isacch, P. Daleo, J. Alberti, O. Iribarne, M. Borges, E. P. Dos Santos, D. Gagliardine, and M. Lasta. 2004. The distribution and ecological effects of the introduced Pacific Oyster *Crassostrea gigas* (Thunberg, 1793) in northern Patagonia. *Journal of Shellfish Research* 23:765–773.

Estabrook, G. F. and A. E. Dunham. 1976. Optimal diet as a function of absolute abundance, relative abundance, and relative value of available prey. *American Naturalist* 110:401–413.

Evans, P. R. and M. W. Pienkowski. 1984. Population dynamics of shorebirds. Pp. 83–123 in J. Burger and B. L. Olla (eds). *Behavior of Marine Animals, Shorebirds: Breeding Behavior and Populations*, vol. 5, Plenum Press: New York.

Evans, S. M., J. Prince, J. Foster-Smith, E. Drew, and R. D. Phillips. 2010. Optimal and anti-predator foraging in the Sand Bubbler Crab *Scopimera inflata* (Decapoda: Ocypodidae). *Journal of Crustacean Biology* 30:194–199.

Evans Ogden, L. J. E., S. Bittman, D. B. Lank, and F. C. Stevenson. 2008. Factors influencing farmland habitat use by shorebirds wintering in the Fraser River Delta, Canada. *Agriculture, Ecosystems and Environment* 124:252–258.

Fedrizzi, C. E. 2003. Abundância sazonal e biologia de aves costeiras na coroa do avião, Pernambuco, Brasil. *M.Sc. Thesis*. Universidade Federal de Pernambuco, Recife, Brazil (in Portuguese).

Fernández, G. and D. B. Lank. 2008. Effects of habitat loss on shorebirds during the non-breeding season: Current knowledge and suggestions for action. *Ornitología Neotropical* 19:633–640.

Ferrari, S., C. Albrieu, and P. Gandini. 2002. Importance of the Rio Gallegos Estuary, Santa Cruz, Argentina, for migratory shorebirds. *Wader Study Group Bulletin* 99:35–40.

Flynn, L., E. Nol, and Y. Zharikov. 1999. Philopatry, nest-site tenacity, and mate fidelity of Semipalmated Plovers. *Journal of Avian Biology* 30:47–55.

Fretwell, S. D. and H. L. Jr. Lucas. 1970. On territorial behaviour and other factors influencing habitat distribution in birds. I. Theoretical Development. *Acta Biotheoretica* 19:16–36.

Gbogbo, F. 2007. Impact of commercial salt production on wetland quality and waterbirds on coastal lagoons in Ghana. *Journal of African Ornithology* 78:81–87.

Gibson, D., M. K. Chaplin, K. L. Hunt, M. J. Friedrich, C. E. Weithman, L. M. Addison, V. Cavalieri, S. Coleman, F. J. Cuthbert, J. D. Fraser, W. Golder, D. Hoffman, S. M. Karpanty, A. Van Zoeren, and D. H. Catlin. 2018. Impacts of anthropogenic disturbance on body condition, survival and site fidelity of nonbreeding Piping Plovers. *Condor* 120:566-580.

Gonzalez, P. M. 1996. Habitat partitioning and the distribution and seasonal abundances of migratory plovers and sandpipers in Los Alamos, Rio Negro, Argentina. *International Wader Studies* 8:93–102.

Gosbell, K. and B. Grear. 2005. The importance of monitoring shorebird utilisation of the Coorong and surrounding wetlands in South Australia. pp 52–61 in P. Straw (ed.), *Status and Conservation of Shorebirds in the East Asian-Australasian Flyway, Proceedings of the Australasian Shorebirds Conference 13–15 December 2003*, Canberra, Australia. Wetland International Global Series 17, Sydney, Australia.

Goss-Custard, J. D. 1996. *The Oystercatcher: From Individuals to Populations*. Oxford Ornithology Series. Oxford University Press: Oxford, UK.

Gratto-Trevor, C., D. Amirault, D. Langlais, D. Catlin, F. Cuthbert, J. Fraser, S. Maddock, E. Roche, and F. Shaffer. 2012. Connectivity in piping plovers: Do breeding populations have distinct winter distributions. *Journal of Wildlife Management* 76:348–355.

Gratto-Trevor, C., S. M. Haig, M. P. Miller, T. D. Mullins, S. Maddock, E. Roche, and P. Moore. 2016. Breeding sites and winter site fidelity of Piping Plovers wintering in The Bahamas, a previously unknown major wintering area. *Journal of Field Ornithology* 87:29–41.

Hamilton, W. J. 1959. Aggressive behavior in migrant Pectoral Sandpipers. *Condor* 61:161–179.

Harebottle, D. M., A. J. Williams, Y. Weiss, and G. B. Tong. 2008. Waterbirds at Paarl waste water treatment works, South Africa, 1994–2004: Seasonality, trends and conservation importance. *Ostrich* 79:147–163.

Hay, J. R. 1984. The behavioural Ecology of the Wrybill Plover *Anarhynchus frontalis* 1984. M.Sc. Thesis, University of Auckland, Auckland, New Zealand.

Hayes, F. E. and J. A. Fox. 1991. Seasonality, habitat use, and flock sizes of shorebirds at the Bahia de Asuncion, Paraguay. *Wilson Bulletin* 103:637–649.

Hayman, P., J. Marchant, and T. Prater. 1986. *Shorebirds: An Identification Guide to Waders of the World*. Croom Helm: London, UK.

Hedenström, A., R. H. Klaassen, and S. Åkesson. 2013. Migration of the Little Ringed Plover *Charadrius dubius* breeding in South Sweden tracked by geolocators. *Bird Study* 60:466–474.

Hobbs, J. N. 1972. Breeding of Red-capped Dotterel at Fletcher's Lake Dareton, NSW. *Emu* 72:121–125.

Hockey, P. A., R. A. Navarro, B. Kalejta, and C. R. Velasquez. 1992. The riddle of the sands: Why are shorebird densities so high in southern estuaries? *American Naturalist* 140:961–79.

Hockey, P. A. R., J. R. Turpie, É. E. Plagányi, and T. E. Phillips. 1999. Scaling patterns in the foraging behaviour of sympatric plovers: Effects of body size and diet. *Journal of Avian Biology* 30:40–46.

Howell, S. N. 2010. *Molt in North American Birds*. Houghton Mifflin Harcourt: Boston, MA.

Hughes, R. A. 1984. Further notes on Puna bird species on the coast of Peru. *Condor* 86:93.

Hughey, K. F. D. 1985. The relationship between riverbed flooding and non-breeding Wrybills on northern feeding grounds in summer. *Notornis* 32:42–50.

Hyde, N. H. and T. H. Worthy. 2010. The diet of New Zealand Falcons (*Falco novaeseelandiae*) on the Auckland Islands, New Zealand. *Notornis* 57:19–26.

Iqbal, M., H. Mulyono, R. Kadarisman, and Surahman. 2010. A new southernmost record of White-faced Plover *Charadrius dealbatus*. *Wader Study Group Bulletin* 117:190–191.

Iqbal, M., I. Taufiqurrahman, K. Yordan, and B. van Balen. 2013. The distribution, abundance and conservation status of the Javan Plover *Charadrius javanicus*. *Wader Study Group Bulletin* 120:75–79.

Isacch, J. P., C. A. Darrieu, and M. M. Martínez. 2005. Food abundance and dietary relationships among migratory shorebirds using grasslands during the non-breeding season. *Waterbirds* 28:238–245.

Isacch, J. P. and M. M. Martínez. 2003. Temporal variance in abundance and the population status of non-breeding Nearctic and Patagonian shorebirds in the flooding pampa grassland of Argentina. *Journal of Field Ornithology* 74:233–242.

Jackson, B. J. and J. A. Jackson. 2000. Killdeer (*Charadrius vociferus*) in A. Poole and F. Gill (eds). *The Birds of North America Online*, No. 517. Cornell Lab of Ornithology: Ithaca, NY. Retrieved from the Birds of North America Online: <https://birdsna.org/Species-Account/bna/species/killdeer> (15 March 2016).

Johnson, C. M. and G. A. Baldassarre. 1988. Aspects of the wintering ecology of Piping Plovers in coastal Alabama. *Wilson Bulletin* 100:214–223.

Johnson, O. W., P. M. Johnson, and D. O'Daniel. 2004. Site fidelity and other features of Pacific Golden-Plovers *Pluvialis fulva* wintering on Johnston Atoll, central Pacific Ocean. *Wader Study Group Bulletin* 104:60–65.

Kalejta, B. 1993. Diets of shorebirds at the Berg River estuary, South Africa: Spatial and temporal variation. *Ostrich* 64:123–133.

Kennerley, P. R., D. N. Bakewell, and P. D. Round. 2008. Rediscovery of a long-lost *Charadrius* plover from South-East Asia. *Forktail* 24:63–79.

Kersten M., L. W. Bruinzeel, P. Wiersma, and T. Piersma. 1998. Reduced basal metabolic rate of migratory waders wintering in coastal Africa. *Ardea* 86:71–80.

Kholil, I. 2014. Predation of Javan Plover *Charadrius javanicus* by Peregrine Falcon *Falco peregrinus* in Pantai Trisik, Yogyakarta, Java, Indonesia. *Stilt* 65:36–37.

Knopf, F. L. 1998. Foods of mountain plovers wintering in California. *Condor* 100:382–384.

Knopf, F. L. and J. R. Rupert. 1995. Habits and habitats of mountain plovers in California. *Condor* 97:743–751.

Kraaijeveld, K. 2008. Non-breeding habitat preference affects ecological speciation in migratory waders. *Naturwissenschaften* 95:347–354.

Kuwae, T. 2007. Diurnal and nocturnal feeding rate in Kentish Plovers *Charadrius alexandrinus* on an intertidal flat as recorded by telescopic video systems. *Marine Biology* 151:663–673.

Lank, D. B. and R. C. Ydenberg. 2003. Death and danger at migratory stopovers: Problems with "predation risk". *Journal of Avian Biology* 34:225–228.

Laredo, C. D. 1996. Observations on migratory and resident shorebirds in lakes in the highlands of north-western Argentina. *International Wader Studies* 8:103–111.

LeDee, O. E., T. W. Arnold, E. A. Roche, and F. J. Cuthbert. 2010. Use of breeding and nonbreeding encounters to estimate survival and breeding-site fidelity of the Piping Plover at the Great Lakes. *Condor* 112:637–643.

Lipshutz, S., M. Remisiewicz, L. G. Underhill, and J. Avni. 2011. Seasonal fluctuations in population size and habitat segregation of Kittlitz's Plover *Charadrius pecuarius* at Barberspan Bird Sanctuary, North West province, South Africa. *Ostrich* 82:207–215.

Lloyd, P. 2008. Adult survival, dispersal and mate fidelity in the White-fronted Plover *Charadrius marginatus*. *Ibis* 150:182–187.

Long, P. R., S. Zefania, R. H. ffrench-Constant, and T. Szekely. 2008. Estimating the population size of an endangered shorebird, the Madagascar Plover, using a habitat suitability model. *Animal Conservation* 11:118–127.

Lourenço, P. M., A. Silva, C. D. Santos, A. C. Miranda, J. P. Granadeiro, and J. M. Palmeirim. 2008. The energetic importance of night foraging for waders wintering in a temperate estuary. *Acta Oecologica* 34:122–129.

Ma, Z. J., K. Jing, S. M. Tang, and J. K. Chen. 2002. Shorebirds in the eastern intertidal areas of Chongming Island during the 2001 northward migration. *Stilt* 41:6–10.

Marchant, S. and P. J. Higgins. 1993. *Handbook of Australasian, New Zealand and Antarctic Birds*, vol. II. Oxford University Press: Melbourne, Australia.

Martínez-Curci, N. S., J. P. Isacch, and A. B. Azpiroz. 2015. Shorebird seasonal abundance and habitat-use patterns in Punta Rasa, Samborombón Bay, Argentina. *Waterbirds* 38:68–76.

Masero, J. A., S. M. Estrella, and J. M. Sánchez-Guzmán. 2007. Behavioural plasticity in foraging mode of typical plovers. *Ardea* 95:259–265.

Masero J. A., A. Perez-Hurtado, M. Castro, and G. M. Arroyo. 2000. Complementary use of intertidal mudflats and adjacent salinas by foraging waders. *Ardea* 88:177–191.

McConkey, K. R. and B. D. Bell. 2005. Activity and habitat use of waders are influenced by time, tide and weather. *Emu* 105:331–340.

Maclean, G. L. 1976. A field study of the Australian Dotterel. *Emu* 76:207–215.

Maclean, G. L. 1977. Comparative notes on black-fronted and red-kneed dotterels. *Emu* 77:199–207.

McNeil, R. 1970. Hivernage et estivage d'oiseaux aquatiques nord-américains dans le Nord-Est du Venezuela (mue, accumulation de graisse, capacité de vol et routes de migration). *Oiseau et la Revue Francaise d'omithologie* 40:185–302 (in French).

Meissner, W., P. Chylarecki, and M. Skakuj. 2010. Ageing and sexing the Ringed Plover *Charadrius hiaticula*. *Wader Study Group Bulletin* 117:99–102.

Meltofte, H. 1996. Are African wintering waders really forced south by competition from northerly wintering conspecifics? Benefits and constraints of northern versus southern wintering and breeding in waders. *Ardea* 84:31–44.

Mendes, L., T. Piersma, M. Lecoq, B. Spaans, and R. E. Ricklefs. 2005. Disease-limited distributions? Contrasts in the prevalence of avian malaria in shorebird species using marine and freshwater habitats. *Oikos* 109:396–404.

Milton, D. A., P. V. Driscoll, and S. B. Harding. 2014. The importance of Bowling Green Bay and Burdekin River Delta, North Queensland, Australia, for shorebirds and waterbirds. *Stilt* 65:3–16.

Minton, C. 1987. Trans-Tasman movements of Double-banded Plovers. *Victoria Wader Study Group Bulletin* 11:27–34.

Mönke, R. and K. -J. Seelig. 2009. Some behavioural observations of wintering Lesser *Charadrius mongolus* and Greater *C. leschenaultii* Sand Plovers in Goa, India. *Indian Birds* 4:110–111.

Morrier, A. and R. McNeil. 1991. Time-activity budget of Wilson's and Semipalmated Plovers in a tropical environment. *Wilson Bulletin* 103:598–620.

Murose, A. 1998. Wintering records of Long-billed Ringed Plover *Charadrius placidus* in Tokachi district, eastern Hokkaido. *Japanese Journal of Ornithology* 47:24 (in Japanese with English Summary).

Musmeci, L., L. O. Bala, and M. De Los Ángeles Hernández. 2013. Dieta del chorlito Doble collar (*Charadrius falklandicus*) en Península Valdés, Patagonia, Argentina. *Hornero* 28:15–21 (in Spanish).

Myers, J. P. 1984. Spacing behavior on nonbreeding shorebirds. pp. 271–321 in J. Burger and B. L. Olla (eds), *Shorebirds: Migration and Foraging Behavior. Behavior of Marine Animals: Current Perspectives in Research*. Plenum Press: New York.

Myers, J. P., P. G. Connors, and F. A. Pitelka. 1979a. Territory size in wintering Sanderlings: The effects of prey abundance and intruder density. *Auk* 96:551–561.

Myers, J. P., P. G. Connors, and F. A. Pitelka. 1979b. Territoriality in non-breeding shorebirds. *Studies in Avian Biology* 2:231–246.

Myers, J. P., J. L. Maron, and M. Sallaberry. 1985. Going to extremes: Why do Sanderlings migrate to the neotropics? *Ornithological Monographs* 36:520–535.

Myers, J. P. and B. J. McCaffery. 1984. Paracas revisited: Do shorebirds compete on their wintering ground? *Auk* 101:197–199.

Myers, J. P., P. D. McLain, R. I. G. Morrison, P. Z. Antas, P. Canevari, B. A. Harrington, T. E. Lovejoy, V. Pulido, M. Sallaberry, and S. E. Senner. 1987. The western hemisphere shorebird reserve network. *Wader Study Group Bulletin* 49:122–124.

Nebel, S., D. B. Lank, P. D. O'Hara, G. Fernández, B. Haase, F. Delgado, F. A. Estela, L. J. Evans Ogden, B. Harrington, B. E. Kus, J. E. Lyons, F. Mercier, B. Ortego, J. Y. Takekawa, N. Warnock, and S. E. Warnock. 2002. Western sandpipers (*Calidris mauri*) during the nonbreeding season: Spatial segregation on a hemispheric scale. *Auk* 119:922–928.

Nebel, S. and R. Ydenberg. 2005. Differential predator escape performance contributes to a latitudinal sex ratio cline in a migratory shorebird. *Behavioral Ecology and Sociobiology* 59:44–50.

Newstead, D. and K. Vale. 2014. Protecting Important Shorebird Habitat Using Piping Plovers as an Indicator Species. Publication CBBEP-96 report to USFWS, Corpus Christi, TX.

Noel, B. L. and C. R. Chandler. 2008. Spatial distribution and site fidelity of non-breeding piping plovers on the Georgia coast. *Waterbirds* 31:241–251.

Nol, E. and M. S. Blanken. 2014. Semipalmated Plover (*Charadrius semipalmatus*). in A. Poole (ed.), *The Birds of North America Online*, No. 444. Cornell Lab of Ornithology: Ithaca, NY. Retrieved from the Birds of North America Online: http://bna.birds.cornell.edu.cat1.lib.trentu.ca:8080/bna/species/444 (18 March 2016).

Nol, E., K. MacCulloch, L. McKinnon, and L. Pollock. 2014. Foraging ecology and time budgets of non-breeding shorebirds in coastal Cuba. *Journal of Tropical Ecology* 30:347–357.

Nol, E., S. Williams, K. Wainio, and A. Storm-Suke. 2013. Plumage dichromatism, wing-mass relationships and assessing the accuracy of field sexing techniques in breeding Semipalmated Plovers. *Wader Study Group Bulletin* 120:114–118.

Norazlimi, N. and R. Ramli. 2014. Temporal variation of shorebirds population in two different mudflats areas. *International Journal of Biological, Biomolecular, Agricultural, Food and Biotechnological Engineering* 8:1265–1271.

Norris, D. R. and C. M. Taylor. 2006. Predicting the consequences of carry-over effects for migratory populations. *Biology Letters* 2:148–151.

Norris, D. R., P. P. Marra, T. K. Kyser, T. W. Sherry, and L. M. Ratcliffe. 2004. Tropical winter habitat limits reproductive success on the temperate breeding grounds in a migratory bird. *Proceedings of the Royal Society of London B* 271:59–64.

Ntiamoa-Baidu, Y., S. K. Nyame, and A. A. Nuoh. 2000. Trends in the use of a small coastal lagoon by waterbirds: Muni Lagoon (Ghana). *Biodiversity and Conservation* 9:527–539.

Ntiamoa-Baidu, Y., T. Piersma, P. Wiersma, M. Poot, P. Battley, and C. Gordon. 1998. Water depth selection, daily feeding routines and diets of waterbirds in coastal lagoons in Ghana. *Ibis* 140:89–103.

Okes, N. C., P. A. Hockey, and G. S. Cumming. 2008. Habitat use and life history as predictors of bird responses to habitat change. *Conservation Biology* 22:151–162.

Ould Aveloitt, M., M. El Morhit, and J. Leyer. 2013. The wintering shorebirds (Aves, Charadrii) in Mauritania: Principal species, and wetlands of major importance. *Revue Marocaine des Sciences Agronomiques et Vétérinaires* 2:67–71.

Owens, N. W. and J. D. Goss-Custard. 1976. The adaptive significance of alarm calls given by shorebirds on their winter feeding grounds. *Evolution* 30:397–398.

Owino, A. 2002. Shoreline distribution patterns of Kittlitz's Plover' *Charadrius pecuarius* Temminck, at Lake Nakuru, Kenya. *African Journal of Ecology* 40:396–398.

Page, G. W., L. E. Stenzel, G. W. Page, J. S. Warriner, J. C. Warriner, and P. W. Paton. [online]. 2009. Snowy Plover (*Charadrius nivosus*). in A. Poole (ed.), *The Birds of North America Online*, No. 145. Cornell Lab of Ornithology: Ithaca, NY. <http://bna.birds.cornell.edu.cat1.lib.trentu.ca:8080/bna/species/154> (20 March 2016).

Page, G. W., M. A. Stern, and P. W. Paton. 1995. Differences in wintering areas of Snowy Plovers from inland breeding sites in western North America. *Condor* 97:258–262.

Pandiyan, J. and S. Asokan. 2016. Habitat use pattern of tidal mud and sand flats by shorebirds (Charadriiformes) wintering in southern India. *Journal of Coastal Conservation* 20:1–11.

Pedro, P. and J. A. Ramos. 2009. Diet and prey selection of shorebirds in salt pans in Mondego Estuary, Western Portugal. *Ardeola* 56:1–11.

Perez-Hurtado, A., J. D. Goss-Custard, and F. Garcia. 1997. The diet of wintering waders in Cádiz Bay, southwest Spain. *Bird Study* 44:45–52.

Piersma, T. and C. Hassell. 2010. Record numbers of grasshopper-eating waders (Oriental Pratincole, Oriental Plover, Little Curlew) on coastal west-Kimberley grasslands of NW Australia in mid February 2010. *Wader Study Group Bulletin* 117:103–108.

Piersma, T., T. Lok, Y. Chen, C. J. Hassell, H.-Y. Yang, A. Boyle, M. Slaymaker, Y.-C. Chan, D. S. Melville, Z.-W. Zhang, and Z. Ma. 2016. Simultaneous declines in summer survival of three shorebird species signals a flyway at risk. *Journal of Applied Ecology* 53:479–490.

Piersma, T., L. Zwarts, and J. H. Bruggemann. 1990. Behavioural aspects of the departure of waders before long-distance flights: Flocking, vocalizations, flight paths and diurnal timing. *Ardea* 78:157–184.

Pomeroy, A. 2006. Tradeoffs between food abundance and predation danger in spatial usage of a stopover site by western sandpipers, *Calidris mauri*. *Oikos* 112:629–637.

Pyle, P. 2008. *Identification Guide to North American Birds: Part II. Anatidae to Alcidae*. Slate Creek Press: Bolinas, CA.

Quaintenne, G., J. A. van Gils, P. Bocher, A. Dekinga, and T. Piersma. 2011. Scaling up ideals to freedom: Are densities of red knots across western Europe consistent with ideal free distribution? *Proceedings of the Royal Society B* 278:2728–2736.

Recher, H. F. 1966. Some aspects of the ecology of migrant shorebirds. *Ecology* 47:393–407.

Ribeiro, P. D., O. O. Iribarne, D. Navarro, and L. Jaureguy. 2004. Environmental heterogeneity, spatial segregation of prey, and the utilization of southwest Atlantic mudflats by migratory shorebirds. *Ibis* 146:672–682.

Robert, M., R. McNeil, and A. Leduc. 1989. Conditions and significance of night feeding in shorebirds and other water birds in a tropical lagoon. *Auk* 106:94–101.

Roche, E. A., J. B. Cohen, D. H. Catlin, D. L. Amirault-Langlais, F. J. Cuthbert, C. L. Gratto-Trevor, J. Felio, and J. D. Fraser. 2010. Range-wide Piping Plover survival: Correlated patterns and temporal declines. *Journal of Wildlife Management* 74:1784–1791.

Rodrigues, A. A. F. 2000. Seasonal abundance of nearctic shorebirds in the Gulf of Maranhão, Brazil. *Journal of Field Ornithology* 71:665–675.

Rogers, K. G., D. I. Rogers, and M. A. Weston. 2014. Prolonged and flexible primary moult overlaps extensively with breeding in beach-nesting Hooded Plovers *Thinornis rubricollis*. *Ibis* 156:840–849.

Rohweder, D. A. 2001. Nocturnal roost use by migratory waders in the Richmond River Estuary, northern New South Wales, Australia. *Stilt* 40:23–28.

Rohweder, D. A., and P. R. Baverstock. 1996. Preliminary investigation of nocturnal habitat use by migratory waders (Order *Charadriformes*) in northern New South Wales. *Wildlife Research* 23:169–183.

Rohweder, D. A. and B. D. Lewis. 2004. Day-night foraging behaviour of banded dotterels (*Charadrius bicinctus*) in the Richmond River estuary, northern NSW, Australia. *Notornis* 51:141–146.

Rose, M. and E. Nol. 2010. Foraging behavior of non-breeding Semipalmated Plovers. *Waterbirds* 33:59–69.

Rose, M., L. Pollock, and E. Nol. 2016. Diet and prey selectivity of Semipalmated Plovers in coastal Georgia. *Canadian Journal of Zoology* 94:727–732.

Round, P. D. 2006. Shorebirds in the inner gulf of Thailand. *Stilt* 50:96–102.

Sánchez, M. I., A. J. Green, and E. M. Castellanos. 2006. Spatial and temporal fluctuations in presence and use of chironomid prey by shorebirds in the Odiel saltpans, south-west Spain. *Hydrobiologia* 567:329–340.

Sandilyan, S., K. Thiyagesan, and R. Nagarajan. 2010. Major decline in species-richness of waterbirds in the Pichavaram mangrove wetlands, southern India. *Wader Study Group Bulletin* 117:91–98.

Sanzenbacher, P. M. and S. M. Haig. 2002. Regional fidelity and movement patterns of wintering Killdeer in an agricultural landscape. *Waterbirds* 25:16–25.

Scherer, A. L. and M. V. Petry. 2012. Seasonal variation in shorebird abundance in the State of Rio Grande Do Sul, Southern Brazil. *Wilson Journal of Ornithology* 124:40–50.

Schneider, D. 1985. Migratory shorebirds: Resource depletion in the tropics. *Ornithological Monographs* 36:546–558.

Senner, S. and M. A. Howe. 1984. Conservation of nearctic shorebirds. Pp. 379–421 in J. Burger and B. L. Olla (eds), *Behavior of Marine Animals: Shorebird Breeding Behavior and Populations*. Plenum Press: New York.

Sillett, T. S. and R. T. Holmes. 2002. Variation in survivorship of a migratory songbird throughout its annual cycle. *Journal of Animal Ecology* 71:296–308.

Simmons, K. E. L. 1956. Territory in the Little Ringed Plover *Charadrius dubius*. *Ibis* 98:390–397.

Sinclair, A. R. E. 1978. Factors affecting the food supply and breeding season of resident birds and movements of Palaearctic migrants in a tropical African savannah. *Ibis* 120:480–497.

Skagen, S. K. and Oman, H. D. 1996. Dietary flexibility of shorebirds in the western hemisphere. *Canadian Field-Naturalist* 110:419–444.

Sprandel, G. L., J. A. Gore, and D. T. Cobb. 2000. Distribution of wintering shorebirds in coastal Florida. *Journal of Field Ornithology* 71:708–720.

Sripanomyom, S., P. D. Round, T. Savini, Y. Trisurat, and G. A. Gale. 2011. Traditional salt-pans hold major concentrations of overwintering shorebirds in Southeast Asia. *Biological Conservation* 144:526–537.

Stillman, R. A., A. D. West, J. D. Goss-Custard, S. McGrorty, N. J. Frost, D. J. Morrisey, A. J. Kenney, and A. L. Drewitt. 2005. Predicting site quality for shorebird communities: A case study on the Humber estuary, UK. *Marine Ecology Progress Series* 305:203–217.

Strauch, Jr. J. G. and L. G. Abele. 1979. Feeding ecology of three species of plovers wintering on the Bay of Panama, Central America. Pp. 217–230 in F. A. Pitelka (ed.), *Shorebirds in Marine Environments. Studies in Avian Biology*, vol. 2, University of California Press: Berkeley, CA.

Storm-Suke, A. 2012. The Use of Stable-Hydrogen Isotopes in Connectivity Studies: A Test of Assumptions and Application with Trace Element Analysis. *Ph.D. Dissertation*. Trent University, Peterborough, Ontario, Canada.

Summers, R. W. and M. Waltner. 1979. Seasonal variations in the mass of waders in southern Africa, with special reference to migration. *Ostrich* 50:21–37.

Summers, R. W., S. Pálsson, C. Corse, B. Etheridge, S. Foster, and B. Swann. 2013. Sex ratios of waders at the northern end of the East Atlantic flyway in winter. *Bird Study* 60:437–445.

Svanbäck, R. and D. I. Bolnick. 2007. Intraspecific competition drives increased resource use diversity within a natural population. *Proceedings of the Royal Society B* 274:839–844.

Taylor, I. R. 2004. Foraging ecology of the Black-fronted Plover on saline lagoons in Australia: The importance of receding water levels. *Waterbirds* 27:270–276.

Thibault, M. and R. McNeil. 1994. Day/night variation in habitat use by Wilson's Plovers in northeastern Venezuela. *Wilson Bulletin* 106:299–310.

Thibault, M. and R. McNeil. 1995. Predator-prey relationship between Wilson's Plovers and fiddler crabs in northeastern Venezuela. *Wilson Bulletin* 107:73–80.

Thomas, D. G. 1972. Moult of the Banded Dotterel (*Charadrius bicinctus*) in winter quarters. *Ostrich* 19:33–35.

Trainor, C. L. 2011. The waterbirds and coastal seabirds of Timor-Leste (East Timor): New site records clarifying residence status, distribution and taxonomy. *Forktail* 27:63–72.

Turpie, J. K. 1995. Non-breeding territoriality: Causes and consequences of seasonal and individual variation in Grey Plover *Pluvialis squatarola* behaviour. *Journal of Animal Ecology* 64:429–438.

Van den Hout, P. J. and G. R. Martin. 2011. Extreme head-tilting in shorebirds: Predator detection and sun avoidance. *Wader Study Group Bulletin* 118:18–21.

Van Den Hout, P. J., B. Spaans, and T. Piersma. 2008. Differential mortality of wintering shorebirds on the Banc d'Arguin, Mauritania, due to predation by large falcons. *Ibis* 150:219–230.

Veitch, C. R. 1978. Waders of the Manukau Harbour and Firth of Thames. *Notornis* 25:1–24.

Velasquez, C. R. 1992. Managing artificial saltpans as a waterbird habitat: Species' responses to water level manipulation. *Colonial Waterbirds* 15:43–55.

Weston, M. A., G. C. Ehmke, and G. S. Maguire. 2009. Manage one beach or two? Movements and space-use of the threatened Hooded Plover (*Thinornis rubricollis*) in south-eastern Australia. *Wildlife Research* 36:289–298.

Weston, M. A., F. J. L. Kraaijeveld-Smit, R. Mcintosh, G. Sofronidis, and M. A. Elgar. 2004. A male-biased sex-ratio in non-breeding Hooded Plovers on a salt-lake in Western Australia. *Pacific Conservation Biology* 9:273–277.

Whitfield, D. P. 1985. Raptor predation on wintering waders in southeast Scotland. *Ibis* 127:544–558.

Withers, K. and B. R. Chapman. 1993. Seasonal abundance and habitat use of shorebirds on an Oso Bay Mudflat, Corpus Christi, Texas. *Journal of Field Ornithology* 64:382–392.

Woodworth, B. K., C. M. Francis, and P. D. Taylor. 2014. Inland flights of young red-eyed vireos *Vireo olivaceus* in relation to survival and habitat in a coastal stopover landscape. *Journal of Avian Biology* 45:387–395.

Wunder, M. B. and F. L. Knopf. 2003. The Imperial Valley of California is critical to wintering Mountain Plovers. *Journal of Field Ornithology* 74:74–80.

Wunderle Jr. J. M., R. B. Waide, and J. Fernandez. 1989. Seasonal abundance of shorebirds in the Jobos Bay Estuary in Southern Puerto Rico. *Journal of Field Ornithology* 60:329–339.

Yasué, M. and P. Dearden. 2009. The importance of supratidal habitats for wintering shorebirds and the potential impacts of shrimp aquaculture. *Environmental Management* 43:1108–1121.

Young, H. G., F. Razafindrajao, R. E. Lewis, and B. A. A. Iahia. 2006. Distribution and status of palearctic shorebirds in Western Madagascar, September-November 2004. *Waterbirds* 29:235–238.

Zefania, S., R. Emilienne, P. J. Faria, M. W. Bruford, P. R. Long, and T. Székely. 2010. Cryptic sexual size dimorphism in Malagasy plovers *Charadrius* spp. *Ostrich* 81:173–178.

Zimmer, C., M. Boos, N. Poulin, A. Gosler, O. Petit, and J. P. Robin, 2011. Evidence of the trade-off between starvation and predation risks in ducks. *PLoS One* 6:e22352.

Zou, F., H. Zhang, T. Dahmer, Q. Yang, J. Cai, W. Zhang, and C. Liang. 2008. The effects of benthos and wetland area on shorebird abundance and species richness in coastal mangrove wetlands of Leizhou Peninsula, China. *Forest Ecology and Management* 255:3813–3818.

CHAPTER NINE

Habitat Ecology and Conservation of Charadrius Plovers*

James D. Fraser and Daniel H. Catlin

Abstract. Habitat loss is a principal cause of species extinction and, therefore, a key conservation problem as the world's human population approaches 8 billion. Plovers worldwide are affected by intensified land use, development, and climate change, with each threat posing unique conservation challenges. Understanding how species use habitats is paramount to population management. Plovers occur in open areas on coasts, rivers, lakeshores, grasslands, and tundra in habitats rich in invertebrate prey (e.g., tidal flats, river shorelines, productive grasslands). Habitat selection theory predicts that habitat choices should enhance a bird's fitness, by increasing foraging rate and decreasing the probability of being killed by a predator. In the breeding season, plovers choose nesting sites near foraging habitat, allowing adults to readily acquire food while tending nests and giving newly hatched chicks access to foraging sites. Plovers also select habitats that allow them to escape predation, which can be difficult in open habitats. They select substrate colors and textures that camouflage them, avoid dense vegetation that can obstruct their view of predators, and select large patches of open area that increase predator search time and effort. Many plovers defend breeding territories, which may set an upper limit on density and limit local population size. Plover population irruptions that occur when new habitat is created by natural or anthropogenic processes are evidence that habitat limits populations. Habitat restoration practices, such as augmenting riverine sandbars or removing vegetation, have been attempted for several plover species with mixed results. Rigorous studies that critically address habitat selection and habitat suitability are needed to complete our understanding of plover habitat ecology. Such studies will provide a scientific underpinning needed to solve increasingly urgent, habitat-related conservation problems.

Keywords: camouflage, foraging, habitat loss, habitat selection, irruption, land use, nesting, population limitation, predation, reproduction, resource selection, restoration, survival.

Extinction due to habitat loss has been called the "signature conservation problem of the 21st century" (Millennium Ecosystem Assessment 2005). Many plover species worldwide are losing habitat, which likely has resulted in a reduction in numbers, and losses of populations, and which may result in future species extinctions. What we already know about loss of shorebird habitats worldwide leads us to expect that continued habitat loss will lead to species losses. Predictions of which species are at greatest risk, how best to mitigate habitat losses, and how to prevent losses of key habitats in the future are hindered by ignorance of the habitat ecology of many species. A clear understanding

* James D. Fraser and Daniel H. Catlin. 2019. Habitat Ecology and Conservation of Charadrius Plovers. Pp. 217–243 in Colwell and Haig (editors). The Population Ecology and Conservation of Charadrius Plovers (no. 52), CRC Press, Boca Raton, FL.

of plover habitat requirements, based on rigorous studies of habitat selection, is vital to effective conservation. Here, we review quantitative and qualitative information available on habitat selection in plovers. We then argue that some plover species are limited by the amount and quality of habitat available, which has consequences for habitat conservation and restoration efforts. We finish with a discussion of the conservation implications of this information and by examining the future of plover habitat conservation and research.

Plovers use a variety of ecoregions year round. At the most general level, most breeding plovers inhabit grasslands and shorelines of oceans, lakes, and rivers (16–20 species), saltpans and alkali wetlands (14 species), and deserts or tundra (three species each, Table 9.1). During the nonbreeding season, habitat use is similar, although only eight species use river habitats in the nonbreeding season, compared to 16 species during the breeding season, and no species inhabit tundra throughout the nonbreeding season. Plover use of tidal flats, salt pans, short grass prairies, pastures, and beaches (Figures 9.1–9.4) is facilitated by a suite of morphological (e.g., large eyes and neural centers; Nol 1984) and behavioral

TABLE 9.1

Ecogeographic habitats used by Charadrius plovers during the breeding (B) and nonbreeding (N) seasons.

	Habitats							
Common name	Ocean shores	Lakeshores	River	Tundra	Grasslands	Desert	Salt pans/ alkali flats	Citations
Red-breasted Dotterel	B,N		B		B,N			Marchant and Higgins (1993)
Lesser Sand-plover	N	B	B	B				Cramp and Simmons (1983)
Greater Sand-plover					B	B	N	Cramp and Simmons (1983)
Caspian Plover	N	N			B,N	B	B	del Hoyo et al. (1996)
Collared Plover	B,N	B,N	B,N		B,N			del Hoyo et al. (1996)
Puna Plover					B,N		B,N	Hayman et al. (1986)
Two-banded Plover	B,N	B	B					Hayman et al. (1986)
Double-banded Plover	B,N	N	B		B			Marchant and Higgins (1993)
Kittlitz's Plover		B,N	B,N		B,N		B,N	Cramp and Simmons (1983)
Red-capped Plover	B,N	B,N					B,N	Marchant and Higgins (1993)
Malay Plover	B,N						B,N	Yasué et al. (2007), Yasué and Dearden (2009)
Kentish Plover	B,N	B					B,N	Cramp and Simmons (1983)
Snowy Plover	B,N	B,N	B				B,N	Page et al. (2009)
Javan Plover	B,N							del Hoyo et al. (1996)
Wilson's Plover	B,N						B	Corbat and Bergstrom (2000)

(Continued)

TABLE 9.1 (*Continued*)

Ecogeographic habitats used by Charadrius plovers during the breeding (B) and nonbreeding (N) seasons.

	Habitats							
Common name	Ocean shores	Lakeshores	River	Tundra	Grasslands	Desert	Salt pans/ alkali flats	Citations
Common Ringed Plover	B	B	B,N	B	B		N	Cramp and Simmons (1983)
Semipalmated Plover	B,N		B,N	B	B,N			Nol and Blanken (2014)
Long-billed Plover		B	B				N	Katayama et al. (2010)
Piping Plover	B,N	B	B				B,N	(Elliott-Smith and Haig 2004)
Black-banded (Madagascar) Plover	B,N	B,N			B,N			Hayman et al. (1986)
Little Ringed Plover	B,N	B,N	B,N					Marchant and Higgins (1993)
Three-banded Plover		B,N			B,N			Marchant and Higgins (1993)
Forbes's Plover		N	B,N		B,N			del Hoyo et al. (1996)
White-fronted Plover		B,N						del Hoyo et al. (1996)
Chestnut-banded Plover		B,N					B,N	del Hoyo et al. (1996)
Killdeer	B,N	B,N	B		B,N		B,N	Jackson and Jackson (2000)
Mountain Plover					B,N		N	Knopf and Wunder (2006)
Oriental Plover	N	B,N			B,N			Marchant and Higgins (1993)
Eurasian Dotterel				B	B,N	N		Cramp and Simmons (1983)
St. Helena Plover					B,N			McCulloch (1991)
Rufous-chested Dotterel	B,N				B,N			Carstairs (1989)
Red-kneed Dotterel		B,N	B,N					Marchant and Higgins (1993)
Hooded Plover	B,N	B,N					B,N	Marchant and Higgins (1993)
Shore Plover	B,N						B,N	Marchant and Higgins (1993)
Black-fronted Dotterel	N	B,N	B,N		B		N	Marchant and Higgins (1993)
Inland Dotterel					B,N	B,N	B,N	Marchant and Higgins (1993)
Wrybill			B,N					Marchant and Higgins (1993)

Figure 9.1. Piping Plover nesting and foraging habitats on sandbars on the Missouri River, South Dakota, USA. (Photo by the U.S. Army Corps of Engineers.)

Figure 9.2. Nesting and foraging habitat of the Eurasian Dotterel, Cairngorms National Park, Scotland. (Photo by Will Boyd-Wallis.)

traits (e.g., vigilance and flocking; Cresswell 1994, Cresswell and Quinn 2011). Their need to forage efficiently while avoiding predators in open habitat has also influenced their breeding biology, including camouflaged eggs, chicks and adults, site selection that allows them to observe their surroundings, stealthy behavior around nests, crouching/freezing when predators threaten, and predator distraction—the "broken wing display"—exhibited by many species (Skutch 1976)

Photograph by Fairchild Aerial Surveys

EVEN THE COASTLINE WAS CHANGED BY THE HURRICANE'S ASSAULT

Where summer cottages once lined the beach, storm waves have cut two new inlets (nearest the camera) into Moriches Bay (left), on Long Island. The third inlet (background) existed previously but was considerably deepened. Just beyond the horizon is Westhampton Beach, where severe damage occurred.

Figure 9.3. Westhampton Island, New York, after it was leveled by the Great Hurricane of 1938. Moriches Bay is to the left, and the Atlantic Ocean is to the right. The town of West Hampton Beach is at the horizon and was severely damaged by the storm. (Photo by Fairchild Aerial Surveys.)

to lead predators away from eggs and chicks. These and other behaviors have evolved in each species and, taken together, represent plovers' solution to the problem of surviving and reproducing in exposed habitats.

HABITAT SELECTION

Early naturalists described habitat use in general terms. For example, Bent (1929) described Snowy Plover habitat as "above the ordinary wash of the tides on ocean beaches. Such places are usually strewn more or less thickly with shells, pebbles and various bits of debris...." More recently, biologists have become interested in using resource selection, the process by which an animal chooses a resource (e.g., habitat) relative to its availability on the landscape (Johnson 1980, Manly et al. 2002), to provide context for animal habitat use and increase our ability to predict habitat use based on its physical characteristics.

Habitat selection theory begins with the assumption that habitat selection evolved via natural selection and that, under normal circumstances, habitat selection is expected to increase an individual's fitness (Fretwell and

Figure 9.4. Old Inlet, Fire Island, New York, USA in April 2013, after Hurricane Sandy. Piping Plovers nested on the sand on both sides of the inlet, and were seen in migration on the tidal flats in the inlet. (Photo by Mike Ferrigno and Rich Giannotti. Additional photos in a time sequence available at http://po.msrc.sunysb.edu/GSB/.) After the island breached, Piping Plovers nested in the sand on both sides of the inlet and both Piping Plovers (6–10), and Semipalmated Plovers (10–159) were counted on the flood shoals during migration. (DeRose-Wilson et al., Kwon et al., unpublished data.)

Lucas 1970, Rosenzweig 1981, Kristan 2003, Morris 2003). From this perspective, habitat quality can be defined as the fitness potential of individuals that use the habitat (Wiens 1989), and to the extent that variation in habitat has effects on individual fitness, it would benefit individuals to be selective. Habitat selection often is measured by comparing patterns of habitat use to measures of the availability of habitat in the area where a species occurs. Use of some feature of the environment disproportionate to its availability is treated as a measure of selectivity by the species for that feature. Unless there is evidence of an ecological trap (Battin 2004, Hale and Swearer 2016), disproportionate use of habitat is assumed to be adaptive, which, in plovers, typically means the habitat chosen is expected to facilitate access to food, reduce the probability of being killed, or both. Although plovers are expected to select habitat that maximizes access to food and minimizes predation risk, habitats that accomplish these goals with equal facility may not always be available, and trade-offs may be necessary.

Resource selection by an animal can be considered at several scales: the geographic range, the home range of an individual, or smaller scales within home ranges, such as nest and feeding sites (Johnson 1980, Hutto 1985, Manly et al. 2002). It can also be assessed at different time scales which is important to migratory species that use habitat in different areas throughout the year.

Trade-offs in Habitat Selection

In all seasons, plovers choose habitats that may involve a trade-off between avoiding predation and obtaining food. Such trade-offs between food supply and predation risk have been noted for a wide variety of taxa (Lima and Dill 1990, Gallagher et al. 2017), including sandpipers. Western Sandpipers (*Calidris mauri*) in British Columbia foraged at sites that balanced the risk of predation with the availability of food (Pomeroy 2006). The addition of an experimental barrier, which prevented sandpipers from seeing potential predators such as Peregrine Falcons (*Falco peregrinus*), changed the density of foraging sandpipers such that densities were lowest closest to the barrier (Pomeroy et al. 2006). Mean body masses and stopover duration of Western Sandpipers declined with increasing numbers of Peregrine Falcons (Ydenberg et al. 2004). Redshanks (*Tringa totanus*) foraged closer to cover and at lower densities during very cold weather, increasing their vulnerability to predation during mid-winter when the risk of starvation was at its seasonal peak (Cresswell and Whitfield 2008). Many plover species form flocks (Chapter Eight, this volume) during the migration and nonbreeding seasons, which may confer benefits to individuals by having more birds to search for predators and also benefits from the dilution effect, such that a bird in a large flock may be less likely to be captured if it is surrounded by other individuals that also are potential prey (Cresswell and Quinn 2011).

Similar trade-offs exist during the breeding season; however, they are complicated by the often-competing interests of adults, eggs, and chicks. After eggs hatch, plover chicks are flightless for several weeks and habitat must be chosen to allow them to obtain food readily, so they can add mass rapidly while avoiding predation. For example, Piping Plover chicks can add more than 1g/day (Catlin et al. 2014). The rapid growth requires rich foraging habitat and is probably the reason that, unlike in the nonbreeding period (Chapter Eight, this volume), active territorial defense is common among plovers during the breeding season (e.g., various species in Marchant and Higgins 1993, Elliott-Smith and Haig 2004, Nol and Blanken 2014). Semipalmated Plovers may defend nest sites, feeding sites, or sites that are used for both (Nol and Blanken 2014). Territory size is related to food supply in many bird species such that the richer the food supply, the smaller the territory (Newton 1998). For example, coastal nests of the Semipalmated Plover averaged 16.5 m apart, where inland nests were at least 50 m apart near Churchill, Manitoba (Nol and Blanken 2014). Nesting densities of Piping Plovers peaked at approximately 1 pair/ha where chicks had access to bayside tidal flats; long-term mean nesting densities were 0.44 pairs/ha where they did not (Cohen et al. 2009).

Habitat Selection that Facilitates Foraging

For plovers, density and availability of food varies across habitats (e.g., Cohen et al. 2009, Cuttriss et al. 2015, Bock et al. 2016). Thus, the selection of habitats in which to forage is vital to plovers' survival and reproduction. It is also important that plovers have good foraging habitats juxtaposed to habitats needed for nesting and roosting; while we often think of selection of habitat by use, that is, foraging habitat or nesting habitat, in fact, birds select habitat complexes where all life needs can be obtained within suitable travel distance. For this reason, the following sections discuss not only foraging habitats *per se*, but also selection of habitats for other purposes (nesting and roosting) that are adjacent to good foraging sites. Where the information is available, we also provide evidence that selecting a particular habitat improves reproduction or survival.

Breeding Season

Hooded Plovers selected breeding sites that had higher invertebrate densities than unused sites (Cuttriss et al. 2015). The Red-breasted Dotterel breeds in areas where freshwater streams meet the sea (Ogden and Dowding 2013), perhaps because of the presence of invertebrates found in this habitat type. Alternatively, the prey obtained from fresh water may have a lower metabolic cost to the plovers than those growing in salt water (Rubega and Oring 2004). Similarly, Ringed Plovers nested close to tidal feeding areas (Pienkowski 1984). On the Atlantic Coast, Piping Plovers nested more frequently where chicks had access to bayside intertidal habitats, or pool habitats that were rich in invertebrates, than on ocean beaches without such access. Fledging success was generally higher in those areas as well (Patterson et al. 1991, Loegering and Fraser 1995, Goldin and Regosin 1998, Elias et al. 2000). In addition to selecting sites where chicks can obtain food, nesting near feeding areas allows access to foraging sites for adults tending nests or defending territories. Kentish Plovers moved chicks from marshes to lakeshores, where they grew faster and had higher survival (Kosztolanyi et al. 2007). Piping Plovers nesting on sandbars in the Missouri River foraged preferentially on low wet edges of sandbars, especially on the low current sides (Le Fer et al. 2008b). Chick mortality increased when high water covered these edges demonstrating the survival value of these habitats (Catlin et al. 2013).

While plovers must select nest sites that allow them to forage from territory establishment to fledging, the microhabitats in which nests are placed may not be productive foraging habitat. For example, Kentish Plovers breeding in saltworks adjacent to Cadiz Bay, Spain, fed primarily on the rag worm (*Hediste diversicolor*) in intertidal habitats. Rag worms comprised more than 80% of their prey intake (biomass) year round. Intake rate (mg ash-free dry mass per second) was ten times higher in the intertidal zone than in nearby saltworks where they nested. Intake rate in the saltworks was about 27% of that required to meet daily energy demands (see equation in Zwarts and Wanink 1993). During the nonbreeding season, low-tide surveys showed approximately 89% of the population foraged on intertidal flats. In contrast, during the breeding season, about 31% of the population foraged on intertidal flats, whereas 59% used saltworks. Therefore, the requirements of breeding apparently prevented these plovers from foraging in the most productive habitat (Castro et al. 2009).

Nonbreeding Season

When unencumbered by nests and young, plovers are faced with simpler habitat choices. Individuals must primarily select foraging habitats that enhance net intake rates while avoiding predation. Choice of roost sites may reduce risk of predation (Chapter Eight, this volume). Foraging locations of Red Knots (*Calidris canutus*) migrating through Virginia were closer to night roosts than expected at random (Cohen et al. 2009). During the nonbreeding season, plovers often roost in groups, presumably as an antipredator behavior, even though they may display some territorial behavior while foraging (Nol et al. 2014). In coastal habitats, plovers often go to roost when tides cover foraging habitats (Rohweder 2001, Mönke and Seelig 2009, Rose and Nol 2010), and some evidence supports the idea that roosting habitat may limit some local populations (Sprandel et al. 2000, Rose and Nol 2010), but not all (Conklin et al. 2008). The choice of roosting habitat for shorebirds is affected by the avoidance of predation (Rogers et al. 2006), as well as the quality and proximity of foraging habitat (Dias et al. 2006, van Gils et al. 2006). Plovers select dry, open habitats for roosting but must balance the energetic costs associated with travel to foraging habitat. For example, Drake et al. (2001) concluded that a mosaic of juxtaposed foraging and roosting habitats contributed to high site fidelity, relatively small home ranges, and high survival in Piping Plovers. Similarly, Piping Plovers preferred spoil islands and flood shoals in North Carolina, but moved to ocean beaches to forage and roost when shoals were covered by the incoming tide (Cohen et al. 2008). In this case, Piping Plovers appeared to be balancing risk with energy intake throughout the tidal cycle, but the habitats that provided greatest energy intake varied with the tide. Foraging activity is likely affected by the tide in all tidal systems (e.g., Fraser et al. 2005, Castro et al. 2009).

In the Laguna Madre, Texas and Tamaulipas, Mexico, most wintering Snowy, Semipalmated, Piping, and Wilson's plovers were observed on algal flats or bayside sand and mudflats (Mabee et al. 2001). Kentish Plovers wintering in Cadiz Bay, Spain, predominately foraged on mudflats at low tide, but were seen more frequently in saltworks at high tide (Castro et al. 2009). Piping Plovers select productive, low-wave-energy bay habitats for foraging (Haig and Oring 1985, Johnson and Baldassarre 1988, Nicholls and Baldassarre 1990, Sprandel et al. 2000, Zonick 2000, Drake et al. 2001). In Tasmania, Hooded Plovers and Red-capped Plovers occurred in greater abundance on beaches with high invertebrate densities (Bock 2014) and on gently sloping beaches with low-wave energy (Bock et al. 2016). In southern California, Snowy Plover abundance was associated with the standing crop of wrack-associated flies (Dugan et al. 2003). Snowy Plovers in northern coastal California selected sites with more brown algae and invertebrates than unoccupied sites (Brindock and Colwell 2011). In Thailand, Greater and Lesser Sand-plovers, Kentish Plovers, and Malay Plovers foraged preferentially on intertidal mudflats (Yasué and Dearden 2009). Double-banded Plovers occupied sites where amphipods and polychaetes were abundant (Pierce 1980). These findings underscore that, across species and regions, nonbreeding plovers select for and are found at greater densities near rich, tidal foraging habitats. See also Piping Plover Case Study, this chapter.

Habitat Selection that Reduces Predation

Nest Substrate and Nearby Debris

Plover eggs are speckled and blend in with gravel or shell substrates. Their breeding plumage breaks up the "bird shape" of an immobile bird. They appear to select habitats that increase the benefits of these physical traits (see also Chapter Six, this volume). Snowy Plovers place nests near driftwood, shells, and stones, and choose areas with greater cover of these objects than found at random points (Hardy and Colwell 2012, Leja 2015). Nests of the Kentish Plover were found on shells more often than expected at random (Valle and Scarton 1999). Mountain Plover nest sites were positively associated with the amount of bare ground (no vegetation) and short vegetation (Goguen 2012). Common Ringed Plovers nested on barren areas in sparse tundra heath and shingle in Greenland and in sand dune gravel flats of fields in the UK where they generally relied on camouflage for protection from aerial predators (Pienkowski 1984). The White-fronted Plover usually placed its nest near an object when nesting on sand, but not when nesting on gravel, suggesting that the objects helped to conceal the nest and/or incubating adult (Maclean and Moran 1965).

Evidence indicates that substrates chosen for nest sites act to camouflage plover nests, which, in turn, improves nest success, chick survival, and presumably lifetime reproductive success (Herman and Colwell 2015). Nest success of Snowy Plovers on gravel bars with numerous egg-sized stones was higher than on nearby sand beaches (Colwell et al. 2011, Hardy and Colwell 2012). Snowy Plover nests subjected to principally avian predation were successful when placed under an object and these nests were more successful when placed on alkali substrate than when placed on sand-gravel substrate, perhaps because nesting density was higher on the former (Page et al. 1983, Page et al. 1985). Survival of Mountain Plover nests was greater when egg coloration most closely matched the substrate (Skrade and Dinsmore 2013). Nest success of Kentish Plovers at Bohai Bay, China, was highest on rocky substrates (Que et al. 2015). In contrast, Snowy Plover nest success in coastal Washington, United States, was not influenced by the amount of debris or vegetation in 1, 5, or 25 m^2 plots centered on nests (Pearson et al. 2016).

Vegetation

Vegetation, or the lack of it, may affect the probability of predation on plovers in several ways. Plovers may use vegetation as cover to avoid detection by predators. Conversely, dense vegetation can afford cover for terrestrial predators, thereby increasing the risk to plovers, which may not detect a threat. Many plovers nest in areas with sparse vegetation. These habitats provide birds with a clear view of the surroundings, so they can observe predators from a distance (e.g., Muir and Colwell 2010). In addition, most implement a "stealthy" departure from the nest (a crouched walk or run) when a predator approaches. These exits presumably reduce the probability of the predator detecting the adult while it still is on or near the nest, and, therefore, determining the nest location. Once the adult has left the nest, the camouflage coloration of the eggs protects them (Skrade and Dinsmore 2013) and the adult often attempts to lure the predator from the nest area by feigning injury.

Many plovers avoid vegetation at or near their breeding locations. Snowy Plovers selected nest sites in areas that were low in overall cover, yet had slightly higher cover than adjacent random sites (Powell and Collier 2000). Muir and Colwell (2010) found that there was less vegetation around Snowy Plover nests and courtship sites than around random points from the same area and that birds flushed from humans at distances similar to the distance by which they avoided vegetation when placing nests. Snowy Plovers avoided breeding in areas between dunes if they were densely vegetated (Webber et al. 2013). Kentish Plover nests had less vegetation with lower height than random sites (Valle and Scarton 1999). Long-billed Plovers avoided forest edges when selecting nest sites and colonized gravel bars newly cleared of vegetation (Katayama et al. 2010). Likewise, Little Ringed Plovers selected nest sites with little vegetation (Parrinder 1989), and Hooded Plovers selected nesting sites with less vegetation on dunes and foredunes compared with random sites (Ehmke et al. 2016). This sample of findings indicates that the relative lack of vegetation is a key feature of plover breeding sites.

The evidence that avoiding vegetation has a positive effect on nest success is mixed. Nest success of Snowy Plovers in Texas was negatively related to the amount of vegetation at the immediate nest site (Saalfeld et al. 2011), and nest success doubled when vegetation was removed from Snowy Plover sites (Dinsmore et al. 2014). However, where predation caused >70% of nest failures, nests near vegetation had higher daily survival than those in barren areas, suggesting a potential fitness benefit to proximity to some vegetation (Ellis et al. 2015). In contrast, nest success of Snowy Plovers in Washington was unaffected by vegetation near nests (Pearson et al. 2016). Mountain Plovers selected prairie dog (*Cynomys* spp.) towns and recently burned areas for nesting, both of which had low vegetation cover (Augustine and Derner 2012), but nest survival was greater in prairie dog towns than in burns, perhaps because of the protection afforded by prairie dog alarms (Augustine and Skagen 2014).

Although protection of nests from predation is important, and in some cases nesting in vegetation may reduce predation (e.g., Ellis et al. 2015), the survival of a breeding adult often has a greater effect on population viability (e.g., Plissner and Haig 2000, Larson et al. 2002). Amat and Masero (2004) showed that incubating Kentish Plovers with obscured vision at the nest took longer to observe approaching predators and were more likely to be killed than those with a clearer view. Such trade-offs undoubtedly affect habitat selection relative to vegetation. These contrasting results suggest that the makeup of local predator communities interacts with habitat selection to affect reproductive output.

The Shore Plover and the Red-kneed Dotterel differ from other plovers in that they build their nests under dense vegetation, likely as protection from avian predators. This is perhaps not surprising for an island endemic like the Shore Plover, which evolved in an environment free of mammalian predators and is consistent with its rapid decline after terrestrial predators were introduced to mainland New Zealand (Davis 1994, Dowding and O'Connor 2013). Although Red-kneed Dotterels would have encountered native terrestrial predators, their nest sites have the added protection from terrestrial predators of being on small islets in freshwater wetlands (Maclean 1977).

Size of Habitat Patch

Because plovers depend on camouflage to protect themselves, their nests, and their chicks, large habitat patches may reduce the probability of predation. Predators passing through or searching large patches are less likely to encounter a plover's nest than predators passing through a small patch. Moreover, small patches of unvegetated nesting habitat have more edge per unit area, and therefore a greater chance of a vegetation-inhabiting predator approaching a nest unseen.

Snowy Plovers selected nest sites on wider beaches than at random points (Patrick and Colwell 2014, Leja 2015). Snowy Plover occupancy of beaches on hypersaline lakes was negatively associated with the amount of shrub cover and distance to water (Ellis et al. 2014). Most nests of the Kittlitz's Plover were on an elevated dune, or next to an open area, which gave incubating birds an unobstructed view of their surroundings (Hall 1958). See also the Piping Plover case study, this chapter.

Selection of wide habitats uncluttered by vegetation, allowing plovers to "hide in plain sight" may extend beyond the breeding season, but the information is meager. Nonbreeding Snowy Plovers occurred on beaches that were wider and had less vegetation than unoccupied sites (Brindock and Colwell 2011). In winter, Mountain Plovers tended to occur in open areas such as alkaline flats, idle fields, burned agricultural fields, or recently grazed pastures (Knopf and Rupert 1995, Hunting et al. 2001, Wunder and Knopf 2003). In contrast, attempts to model the distribution and abundance of Killdeer using agricultural habitats found no association with food abundance and landscape context (availability of wet, less-vegetated habitat nearby; Taft and Haig 2006a, 2006b). This may have been because Killdeer used a wide variety of sites (Sanzenbacher and Haig 2002) or because of a lack of selection for wet, open habitats (Taft and Haig 2006b).

Trade-offs in Site Selection

When several factors affect a bird's reproduction or survival, site selection may involve trade-offs. For visual foragers, it would seem most advantageous to hunt throughout the daylight hours, assuming preys are equally available day and night. But Thibault and McNeil (1994) found that Wilson's Plovers foraged more at night than during the day, which they attributed to avoidance of predation by Peregrine Falcons (*Falco peregrinus*). Moreover, during the seasonal peak of falcon abundance, plovers favored daytime roosts that provided better concealment from aerial predators. Kentish Plovers may trade selection for a clear view of the area around a nest site with the crypsis afforded by dense vegetation (Gomez-Serrano and Lopez-Lopez 2014). It is possible that these trade-offs are mediated by an individual's experience at particular sites. Experience with aerial predators, for example, might increase the probability that a plover would select cover for crypsis, but where terrestrial predators are more common, plovers may favor selection for open vistas. Similarly, when the nesting substrate is comparable in color and pattern to eggs, open nesting may be a better choice than when the eggs contrast with the substrate. These ideas require further study. Piping Plovers nesting on shorelines selected areas with less vegetation and more cobble, but at lower elevation relative to the water than randomly selected nearby areas, suggesting a trade-off of a greater risk in flooding for a substrate that increases crypsis or a better view of oncoming predators (Anteau et al. 2012).

Nesting on Islands

Many plovers nest on islands which likely have fewer predators than nearby mainland areas (e.g., Dowding 1999, Ogden and Dowding 2013). Nearly half of plover species have been recorded nesting on rivers, typically on sandbars (Table 9.1). Moreover, many of the coastal species nest on coastal islands.

HABITAT AS A LIMITING FACTOR

Design of effective conservation programs requires an understanding of the factors limiting populations. All wildlife species are *potentially* limited by habitat, but not all populations are *actually* limited by habitat. Some populations are kept lower than the level the habitat can support (the carrying capacity) by environmental factors that affect reproduction and/or survival. A well-known example of such a demographically driven limitation was the long-term decline of bird-eating and fish-eating birds in the USA and Europe in the middle of the 20th century. This decline was caused by reduced reproduction and increased mortality due to DDT and other organochlorine pesticides. Populations declined substantially, but when the contaminants were reduced in their environments, and reproduction and survival improved, populations rebounded (Holm et al. 2003, Bierregaard et al. 2014, Eakle et al. 2015). We distinguish demographically driven limitation from habitat limitation, because the two kinds of problems require different conservation interventions, so confounding one with the other can result in ineffective management. Much of the conservation for the most intensively managed plovers has been aimed at increasing reproductive output (Hecht and Melvin 2009). However, we believe the evidence indicates that many species generally are limited by the amount of breeding habitat available (see the Piping Plover case study, this chapter).

The best evidence that plover populations are habitat limited is the population irruptions that occur when habitat is created or improved. For example, numbers of breeding Long-billed Plovers increased after vegetation was removed from a river gravel bar and the bar was covered with sand and gravel (Katayama et al. 2010). Similarly, a Little Ringed Plover population increased when a channelized river was restored to a braided condition (Arlettaz et al. 2011). Other authors have suggested habitat limitation for Red-breasted Dotterel (Ogden and Dowding 2013) and Little Ringed Plover (Parrinder 1989) based on constant population numbers. See also the Piping Plover case study, this chapter.

The factors determining the carrying capacity for many plover species are poorly understood. However, it seems likely that, for many, carrying capacity is determined by food supply and the size of the territories defended (Newton 1998). For example, the St. Helena Plover inhabits a single, isolated island approximately 120 km^2, where they use only a quarter of that area (McCulloch 2009). The population recently decreased by more than 43% over 5 years associated with the loss of their preferred grassland nesting habitat through a reduction in cattle grazing that had maintained short grass heights. Densities are positively correlated with their invertebrate prey and negatively associated with vegetation height (McCulloch 1991), suggesting that their distribution and carrying capacity are determined by food and predation risk from introduced mammalian predators, but further study of the relative impacts of habitat and predation are warranted (McCulloch 2009).

The Piping Plover: A Case Study of Breeding Habitat Limitation

Piping Plovers breed in three areas of North America: the north Atlantic Coast, where they nest primarily on barrier islands; the Great Lakes, where they breed on lakeshores; and the Great Plains, where they nest on alkali wetlands and sandbars in large rivers (Elliott-Smith and Haig 2004). The species suffered a decline driven by unprecedented hunting from the mid-19th century to the early 20th century (Forbush 1913). A second decline in the late 20th century was attributed to habitat loss and degradation, predation on nests and chicks, and human disturbance (USFWS 2009), which led to the species' listing under the United States Endangered Species Act. The listing stimulated research to understand Piping Plover demography and habitat ecology (reviewed in Catlin et al. 2015), making it one of the best-studied of the plovers.

Piping Plovers breed on substrates with materials that vary from fine sand to mixtures of sand, gravel, cobble, and shell fragments (Cairns 1982, Patterson 1988, Flemming et al. 1992, Espie et al. 1996, Cohen et al. 2009). They select unvegetated or sparsely vegetated nesting habitat (Maslo et al. 2011, Anteau et al. 2012) on relatively wide beaches (approximately 27–150 m, Patterson 1988, Prindiville-Gaines and Ryan 1988, Gieder et al. 2014). As for other plovers, selection of these habitats appears to reduce the risk of predation (this chapter; Chapter Six, this volume).

Nest sites often are clumped near productive foraging areas (e.g., Patterson et al. 1991, Cohen et al. 2009). Adults typically arrive on breeding sites in early spring when the weather may be inhospitable, making access to consistently high-quality foraging habitat essential. There is an advantage to arriving early, as early arrival affords a greater likelihood of reestablishing a prior territory and pairing with the previous year's mate (Friedrich et al. 2015), a pattern seen in other shorebirds (Lank and Oring 1982, Flynn et al. 1999). Moreover, earlier hatched chicks have a higher survival rate than later chicks (Roche et al. 2008, Saunders et al. 2014, Catlin et al. 2015).

On barrier islands, bayside intertidal habitats and ephemeral pools provide more food than other habitats (Elias et al. 2000, Cohen et al. 2009). The bayside flats are particularly important for early arrivers who are on their breeding sites before dipterans and other flying arthropods are available (Cohen et al. 2009) and plovers use them extensively early in the season (Fraser et al. 2005, Cohen and Fraser 2010). Piping Plover nests on Assateague Island, Maryland, were clumped on either end of the island, close to bayside intertidal flats, ephemeral pools on the beach, and in the center of the island adjacent to a large managed wetland (Patterson et al. 1991, Loegering and Fraser 1995).

Choosing nesting habitats with adjacent foraging opportunities that were accessible to flightless chicks was adaptive. Bayside tidal flats, ephemeral pools, and protected sandbar edges typically supported more arthropods than other habitats. Moreover, plovers had higher foraging rates, and, in general, higher chick growth rates and higher survival in these habitats (Loegering and Fraser 1995, Goldin and Regosin 1998, Elias et al. 2000, Cohen et al. 2009). After chicks fledged, adults and young spent considerable time foraging in these habitats (Fraser et al., unpublished data). Chicks weigh approximately 34 g when they begin to fly (Catlin et al. 2013, Catlin et al. 2014) compared to average masses of 52 g for nesting adults (Catlin et al. 2014). Fledgling chicks may, therefore, increase their weight by as much as 50% while foraging in these or similar nearby habitats.

The primary evidence that Piping Plovers are habitat limited is that, when habitat was improved or new habitat was created, local populations irrupted. In the examples below, local populations more than doubled in 2–9 years. During these local population increases, regional populations also grew, indicating that the local gains were not caused by adjacent adult breeders shifting short distances to the new habitat without replacement. Substantial increases in both the local and regional populations in response to habitat creation and/or improvement (Wilcox 1959, Cohen et al. 2009, Catlin et al. 2015, Hunt 2016), suggest that population regulation acts through habitat limitation. We discuss the changes in vital rates that regulate Piping Plover populations below.

The first recorded irruption occurred after the Great New England Hurricane of 1938. This storm cleared vegetation, roads, and buildings from Westhampton Island, New York, creating uninterrupted sand from the Atlantic Ocean to Moriches Bay (Figure 9.3). Following the hurricane, the local population tripled in 2 years (Figure 9.5a), but later decreased. Wilcox (1959) hypothesized that the decrease was caused by invasion of native beach grass (*Ammophila breviligulata*) on dredge-created dunes that reduced the suitability of the habitat.

In the same location, winter storms of 1992–1993 breached Westhampton Island again. The breach was filled by the U.S. Army Corps of Engineers,

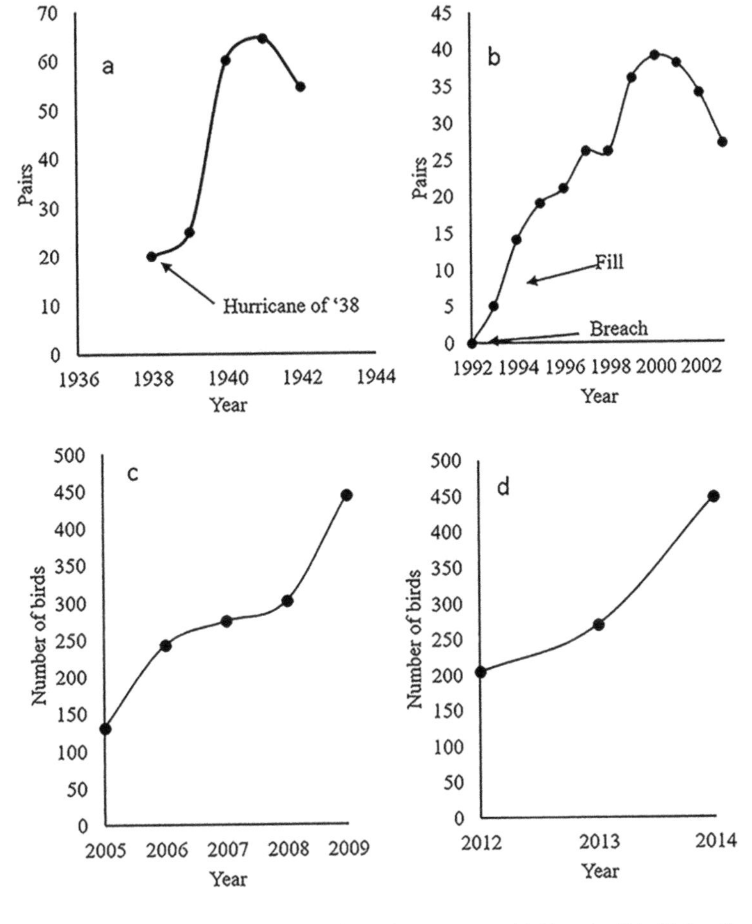

Figure 9.5. Piping Plover irruptive population growth after habitat creation. (a) Growth of the Piping Plover population on Westhampton Island, New York, after habitat changes caused by the great Hurricane of 1938. (Figure created from data in Wilcox 1959.) (b) Piping Plover population at West Hampton Dunes, New York. The population was established following a breach of the barrier island that subsequently was filled by the U.S. Army Corps of Engineers. (From Cohen et al. (2009).) (c) Piping Plover population on the Gavins Point Reach and Lewis and Clark Reservoir on the Missouri River (2005–2009). Habitat (240 ha) was created by the U.S. Army Corps of Engineers from 2004 to 2008. (Data from Catlin et al. (2015) and unpublished data.) (d) Piping Plover population on the Gavins Point Reach and Lewis and Clark Lake on the Missouri River, following habitat creating floods in 2010 and 2011. (Data from Catlin et al., unpublished.)

which created wide beaches with no vegetation or buildings between ocean and bay. The population grew from 0 pairs in 1992 to 39 pairs in 1999 (Figure 9.5; Cohen et al. 2009). The growth was not merely a response to increased nesting habitat, but also to new, productive foraging areas. Nesting densities reached 1.05 pairs/ha near these habitats, compared to a 13-year average of 0.44 pairs/ha on adjacent beaches that lacked productive foraging habitats. The population declined when the quality of bayside habitat declined due to vegetation encroachment (Cohen et al. 2009).

Other examples provide additional evidence that habitat area and quality limit plover populations. On the Missouri River, the U.S. Army Corps of Engineers built about 240 ha of sandbars from 2005 to 2008 (Catlin et al. 2015). Plovers colonized the sandbars and the population on a segment of the river grew from 132 to 443 individuals (Figure 9.5c). In 2010 and 2011, the Missouri River flooded, depositing sand throughout the system, creating nearly 800 ha of additional habitat during the 2012 breeding season. Plover numbers along this stretch of river more than doubled in 3 years (Figure 9.5d).

During population irruptions, studies with marked birds show that new recruits to the population often settled in sites unoccupied by competitors. Because many new recruits are yearlings, this probably is a mechanism for avoiding competition with more experienced birds for a breeding territory (Haig and Oring 1988, Friedrich et al. 2015). On average, yearlings arrive later than older birds (Catlin et al. 2015), which is likely an adaptation that prevents them from spending energy competing with older, more experienced adults for territories. At the start of irruptions (when density is low), yearling philopatry is high (Catlin et al. 2015). The site at which a bird settles may be influenced by prospecting that occurred during fledgling dispersal (Davis et al. 2017) or following nest failure for adults (Rioux et al. 2011). On the Missouri River, Piping Plover yearlings selected sandbars that they prospected as fledglings (Davis et al. 2017). These sites had higher average nest success and lower density per unit of foraging habitat than other randomly selected sandbars, suggesting that they were able to glean important information about habitat quality to influence their subsequent choice of a breeding site.

Determining the mechanisms driving population regulation is challenging (Newton 1994, 1998). However, during the irruptions on the Missouri River, most Piping Plovers were individually marked, facilitating detailed analyses of demographic rates from 2005 to 2014. The changes in abundance were driven by density-dependent chick survival (Hunt 2016), density-dependent recruitment (some immigration but primarliy local reproduction, Catlin et al. 2015), and emigration (Catlin et al. 2016). Although some birds immigrated from other populations, rates were low (<5%, Catlin et al. 2016). The increases were largely driven by high reproductive output (>2 chicks fledged per female), by double-brooding (production of a second clutch after a successful first attempt), and by high chick survival when density was low. In addition, yearling philopatry was high, resulting in increased recruitment (Catlin et al. 2015, Hunt et al. 2015, Hunt 2016). As irrupting Piping Plover populations approach carrying capacity, it is likely that the high densities lead to competition, forcing young birds to establish territories elsewhere. Moreover, as density increases, density-dependent predation and starvation of hatchlings lowers reproductive output (Catlin et al. 2015, Hunt 2016).

These examples are not unique to Piping Plovers (see above), but they represent well-studied illustrations. The evidence suggests that population irruptions following habitat creation events was a natural part of Piping Plover population ecology, and that Piping Plover populations are usually at or close to the carrying capacity of their habitats.

HABITAT LOSS AND DEGRADATION

Plover habitats are being lost or degraded throughout the world. The number of hectares lost each year is unknown. Below we provide some indication of the mechanisms of loss and, qualitatively, the magnitude of that loss.

Coastal Habitats

About two-thirds of humans live within 60 km of the ocean (UN Atlas of Oceans 2016), and coastal areas are being converted to housing, industrial sites, and recreation areas. This habitat conversion results in destruction of plover breeding, foraging, and roosting habitats.

Barrier Islands

Barrier Islands are long, low, sandy islands, oriented parallel to the coastline and are largely unstable and subject to erosion and movement. They are rare, associated with only 6.5% of the world's open ocean coastline (Stutz and Pilkey 2001). On the Atlantic and Gulf coasts of the USA, 156 barrier islands extend 3,605 km from New England to the Mexican border. These islands amount to 10.4% of the world's barrier islands, and 23.9% of the world's total barrier island length (Stutz and Pilkey 2001). They provide key plover habitats on the North American Atlantic Flyway.

Atlantic and Gulf barrier islands have been extensively manipulated, with beach nourishment, dune building, and construction of seawalls, groins, and jetties. More than 6,000 km of beach-front was treated from 1922 to 2003 on the U.S. Atlantic and Gulf shores (Peterson and Bishop 2005). This is nearly equivalent to the coastline distance from Maine to Mexico and back again. A typical beach nourishment/dune building project might include a dune 4–5 m above sea level with a backshore extending 30 m between the dune and the high tide line. This often is achieved by dredging sand from the sea floor and placing it in the treatment area. Beach nourishment kills infauna, and while these organisms may recover in a year (Schlacher et al. 2014), in the interim, plovers may have a reduced prey base.

The direct mortality of infauna is not the greatest impact of beach engineering on plover habitat. Two key barrier island plover habitats are bayside intertidal flats and flood shoals, where plovers forage. Bayside flats form when waves overwash barrier islands, flattening the islands and depositing sand in the bay (Figure 9.3). Flood shoals occur when storm waves breach islands, and sediment is deposited by the flood tide (Figures 9.4 and 9.6). Widening the backshore and building dunes prevent these habitat-forming events.

Coasts of East and Southeast Asia

Wetlands on the coasts of East Asia and Southeast Asia are being lost at an alarming rate. In a 50-year period, more than 40% of wetlands were lost in China, Japan, the Republic of Korea, and Singapore (An et al. 2007, Yee et al. 2010). In Thailand, the Malay Plover is threatened by loss of beach habitat (Yasué et al. 2007). By one estimate, 1.19 million ha of China's coastal tidal flats and 51% of coastal wetlands have been lost (MacKinnon et al. 2012). The main cause of loss is the conversion of intertidal habitats ("land claiming") for agriculture, aquaculture, salt production, construction of ports and other urban development, and tidal energy projects (MacKinnon et al. 2012, Melville et al. 2016). In addition, river damming has trapped sand and silt that otherwise would have been washed to the sea to replenish intertidal flats (MacKinnon et al. 2012). In some cases, substantial tidal flats still exist seaward of seawalls, but the walls have been built so far downslope that the flats are exposed for a short interval of the tidal cycle, restricting feeding time for birds. The development in this area has been linked to the decline of the Lesser Sand-plover and six other shorebird species (Studds et al. 2017). This problem will be exacerbated as sea levels rise (Galbraith et al. 2002, Galbraith et al. 2014). For example, 23%–40% of the intertidal habitat used by migrating shorebirds in the East Asian-Australasian Flyway will be inundated by rising seas (Iwamura et al. 2013).

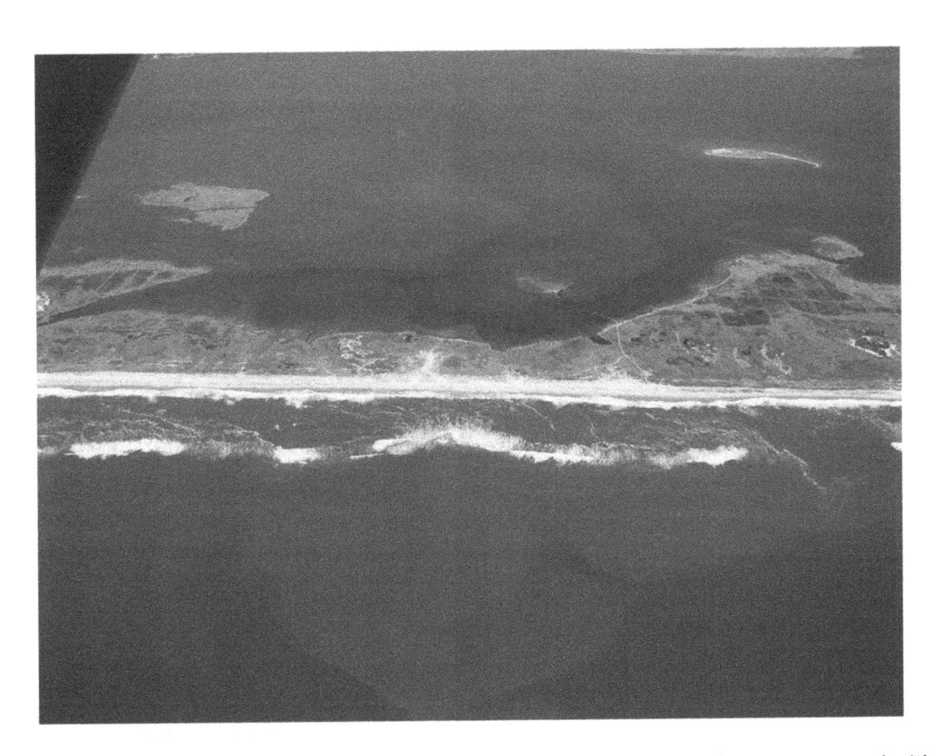

Figure 9.6. View of Old Inlet, Fire Island, New York, USA before the breach in April 2005 from Hurricane Sandy. (Photo by Charlie Flagg.)

Rivers and River Sandbars

Many plovers breed and forage on river sandbars. Reservoirs created by dams inundate these sandbars. Moreover, river sandbars are renewed when high waters deposit sand. Thus, when dams reduce flooding, sandbars erode and become vegetated, decreasing the area suitable for plovers.

More than 58% of the world's river systems have been dammed, some of them in multiple locations (Nilsson et al. 2005). Dams and sandbar degradation have affected Piping Plovers on the Missouri and Platte rivers (Catlin et al. 2015, Catlin et al. 2016, Hunt 2016), and likely Snowy Plovers and Killdeer that co-occurred with them. Long-billed Plover habitat was reduced by damming of the Tama River in Japan (Katayama et al. 2010). Similarly, populations of the Little Ringed Plover were reduced by channelization of the Rhone River in Switzerland (Arlettaz et al. 2011). Damming and channelization are expected to negatively affect the Little Ringed Plovers on the Mekong, Sekong, and Sesan Rivers, Cambodia (A. Claassen, University of Minnesota, pers. comm.) and the Double-banded Plover on the South Island of New Zealand (Maloney 1999). The flow of the Kuiseb River in Namibia has been so compromised by dams that it now rarely reaches the ocean so that less sediment is available to create plover habitat (Wearne and Underhill 2005).

In addition to habitat loss associated with damming rivers and streams, foraging opportunities for plovers may be compromised by dams. On the Missouri River, United States, prey abundance and biomass were lower downstream of a dam with a hypolimnetic release (cold water released from deep below the surface) than downstream of a dam releasing warmer water (Le Fer et al. 2008a).

Temperate Lakeshores

As for all other habitats, it is unknown how much lakeside plover habitat has been lost. However, the loss probably is correlated with the loss of freshwater wetlands, estimated at 60% of historic coverage worldwide (Davidson 2014).

The Prairie Pothole Region in the USA encompasses some 715,000 km^2 of mid- and tall-grass prairies. The potholes are small lakes created during glacial retreat. Piping Plovers, Snowy Plovers, and Killdeer nest on these lake shores, and Semipalmated Plovers use them on migration. Some 50% of these potholes have been drained to create more arable land (Euliss et al. 1999). Wetland loss from the 1780s to 1980s in Montana, North Dakota, South Dakota, and Nebraska, USA, ranged from 27% to 49% (Dahl 1990), though those losses have slowed and may have begun to reverse recently (Dahl 2005, 2011).

China lost 16% of freshwater lakes in the 50 years prior to 2007, much of this due to conversion for agriculture (An et al. 2007). Parts of the larger lakes, like Poyang and Dongting, also were converted (Wang et al. 2016). Fortunately, the Chinese government has halted land claiming in the Lower and Middle Yangtze flood plain, but there is a new threat to shorebird habitat there. Wetlands are being reclaimed to plant the Italian poplar (*Populous euramericana*), which is used for pulp and fiberboard (Wang et al. 2016). In semiarid northern China, lakes have decreased in size or disappeared due to climate change. Decreased precipitation seems to have played the major role here, but increasing temperatures and human water use may also have contributed (Wang et al. 2016).

Grasslands

In the last 300 years, 49% of the original grasslands of the world have been transformed into agricultural land; as of 1990, ~17.5 million km^2 of grassland remained (Goldewijk 2001). While perhaps 60% of these lands were converted from native grasslands to pasture, which may retain some value for plovers, this still represents a large loss in grasslands worldwide.

Tundra

Perhaps the greatest threat to tundra-breeding plovers is loss of the open habitat they require. Numerous studies have indicated that tundra habitats are being invaded by shrubs as the earth continues to warm. In the Chugach Mountains of Alaska, for example, tundra vegetation is retreating upslope at 1.2 m/year, while shrubs are gaining in elevation at 1.3 m/year (Dial et al. 2016). Such changes have been observed around the Arctic (Myers-Smith et al. 2011). We have not found studies of the response of plovers to these changes. However, Ballantyne and Nol (2015) showed a decline in breeding Whimbrels (*Numenius phaeopus*) near Churchill, Manitoba,

Canada, coincident with an increase in tree and shrub cover.

CONSERVATION IMPLICATIONS

Where plovers are habitat limited, loss of habitat or degradation that leads to lowered carrying capacity will reduce population numbers. Where the physical habitat is intact, human use may compromise the habitat quality or preclude use by plovers (Chapter Eleven, this volume). The population-level effects of consistently precluding plovers from an area are equivalent to habitat loss, though potentially reversible. When habitat-limited species are below carrying capacity for a time, population management actions that increase survival or reproduction may speed population recovery to carrying capacity, but will not result in the population surpassing that level for a significant length of time, as density-dependent forces will intervene (Newton 1998).

Anthropogenic habitat loss will continue (Nilsson et al. 2005, Peterson and Bishop 2005, Myers-Smith et al. 2011, MacKinnon et al. 2012, Davidson 2014). Local action will be required to protect habitats where possible, and protective efforts will need to be tailored to local political and social systems. The least expensive way to protect habitats will, in many cases, be to allow natural processes, such as river flows and coastal overwashes, to maintain the early successional landscapes needed by plovers. However, conservation externalities, such as destruction of human infrastructure, will be undesirable to some. For some species, features other than the physical habitat, such as disturbance (Chapter Eleven, this volume), or the presence of predators (Pomeroy 2006, Pomeroy et al. 2006) can prevent birds from using otherwise suitable habitat.

Some species have such limited habitat that they will be of conservation concern for the foreseeable future. The St. Helena Plover, for example, only regularly uses a 30 km^2 area (McCulloch 2009), and that area may be shrinking due to development and reductions in grazing that maintains the sparse vegetation they prefer (McCulloch and Norris 2001, McCulloch 2009). The Shore Plover, a New Zealand endemic, was extirpated from the South Island in the 19th century, probably by introduced predators, but remnant populations persisted on four of the Chatham Islands. In 2013, approximately 63 breeding pairs existed on five islands (Davis 1994, Dowding and O'Connor 2013). The entire area of these islands is only about 2.2 km^2 and likely only a small portion of this is suitable breeding habitat. The Madagascar Plover also has limited habitat; although Madagascar is large (>580,000 km^2), only about 139 km^2 is suitable habitat (Long et al. 2008), and demographic modeling of this species suggested that low reproductive output could lead to species extinction (Zefania et al. 2008).

For most species, habitat loss is a continuing, chronic threat. In addition to habitat destruction by conversion for human uses, for many species, habitat may be threatened by climate change and rising sea levels (Galbraith et al. 2002, 2014; Chapter Three, this volume). Increasing uncertainty in future scenarios suggests that we need to adopt methods that account for a changing climate and uncertainty when modeling habitat (e.g., Gieder et al. 2014) and the demographic interactions with habitat (e.g., Zeigler et al. 2017).

HABITAT RESTORATION

Because of shorebird habitat declines due to anthropogenic causes, and their links to declines in shorebird populations throughout the globe, habitat restoration has been proposed as a means to protect populations (e.g., Davidson and Evans 1987, Erwin 1996, Atkinson 2003, Caruso 2006). Shorebirds often use human-created or modified habitats such as sand and gravel mines (Davidson and Evans 1987, DeVault et al. 2002, Catlin et al. 2016), impoundments (Erwin 1996), grazed and burned pastures (Fletcher and Koford 2003, Augustine 2011, Augustine and Derner 2012), and dredge-spoil islands for feeding and nesting (Powell and Collier 2000, Collis et al. 2001, Erwin et al. 2003, Spear et al. 2007, Catlin et al. 2011). These observations suggest that restoration of habitats could be successful.

Plovers readily use human-placed substrates for breeding and foraging (Powell and Collier 2000). Sometimes sand is available as a by-product of dredging (Yozzo et al. 2004, Guilfoyle et al. 2006), but in other cases it has been dredged for habitat creation. Snowy Plover nesting at a lagoon in southern California increased following restoration with dredge material for California Least Terns (*Sternula antillarum browni*), and reproductive output was initially higher at these sites than in natural habitat (Powell and Collier 2000).

The same pattern occurred on created islands in the Missouri River, where Piping Plovers selected newly created habitat, though reproductive output declined after 2 years (Catlin et al. 2011, Catlin et al. 2015). Both studies pointed to vegetation encroachment and discovery by predators as likely causes of reproductive declines (Powell and Collier 2000, Catlin et al. 2015).

Vegetation management to create open areas favored by plovers may be useful (Katayama et al. 2010, Augustine 2011, Catlin et al. 2011, Augustine and Derner 2012). Long-billed Plovers used gravel bars where vegetation was removed (Katayama et al. 2010), but Piping Plovers avoided habitat where vegetation was removed (Catlin et al. 2011). Snowy Plover's nest survival more than doubled when vegetation was removed from sites, but this did not lead to increases in fledging success and perhaps not to population growth (Dinsmore et al. 2014). Mountain Plovers breed in a variety of sparsely vegetated areas such as prairie dog towns and burned and grazed pasture (Knopf and Wunder 2006), but they bred in higher densities in the burned areas and prairie dog towns than in grazed habitats (Augustine 2011, Augustine and Derner 2012). Thus, burning could be effective for habitat restoration, but plover densities declined within a few years of burning (Augustine and Skagen 2014). For this and other systems where vegetation is removed, removal may have to be repeated (Powell and Collier 2000, Erwin et al. 2003, Catlin et al. 2015).

Given that it can be difficult to create high-quality habitat, particularly in the absence of detailed habitat studies (Atkinson 2003), it is imperative that we work to protect habitat where we can (Davidson and Evans 1987), while pushing for ecosystem-level restoration (Arlettaz et al. 2011, Hunt 2016). For example, after restoration of the Rhone River in Switzerland, habitat diversity and natural function increased which led to a long-term increase in Little Ringed Plovers and Common Sandpipers (*Actitis hypoleucos*; Arlettaz et al. 2011). Efforts to create sandbar habitat for Piping Plovers on the Missouri River were only moderately successful because reproductive success plummeted soon after creation (Catlin et al. 2015). In contrast, a system-wide flood that mimicked pre-dam conditions increased reproductive output and population growth for several years (Hunt 2016). The USFWS (1996) has long called for restoration of ecosystem processes (barrier island overwash) to create and maintain habitat for Piping Plovers, but that has been met with resistance because of concerns about infrastructure protection. As a substitute, USFWS (2014) recently required habitat restoration and maintenance as part of a coastal stabilization project. Long-term plover conservation will likely require protection, allowing natural processes to continue where feasible, and local and ecosystem restoration where natural processes damage infrastructure. Success will depend on an expansion of the literature that critically addresses the effects of habitat restoration on plovers.

Evaluating Habitat Restoration Projects

The effectiveness of restoration projects should be evaluated (Wortley et al. 2013). Because some kinds of plover habitat restoration can be very expensive, it is important that past success guide future conservation efforts. Here we offer a few guidelines for restoration evaluation.

To be properly evaluated, the objectives of the project must be clearly stated during the planning state (Doran 1981), and these objectives need to be expressed to managers, politicians, and engineers (Atkinson 2003). The primary objectives of restoration projects should be stated in terms of number of birds added or population growth. There is a temptation to limit the objectives to statistics about the habitat (e.g., number of ha of vegetation removed). While such objectives are useful in evaluating the manipulation itself, they do not address the main point—increasing population size. Thus, we recommend that evaluation of a project includes monitoring changes in target species population and selection of new habitat compared to selection of preexisting habitat. Moreover, it is important to evaluate the fitness of birds using restored versus preexisting habitat by comparing a suite of vital rates (e.g., nest success, chick survival, reproductive output, adult survival, immigration, and emigration), and these rates should be evaluated over time to highlight temporal variation in the results (Powell and Collier 2000). The ultimate test of the usefulness of a habitat restoration project will be in population growth, as increases in one or more vital rate may not lead to growth (Dinsmore et al. 2014, Catlin et al. 2015, Hunt 2016). Monitoring food resources within habitats, and sources of mortality, often can help to explain the vital rates observed. Many plover habitats are ephemeral being lost to vegetative succession or erosion. Monitoring programs should continue throughout the life cycle of the habitat created. Therefore, if vegetation is removed that can completely regrow in, say, 7 years, monitoring should be ongoing for at least that long.

THE FUTURE

Conservation of plovers will require multidimensional strategies that include both protecting and restoring habitats. Planning these strategies will be complicated by climate change and sea level rise.

Opportunity abounds for biologists wishing to expand our understanding of plovers and their habitat use. The specificity with which habitat use is described in the literature varies widely across species reflecting the detail and focus of research efforts on individual species. For most species, habitats have only been described qualitatively. Habitat studies evaluating selection, except at the microhabitat level for nest sites, are rare. Such studies are needed to explain why some habitats that appear similar to occupied habitats remain unoccupied by plovers. For example, Ehmke et al. (2016) cite the case of Ninety Mile Beach in Victoria, Australia, one of the longest contiguous beaches in the world, which appears suitable for the Hooded Plover, but which is uninhabited by that species. While this may be an extreme case, plover biologists in other places are called upon to explain how plovers could possibly be habitat limited when there is so much apparently suitable habitat available. We believe the solution to this problem is to be found in more detailed, statistical studies of habitat selection. It will be hard to make the case for habitat conservation without specific understanding of the habitat needs of each species.

Much work needs to be done on the breeding habitat of many species, but resource selection studies are even less common for nonbreeding plovers. Habitats used by fledglings before departing for the wintering areas are less well-studied, though research on Piping Plovers suggests they are prospecting for future territories (Davis et al. 2017). Evidence to date shows that the quality of chick foraging habitat can have lifelong effects on the fitness of chicks raised in different habitats (Catlin et al. 2014). Thus, there is a need to learn how food supplies differ across brood-rearing habitats of different species and how this affects chick fitness. For many species, winter food resources have not been studied, and the question of roost-site selection is relatively untouched. In addition to simply studying resource selection, it will be paramount to link habitats with vital rates and ultimately to the carrying capacity of various habitats.

Fortunately, the toolbox has never been better equipped for this work. Imagery that can be used to assess habitat now is available at very high resolution. For the biologist who hopes to study plovers in the wild, some sensors can be mounted on lightweight drones. Such high-resolution images allow for detailed studies of habitat use and availability at a scale meaningful to a plover. Complementary to that, new methods of calculating population vital statistics are available to link reproductive output, survival, immigration, and emigration to habitat types (e.g., Catlin et al. 2015, McCaffery and Lukacs 2016; Chapter Eleven, this volume).

Habitat conservation will remain a fundamental challenge in plover conservation. While it is important to maximize survival and reproduction, this will not allow plover populations to exceed the carrying capacity of their habitat. Given competing human desires for development and recreation, it will be imperative for plover conservationists to articulate exactly what habitat is needed, when, and why.

LITERATURE CITED

Amat, J. A. and J. A. Masero. 2004. Predation risk on incubating adults constrains the choice of thermally favourable nest sites in a plover. *Animal Behaviour* 67:293–300.

An, S. Q., H. B. Li, B. H. Guan, C. F. Zhou, Z. S. Wang, Z. F. Deng, Y. B. Zhi, Y. H. Liu, C. Xu, S. B. Fang, J. H. Jiang, and H. L. Li. 2007. China's natural wetlands: Past problems, current status, and future challenges. *Ambio* 36:335–342.

Anteau, M. J., M. H. Sherfy, and M. T. Wiltermuth. 2012. Selection indicates preference in diverse habitats: A ground-nesting bird (*Charadrius melodus*) using reservoir shoreline. *Plos One* 7:e30347.

Arlettaz, R., A. Ligon, A. Sierro, P. Werner, M. Kery, and P. A. Oggier. 2011. River bed restoration boosts habitat mosaics and the demography of two rare non-aquatic vertebrates. *Biological Conservation* 144:s2126–2132.

Atkinson, P. W. 2003. Can we recreate or restore intertidal habitats for shorebirds? *Wader Study Group Bulletin* 100:67–72.

Augustine, D. J. 2011. Habitat selection by Mountain plovers in shortgrass steppe. *Journal of Wildlife Management* 75:297–304.

Augustine, D. J. and J. D. Derner. 2012. Disturbance regimes and Mountain plover habitat in shortgrass steppe: Large herbivore grazing does not substitute for prairie dog grazing or fire. *Journal of Wildlife Management* 76:721–728.

Augustine, D. J. and S. K. Skagen. 2014. Mountain plover nest survival in relation to prairie dog and fire dynamics in shortgrass steppe. *Journal of Wildlife Management* 78:595–602.

Ballantyne, K. and E. Nol. 2015. Localized habitat change near Churchill, Manitoba and the decline of nesting Whimbrels (*Numenius phaeopus*). *Polar Biology* 38:529–537.

Battin, J. 2004. When good animals love bad habitats: Ecological traps and the conservation of animal populations. *Conservation Biology* 18:1482–1491.

Bent, A. C. 1929. Life Histories of North American Shorebirds, part 2. Smithsonian Institution United States National Museum Bulletin 146.

Bierregaard, R. O., A. Ben David, L. Gibson, R. S. Kennedy, A. F. Poole, M. S. Scheibel, and J. Victoria. 2014. Post-ddt recovery of Osprey (*Pandion Haliaetus*) populations in southern New England and Long Island, New York, 1970–2013. *Journal of Raptor Research* 48:361–374.

Bock, A. 2014. Characteristics of Sandy Beaches Used by Resident Shorebirds in Tasmania. M.S. Thesis, University of Akureyi, Iceland.

Bock, A., M. R. Phillips, and E. Woehler. 2016. The role of beach and wave characteristics in determining suitable habitat for three resident shorebird species in Tasmania. *Journal of Coastal Research* 75:358–362.

Brindock, K. M. and M. A. Colwell. 2011. Habitat selection by western Snowy plovers during the nonbreeding season. *Journal of Wildlife Management* 75:786–793.

Cairns, W. E. 1982. Biology and behavior of breeding piping plovers. *Wilson Bulletin* 94:531–545.

Carstairs, D. N. 1989. The status of the rufous-chested dotterel *Zonibyx modestus* in the Falkland Islands. *Bulletin of the British Ornithologist's Club* 109:166–170.

Caruso, B. S. 2006. Effectiveness of braided, gravel-bed river restoration in the Upper Waitaki Basin, New Zealand. *River Research and Applications* 22:905–922.

Castro, M., J. A. Masero, A. Perez-Hurtado, J. A. Amat, and C. Megina. 2009. Sex-related seasonal differences in the foraging strategy of the Kentish plover. *Condor* 111:624–632.

Catlin, D. H., J. H. Felio, and J. D. Fraser. 2013. Effects of water discharge on fledging time, growth, and survival of piping plovers on the Missouri River. *Journal of Wildlife Management* 77:525–533.

Catlin, D. H., J. D. Fraser, and J. H. Felio. 2015. Demographic responses of piping plovers to habitat creation on the Missouri River. *Wildlife Monographs* 192:1–42.

Catlin, D. H., J. D. Fraser, J. H. Felio, and J. B. Cohen. 2011. Piping plover habitat selection and nest success on natural, managed, and engineered sandbars. *Journal of Wildlife Management* 75:305–310.

Catlin, D. H., O. Milenkaya, K. L. Hunt, M. J. Friedrich, and J. D. Fraser. 2014. Can river management improve the Piping plover's long-term survival on the Missouri River? *Biological Conservation* 180:196–205.

Catlin, D. H., S. L. Zeigler, M. B. Brown, L. R. Dinan, J. D. Fraser, K. L. Hunt, and J. G. Jorgensen. 2016. Metapopulation viability of an endangered shorebird depends on dispersal and human-created habitats: Piping plovers (*Charadrius melodus*) and prairie rivers. *Movement Ecology* 4:6.

Cohen, J. B. and J. D. Fraser. 2010. Piping plover foraging distribution and prey abundance in the pre-laying period. *Wilson Journal of Ornithology* 122:578–582.

Cohen, J., L. Houghton, and J. Fraser. 2009. Nesting density and reproductive success of piping plovers in response to storm- and human-created habitat changes. *Wildlife Monographs* 173:1–24.

Cohen, J. B., S. M. Karpanty, D. H. Catlin, J. D. Fraser, and R. A. Fischer. 2008. Winter ecology of piping plovers at Oregon Inlet, North Carolina. *Waterbirds* 31:472–479.

Collis, K., D. D. Roby, D. P. Craig, B. A. Ryan, and R. D. Ledgerwood. 2001. Colonial waterbird predation on juvenile salmonids tagged with passive integrated transponders in the Columbia River estuary: Vulnerability of different salmonid species, stocks, and rearing types. *Transactions of the American Fisheries Society* 130:385–396.

Colwell, M. A., J. J. Meyer, M. A. Hardy, S. E. Mcallister, A. N. Transou, R. R. Levalley, and S. J. Dinsmore. 2011. Western Snowy plovers *Charadrius alexandrinus nivosus* select nesting substrates that enhance egg crypsis and improve nest survival. *Ibis* 153:303–311.

Conklin, J. R., M. A. Colwell, and N. W. Fox-Fernandez. 2008. High variation in roost use by Dunlin wintering in California: Implications for habitat limitation. *Bird Conservation International* 18:275–291.

Corbat, C. A. and P. W. Bergstrom. 2000. Wilson's Plover (*Charadrius wilsonia*) in Rodewald, P. G. (ed.), *The Birds of North America Online*. Cornell Lab of Ornithology: Ithaca, NY.

Cramp, S. and K. E. L. Simmons. 1983. *Handbook of the Birds of Europe, the Middle East, and North Africa: The Birds of the Western Palearctic, Volume 3: Waders to Gulls*. Oxford University Press: New York.

Cresswell, W. 1994. Flocking is an effective anti-predation strategy in redshanks, *Tringa totanus*. *Animal Behaviour* 47:433–442.

Cresswell, W. and D. P. Whitfield. 2008. How starvation risk in Redshanks *Tringa totanus* results in predation mortality from Sparrowhawks *Accipiter nisus*. *Ibis* 150:209–218.

Cresswell, W. and J. L. Quinn. 2011. Predicting the optimal prey group size from predator hunting behaviour. *Journal of Animal Ecology* 80:310–319.

Cuttriss, A., G. S. Maguire, G. Ehmke, and M. A. Weston. 2015. Breeding habitat selection in an obligate beach bird: A test of the food resource hypothesis. *Marine and Freshwater Research* 66:841–846.

Dahl, T. E. 1990. *Wetland Losses in the United States 1780s to 1980s*. U.S. Fish and Wildlife Service: Washington, DC.

Dahl, T. E. 2005. *Status and Trends of Wetlands in the Conterminus United States from 1998 to 2004*. U.S. Fish and Widlife Service: Washington, DC.

Dahl, T. E. 2011. *Status and Trends of Wetlands in the Conterminus United States from 2004 to 2009*. U.S. Fish and Wildlife Service: Washington, DC.

Davidson, N. C. 2014. How much wetland has the world lost? Long-term and recent trends in global wetland area. *Marine and Freshwater Research* 65:934–941.

Davidson, N. C. and P. R. Evans. 1987. Habitat restoration and creation: Its role and potential in the conservation of waders. *Wader Study Group Bulletin* 49:139–145.

Davis, A. 1994. Breeding biology of the New Zealand Shore plover *Thinornis novaeseelandiae*. *Notornis* 41:195–208.

Davis, K. L., K. L. Schoenemann, D. H. Catlin, K. L. Hunt, M. J. Friedrich, S. J. Ritter, J. D. Fraser, and S. M. Karpanty. 2017. Hatch-year Piping plover (*Charadrius melodus*) prospecting and habitat quality influence second-year nest site selection. *Auk* 134:92–103.

del Hoyo, J., A. Elliot, and J. Sargatal. 1996. *Handbook of the Birds of the World, Volume 3: Hoatzin to Auks*. Lynx Edicions: Barcelona, Spain.

DeVault, T. L., P. E. Scott, R. A. Bajema, and S. L. Lima. 2002. Breeding bird communities of reclaimed coal-mine grasslands in the American midwest. *Journal of Field Ornithology* 73:268–275.

Dial, R. J., T. S. Smeltz, P. F. Sullivan, C. L. Rinas, K. Timm, J. E. Geck, S. C. Tobin, T. S. Golden, and E. C. Berg. 2016. Shrubline but not treeline advance matches climate velocity in montane ecosystems of south-central Alaska. *Global Change Biology* 22:1841–1856.

Dias, M. P., J. P. Granadeiro, M. Lecoq, C. D. Santos, and J. M. Palmeirim. 2006. Distance to high-tide roosts constrains the use of foraging areas by Dunlins: Implications for the management of estuarine wetlands. *Biological Conservation* 131:446–452.

Dinsmore, S. J., D. J. Lauten, K. A. Castelein, E. P. Gaines, and M. A. Stern. 2014. Predator exclosures, predator removal, and habitat improvement increase nest success of Snowy plovers in Oregon, USA. *Condor* 116:619–628.

Doran, G. T. 1981. There is a S.M.A.R.T. way to write management goals and objectives. *Management* 70:35–36.

Dowding, J. E. 1999. Past distribution and decline of the New Zealand Dotterel (*Charadrius obscurus*) in the South Island of New Zealand. *Notornis* 46:167–180.

Dowding, J. E. and C. M. O'Connor. 2013. Reducing the risk of extinction of a globally threatened shorebird: Translocations of the New Zealand Shore plover (*Thinornis novaeseelandiae*), 1990–2012. *Notornis* 60:70–84.

Drake, K. R., J. E. Thompson, K. L. Drake, and C. Zonick. 2001. Movements, habitat use, and survival of nonbreeding piping plovers. *Condor* 103:259–267.

Dugan, J. E., D. M. Hubbard, M. D. McCrary, and M. O. Pierson. 2003. The response of macrofauna communities and shorebirds to macrophyte wrack subsidies on exposed sandy beaches of southern California. *Estuarine Coastal and Shelf Science* 58:25–40.

Eakle, W. L., L. Bond, M. R. Fuller, R. A. Fischer, and K. Steenhof. 2015. Wintering Bald Eagle count trends in the conterminous United States, 1986–2010. *Journal of Raptor Research* 49:259–268.

Ehmke, G., G. S. Maguire, T. Bird, D. Ierodiaconou, and M. A. Weston. 2016. An obligate beach bird selects sub-, inter- and supra-tidal habitat elements. *Estuarine Coastal and Shelf Science* 181:266–276.

Elias, S. P., J. D. Fraser, and P. A. Buckley. 2000. Piping Plover brood foraging ecology on New York barrier islands. *Journal of Wildlife Management* 64: 346–354.

Elliott-Smith, E. and S. M. Haig. 2004. Piping Plover (*Charadrius melodus*) in Poole, A. (ed.), *The Birds of North America Online*. Cornell Lab of Ornithology: Ithaca, NY.

Ellis, K. S., J. F. Cavitt, and R. T. Larsen. 2015. Factors influencing Snowy plover (*Charadrius nivosus*) nest survival at Great Salt Lake, Utah. *Waterbirds* 38:58–67.

Ellis, K. S., R. T. Larsen, R. N. Knight, and J. F. Cavitt. 2014. Occupancy and detectability of Snowy plovers in western Utah: An application to a low density population. *Journal of Field Ornithology* 85:355–363.

Erwin, R. M. 1996. Dependence of waterbirds and shorebirds on shallow-water habitats in the mid-Atlantic coastal region: An ecological profile and management recommendations. *Estuaries* 19:213–219.

Erwin, R. M., D. H. Allen, and D. Jenkins. 2003. Created versus natural coastal islands: Atlantic waterbird populations, habitat choices, and management implications. *Estuaries* 26:949–955.

Espie, R. H. M., R. M. Brigham, and P. C. James. 1996. Habitat selection and clutch fate of piping plovers (*Charadrius melodus*) breeding at Lake Diefenbaker, Saskatchewan. *Canadian Journal of Zoology-Revue Canadienne De Zoologie* 74:1069–1075.

Euliss, Jr. N. H., D. M. Mushet, and D. A. Wrubleski. 1999. Wetlands of the prairie potholde region: Invertebrate species composition, ecology, and management. In Batzer, D. P., Rader, R. B., and Wissinger, S. A. (eds), *Invertebrates in Freshwater Wetlands of North America: Ecology and Management*. John Wiley and Sons: New York.

Flemming, S. P., R. D. Chiasson, and P. J. Austinsmith. 1992. Piping plover nest site selection in New-Brunswick and Nova-Scotia. *Journal of Wildlife Management* 56:578–583.

Fletcher, R. J. and R. R. Koford. 2003. Changes in breeding bird populations with habitat restoration in northern Iowa. *American Midland Naturalist* 150:83–94.

Flynn, L., E. Nol, and Y. Zharikov. 1999. Philopatry, nest-site tenacity, and mate fidelity of Semipalmated plovers. *Journal of Avian Biology* 30:47–55.

Forbush, E. H. 1913. *Useful Birds and Their Protection*. 4th ed., Massachusetts State Board of Agriculture, Wright and Potter Printing Company, State Printers: Boston, MA.

Fraser, J. D., S. E. Keane, and P. A. Buckley. 2005. Prenesting use of intertidal habitats by piping plovers on South Monomoy Island, Massachusetts. *Journal of Wildlife Management* 69:1731–1736.

Fretwell, S. D. and H. L. Lucas. 1970. On territorial behavior and other factors influencing habitat distribution in birds. I. Theirtical development. *Acta Biotheoretica* 19:16–36.

Friedrich, M. J., K. L. Hunt, D. H. Catlin, and J. D. Fraser. 2015. The importance of site to mate choice: Mate and site fidelity in piping plovers. *Auk* 132:265–276.

Galbraith, H., R. Jones, R. Park, J. Clough, S. Herrod-Julius, B. Harrington, and G. Page. 2002. Global climate change and sea level rise: Potential losses of intertidal habitat for shorebirds. *Waterbirds* 25:173–183.

Galbraith, H., D. W. DesRochers, S. Brown, and J. M. Reed. 2014. Predicting vulnerabilities of North American Shorebirds to climate change. *Plos One*:e108899.

Gallagher, A. J., S. Creel, R. P. Wilson, and S. J. Cooke. 2017. Energy landscapes and the landscape of fear. *Trends in Ecology & Evolution* 32:88–96.

Gieder, K. D., S. M. Karpanty, J. D. Fraser, D. H. Catlin, B. T. Gutierrez, N. G. Plant, A. M. Turecek, and R. Thieler. 2014. A Bayesian network approach to predicting nest presence of the federally-threatened piping plover (*Charadrius melodus*) using barrier island features. *Ecological Modelling* 276:38–50.

Goguen, C. B. 2012. Habitat use by Mountain plovers in prairie dog colonies in northeastern New Mexico. *Journal of Field Ornithology* 83:154–165.

Goldewijk, K. K. 2001. Estimating global land use change over the past 300 years: The HYDE Database. *Global Biogeochemical Cycles* 15:417–433.

Goldin, M. R. and J. V. Regosin. 1998. Chick behavior, habitat use, and reproductive success of piping plovers at Goosewing Beach, Rhode Island. *Journal of Field Ornithology* 69:228–234.

Gomez-Serrano, M. A. and P. Lopez-Lopez. 2014. Nest site selection by Kentish Plover suggests a trade-off between nest-crypsis and predator detection strategies. *PLoS One* 9:e107121.

Guilfoyle, M. P., R. A. Fischer, D. N. Pashley, and C. A. Lott. 2006. Summary of first regional workshop on dredging, beach nourishment, and birds on the south Atlantic Coast. US Army Corps of Engineers Report 06–10.

Haig, S. M. and L. W. Oring. 1985. Distribution and status of the Piping plover throughout the annual cycle. *Journal of Field Ornithology* 56:334–345.

Haig, S. M. and L. W. Oring. 1988. Mate, site, and territory fidelity in piping plovers. *Auk* 105:268–277.

Hale, R. and S. E. Swearer. 2016. Ecological traps: Current evidence and future directions. *Proceedings of the Royal Society B* 283: 20152647.

Hall, K. R. L. 1958. Observations on the nesting sites and nesting behaviour of the Kittlitz's Sandplover *Charadrius pecuarius*. *Ostrich* 29:113–125.

Hardy, M. A. and M. A. Colwell. 2012. Factors influencing Snowy plover nest survival on ocean-fronting beaches in coastal northern California. *Waterbirds* 35:503–511.

Hayman, P., J. Marchant, and T. Prater. 1986. *Shorebirds: An Identification Guide to the Waders of the World*. Houghton Mifflin Company, Boston, MA.

Hecht, A. and S. Melvin. 2009. Expenditures and effort associated with recovery of breeding Atlantic coast piping plovers. *The Journal of Wildlife Management* 73:1099–1107.

Herman, D. M. and M. A. Colwell. 2015. Lifetime reproductive success of Snowy plovers in coastal northern California. *Condor* 117:473–481.

Holm, G. O., T. J. Hess, D. Justic, L. McNease, R. G. Linscombe, and S. A. Nesbitt. 2003. Population recovery of the Eastern Brown Pelican following its extirpation in Louisiana. *Wilson Bulletin* 115:431–437.

Hunt, K. L. 2016. Management and Mother Nature: Piping Plover Demography and Condition in Response to Flooding on the Missouri River. M.S. *Thesis*, Virginia Polytechnic Institute and State University, Blacksburg, VA.

Hunt, K. L., L. R. Dinan, M. J. Friedrich, M. Bomberger Brown, J. G. Jorgensen, D. H. Catlin, and J. D. Fraser. 2015. Density dependent double brooding in piping plovers (*Charadrius melodus*) in the northern Great Plains, USA. *Waterbirds* 38:321–329.

Hunting, K. W., S. Fitton, and L. Edson. 2001. Distribution and habitat associations of the Mountain plover (*Charadirus montanus*) in California. *Transactions of the Western Section of the Widlife Society* 37:37–42.

Hutto, R. L. 1985. Habitat selection by nonbreeding, migratory land birds. pp. 455–476 in Cody, M. L. (ed.), *Habitat Selection in Birds*, Academic Press: New York.

Iwamura, T., H. P. Possingham, I. Chades, C. Minton, N. J. Murray, D. I. Rogers, E. A. Treml, and R. A. Fuller. 2013. Migratory connectivity magnifies the consequences of habitat loss from sea-level rise for shorebird populations. *Proceedings of the Royal Society B* 280:1–8.

Jackson, B. J. and J. A. Jackson. 2000. Killdeer (*Charadrius vociferus*) in Rodewald, P. G. (ed.), *The Birds of North America Online*. Cornell Lab of Ornithology, Ithaca, NY.

Johnson, D. H. 1980. The comparison of usage and availability measurements for evaluating resource preference. *Ecology* 61:65–71.

Johnson, C. M. and G. A. Baldassarre. 1988. Aspects of the wintering ecology of piping plovers in coastal Alabama. *Wilson Bulletin* 100:214–223.

Katayama, N., T. Amano, and S. Ohori. 2010. The effects of gravel bar construction on breeding Long-billed plovers. *Waterbirds* 33:162–168.

Knopf, F. L. and J. R. Rupert. 1995. Habits and habitats of Mountain plovers in California. *Condor* 97:743–751.

Knopf, F. L. and M. B. Wunder. 2006. Mountain plover (*Charadrius montanus*) in Rodewald, P. G. (ed.), *The Birds of North America Online*. Cornell Lab of Ornithology, Ithaca, NY.

Kosztolanyi, A., T. Szekely, and I. C. Cuthill. 2007. The function of habitat change during brood-rearing in the precocial Kentish plover *Charadrius alexandrinus*. *Acta Ethologica* 10:73–79.

Kristan, W. B. 2003. The role of habitat selection behavior in population dynamics: Source-sink systems and ecological traps. *Oikos* 103:457–468.

Lank, D. B. and L. W. Oring. 1982. Sexual selection, arrival times, philopatry and site fidelity in the polyandrous Spotted Sandpiper. *Behavioral Ecology and Sociobiology* 10:185–191.

Larson, M. A., M. R. Ryan, and R. K. Murphy. 2002. Population viability of piping plovers: Effects of predator exclusion. *Journal of Wildlife Management* 66:361–371.

Le Fer, D., J. D. Fraser, and C. D. Kruse. 2008a. Piping plover chick foraging, growth, and survival in the Great Plains. *Journal of Wildlife Management* 72:682–687.

Le Fer, D., J. D. Fraser, and C. D. Kruse. 2008b. Piping plover foraging-site selection on the Missouri River. *Waterbirds* 31:587–592.

Leja, S. D. 2015. Habitat Selection and Response to Restoration by Breeding Western Snowy Plovers in Coastal Northern California. M.S. *Thesis*, Humboldt State University, Arcata, CA.

Lima, S. L. and L. M. Dill. 1990. Behavioral decisions made under the risk of predation—A review and prospectus. *Canadian Journal of Zoology* 68:619–640.

Loegering, J. P. and J. D. Fraser. 1995. Factors affecting Piping plover chick survival in different brood-rearing habitats. *Journal of Wildlife Management* 59:646–655.

Long, P. R., S. Zefania, R. H. French-Constant, and T. Szekely. 2008. Estimating the population size of an endangered shorebird, the Madagascar plover, using a habitat suitability model. *Animal Conservation* 11:118–127.

Mabee, T. J., J. H. Plissner, S. M. Haig, and J. P. Goosen. 2001. Winter distributions of North American plovers in the Laguna Madre regions of Tamaulipas, Mexico and Texas, USA. *Wader Study Group Bulletin* 94:39–43.

MacKinnon, J., Y. I. Verkuil, and N. Murray. 2012. *IUCN Situation Analysis on East and Southeast Asian Intertidal Habitats, with Particular Reference to the Yellow Sea (Including the Bohai Sea)*. vol. 47. IUCN, Gland, Switzerland and Cambridge, UK.

Maclean, G. L. and V. C. Moran. 1965. The choice of nest site in the White-fronted Sandplover *Charadrius marginatus vieillot*. *Ostrich* 36:63–72.

Maclean, G. L. 1977. Comparative notes on the Black-fronted and Red-kneed dotterels. *Emu* 77: 199–207.

Maloney, R. F. 1999. Bird populations in nine braided rivers of the Upper Waitaki Basin, South Island, New Zealand: Changes after 30 years. *Notornis* 46:243–256.

Manly, B. F. J., L. L. McDonald, D. L. Thomas, T. L. McDonald, and W. P. Erickson. 2002. *Resource Selection by Animals: Statistical Design and Analysis for Field Studies*. 2nd ed., Kluwer Academic Publishers, New York.

Marchant, S. and P. J. Higgins. 1993. *Handbook of Australian, New Zealand, and Antarctic Birds: Volume 2 Raptors to Lapwings*. Oxford University Press, New York.

Maslo, B., S. N. Handel, and T. Pover. 2011. Restoring beaches for Atlantic coast piping plovers (*Charadrius melodus*): A classification and regression tree analysis of nest-site selection. *Restoration Ecology* 19:194–203.

McCaffery, R. and P. M. Lukacs. 2016. A generalized integrated population model to estimate greater sage-grouse population dynamics. *Ecosphere* 7:e01585.

McCulloch, M. N. 1991. Status, habitat and conservation of the St. Helena Wirebird *Charadrius sanctahelenae*. *Bird Conservation International* 1:361–392.

McCulloch, N. 2009. Recent decline of the St. Helena Wirebird *Charadrius sanctaehelenae*. *Bird Conservation International* 19:33–48.

McCulloch, N. and K. Norris. 2001. Diagnosing the cause of population changes: Localized habitat change and the decline of the endangered St Helena Wirebird. *Journal of Applied Ecology* 38:771–783.

Melville, D. S., Y. Chen, and Z. J. Ma. 2016. Shorebirds along the Yellow Sea coast of China face an uncertain future—A review of threats. *Emu* 116:100–110.

Milennium Ecosystem Assessment. 2005. *Ecosystems and Human Well-Being: Biodiversity Synthesis*. World Resources Institute, Washington, DC.

Mönke, R. and K. J. Seelig. 2009. Some behavioural observations of wintering Lesser *Charadrius mongolus* and Greater *C. leschenaultii* sand plovers in Goa, India. *Indian Birds* 4:110–111.

Morris, D. W. 2003. Toward an ecological synthesis: A case for habitat selection. *Oecologia* 136: 1–13.

Muir, J. J. and M. A. Colwell. 2010. Snowy plovers select open habitats for courtship scrapes and nests. *Condor* 112:507–510.

Myers-Smith, I. H., B. C. Forbes, M. Wilmking, M. Hallinger, T. Lantz, D. Blok, K. D. Tape, M. Macias-Fauria, U. Sass-Klaassen, E. Levesque, S. Boudreau, et al. 2011. Shrub expansion in tundra ecosystems: Dynamics, impacts and research priorities. *Environmental Research Letters* 6.

Newton, I. 1994. Experiments on the limitation of bird breeding densities: A review. *Ibis* 136:397–411.

Newton, I. 1998. *Population Limitation in Birds*. Academic Press, New York.

Nicholls, J. L. and G. A. Baldassarre. 1990. Habitat associations of piping plovers wintering in the United-States. *Wilson Bulletin* 102:581–590.

Nilsson, C., C. A. Reidy, M. Dynesius, and C. Revenga. 2005. Fragmentation and flow regulation of the world's large river systems. *Science* 308:405–408.

Nol, E. 1984. Reproductive Strategies in the Oystercatchers (Aves: Haematopodidae). *Ph.D. Dissertation*, University of Toronto, Toronto, Canada.

Nol, E. and M. S. Blanken. 2014. Semipalmated plover (*Charadrius semipalmatus*). *In* Rodewald, P. G. (ed.), *The Birds of North America Online*. Cornell Lab of Ornithology, Ithaca, NY.

Nol, E., K. MacCulloch, L. Pollock, and L. McKinnon. 2014. Foraging ecology and time budgets of nonbreeding shorebirds in coastal Cuba. *Journal of Tropical Ecology* 30:347–357.

Ogden, J. and J. E. Dowding. 2013. Population estimates and conservation of the New Zealand Dotterel (*Charadrius obscurus*) on Great Barrier Island, New Zealand. *Notornis* 60:210–223.

Page, G. W., L. E. Stenzel, and C. A. Ribic. 1985. Nest site selection and clutch predation in the Snowy plover. *Auk* 102:347–353.

Page, G. W., L. E. Stenzel, J. S. Warriner, J. C. Warriner, and P. W. C. Paton. 2009. Snowy plover (*Charadrius alexandrinus*). In Poole, A. (ed.), *The Birds of North America Online*. Cornell Lab of Ornithology, Ithaca, NY.

Page, G. W., L. E. Stenzel, D. W. Winkler, and C. W. Swarth. 1983. Spacing out at Mono Lake: Breeding success, nest density, and predation in the Snowy plover. *Auk* 100:13–24.

Parrinder, E. D. 1989. Little ringed plovers *Charadrius dubius* in Britain in 1984. *Bird Study* 36:147–153.

Patrick, A. M. and M. A. Colwell. 2014. Snowy plovers select wide beachs for nesting. *Wader Study Group Bulletin* 121:17–20.

Patterson, M. E. 1988. Piping Plover Breeding Biology and Reproductive Success on Assateague Island. *M.S. Thesis*, Virginia Polytechnic Institute and State University, Blacksburg, VA.

Patterson, M. E., J. D. Fraser, and J. W. Roggenbuck. 1991. Factors affecting piping plover productivity on Assateague Island. *Journal of Wildlife Management* 55:525–531.

Pearson, S. F., S. M. Knapp, and C. Sundstrom. 2016. Evaluating the ecological and behavioural factors influencing Snowy plover *Charadrius nivosus* egg hatching and the potential benefits of predator exclosures. *Bird Conservation International* 26:100–118.

Peterson, C. H. and M. J. Bishop. 2005. Assessing the environmental impacts of beach nourishment. *Bioscience* 55:887–896.

Pienkowski, M. W. 1984. Breeding biology and population-dynamics of Ringed plovers *Charadrius hiaticula* in Britain and Greenland: Nest-predation as a possible factor limiting distribution and timing of breeding. *Journal of Zoology* 202:83–114.

Pierce, R. J. 1980. Habitats and feeding of the Auckland Island banded dotterel (*Charadrius bicinctus exilis* Falla 1978) in autumn. *Notornis* 27:309–324.

Plissner, J. and S. Haig. 2000. Viability of piping plover *Charadrius melodus* metapopulations. *Biological Conservation* 92:163–173.

Pomeroy, A. C. 2006. Tradeoffs between food abundance and predation danger in spatial usage of a stopover site by Western Sandpipers, *Calidris mauri*. *Oikos* 112:629–637.

Pomeroy, A. C., R. W. Butler, and R. C. Ydenberg. 2006. Experimental evidence that migrants adjust usage at a stopover site to trade off food and danger. *Behavioral Ecology* 17:1041–1045.

Powell, A. N. and C. L. Collier. 2000. Habitat use and reproductive success of western Snowy plovers at new nesting areas created for California Least Terns. *Journal of Wildlife Management* 64:24–33.

Prindiville-Gaines, E. P. and M. R. Ryan. 1988. Piping plover habitat use and reproductive success in North Dakota. *Journal of Wildlife Management* 52:266–273.

Que, P. J., Y. J. Chang, L. Eberhart-Phillips, Y. Liu, T. Szekely, and Z. W. Zhang. 2015. Low nest survival of a breeding shorebird in Bohai Bay, China. *Journal of Ornithology* 156:297–307.

Rioux, S., D. Amirault-Langlais, and F. Shaffer. 2011. Piping plover make decisions regarding dispersal based on personal and public information in a variable coastal ecosystem. *Journal of Field Ornithology* 82:32–43.

Roche, E. A., F. J. Cuthbert, and T. W. Arnold. 2008. Relative fitness of wild and captive-reared piping plovers: Does egg salvage contribute to recovery of the endangered Great Lakes population? *Biological Conservation* 141:3079–3088.

Rogers, D. I., P. F. Battley, T. Piersma, J. A. Van Gils, and K. G. Rogers. 2006. High-tide habitat choice: Insights from modelling roost selection by shorebirds around a tropical bay. *Animal Behaviour* 72:563–575.

Rohweder, D. A. 2001. Nocturnal roost use by migratory waders in the Richmond River Estuary, northern New South Wales, Australia. *Stilt* 40:23–28.

Rose, M. and E. Nol. 2010. Foraging behavior of non-breeding Semipalmated plovers. *Waterbirds* 33:59–69.

Rosenzweig, M. L. 1981. A theory of habitat selection. *Ecology* 62:327–335.

Rubega, M. A. and L. W. Oring. 2004. Excretory organ growth and implications for salt tolerance in hatchling American Avocets *Recurvirostra americana*. *Journal of Avian Biology* 35:13–15.

Saalfeld, S. T., W. C. Conway, D. A. Haukos, and W. P. Johnson. 2011. Nest success of Snowy plovers (*Charadrius nivosus*) in the southern high plains of Texas. *Waterbirds* 34:389–399.

Sanzenbacher, P. M. and S. M. Haig. 2002. Regional fidelity and movement patterns of wintering Killdeer in an agricultural landscape. *Waterbirds* 25:16–25.

Saunders, S. P., T. W. Arnold, E. A. Roche, and F. J. Cuthbert. 2014. Age-specific survival and recruitment of piping plovers *Charadrius melodus* in the Great Lakes region. *Journal of Avian Biology* 45:1–13.

Schlacher, T. A., D. S. Schoeman, A. R. Jones, J. E. Dugan, D. M. Hubbard, O. Defeo, C. H. Peterson, M. A. Weston, B. Maslo, A. D. Olds, F. Scapini, et al. 2014. Metrics to assess ecological condition, change, and impacts in sandy beach ecosystems. *Journal of Environmental Management* 144:322–335.

Skalski, J. R. 1996. Regression of abundance estimates from mark-recapture surveys against environmental covariates. *Canadian Journal of Fisheries and Aquatic Sciences* 53:196–204.

Skrade, P. D. B. and S. J. Dinsmore. 2013. Egg crypsis in a ground-nesting shorebird influences nest survival. *Ecosphere* 4.

Skutch, A. F. 1976. *Parent Birds and Their Young*. University of Texas Press: Austin, TX.

Spear, K. A., S. H. Schweitzer, R. Goodloe, and D. C. Harris. 2007. Effects of management strategies on the reproductive success of Least Terns on dredge spoil islands in Georgia. *Southeastern Naturalist* 6:27–34.

Sprandel, G. L., J. A. Gore, and D. T. Cobb. 2000. Distribution of wintering shorebirds in coastal Florida. *Journal of Field Ornithology* 71:708–720.

Studds, C. E., B. E. Kendall, N. J. Murray, H. B. Wilson, D. I. Rogers, R. S. Clemens, K. Gosbell, C. J. Hassell, R. Jessop, D. S. Melville, D. A. Milton, et al. 2017. Rapid population decline in migratory shorebirds relying on Yellow Sea tidal mudflats as stopover sites. *Nature Communications* 8:14895.

Stutz, M. L. and O. H. Pilkey. 2001. A review of global barrier island distribution. *Journal of Coastal Research SI* 34:15–22.

Taft, O. W. and S. M. Haig. 2006a. Landscape context mediates influence of local food abundance on wetland use by wintering shorebirds in an agricultural valley. *Biological Conservation* 128:298–307.

Taft, O. W. and S. M. Haig. 2006b. Importance of wetland landscape structure to shorebirds wintering in an agricultural valley. *Landscape Ecology* 21:169–184.

Thibault, M. and R. Mcneil. 1994. Day-night variation in habitat use by Wilsons plovers in northeastern Venezuela. *Wilson Bulletin* 106:299–310.

UN Atlas of Oceans. 2016. www.oceansatlas.org/.

USFWS. 1996. *Piping Plover (Charadrius melodus) Atlantic Coast Population: Revised Recovery Plan*. USFWS: Hadley, MA.

USFWS. 2009. *Piping Plover (Charadrius melodus) 5-Year Review: Summary and Evaluation*. USFWS: Hadley, MA.

USFWS. 2014. *Biological Opinion and Conference Opinion: Fire Island Inlet to Moriches Inlet, Fire Island Stabilization Project*. Suffolk County, New York. Hadley, MA: USFWS.

Valle, R. and F. Scarton. 1999. Habitat selection and nesting association in four species of Charadriiformes in the Po Delta (Italy). *Ardeola* 46:1–12.

van Gils, J. A., B. Spaans, A. Dekinga, and T. Piersma. 2006. Foraging in a tidally structured environment by Red Knots (*Calidris canutus*): Ideal, but not free. *Ecology* 87:1189–1202.

Wang, W., J. D. Fraser, and J. Chen. 2016. Wintering waterbirds in the middle and lower Yangtze River floodplain: Changes in abundance and distribution. *Bird Conservation International* 27:167–186.

Wearne, K. and L. G. Underhill. 2005. Walvis Bay, Namibia: A key wetland for waders and other coastal birds in sourthern Africa. *Wader Study Group Bulletin* 107:24–30.

Webber, A. F., J. A. Heath, and R. A. Fischer. 2013. Human disturbance and stage-specific habitat requirements influence Snowy plover site occupancy during the breeding season. *Ecology and Evolution* 3:853–863.

Wiens J. A. 1989. *The ecology of bird communities: foundations and patterns*. Cambridge University Press, Cambridge, UK.

Wilcox, L. 1959. A twenty year banding study of the piping plover. *Auk* 76:129–152.

Wortley, L., J. M. Hero, and M. Howes. 2013. Evaluating ecological restoration success: A review of the literature. *Restoration Ecology* 21:537–543.

Wunder, M. B. and F. L. Knopf. 2003. The Imperial Valley of California is critical to wintering Mountain plovers. *Journal of Field Ornithology* 74:74–80.

Yasué, M. and P. Dearden. 2009. The Importance of supratidal habitats for wintering shorebirds and the potential impacts of shrimp aquaculture. *Environmental Management* 43:1108–1121.

Yasué, M., A. Patterson, and P. Dearden. 2007. Are saltflats suitable supplementary nesting habitats for Malaysian plovers *Charadrius peronii* threatened by beach habitat loss in Thailand? *Bird Conservation International* 17:211–223.

Ydenberg, R. C., R. W. Butler, D. B. Lank, B. D. Smith, and J. Ireland. 2004. Western Sandpipers have altered migration tactics as peregrine falcon populations have recovered. *Proceedings of the Royal Society B* 271:1263–1269.

Yee, A. T. K., W. F. Ang, S. Teo, S. C. Liew, and H. T. W. Tan. 2010. The present extent of mangrove forests in Singapore. *Nature in Singapore* 3:139–145.

Yozzo, D. J., P. Wilber, and R. J. Will. 2004. Beneficial use of dredged material for habitat creation, enhancement, and restoration in New York-New Jersey Harbor. *Journal of Environmental Management* 73:39–52.

Zefania, S., R. French-Constant, P. R. Long, and T. Szekely. 2008. Breeding distribution and ecology of the threatened Madagascar plover *Charadrius thoracicus*. *Ostrich* 79:43–51.

Zeigler, S. L., D. H. Catlin, M. Bomberger-Brown, J. D. Fraser, L. R. Dinan, K. L. Hunt, J. G. Jorgensen, and S. M. Karpanty. 2017. Effects of climate change and anthropogenic modification on a disturbance-dependent species in a large riverine system. *Ecosphere* 8:e01653.

Zonick, C. 2000. The Winter Ecology of the Piping Plover (*Charadius melodus*) Along the Texas Gulf Coast. *Ph.D. Dissertation*, University of Missouri, Columbia, MO.

Zwarts, L. and J. H. Wanink. 1993. How the food-supply harvestable by waders in the Wadden Sea depends on the variation in energy density, body-weight, biomass, burying depth and behavior of tidal-flat invertebrates. *Netherlands Journal of Sea Research* 31:441–476.

Ram Papish

CHAPTER TEN

Population Biology*

Stephen J. Dinsmore

Abstract. Demography is the foundation of conservation directed at influencing population growth. A review of 39 species of plover revealed the disparity in our understanding of their vital rates and population biology. A few species have been well studied (Kentish, Snowy, Piping, and Mountain plovers), which contrasts with the dearth of knowledge for most others. Nest survival is the best-studied vital rate and many species have an approximately 50% chance of producing young from a single nesting attempt. Nest survival patterns vary in response to intrinsic factors such as nest age, season, adult fitness, and egg quality and extrinsic factors such as habitat at the nest site, and anthropogenic influences. Many studies examined the role of predator control and exclusion and found that, while exclosures lead to increased hatching success, the contribution to long-term population growth was less certain. Age-specific survival rates have been well studied in just a few species and reveal a pattern of low chick survival (often <0.10 for some species, but highly variable), moderate juvenile survival (0.20–0.50 for most species), and relatively high adult survival (0.60–0.80 for most species). No strong patterns appear to exist between clutch size and either nest survival or adult annual survival. Detailed population models exist for a few species of plover, providing a valuable conservation planning tool. Future population-level work on plovers should emphasize studies of age-specific annual survival (chicks, juveniles, and adults) and development of population models to address management and conservation concerns.

Keywords: age at maturity, breeding, *Charadriidae*, clutch size, nest survival, plover, population biology, population model, survival.

Population biology is the study of vital rates with a focus on factors that affect population growth and decline; it forms a basis for many conservation actions. Key vital rates include survival (and its complement, mortality), components of fecundity, and immigration and emigration, plus key drivers such as sex ratio and age structure (Cole 1957). Survival is broadly defined and often emphasizes one of the many components (nest, chick, adult, etc.) of an individual's life. Many questions center on the effect of limiting factors (e.g., competition, predation, weather patterns) on vital rates and, ultimately, population growth. Vital rates are often integrated with estimates of population size in a variety of population models that are designed to estimate the rate of population change, lambda (λ). The analysis of animal populations has a long history and there are many good resources on the topic (e.g., Seber 1982, Williams et al. 2002, Dinsmore and Johnson 2012).

* Stephen J. Dinsmore. 2019. Population Biology. Pp. 245–274 in Colwell and Haig (editors). The Population Ecology and Conservation of Charadrius Plovers (no. 52), CRC Press, Boca Raton, FL.

Evans and Pienkowski (1984) provided the first comprehensive review of the population biology of shorebirds. Subsequently, some species have been particularly well studied, including the Eurasian Oystercatcher (*Haematopus ostralegus*; van de Pol et al. 2010), Black-tailed Godwit (*Limosa*; Groen et al. 2012, Kentie et al. 2015), and Red Knot (*Calidris canutus*; McGowan 2015, Bijleveld et al. 2016). For plovers, details from a small percentage of species contribute to generalizations of demographic patterns within this group. Studying the population biology of shorebirds has its own set of challenges, beyond those common to other birds. Sandercock (2003) summarized approaches to estimating survival in shorebirds using capture–recapture techniques. In particular, he noted the difficulty in interpreting return rates, which were commonly reported in early shorebird studies, and the more recent shift to using mark-recapture models for estimating apparent survival when capture probabilities are <1.0. Despite these challenges, studies addressing population biology in shorebirds are receiving greater attention and our knowledge in this area has grown considerably in the last decade.

As a group, plovers are better studied than many other shorebird groups, although there are still significant information gaps (Colwell 2010). Collectively, >75% of published studies on population biology in this group have focused on just four species: Kentish, Snowy, Piping, and Mountain plovers. At the other extreme, more than half the species are represented by at most a single study investigating some element of population biology. Thus, our understanding of plover population biology is greatly influenced by findings from just a handful of species, most of which breed only in North America. In this chapter, I summarize what is known about plover population biology, focusing on studies of vital rates (nest survival, age-specific annual survival, immigration, and emigration) and population models that integrate information to better understand population change (e.g., Population Viability Analyses [PVAs]). I finish by highlighting gaps in information that, when filled, should improve conservation efforts.

COMPONENTS OF POPULATION BIOLOGY

Basic needs in population biology are to estimate vital rates for a population, and to understand how these rates are affected by individual attributes, the surrounding habitat, competition, density dependence, and other environmental factors. Once these vital rates are estimated they can be integrated into population models to gain a more holistic understanding of population status. The information presented here is organized by vital rate followed by an overview of population models in the context of conservation. I have summarized the key vital rates by species (Table 10.1) and provide a summary of key causal factors and patterns, below.

Clutch Size

The number of eggs in a clutch plays an important role in fecundity and, ultimately, population growth. This information is covered in detail elsewhere (Chapter Six, this volume), so only the key points that relate explicitly to demography are mentioned here. Shorebirds have a determinate clutch that consists of two to four eggs; clutch size is rarely outside this narrow range (Table 10.1; Chapter Five, this volume). Many species are capable of just a single successful nesting attempt per year, but will renest if a clutch is lost early in incubation (Chapter Five, this volume). The combined effects of clutch size and number of nesting attempts, as well as information about the mating system (see next section), provide a baseline for productivity potential that is mediated by survival at each stage.

Nest Survival

Understanding the success of a nesting attempt is one of the most frequently studied aspects of plover biology. From a population perspective, the success of nesting attempts is a key demographic rate that ultimately informs estimates of productivity and population growth (Sandercock 2003). Information about the success of nesting attempts is vital because plovers have small clutch sizes. Most species breeding in temperate regions have a clearly defined nesting season whereas those nesting in tropical environments may be capable of year-round nesting (Chapter Five, this volume). The length of the nesting season can be quite variable, e.g., Collared Plovers in Brazil nest in an 8-month period between May and December (Efe et al. 2001) while the Mountain Plover nests in a 3-month period between April and July (Knopf and Wunder 2006). Another important consideration is the variation in mating systems found in plovers (Chapter Four, this volume). For example, there might be differences between species in terms of uniparental versus biparental care, or between male- and female-attended nests as in the Mountain Plover

TABLE 10.1

Summary of key vital rates for plovers.

Species	Region	Clutch size	Nest survival	Chick survival	Juvenile survival	Adult survival		Source(s)
						(Male)	(Female)	
Red-breasted Dotterel	New Zealand	3				0.93		Marchant and Higgins (1993)
Kittlitz's Plover	South Africa	3	0.423[a]					Tulp (1998)
Malay Plover	Thailand	3	0.39–0.61[a]	0.51–0.53				Yasué and Dearden (2006)
Kentish Plover	Hungary	3		0.011–0.655				Noszály et al. (1995)
	China	3	0.575[b]					Yu and Pei (1996)
	Spain	3	0.12–0.54[b]					Amat et al. (1999)
	Turkey	3		0.08	0.15[c]	0.59–0.64		Sandercock et al. (2005)
	Netherlands	3			0.28[d]	0.65	0.73–0.91	Foppen et al. (2006)
	South Korea	3	0.369[a]					Hong and Higashi (2008)
	Abu Dhabi	3	0.217[a]					Kosztolányi et al. (2009)
	China	3	0.138[a]					Lei (2010)
	Spain	3	0.49–0.66[a]					Toral and Figuerola (2012)
	China	3	0.131[a]					Que et al. (2015)
Snowy Plover	California, USA	3	0.40–0.68[a]		0.643[d]	0.743		Page et al. (1983)
	California, USA	3	0.58[b]	0.39, 0.42				Warriner et al. (1986)
	Utah, USA	3			0.385[d]	0.578–0.880		Paton (1994)
	Utah, USA	3	0.054–0.492[a]					Paton (1995)
	California, USA	3	0.40–0.53[b]					Colwell et al. (2005)
	California, USA	3			0.283–0.575[c]			Stenzel et al. (2007)
	Texas, USA	3	0.03–0.46[a]					Hood and Dinsmore (2007)
	California, USA	3		0.06–0.80				Colwell et al. (2007)
	California, USA	3			0.40[c]	0.61	0.50	Mullin et al. (2010)
	California, USA	3	0.55[a]					Colwell et al. (2011)

(Continued)

TABLE 10.1 (*Continued*)
Summary of key vital rates for plovers.

Species	Region	Clutch size	Nest survival	Chick survival	Juvenile survival	Adult survival (Male)	Adult survival (Female)	Source(s)
	Texas, USA	3	0.07–0.33[a]					Saalfield et al. (2011)
	California, USA	3				0.734	0.693	Stenzel et al. (2011)
	Washington, USA	3	0.25[a,e]					Pearson et al. (2016)
	Oregon, USA	3		0.57				Dinsmore et al. (2017)
	Washington, USA	3		0.27–0.67				Dinsmore et al. (2017)
Wilson's Plover	Texas, USA	3	0.58[a]					Hood and Dinsmore (2007)
	St. Martin	3	0.371[b]					Brown and Brindock (2011)
	North Carolina, USA	3	0.35[a]	0.74	0.42[c]	0.77		DeRose-Wilson et al. (2013)
Common Ringed Plover	Europe	4				0.58		Boyd (1962)
	United Kingdom	4		0.40–0.60	0.57[c]	0.80		Pienkowski (1984)
	Netherlands	4			0.27–0.35[d]	0.61–0.82		Foppen et al. (2006)
Semipalmated Plover	Manitoba, Canada	4			0.048[d]	0.866		Nol et al. (2010)
	Manitoba, Canada	4				0.71		Nol and Blanken (2014)
Piping Plover	New York, USA	4	0.91[b]					Wilcox (1959)
	Manitoba, Canada	4	0.362[b]					Haig and Oring (1988)
	Maryland, USA	4	0.057–0.880[a]	0.08–0.60				Patterson et al. (1991)
	North Dakota, USA	4				0.664		Root et al. (1992)
	North Dakota, USA	4			0.318[d]	0.737		Larson et al. (2000)
	Saskatchewan, Canada	4	0.75–0.88[a]					Harris et al. (2005)
	Southern Nova Scotia, Canada	4			0.328[c]	0.732		Calvert et al. (2006)
	Gulf of St. Lawrence, Canada	4			0.240[c]	0.733		Calvert et al. (2006)
	New York, USA	4				0.703		Cohen et al. (2006)
	New York, USA	4	0.46–0.61[a]	~0.28–0.75				Cohen et al. (2009)
	USA	4				0.56–0.81		Roche et al. (2010)

(*Continued*)

TABLE 10.1 (*Continued*)
Summary of key vital rates for plovers.

Species	Region	Clutch size	Nest survival	Chick survival	Juvenile survival	Adult survival (Male)	Adult survival (Female)	Source(s)
	Prince Edward I., Canada	4	0.645[b,f], 0.339[b,g]					Barber et al. (2010)
	Great Plains, USA	4	0.47[a,h], 0.66[a,i]					Catlin et al. (2011b)
	Saskatchewan, Canada	4			0.57[d]	0.80		Cohen and Gratto-Trevor (2011)
	North Dakota, USA	4	0.08–0.35[a]					Anteau et al. (2012)
	Great Lakes, USA	4		0.556				Brudney et al. (2013)
	Great Lakes, USA	4	~0.40–0.75[a]					Claassen et al. (2014)
	Great Lakes, USA	4			0.284[d]	0.742	0.725	Saunders et al. (2014)
Madagascar Plover	Madagascar	2	0.229[a]	0.414				Zefania et al. (2008)
Little Ringed Plover	Finland	4				0.72		Pakanen et al. (2015)
White-fronted Plover	South Africa	2	0.284[b]			0.876		From Summers and Hockey (1980)
	South Africa	2	0.430[a]					Tulp (1998)
	South Africa	2	0.038–0.291[a]					Lloyd and Plagányi (2002)
	South Africa	2				0.865	0.928	Lloyd (2008)
Killdeer	Michigan, USA	4		0.27				Powell (1992)
Mountain Plover	Colorado, USA	3		0.466				Calculated from Miller and Knopf (1993)
	Colorado, USA	3	0.26–0.37[a]					Knopf and Rupert (1996)
	Montana, USA	3	0.33[a,j], 0.49[a,k]					Dinsmore et al. (2002)
	Montana, USA	3			0.46, 0.49[d]	0.68		Dinsmore et al. (2003)
	Montana, USA	3		0.01–0.26				Calculated from Dinsmore and Knopf (2005)
	Wyoming, USA	3	0.64[b]					Plumb et al. (2005)
	Colorado, USA	3	0.56[b]					Mettenbrink et al. (2006)
	Montana, USA	3		0.06		0.74–0.96		Dinsmore (2008)

(*Continued*)

TABLE 10.1 (*Continued*)
Summary of key vital rates for plovers.

Species	Region	Clutch size	Nest survival	Chick survival	Juvenile survival	Adult survival (Male)	Adult survival (Female)	Source(s)
	Colorado, USA	3	0.272[a]					Dreitz et al. (2012)
	Montana, USA	3	0.01–0.63[a]					Dinsmore (2013)
	Colorado, USA	3	0.17–0.81[a]					Augustine and Skagen (2014)
Eurasian Dotterel	Norway	3	0.22–0.67[b]					Kålås and Byrkjedal (1984)
St. Helena Plover	St. Helena	2	0.227–0.238[a]					Burns et al. (2013)
Hooded Plover	Australia	3		0.196				Weston (2000)
	Australia	3	0.22[b]	0.40				Baird and Dann (2003)
Shore Plover	New Zealand	3	0.79, 0.89[b]	0.07, 0.43				Davis (1994a)
Wrybill	New Zealand	2				0.83		Hay (1984)
	New Zealand	2			0.45–0.50[d]	0.80		Dowding and Murphy (2001), Riegen and Dowding (2003)

When multiple estimates were available, the range of values is given. Nest survival estimates pertain to the incubation period, usually excluding the egg-laying period. Estimates of adult survival are annual estimates. Most estimates of chick survival pertain to the mobile fledgling period, although some estimates may be biased because all chicks are not always marked at hatching. Finally, most estimates of juvenile survival pertain to the period from fledging to first birthday (a roughly 11-month period), although some estimates cover the time from hatch to first birthday.

[a] Estimated using a model-based approach (Mayfield, others).

[b] Apparent nest success (the proportion of nests in a sample that were successful).

[c] Survival is from fledging to age one.

[d] Survival is estimated from chicks banded throughout the fledgling period to age one.

[e] Not exclosed nests only.

[f] Exclosed nests.

[g] Not exclosed nests.

[h] Nests on natural sandbars.

[i] Nests on engineered sandbars.

[j] Female-tended nests.

[k] Male-tended nests.

(Dinsmore et al. 2002). Here, I cover the survival of plover nests in detail; productivity is covered elsewhere (Chapter Five, this volume).

Most studies of nest survival aim to estimate the probability that a nest will be successful, usually defined as nests where ≥1 egg hatches (Mayfield 1961, 1975, Dinsmore et al. 2002). Many studies use the term "nest success," which is probably synonymous with nest survival (see Dinsmore et al. 2002) because most plovers are single brooded. Other terms such as hatching success are sometimes used, although this term typically refers to the proportion of eggs that hatch and not the overall outcome of a nesting attempt (Dinsmore et al. 2002). Methods to estimate nest survival vary. The simplest approach is to use the proportion of a sample of nests that were successful, often called apparent nest success (Mayfield 1961). Such estimates tend to be biased high because they fail to account for nest losses prior to nest discovery (Mayfield 1961); consequently, Mayfield estimators or more recently developed likelihood-based models (e.g., Dinsmore et al. 2002, Rotella et al. 2004, Shaffer 2004, Shaffer and Thompson 2007) have been used increasingly often to estimate the daily nest survival rate. In any case, I have attempted to use the best estimates whenever possible (e.g., a daily survival rate from a formal nest survival analysis; Table 10.1).

Nest survival may be the easiest population parameter to estimate because plover nests are relatively straightforward to find and monitor. The sources of nest loss are many, although most can be attributed to predation, abandonment, or weather events (see Chapter Six, this volume, for details on predation). A thorough examination of published studies with information on nest losses revealed that, across all species, >50% of nest losses could be attributed to predation (Chapter Six, this volume). In most species, the primary predators were mammals, followed closely by other birds (pers. obs.). In several studies, the primary avian predators were members of the Corvidae (Lloyd and Plagányi 2002, Colwell et al. 2011, DeRose-Wilson et al. 2013, Pearson et al. 2016). Nest abandonment was also frequent and resulted from human disturbance, adult death, and other causes. One of the challenges when estimating nest survival probabilities is how to treat abandoned nests. Some nest abandonment happens naturally and should be reflected in the estimates whereas losses caused by researchers' activities should be discarded from analyses.

Considerable effort has been devoted to understanding how environmental effects, habitat, characteristics of the tending adult(s), and other factors affect nest survival in plovers. The effects of nest age are often included in models because nests can be aged by the use of field techniques such as egg flotation (Mabee et al. 2006). For most species, survival increases as the nest ages (Claassen et al. 2014), although there are exceptions where survival is not related to nest age (Toral and Figuerola 2012) or where it declines with age (DeRose-Wilson et al. 2013). This pattern of nest survival increasing with nest age presumably results because of the early loss of nests that are placed in vulnerable locations. Daily and seasonal patterns in nest survival have been well documented in some studies (Dinsmore et al. 2002, Brown and Brindock 2011, Toral and Figuerola 2012) and are often shown to respond to changing weather or habitat conditions or changes in predator pressure. Many studies have examined the influence of habitat attributes on nest survival at varying spatial scales. Not surprisingly, nest survival is often positively correlated with sparse vegetation and more open habitats within this group (Yasué and Dearden 2006, DeRose-Wilson et al. 2013). Claassen et al. (2014) found that nest survival increased with male age in the Piping Plover. Anthropogenic effects on nest survival have received some attention, especially for a few species. This is an important topic because we can often manage human access to plover nests with nesting area closures or fewer visits to nests for monitoring efforts. Lloyd and Plagányi (2002) demonstrated that observer visits to nests can lower daily survival and this has implications for future studies that involve nest checks. Schneider and McWilliams (2007) used small data loggers to monitor nests and demonstrated that the devices did not affect hatching success or predation rates, suggesting that use of such loggers might be useful for future studies where reducing nest disturbance is a priority. Human disturbance often results in negative effects during the nesting season as documented for Malay Plover (Yasué and Dearden 2006), Piping Plover (Cohen et al. 2009), Snowy Plover (Lafferty et al. 2006), and Hooded Plover (Weston et al. 2011). However, equivocal findings for the Wilson's Plover (DeRose-Wilson et al. 2015) and White-fronted Plover (Baudains and Lloyd 2007) indicate the complexity of behavioral responses by nesting plovers to human disturbance. The effects of human-created habitat on multiple demographic responses have been studied in the Piping Plover (Cohen et al. 2009, Catlin et al. 2015). Intrinsic factors should not be ignored in this discussion and factors such as the quality of adults (Catlin et al. 2014) or eggs (Skrade and Dinsmore 2013) may also play a role in the success of a nesting attempt.

One general technique that has been used to improve plover nest survival is predator management, either by direct removal of predators or the use of exclosures or other deterrents (see Chapter Six, this volume). Predator exclosures have been used to enhance the nesting success of several species of plovers and has been a common practice with endangered ground-nesting birds (Mabee and Estelle 2000, Maslo and Lockwood 2009). Exclosures have been frequently used with Snowy Plovers in the western United States (Neuman et al. 2004, Hardy and Colwell 2008, Dinsmore et al. 2014), but also in the St. Helena Plover (Burns et al. 2013), Hooded Plover (Maguire et al. 2011), and Piping Plover (Cohen et al. 2009, Barber et al. 2010). In general, predator exclosures are effective at increasing the number of eggs hatched, but their effect on recruitment is less well understood (Neuman et al. 2004). One serious drawback of exclosures is that they can increase mortality rates of the attending adults (Hardy and Colwell 2008, Barber et al. 2010, Burns et al. 2013, Pearson et al. 2016); this is especially problematic because adult survival is arguably the most important component of population growth in long-lived birds such as plovers (Dinsmore et al. 2010). Watts et al. (2012) predicted that exclosures would have a net negative effect on a population of Snowy Plovers if adult survival were reduced to 90% of the rate without exclosures. Similarly, Calvert and Taylor (2011) found that the benefit of exclosures for increasing chick production at best compensated for a corresponding increase in adult mortality. The use of shelters to protect chicks was studied in the Hooded Plover and shown to increase fledging probability by more than 70% (Maguire et al. 2011). A pattern that emerges is that exclosures can improve nesting success in plovers, but that more work is needed to fully assess whether this translates into increased recruitment and enhanced population growth.

Few studies have examined the role of conspecifics or other bird species on plover nest survival. Que et al. (2015) found evidence for density-dependent nest survival in the Kentish Plover, a semicolonial breeder, such that nests located along colony edges had lower survival than those in the center of the colony. Mayer and Ryan (1991) used artificial nests to test the prediction that Piping Plover nests within American Avocet colonies would have greater survival probabilities because of the mobbing behavior of avocets to deter predators; this hypothesis was not supported in their study. Semipalmated Plovers nesting within Arctic Tern (*Sterna paradisaea*) colonies benefited from aggressive nest predator protection by the terns (Nguyen et al. 2006).

Snowy, Piping, and Mountain plovers have been the subject of considerable study with respect to nest survival. Below, I provide a short summary of the relevant work for each species as a means of illustrating the similarities and differences.

Snowy Plover

The Snowy Plover has been well studied with respect to nest survival patterns, especially in the western United States, although fewer published works are from populations in the Great Plains and Gulf Coast regions. Colwell et al. (2005) found higher nesting success in riverine habitats when compared to sandy beaches, presumably because river bars contained lower quality substrate for camouflaging the eggs. Colwell et al. (2011) then documented the role of egg crypsis by showing that nest survival was highest in less heterogeneous substrates with a greater number of egg-sized stones. Hood and Dinsmore (2007) found that the survival of plover nests along the Texas Gulf Coast increased with nest age and the presence of a conspicuous object at the nest site; nests along the immediate coast survived better than those at a short-distance inland. A study at inland saline lakes in Texas estimated nest success at 22% and found it was negatively influenced by the number of plants and positively influenced by the amount of surface water in close proximity (Saalfield et al. 2011). This study also noted a 31% decline in nest success when compared to the previous decade of work in this area. A long-term study in coastal Oregon found that Snowy Plovers responded positively to habitat management, nest exclosures, and lethal predator control (Dinsmore et al. 2014). Reproductive success is also linked to dispersal patterns. Stenzel et al. (1994) documented long-distance dispersal along the Pacific Coast (up to 1,140 km), found that it was unrelated to breeding success, noted that dispersal patterns of males and females differed, and concluded that there was no relationship between site fidelity and nesting success. In coastal California, both sexes dispersed farther when switching mates or when following a nest failure; dispersal distances ranged from 0.9 to 13.0 km (Pearson and Colwell 2013).

Piping Plover

Unlike other species, some information is available from each of the major subpopulations of Piping Plover, including those inhabiting the Atlantic Coast, Great Lakes, and northern Great Plains. A study in New York found that nest survival benefited from predator control and nest exclosures, and was greater in a newly established site that was adjacent to a beach with considerable human development than at an existing site with a long history of plover use (Cohen et al. 2009). In the Great Lakes population, with fewer than 100 pairs (Saunders and Cuthbert 2015), nest success was 45% during a 22-year period (Barber et al. 2010), with a large difference between exclosed (65%) and not exclosed nests (34%). Another study making use of many of the same nests over an 18-year period estimated that nest survival for the 28-day incubation period was 0.70 (Claassen et al. 2014). Several studies of nest survival have been conducted for plover populations in the Great Plains. Nest survival in this region appears lower with estimates of 0.08–0.35 for a 4-year study in North Dakota (Anteau et al. 2012), 0.75–0.88 in Saskatchewan (Harris et al. 2005), and 0.47 on natural sandbars and 0.66 on man-made sandbars (Catlin et al. 2011b).

Mountain Plover

Nest survival patterns in the Mountain Plover have been particularly well studied. Nest success averaged about 50% rangewide, although it varied between sites and years. Hatching success was 26%–37% in Weld County, Colorado (Knopf and Rupert 1996), 50% across eastern Colorado (Mettenbrink et al. 2006), 58% in Phillips County, Montana (Dinsmore et al. 2002), 64% in Wyoming (Plumb et al. 2005), 47% on black-tailed prairie dog (*Cynomys ludovicianus*) colonies treated with deltamethrin (used to control plague), and 66% on control colonies in Phillips County, Montana (Dinsmore 2013). Dinsmore et al. (2002) found that male-tended nests had greater survival than nests tended by females, and that daily nest survival increased as nests aged and was negatively influenced by daily precipitation. Treatments of the chemical deltamethrin on prairie dog burrows had a strong negative effect on nest survival, presumably because of a reduction in food supply for the parents (Dinsmore 2013). Dreitz et al. (2012) found that nest survival was positively correlated with drought conditions and cooler temperatures in Colorado. Nest survival is higher on prairie dog colonies compared to managed burns (Augustine and Skagen 2014), a further hint at the importance of habitat and disturbance to vegetation in this species. Plovers prefer open areas for nesting, but distance to anthropogenic edges was not correlated with nest success (Mettenbrink et al. 2006). Dreitz and Knopf (2007) documented the causes of nest mortality in Colorado, where nests on rangeland were mostly depredated while those in agricultural fields failed owing to farming practices such as tilling. The influence of nest survival on population growth rate in this species is less certain, with one study suggesting a positive effect related to nest protection strategies (Post van der Burg and Tyre 2011), whereas another study found that possible benefits were swamped by a negative influence of adult survival (Dinsmore et al. 2010).

These three species provide interesting contrasts with respect to nest survival patterns. There does not appear to be any pattern in nest survival as a function of clutch size or migratory strategy (all three are mid-distance Northern Hemisphere migrants; see Chapter Seven, this volume). Nest survival estimates were variable for each species, although on average it appears that a nest of any of the three species has ~50% chance of producing chicks. Exclosures and predator control have been used with the two listed species (Snowy and Piping plovers) and have increased hatching success, but contributions to population growth have not been well documented. Studies that examined habitat at the nest site found correlates with nest survival that are consistent with the habitat preferences of the species. Piping Plover's use of constructed nesting habitat and the consequences of the unique mating system of the Mountain Plovers on nest survival are perhaps the most novel contributions of these studies.

BOX 10.1 Case Study: Mountain Plover

The Mountain Plover is a patchily distributed shorebird of the North American Great Plains (Knopf and Wunder 2006). The species often associates with disturbance in arid grasslands and shrublands, especially sites created by herbivores such as the black-tailed prairie dog (*Cynomys ludovicianus*) or domestic cattle and sheep (Knopf and Miller 1994). The species is of heightened conservation status throughout much of its range, although listing was denied under the U.S. Endangered Species Act (U.S. Department of the Interior 2002). The Mountain Plover also has an unusual mating system where males and females tend separate nests simultaneously (Graul 1975, Dinsmore et al. 2002), offering a unique opportunity to study male and female investments in reproduction.

The northern limit of the breeding range of the Mountain Plover lies in north-central Montana, an area of high-density prairie dog colonies in Phillips County (Knowles et al. 1982, Olson-Edge and Edge 1987). This ecoregion may also represent one of the most intact natural ecosystems inhabited by the plover, in contrast to its use of agricultural fields elsewhere in its range (Shackford et al. 1999, Knopf and Wunder 2006). Its population biology has been extensively studied in southern Phillips County since the 1980s, making it one of the best-studied plover species. Within this region, prairie dog numbers fluctuate partly in response to sylvatic plague (*Yersinia pestis*), an introduced flea-borne disease that causes >95% mortality in prairie dogs; plague-affected colonies lack the continual grazing disturbance necessary to create plover nesting habitat (Dinsmore and Smith 2010). The plover selectively uses active black-tailed prairie dog colonies (Knowles et al. 1982, Olson-Edge and Edge 1987, Dinsmore et al. 2003), especially larger colonies (>50 ha) that are not affected by plague and where annual occupancy rates by breeding plovers are approximately 0.75 (Dinsmore and Smith 2010). Birds select nest sites on prairie dog colonies with sparser vegetation and less bare ground than random sites within those colonies (Olson and Edge 1985). Childers (2006) found that shrub and bare ground cover within 25 m of nests was positively correlated with nest survival while the habitat heterogeneity within this region was negatively correlated with nest survival. Many of these habitat attributes disappear when colonies are lost with the onset of plague. Olson-Edge and Edge (1987) estimated a density of 17.5 plovers/100 ha of prairie dog colony habitat using simple colony visit surveys. The density of breeding plovers on these same prairie dog colonies was estimated at 7.20 individuals/km^2 in 2003–2004, which translates into a breeding population of approximately 750 adults (Childers and Dinsmore 2008); the population is thought to be much lower today (pers. obs.). Trend analyses document population fluctuations early in the study (Dinsmore et al. 2005) and a rangewide analysis revealed that this population was most susceptible to declines in adult annual survival (Dinsmore et al. 2010).

The demography of the Mountain Plover has been studied extensively in Montana through a color-banding study begun in 1995. As part of that study, >1,800 nests have been monitored to understand nest survival patterns. Nest survival is greater for male-tended nests, presumably because males did not lay eggs and were in better body condition (pers. obs.); nest survival also shows considerable seasonal variation and is negatively affected by rainfall (Dinsmore et al. 2002). On average, a nest has a roughly 50% chance of having ≥1 egg hatch. A total of >1,900 plovers (roughly equal number of juveniles and adults) has been uniquely banded and monitored during the study. The annual adult survival rate averages ~0.80 and is positively correlated with drought severity, suggesting this is a drought-adapted species (Dinsmore et al. 2003, Dinsmore 2008). The juvenile survival rate, from hatch to age one, is relatively low and has been estimated at 0.06 with most mortality occurring during the brood-rearing period (Dinsmore and Knopf 2005, Dinsmore 2008). The extremely low juvenile survival rate may be partly explained by natal dispersal, which probably occurs but has not been estimated. Prairie dog colonies are dynamic habitats, and the plovers in this region are mobile and will disperse within short distances to

(Continued)

BOX 10.2 (Continued) Case Study: Mountain Plover

breed. Skrade and Dinsmore (2010) found that mean dispersal distances were <5 km for both sexes with successful breeders being less likely to move in a subsequent year.

As this example illustrates, the Mountain Plover is a moderately long-lived shorebird that is adapted to a terrestrial, herbivore-dominated ecosystem. Breeding pairs have two separate nests, thereby increasing nest survival to nearly 50%. Dependent chicks have extremely low survival, which is balanced by high survival of adults. This study also clearly illustrates the value of long-term research with plovers. In this relatively long-lived species, the key vital rates cannot be reliably estimated from studies lasting just a few years. Many key vital rates vary temporally and long-term studies may allow such patterns to be more easily discerned. This field study provides a template for future research to better understand patterns of nest, chick, and annual survival in plovers.

Chick Survival

Fewer studies have been conducted on the survival of plover chicks during the period from hatching to fledging than for nest survival. Such studies are more difficult in practice because broods are mobile and chicks must be marked immediately at hatching using color bands or radio telemetry. Brood checks must be frequent (often every 1–2 days) because mortality during this life history stage is high and often occurs during the first week post-hatch. Technological advances leading to smaller tracking devices have helped, but they are still expensive (>$100 each) and moderate numbers of individuals must be marked and tracked to get meaningful estimates of chick survival.

Detailed studies of chick survival have been conducted on just six species. For many, the mortality curve confirms that most deaths happen soon after hatching, often within the first week, presumably because young chicks are less mobile and are more vulnerable to predators. Species for which studies that found high mortality during the early chick stage included Shore Plover (50% in the first 10 days; Davis 1994a), Snowy Plover (Warriner et al. 1986, Colwell et al. 2007, Dinsmore et al. 2017), Piping Plover (Loegering and Fraser 1995), and Mountain Plover (Dinsmore and Knopf 2005). Amat et al. (2001) found that egg mass and chick mass at hatching were positively correlated in the Kentish Plover, and that heavier chicks survived better and were more likely to recruit into the population. Dreitz (2009) investigated survival patterns in Mountain Plover chicks in response to habitat type and found that survival was greatest on prairie dog colonies and lowest on agricultural land and grasslands. DeRose-Wilson et al. (2013) found that Wilson's Plover chick survival was weakly correlated with hatch date and did not vary seasonally. Catlin et al. (2011a) found that removal of Great Horned Owls (*Bubo virginianus*), a major predator of Piping Plover chicks, increased daily chick survival, but the effect decreased as chicks aged. In Great Lakes Piping Plovers, first-year survival was positively correlated with earlier hatching date, older age at banding, greater number of fledglings at the site, and better body condition at banding (Saunders et al. 2014). Saunders and Cuthbert (2015) found evidence for sex-biased chick survival in Piping Plovers; higher male survival was responsible for a skewed sex ratio in this species. Sandercock et al. (2005) found similar evidence for sex bias in the Kentish Plover. Human disturbance can also affect chick survival, as documented in the Snowy Plover in California (Ruhlen et al. 2003). See Chapter Four, this volume, for a detailed discussion of plover breeding systems.

Published chick survival rates in plovers are highly variable (Table 10.1). The probability of surviving the fledgling stage (4–6 weeks in most plover species) ranged from <0.10 in several species to as high as 0.80 in the Snowy Plover (Colwell et al. 2007) and 0.74 in the Wilson's Plover (DeRose-Wilson et al. 2013). Dinsmore et al. (2017) illustrated that chick survival estimates in a Snowy Plover population differed depending on the analysis used; they suggested that care must be taken when considering assumptions such as the detectability of chicks in broods. The method developed by Lukacs et al. (2004) can be used to estimate chick survival when only the adult is marked, which is advantageous because transmitters that may affect survival do not need to be placed on chicks. Some literature on Piping Plovers is confusing because a common metric of monitoring programs for this federally listed species is the fledging rate expressed as number of young per breeding pair per season, leading some authors to fail to report the actual chick survival rates. Nonetheless, a paucity of information about chick survival patterns in many plovers suggests that this stage should be a focus of further research within this group.

Probability of Breeding

The age of first reproduction, which can also be summarized as an age-specific probability of breeding, is an important characteristic of any population. In birds, age often explains much of the variation in reproductive success between individuals; success often increases during early age to a peak in middle age, followed by senescence (Forslund and Pärt 1995). Thus, some individuals may choose to delay their first reproductive attempt. Much of the published literature on this topic for plovers consists of generalizations about the typical age of first reproduction; few studies have developed a probabilistic model of breeding by age. Saunders et al. (2014) developed such a model for Great Lakes Piping Plovers and found a strong sex bias in the probability of breeding at age one (0.56 for females and 0.35 for males), and that almost all individuals bred by age three. Age of first reproduction has not been explicitly studied in most plover species. Most references to this topic indicate that individuals often breed at age one, although there is undoubtedly some variation in timing. Nol et al. (2010) noted that the Semipalmated Plover first breeds at age two or three (rarely at age one) and in the Mountain Plover some individuals may delay breeding until age two while others breed at age one (pers. obs.).

Annual Age-Specific Survival

Many studies of age-specific survival include two age classes (juveniles and adults) and estimate survival on an annual basis. Like dependent chicks, survival has been monitored with both color-banding studies and radio telemetry, the latter mostly used for studies of seasonal (rather than annual) survival. Importantly, color-banding studies allow estimates of apparent survival, a product of true survival and site fidelity; thus, apparent survival is biased low with respect to true survival (Cormack 1964, Pollock 1982). Often, the timing of marking chicks after hatching precludes estimating chick survival to independence, but will allow survival to be estimated from marking to 1 year of age; this is juvenile survival. Because young birds are marked at different ages (some right after hatch versus some just before fledging) juvenile survival rates within and between species are difficult to compare directly. For example, a study using information only from individuals marked right after hatching would produce lower estimates of juvenile survival than a study of the same species where the same birds were marked immediately prior to fledging, simply because of high chick mortality rates immediately post-hatch.

Few estimates of juvenile survival are available for plovers. Published estimates exist for seven species (Table 10.1). As expected, estimates of juvenile survival in this group were highly variable and often fall between estimates of chick and adult survival. The time period for these estimates is inconsistent, although most are for the approximately 11-month period from fledging to first birthday. For many studies, the estimates often range from 0.20 to 0.50. No study has documented a strong effect of sex on juvenile survival, although it has been investigated in only a few species. In the Mountain Plover, body mass was positively correlated with juvenile survival, presumably because body mass was also correlated with chick age (Dinsmore et al. 2003, Dinsmore 2008). Many studies also noted that estimates of juvenile survival may be biased low because of dispersal and permanent emigration of this age class, providing further justification for the use of open population models at large geographic scales to estimate this parameter.

A greater number of studies have estimated adult survival. Estimates of adult annual survival exist for 11 plovers and for most they range from 0.60 to 0.80 (Table 10.1). Notable exceptions are three Southern Hemisphere species (Red-breasted Dotterel, White-fronted Plover, and Wrybill), all of which have adult survival estimates ≥0.80, peaking at 0.93 in the long-lived Red-breasted Dotterel (Marchant and Higgins 1993). These species are nonmigratory or short-distance migrants, suggesting a trade-off between migration and survival is present (or site fidelity; see Chapter Seven, this volume). In single studies of Kentish and Common Ringed plovers the estimated adult annual survival rate was <0.60, comprising the lowest estimates within the group (Table 10.1). In the well-studied Snowy and Piping plovers, adult survival rates from many studies seem to be consistent at ~0.70.

Many studies have investigated the factors affecting adult survival. The most common question dealt with possible differences between males and females (see Chapter Four, this volume). Sex differences in adult survival have been found in Kentish (Foppen et al. 2006), Snowy (Mullin et al. 2010, Stenzel et al. 2011, Colwell et al. 2013), Piping (Saunders et al. 2014), and White-fronted (Lloyd 2008) plovers. In Kentish and White-fronted plovers, the survival of females was greater than survival of males; the opposite was true for the other two species. In the Mountain Plover there is a suggestion that males may have greater annual survival than females (Dinsmore 2008). In addition to sex, several studies have investigated the influence of other factors on adult survival. Dinsmore et al. (2003) found no effect of the extent of breeding habitat (prairie dog colonies) on the adult survival of Mountain Plovers. Dinsmore (2008) found that annual survival in adult Mountain Plovers was positively correlated with drought conditions, consistent with a similar correlation with nest survival in this species. Adult survival did not differ for Piping Plovers nesting in engineered and natural habitats in the U.S. Great Plains (Catlin et al. 2015). Annual survival of Snowy Plovers was positively affected by current or cumulative reproductive effort, which is contrary to predicted trade-offs between survival and reproduction (Colwell et al. 2013). Annual variation in adult survival was found in some long-term studies (Dinsmore 2008, Stenzel et al. 2011), but not in others, suggesting that this is not a consistent pattern or that too few studies have been conducted long enough to detect such patterns.

Little work has been done on nonbreeding season survival in plovers. Drake et al. (2001) studied the nonbreeding survival of 49 Piping Plovers in Texas between August and April and documented no mortality among radio-marked individuals, suggesting that the nonbreeding portion of the annual cycle was not contributing to population declines. In the Mountain Plover, breeding season survival in Colorado was 1.0 (Miller and Knopf 1993) whereas overwinter survival in California was 0.947 (1 November–15 March; Knopf and Rupert 1995). Roche et al. (2010) suggested that wintering ground differences might explain annual variation in Piping Plover survival, but they did not directly measure overwinter survival.

Summary of Limiting Factors on Survival

Patterns of plover survival (nest, chick, juvenile, and adult) are linked to several limiting factors, some of which can be addressed through specific management actions. On the basis of the information reviewed here, human disturbance, predation, and the extent of suitable breeding habitat emerge as important limiting factors. Human disturbance has pervasive negative effects on all vital rates (see Chapter Eleven, this volume), but especially nest survival where it can result in a direct loss of the nest or have an indirect effect by altering adult behavior and making the nest more vulnerable to predation. Predation is also a factor at all life history stages although it is probably most important at the nest (losses average ~0.43 across all species) and chick stages (losses average ~0.37 across all species; see Chapter Six, this volume). Adult plovers generally have moderate to high (0.70–0.90) annual survival, and predation is not known to be a major influence.

The extent of nesting habitat is the third primary limiting factor with examples affecting the chick and adult life history stages (see Chapter Five, this volume). Many studies note that nest survival is correlated with sparse to no vegetation; our understanding of habitat influences on chick or adult survival is limited to just a couple of well-studied species. Habitat use during migration or in the nonbreeding season has not been identified as a limiting factor for any plover species. Studds et al. (2017) estimated a 6% annual decline in Lesser Sand-plovers using the Yellow Sea in China as a stopover site during a 20-year period, a trend that was attributed to the loss of mudflat habitat. However, migratory species depend upon one or more sites (habitats), each with precisely timed resources, to complete their annual cycle (see Chapter Seven, this volume).

The conservation of many plover species may depend upon our ability to manage these limitations, which can be accomplished using such diverse tools as habitat restoration/creation or predator control. Weather is most often identified as a limiting factor on nest survival with many studies hinting at the negative effects of too much precipitation or extremely hot temperatures. Abnormal weather patterns have been identified as a limiting factor on adult survival in two species (cold winters for the Snowy Plover [Eberhart-Phillips et al. 2016] and non-drought years for the Mountain Plover [Dinsmore 2008]). Other limiting factors such as disease or competition have not been identified as influencing demographic processes in plovers, although this is certainly an area where further work is needed.

Longevity

Longevity, or maximum life span, is a way to represent the consequences of the estimates of survival at each life history stage. A simple method to calculate a mean expected life span uses an estimate of annual survival (Lifespan $= -\frac{1}{\ln(\phi)}$ where ϕ is the estimated annual survival rate; Brownie et al. 1985). Longevity, on the other hand, is documented as a consequence of having a population that is marked and well monitored. Estimates of longevity are often positively correlated with the length of a study and the number of individuals banded. I found reliable longevity estimates for 12 species of plover, although in at least one species (Wilson's Plover) the lack of a long-term banding study means that longevity is probably underestimated. As a group, plovers appear to be relatively long-lived, although this conclusion is based on findings from only the best-studied species. The longevity estimates reveal a maximum life expectancy of about 12–17 years for most species (Table 10.2). Less has been published about population-specific age distributions, although there are two examples for the group. In a New Zealand population of Wrybill, there were 165 individuals that reached at least 12 years of age (A. Riegen, pers. comm.). In a well-studied population of the Mountain Plover, the known age distribution from individuals marked as chicks revealed that just seven individuals (0.4% of total) reached 10 years of age (pers. obs.). Herman and Colwell (2015) documented a highly skewed pattern of lifetime reproductive success in a well-studied Snowy Plover population, in which 13% of individuals produced half of all fledglings. A few studies have used estimates of survival rates to estimate mean lifespan, as for adult Snowy Plovers in Utah (2.7 years; Paton 1994) and Mountain Plovers at initial capture in Montana (1.92 years; Dinsmore et al. 2003).

Immigration and Emigration

The dynamics of a population fluctuate in response to many changes that include the consequences of reproduction (births), losses due to deaths (the complement of survival), and physical movements of individuals that results from

TABLE 10.2
Summary of longevity records (in years and months) for plovers.

Species	Longevity record	Source
Red-breasted Dotterel	41 years	Marchant and Higgins (1993)
Kentish Plover	19 years	Fransson et al. (2010)
Snowy Plover	15 years and 11 months	Colwell et al. (2017b)
Wilson's Plover	6 years	U.S. Geological Survey (2016)
Common Ringed Plover	20 years and 10 months	Fransson et al. (2010)
Semipalmated Plover	10 years and 8 months	Nol and Blanken (2014)
Piping Plover	17 years	Piping Plover Conservation in Nova Scotia (2017)
Little Ringed Plover	13 years	Fransson et al. (2010)
Killdeer	10 years and 11 months	U.S. Geological Survey (2016)
Mountain Plover	12 years and 1 month	pers. obs.
Shore Plover	17 years	Davis (1994b)
Wrybill	22 years and 4 months	Adrian Riegen, pers. comm.

immigration and emigration. Immigration and emigration are often hard to estimate because they (1) occur infrequently or (2) encompass great distances that make it difficult to relocate dispersers. It is also worth noting that both movements can be permanent, such as the permanent emigration of a juvenile plover to a population different from where it hatched, or temporary, when an individual moves between populations annually. Unfortunately, the study of immigration and emigration has received almost no attention in the plovers. Catlin et al. (2016) used multistate models to examine movements in Great Plains Piping Plovers and found that juveniles were more likely to disperse than adults, and that emigration and immigration rates dropped with increasing distance between populations and were further affected by environmental conditions (river flows and habitat creation). Colwell et al. (2017a) documented the role of immigration in maintaining a population of Snowy Plovers, noting that >60% of the breeding adults were immigrants. Hillman et al. (2012) documented the first interpopulation dispersal event for the Piping Plover that resulted in successful breeding, when an individual originating in the Great Lakes population emigrated to the Atlantic Coast population. Population genetic studies can also be used to infer rates of immigration and emigration, as in a study of the Mountain Plover where inferred gene flow resulted in a lack of genetic differentiation across the breeding range (Oyler-McCance et al. 2005). As these few examples suggest, estimates of population-level rates of emigration and immigration are needed for most plovers, yet they provide information that is critical to fully understanding plover population dynamics.

Population Models

A population model is typically used to assess past and future population growth, and can be accomplished using multiple approaches. In some cases, actual population counts may be available, from which the rate of population change (λ) can be estimated as a ratio of counts. A different approach is to use age-specific estimates of survival and fecundity to make predictions about the rate of population growth. From either approach, inferences are made by comparing the actual estimate of lambda to a scenario where the population is stable ($\lambda = 1.0$).

Demographic models require detailed information about the population of interest, and there are examples from just a few species of plover. Some are simple estimates of population change during a specified time period whereas others use a model-based approach to estimate trends. McCulloch and Norris (2001) used surveys to estimate a 21% decline in the St. Helena Plover during a 10-year period. Several studies of local segments of the Pacific Coast Snowy Plovers have documented long-term population declines (Mullin et al. 2010, Colwell et al. 2010), which provided an incentive to more fully assess population status. Nur et al. (1999) developed a detailed PVA that evaluated how changes in demographic rates affected population growth; this model is integral to the species' recovery plan (U.S. Fish and Wildlife Service 2007). Mullin et al. (2010) used multiple approaches to assess the stability of a northern California population and concluded it was a sink that was maintained largely by immigration from surrounding regions. Eberhart-Phillips et al. (2014) modeled the viability of this same population and confirmed that it was a sink being maintained by immigrants, that local management actions should emphasize increasing productivity (via managing predators and humans without the use of nest exclosures), and that cold winter weather was an important limiting factor. Lastly, Colwell et al. (2017a) used 16 years of data from the same population and noted that the population is composed of ~63% immigrants, it is a demographic sink, and that local management efforts alone are insufficient to grow the population. The examples with the Snowy Plover illustrate both the complexity and usefulness of detailed population models and demonstrate how they can inform future management decisions.

In other species, sophisticated models have been used to integrate vital rates and population estimates to make predictions about growth. Population trends of the federally endangered Piping Plover have received considerable attention in North America. Larson et al. (2002) modeled growth rates in the Great Plains population and suggested that declines of 4.3% per year could be expected under existing management plans; management actions to increase reproductive success were needed to stabilize or increase this population. Larson et al. (2002) concluded that a rangewide fledging rate of 1.10 fledglings per pair was needed to stabilize the population. Wemmer et al. (2001) developed a habitat-based model for the Great Lakes population and concluded that measures focused only on increasing productivity were unlikely to be successful; more nesting sites are necessary, and the model was most sensitive to changes in adult survival. Catlin et al. (2015) concluded that a fledging rate of 1.25 chicks per pair was needed for population stability in the Great Plains. Catlin et al. (2016), also working in the Great Plains, concluded that population persistence was enhanced by the presence of anthropogenically created habitats that were an important and consistent source of nesting sites and dispersing individuals. For the Atlantic Coast population, abundance of Piping Plovers increased substantially between 1989 and 2006 (λ = 1.83), despite the appearance of a decline in survival rates with increasing latitude (Hecht and Melvin 2009). In eastern Canada, Calvert et al. (2006) estimated that stochastic population growth rates were <1.0 and most affected by low juvenile survival, although they noted that there was considerable uncertainty in the growth rate estimates. In prairie Canada, Cohen and Gratto-Trevor (2011) estimated that a fledging rate of 0.75 chicks per pair was needed to maintain a stable population.

In the Mountain Plover, population trends have been assessed locally and rangewide. Dinsmore et al. (2005) monitored a local population for 6 years and found that recruitment rates and rate of population change were correlated with changes in the amount of prairie dog habitat; the population initially declined but then stabilized during this period. Dinsmore et al. (2010) used a stage-based matrix model to assess the influence of vital rates (clutch size, hatching success, nest survival, and chick, juvenile and adult survival) on the rate of population change at three key breeding sites. They found that lambda was most influenced by adult survival, which is consistent with life history theory for a long-lived bird that lays two clutches of three eggs in rapid succession. Estimates of lambda were >1.0 at all three sites, although there was variation in age-specific fecundity between the sites. Dreitz et al. (2006) discussed the feasibility of using an occupancy-abundance (N-mixture) model as a monitoring tool and noted its usefulness at providing insight into population patterning among, for example, different habitat types such as those occurring on prairie dog colonies versus those using agricultural fields.

Demographic Patterns

In this chapter, I have summarized estimates of multiple demographic parameters related to the understanding of plover population biology. Within the group we have a better understanding of nest survival than any other vital rate; this has implications for managing limiting factors such as predation or human activities. Age-specific survival patterns are less well understood. Population models that integrate these vital rates to predict future population growth are lacking for nearly all species. Despite the sparseness of information, several patterns emerge. First, three Southern Hemisphere plovers (Red-breasted Dotterel, White-fronted Plover, and Wrybill) experience high annual survival (adult survival >0.80), consistent with their smaller clutch size. The Double-banded Plover, another Southern Hemisphere species, has been poorly studied but may also be long-lived. This pattern contrasts with many well-studied Northern Hemisphere species in which adult survival rates are lower and generally better estimated. It is telling that many long-term studies, involving many thousands of uniquely marked individuals, have rarely documented a Snowy, Piping, or Mountain Plover older than 15 years.

Conversely, I found differing patterns between clutch size and two vital rates (nest and adult survival) in plovers. Modal clutch size ranges from two to four eggs by species; most have a typical clutch of three eggs. Using all published information (Table 10.3), and the mean of the estimates from studies where a range was reported, I calculated estimates of nest survival and adult survival as a function of clutch size (Figure 10.1). Estimates of nest survival were highly variable, especially for three- and four-egg clutch sizes, and there was no statistical difference between mean nest survival estimates of differing clutch sizes (Figure 10.1a). Nest survival seems to increase with increasing clutch size. Thus, nest survival for species with two-egg clutches may have been lowest; this makes evolutionary sense because two of the species with high adult survival fall into this group (White-fronted Plover and Wrybill). The relationship between clutch size and adult annual survival (Figure 10.1b) was more revealing and suggests that adult survival is greatest for species with two-egg clutches, which is consistent with the pattern noted earlier for nest survival. Adult survival for

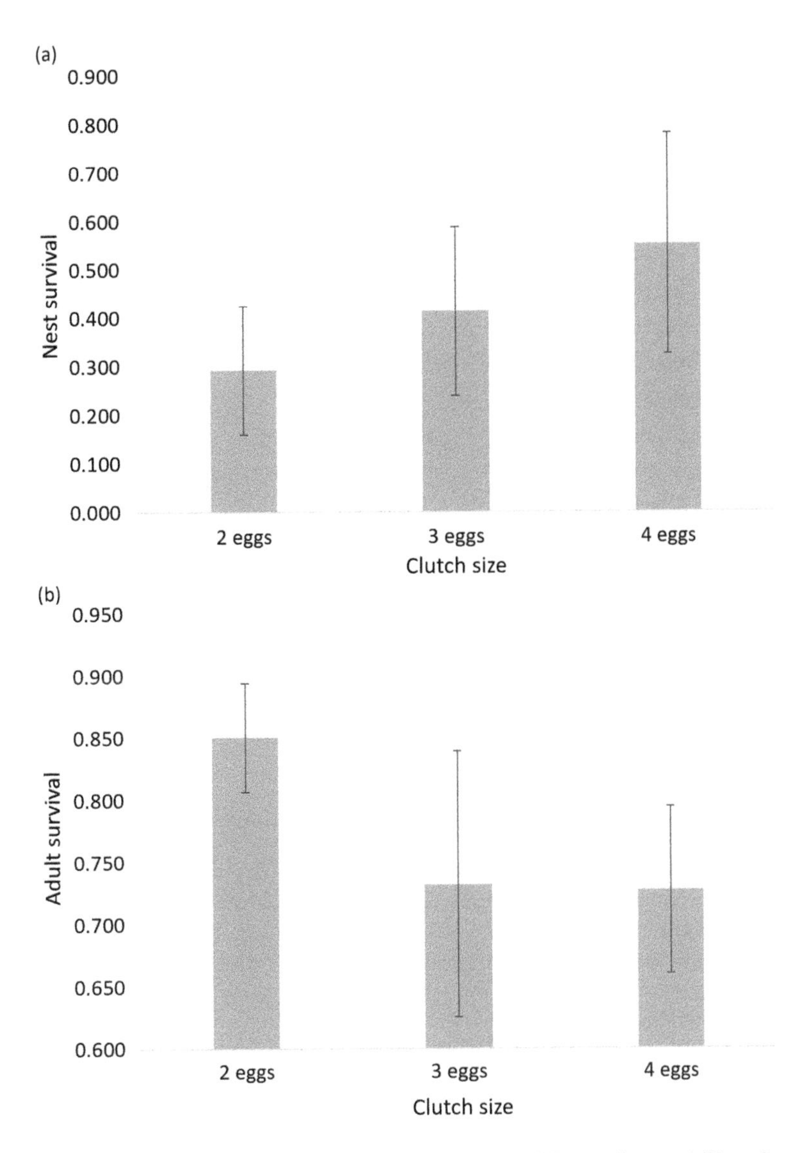

Figure 10.1. The relationship between clutch size and nest survival (a) and adult annual survival (b) in plovers. The estimates for each clutch size were calculated using the data in Table 10.3 and using the mean for any estimates that included a range in the result. Vertical bars illustrate one standard deviation from the mean. Sample sizes for nest survival [2 eggs (n = 3), 3 eggs (n = 27), 4 eggs (n = 8)] differed from those for adult survival [2 eggs (n = 4), 3 eggs (n = 10), 4 eggs (n = 14)].

species with three- and four-egg clutches did not appear to differ but was highly variable within each group. I did not investigate a pattern between clutch size and chick or juvenile survival because there were too few estimates to make such an exercise meaningful.

Population Estimates

Population estimates can aid conservation planning at local, regional, and rangewide spatial scales by providing a benchmark from which to gauge recovery goals and provide an assessment tool for gauging the success of management practices such as predator control or habitat restoration aimed at achieving conservation objectives. Population estimates, when combined with demographic information, also provide the most comprehensive look at current population status. The term "estimate" is used loosely here because of the varied methods that are used to count shorebirds. Some are a result of rigorous surveys

TABLE 10.3
Local population estimates (number of adults) for selected plovers that have been derived from formal sampling efforts.

Species	Site	Estimate	Source
Snowy Plover	Lower Laguna Madre, Texas, USA	416 (394, 438)	Hood and Dinsmore (2007)
	North America	23,555 (17,299, 29,859)	Thomas et al. (2012)
Wilson's Plover	Lower Laguna Madre, Texas, USA	279 (262, 296)	Hood and Dinsmore (2007)
Piping Plover	North America (1991)	5,484	Haig and Plissner (1993)
	North America (1996)	5,913	Plissner and Haig (2000)
	North America (2001)	5,945	Haig et al. (2005)
	North America (2006)	8,092	Elliott-Smith et al. (2009)
	North America (2011)	5,723	Elliott-Smith et al. (2015)
Killdeer	Eastern Rainwater Basin, Nebraska, USA	101,883 (87,450, 116,361)	Jorgensen et al. (2009)
Mountain Plover	Northeastern Montana, USA	1,028 (903, 1,153)	Childers and Dinsmore (2008)
Mountain Plover	Wyoming, USA	3,393	Plumb et al. (2005)
Mountain Plover	Nebraska, USA	1,568–1,650	Post van der Burg et al. (2010)
Mountain Plover	South Park, Colorado, USA	2,310	Wunder et al. (2003)
Mountain Plover	Eastern Colorado, USA	8,577 (7,511, 35,130)	Tipton et al. (2009)
Mountain Plover	Oklahoma, USA	68–91	McConnell et al. (2009)

When available, the 95% confidence limits are given in parentheses.

and include estimates of precision while others range from minimum counts to rough estimates based on professional opinion.

Considerable variation exists in the estimated size of plover populations (Appendix 10.1). Of the 39 species in this list, 26 have "Least Concern" status and population sizes that are known, or thought to be, sufficiently "large." A few species have populations that are thought to exceed 100,000 individuals. At the other extreme are species whose population status is elevated (two Vulnerable, five Near Threatened, one Endangered, and one Critically Endangered; IUCN 2016) and for which the estimated population is either small or vulnerable to a key threat. The St. Helena Plover and Shore Plover are the rarest members of this group, with bird populations numbering <500. Notably, population estimates are missing for 17 species, almost half the group; some of these could be in greater danger than believed but cannot be critically assessed because basic population data are unavailable. The distribution of plover population sizes, plotted on a logarithmic scale, reveals that known estimates are rather evenly distributed from rare (<1,000 individuals) to abundant (>100,000 individuals) (Figure 10.2).

Detailed population estimates are available for some plover species and span everything from local populations to rangewide estimates. Rangewide estimates for most species are summarized by several sources (Delaney et al. 2009, IUCN 2016; Appendix 10.1). Rangewide estimates aggregate information from local studies and make extrapolations to unsampled portions of a species' range to derive the final estimate. One unique approach was to map habitat suitability for the Madagascar Plover and then use known plover densities at specific sites to extrapolate a total population estimate (Long et al. 2008). A possible limitation to this approach is that some plovers have clumped distributions as semicolonial species that can result in their absence from seemingly "suitable" breeding habitat. Thus, such habitat-based extrapolations are useful but may lead to biased population estimates for some species. For many species, there are good estimates of the size of particular local populations. Some "estimates" are the result of coordinated surveys of suitable habitat such as the International Piping Plover Census (Haig and Plissner 1993, Haig et al. 2005); more localized efforts include those of Edgar (1969) with Red-breasted Dotterel.

APPENDIX 10.1

List of plover species covered in this book including subspecies information, breeding range, population size, population status, and migratory status.

Common name	Latin name/subspecies[a]	Breeding range[a]	Population size[b]	Population status[b]	Migratory status[b]
Red-breasted Dotterel	*C. o. obscurus*	Steward Island, New Zealand			
	C. o. aquilonius	North Island, New Zealand			
Lesser Sand-plover	*C. m. mongolus*	E Siberia, Russian Far East	310,000–390,000	Least Concern	Migratory
	C. m. pamirensis	Pamirs to w China			
	C. m. atrifrons	Himalayas, s Tibet			
	C. m. schaeferi	E Tibet to s Mongolia			
	C. m. stegmanni	Kamchatka to Chukotsk Peninsula			
Greater Sand-plover	*C. l. leschenaultii*	W. China to s Mongolia, s Siberia	180,000–360,000	Least Concern	Migratory
	C. l. columbinus	Turkey to s Afghanistan			
	C. l. scythicus	Transcaspia to se Kazakstan			
Caspian Plover	*C. asiaticus*	Caspian Sea to w China	No data	Least Concern	Migratory
Collared Plover	*C. collaris*	Mexico to n Argentina, c Chile	<10,000	Least Concern	No data
Puna Plover	*C. alticola*	Andes of Peru to nw Argentina, n Chile	No data	Least Concern	No data
Two-banded Plover	*C. falklandicus*	S. Chile, Argentina and Falkland Islands	No data	Least Concern	No data
Double-banded Plover	*C. b. bicinctus*	New Zealand and Chatham Islands	No data	Least Concern	No data
	C. b. exilis	Auckland Islands			
Kittlitz's Plover	*C. pecuarius*	Africa south of the Sahara, ne Egypt, Madagascar	No data	Least Concern	Non-migratory
Red-capped Plover	*C. ruficapillus*	Australia and Tasmania	No data	Least Concern	No data
Malay Plover	*C. peronii*	SE Asia to Philippines, Indonesia	No data	No data	No data
Kentish Plover	*C. a. alexandrinus*	W Palearctic to e China, s Japan	No data	Least Concern	Migratory; sedentary
	C. a. seebohmi	SE India, Sri Lanka			
	C. a. dealbatus	SE China			
Snowy Plover	*C. n. nivosus*	U.S. to Mexico, West Indies	25,869	Near Threatened	Migratory; sedentary
	C. n. occidentalis	Coastal Peru to south-central Chile	No data		
Javan Plover	*C. javanicus*	Coastal Java, Bali, Kangean Islands	No data	No data	No data

(*Continued*)

APPENDIX 10.1 (*Continued*)

List of plover species covered in this book including subspecies information, breeding range, population size, population status, and migratory status.

Common name	Latin name/subspecies[a]	Breeding range[a]	Population size[b]	Population status[b]	Migratory status[b]
Wilson's Plover	*C. w. wilsonia*	E USA to Belize, West Indies	No data	Least Concern	No data
	C. w. beldingi	Baja California, Mexico to s Peru			
	C. w. cinnamominus	Colombia to French Guiana			
	C. w. crassirostris	NE Brazil			
Common Ringed Plover	*C. h. hiaticula*	NE Canada, Greenland to Scandinavia	360,000–1,300,000	Least Concern	Migratory
	C. h. tundrae	Russia, Siberia			
Semipalmated Plover	*C. semipalmatus*	North America	>150,000	Least Concern	Migratory
Long-billed Plover	*C. placidus*	E Asia	1,000–25,000	Least Concern	No data
Piping Plover	*C. m. melodus*	North America	8,092	Near Threatened	Migratory
Madagascar Plover	*C. thoracicus*	SW Madagascar	2,700–3,500	Vulnerable?	Sedentary
Little Ringed Plover	*C. d. dubius*	Philippines to New Guinea, Bismarck Archipelago	280,000–530,000	Least Concern	Migratory
	C. d. curonicus	Palearctic			
	C. d. jerdoni	India, SE Asia			
Three-banded Plover	*C. t. tricollaris*	Ethiopia to Tanzania, Gabon, Chad and S. Africa	70,000–140,000	Least Concern	Migratory (?)
	C. t. bifrontalis	Madagascar	10,000–30,000	Least Concern	No data
Forbe's Plover	*C. forbesi*	W. and C. Africa	No data	No data	No data
White-fronted Plover	*C. m. marginatus*	S Angola to sw Cape Province	No data	Least Concern	Migratory; sedentary
	C. m. mechowi	Africa s of Sahara to n Angola, Botswana, Mozambique			
	C. m. arenaceus	S Mozambique to s Cape Province			
	C. m. tenellus	Madagascar			
Chestnut-banded Plover	*C. p. pallidus*	S Africa	11,000–16,000	Near Threatened	Migratory; sedentary
	C. p. venustus	Rift Valley of Kenya/Tanzania border	6,500		

(Continued)

APPENDIX 10.1 (*Continued*)

List of plover species covered in this book including subspecies information, breeding range, population size, population status, and migratory status.

Common name	Latin name/subspecies[a]	Breeding range[a]	Population size[b]	Population status[b]	Migratory status[b]
Killdeer	*C. v. vociferus*	N America	1,000,000	Least Concern	Migratory; sedentary
	C. v. temominatus	Greater Antilles			
	C. v. peruvianus	Peru and NW Chile			
Mountain Plover	*C. montanus*	Great Plains of W Canada and United States	11,000–14,000	Near Threatened	Migratory
Oriental Plover	*C. veredus*	Siberia to Manchuria and Mongolia	160,000	Least Concern	Migratory
Eurasian Dotterel	*C. morinellus*	W Alaska and n Palearctic	50,000–220,000	Least Concern	Migratory
St. Helena Plover	*C. sanctaehelanae*	St. Helena Island	<500	Critically Endangered	Sedentary
Rufous-chested Dotterel	*C. modestus*	Tierra del Fuego and Falkland Islands	No data	Least Concern	No data
Red-kneed Dotterel	*Erythrogonys cinctus*	Australia and s New Guinea	No data	Least Concern	No data
Hooded Plover	*Thinornis cucullatus*	S Australia, Tasmania, New Zealand	~10,000	Vulnerable	No data
Shore Plover	*Thinornis novaeseelandiae*	Rangitara Island, New Zealand	~300	Endangered	Migratory; sedentary
Black-fronted Dotterel	*Elseyornis melanops*	Australia, Tasmania, New Zealand	No data	Least Concern	No data
Inland Dotterel	*Peltohyas australis*	S Australia	No data	Least Concern	No data
Wrybill	*Anarhynchus frontalis*	N South Island, New Zealand	4,500–5,000	Vulnerable	Migratory
Diademed Sandpiper-Plover	*Phegornis mitchellii*	S Peru, Bolivia, W Argentina, Chile	<10,000	Near Threatened	Migratory; sedentary
Tawny-throated Dotterel	*Oreopholus r. ruficollis*	Peru to c Tierra del Fuego	1,000–10,000	Least Concern	Migratory
	O. r. pallidus	SW Ecuador and N Peru			

[a] Clements et al. (2016).
[b] IUCN website (www.iucnredlist.org/); Delaney et al. (2009) An atlas of wader populations in Africa and Western Eurasia. Wetlands International and Wader Study Group; population size estimates are broken down by species unless otherwise noted.

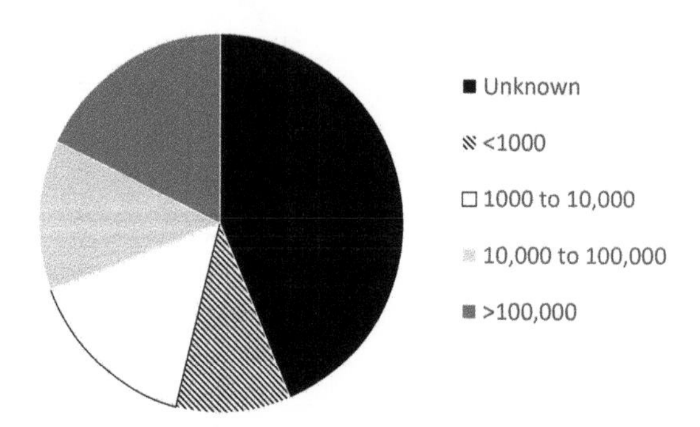

Figure 10.2. Distribution of estimated population sizes of plovers. Estimates come from Appendix 10.1 and the low end of a range of estimates is used in this figure. Species for which there are no reliable population estimates are termed "Unknown."

The other general approach to estimating population size is with the use of line or transect surveys or from capture–recapture data (Table 10.3). Systematic surveys have an added advantage of estimating detection probability and provide a more robust estimate of precision from which to calculate confidence limits. An example with Mountain Plovers illustrates some of the advances in estimating population size for shorebirds. Graul and Webster (1976) used extrapolation of density from the best habitats to estimate the continental population at a minimum of 214,200 individuals. The estimate was obviously biased high because the best density estimate was applied to all suitable habitats within the species' breeding range, which is unrealistic in practice. More recent estimates, derived by aggregating rigorous estimates of density and abundance at key breeding sites, place the population numbers at 11,000–14,000 individuals (Plumb et al. 2005). The estimates differ by more than an order of magnitude, which can be partially explained by population declines in the last several decades but also by differences in the estimation technique.

Conservation Issues

As a group, plovers face many conservation challenges that have resulted in elevated concern for some species (Appendix 10.1). Biologists have responded to these conservation concerns using varied techniques to stabilize or increase populations. Some plans take a hands-on approach such as the techniques described earlier to manage predators (see also Chapter Six, this volume). Strategies include manipulating or restoring habitat, especially for nesting (Dinsmore et al. 2014), cross-fostering chicks, or releasing captive-reared individuals, as has been done with Piping Plovers (Powell and Cuthbert 1993, Powell et al. 1997) and Shore Plovers (Davis 1994b). Translocation of independent young has been tried unsuccessfully in the Mountain Plover (Ptacek and Schwilling 1983), and is a strategy that has met with relatively little success in many threatened or endangered species (Griffith et al. 1989).

Future Perspectives

In this chapter, I have attempted to highlight what is known about the population biology of plovers. The body of work with this group is indeed impressive with >150 published works on this topic, although the bulk of the work has focused on just four species. Thus, there is a need for even the most basic demographic work for the majority of plover species. Most research has involved studies of nest survival, probably because nests are easier to monitor than other life history stages; fewer studies have specifically addressed questions relating to age-specific survival (chick, juvenile, or adult). Likewise, it is relatively easy to implement nest protection programs, but much harder to protect mobile chicks or increase the survival of juveniles or adults. Population models that integrate information from multiple life history stages to make predictions about future population change are scarce in this group and should be a priority for future research. One promising area is the development of Integrated Population Models (IPMs), which aggregate demographic data and population counts under a single modeling framework (Besbeas et al. 2002).

Unfortunately, much of the research on population biology in plovers has focused on just a handful of species, and more work is needed on the other species in this group.

Three topics should be the focus of future population biology studies: chick/juvenile survival, adult survival, and population models. Embedded within each of these topics is a need for basic science in addition to studies that provide links to the effects of specific management strategies. First, it is apparent from the relatively few published studies that chick/juvenile survival in this group is often low, and may be a key life history stage for many species. Thus, rigorous estimates of chick/juvenile survival, making use of radio telemetry or individual color bands, are needed for many species. Having more information on this topic would also allow future comparisons of survival patterns as they relate to differences in clutch size, migratory strategy, and other factors of interest. A related need is to better understand adult survival patterns, again by making use of radio telemetry or individual color-banding studies. Survival studies should include relevant covariates such as age (or age class), habitat, body condition, sex, or possibly others to better explain observed patterns. For example, covariates to explain body condition, sex, age (or even age class), habitat, or other factors could all contribute to a greater understanding of survival patterns in this group. These studies could also include specific management actions that are measured in terms of the demographic response such as an increase in chick survival to fledge. Lastly, there is a need to better integrate information about vital rates to make predictions about future population trajectories. This is especially important for this group because nine of the 39 species are thought to be at risk (Appendix 10.1; IUCN 2016). One final point is the need for consistency in the time intervals used to estimate the various survival rates. Most estimates of adult survival are on an annual basis, making comparisons between estimates straightforward. However, estimates of chick and juvenile survival are sometimes more difficult to interpret because the timing of marking (always at hatch, or throughout the pre-fledging stage) is not consistent across studies, or in some cases is not specified.

In addition to these broad needs there are many other specific questions, often related to a single species, that beg for a good answer. The Killdeer is the most widely distributed plover in North America, yet almost nothing is known about its population biology except for a single study (Powell 1992). Hopefully, someone will start marking individual Killdeer to estimate vital rates and monitor their reproductive success to fill information gaps for this common plover. The Wrybill, an enigmatic plover endemic to New Zealand, has been the focus of long-term banding efforts that have documented a longevity record of 22 years (A. C. Riegen, pers. comm.). However, the only published information on adult survival rates (estimated at 0.80–0.83 in two studies) seem at odds with a known age distribution, and I suspect adult survival rates may exceed 0.90. Lastly, there is a need to fill in the huge geographic gaps in knowledge within this group. Species in Europe and North America are generally well studied, although species such as the Wilson's Plover and Eurasian Dotterel could benefit from greater attention. There are more information gaps with the species inhabiting Africa and Australia/New Zealand, but in each region there is at least one well-studied species. In general, almost nothing is known about plover species occurring in Asia (seven species) and South America (six species). Despite these gaps in knowledge, much is known about the demography of the plovers, albeit with an emphasis on four species. Most are moderately long-lived with low chick/juvenile survival compensated by high adult survival. Nest survival increases with increasing clutch size, but annual adult survival declines with increasing clutch size. There are no clear patterns between clutch size and chick or juvenile survival, largely because of a paucity of estimates of survival for these age classes. Several plovers have a small population size or restricted geographic range that raise conservation concerns while other species are more numerous and widespread. The variation in demographic patterns highlighted in this chapter raises many unanswered questions, but at the same time it is what makes this group so fascinating.

ACKNOWLEDGMENTS

I thank A.C. Riegen for information on Wrybill and B. Vieira for information on Collared Plovers. I also thank B.K. Sandercock and an anonymous reviewer for their thoughtful critiques on this chapter and suggestions for improvement.

REFERENCES

Amat, J. A., R. M. Fraga, and G. M. Arroyo. 1999. Brood desertion and polygamous breeding in the Kentish plover *Charadrius alexandrinus*. *Ibis* 141:596–607.

Amat, J. A., R. M. Fraga, and G. M. Arroyo. 2001. Intraclutch egg-size variation and offspring survival in the Kentish plover *Charadrius alexandrinus*. *Ibis* 143:17–23.

Anteau, M. J., T. L. Shaffer, M. H. Sherfy, M. A. Sovada, J. H. Stucker, and M. T. Wiltermuth. 2012. Nest survival of piping Plovers at a dynamic reservoir indicates an ecological trap for a threatened population. *Oecologia* 170:1167–1179.

Augustine, D. J. and S. K. Skagen. 2014. Mountain plover nest survival in relation to prairie dog and fire dynamics in shortgrass steppe. *Journal of Wildlife Management* 78:595–602.

Baird, B. and P. Dann. 2003. The breeding biology of Hooded Plovers, *Thinornis rubricollis*, on Phillip Island, Victoria. *Emu* 103:323–328.

Barber, C., A. Nowak, K. Tulk, and L. Thomas. 2010. Predator exclosures enhance reproductive success but increase adult mortality of Piping Plovers (*Charadrius melodus*). *Avian Conservation and Ecology* 5: art6.

Baudains, T. P. and P. Lloyd. 2007. Habituation and habitat changes can moderate the impacts of human disturbance on shorebird breeding performance. *Animal Conservation* 10:400–407.

Besbeas, P., S. N. Freeman, B. J. T. Morgan, and E. A. Catchpole. 2002. Integrating mark-recapture-recovery and census data to estimate animal abundance and demographic parameters. *Biometrics* 58:540–547.

Bijleveld, A. I., R. B. MacCurdy, Y.-C. Chan, E. Penning, R. M. Gabrielson, J. Cluderay, E. L. Spaulding, A. Dekinga, S. Holthuijsen, J. ten Horn, M. Brugge, J. A. van Gils, D. W. Winkler, and T. Piersma. 2016. Understanding spatial distributions: Negative density-dependence in prey causes predators to trade-off prey quantity with quality. *Proceedings of the Royal Society B* 283:20151557.

Boyd, H. 1962. Mortality and fertility of European Charadrii. *Ibis* 104:368–387.

Brown, A. C. and K. Brindock. 2011. Breeding success and nest site selection by a Caribbean population of Wilson's Plovers. *Wilson Journal of Ornithology* 123:814–819.

Brownie, C., D. R. Anderson, K. P. Burnham, and D. S. Robson. 1985. *Statistical Inference from Band-Recovery Data: A Handbook*, 2nd ed., Resource Publication No. 156. U.S. Fish and Wildlife Service: Washington, DC, 305.

Brudney, L. J., T. W. Arnold, S. P. Saunders, and F. J. Cuthbert. 2013. Survival of Piping plover (*Charadrius melodus*) chicks in the Great Lakes region. *Auk* 130:150–160.

Burns, F., N. McCulloch, T. Székely, and M. Bolton. 2013. No overall benefit of predator exclosure cages for the endangered St. Helena plover *Charadrius sanctaehelenae*. *Ibis* 155:397–401.

Calvert, A. M., D. L. Amirault, F. Shaffer, R. Elliot, A. Hanson, J. McKnight, and P. D. Taylor. 2006. Population assessment of an endangered shorebird: The Piping Plover (*Charadrius melodus melodus*) in Eastern Canada. *Avian Conservation and Ecology* 1:art4.

Calvert, A. M. and P. D. Taylor. 2011. Measuring conservation trade-offs: Demographic models provide critical context to empirical studies. *Avian Conservation and Ecology* 6:art2.

Catlin, D. H., J. H. Felio, and J. D. Fraser. 2011a. Effect of great horned owl trapping on chick survival in Piping Plovers. *Journal of Wildlife Management* 75:458–462.

Catlin, D. H., J. D. Fraser, J. H. Felio, and J. B. Cohen. 2011b. Piping plover habitat selection and nest success on natural, managed, and engineered sandbars. *Journal of Wildlife Management* 75:305–310.

Catlin, D. H., J. D. Fraser, and J. H. Felio. 2015. Demographic responses of Piping Plovers to habitat creation on the Missouri River. *Wildlife Monographs* 192:1–42.

Catlin, D. H., O. Milenkaya, K. L. Hunt, M. J. Friedrich, and J. D. Fraser. 2014. Can river management improve the Piping Plover's long-term survival on the Missouri River? *Biological Conservation* 180:196–205.

Catlin, D. H., S. L. Zeigler, M. B. Brown, L. R. Dinan, J. D. Fraser, K. L. Hunt, and J. G. Jorgensen. 2016. Metapopulation viability of an endangered shorebird depends on dispersal and human-created habitats: Piping Plovers (*Charadrius melodus*) and prairie rivers. *Movement Ecology* 4:6.

Childers, T. M. 2006. Mountain Plover Abundance and Nest Survival in Northeastern Montana. M.S. Thesis, Mississippi State University, Starkville, MS.

Childers, T. M. and S. J. Dinsmore. 2008. Density and abundance of Mountain Plovers in northeastern Montana. *Wilson Journal of Ornithology* 120:700–707.

Claassen, A. H., T. W. Arnold, E. A. Roche, S. P. Saunders, and F. J. Cuthbert. 2014. Factors influencing nest survival and renesting by Piping Plovers in the Great Lakes region. *Condor* 116:394–407.

Clements, J. F., T. S. Schulenberg, M. J. Iliff, D. Roberson, T. A. Fredericks, B. L. Sullivan, and

C. L. Wood. [online]. 2016. The eBird/Clements Checklist of Birds of the World: v2016. <www.birds.cornell.edu/clementschecklist/download/> (23 April 2017).

Cohen, J. B., J. D. Fraser, and D. H. Catlin. 2006. Survival and site fidelity of Piping Plovers on Long Island, New York. *Journal of Field Ornithology* 77:409–417.

Cohen, J. B. and C. Gratto-Trevor. 2011. Survival, site fidelity, and the population dynamics of Piping Plovers in Saskatchewan. *Journal of Field Ornithology* 82:379–394.

Cohen, J. B., L. M. Houghton, and J. D. Fraser. 2009. Nesting density and reproductive success of Piping Plovers in response to storm- and human-created habitat changes. *Wildlife Monographs* 173:1–24.

Cole, L. C. 1957. Sketches of general and comparative demography. *Quantitative Biology* 22:1–15.

Colwell, M. A. 2010. *Shorebird Ecology, Conservation, and Management*. University of California Press, Berkeley, CA.

Colwell, M. A., N. S. Burrell, M. A. Hardy, K. Kayano, J. J. Nuir, W. J. Pearson, S. A. Peterson, and K. A. Sesser. 2010. Arrival times, laying dates, and reproductive success of Snowy Plovers in two habitats in coastal northern California. *Journal of Ornithology* 81:349–360.

Colwell, M. A., E. J. Feucht, M. J. Lau, D. J. Orluck, S. E. McAllister, and A. N. Transou. 2017a. Recent Snowy Plover population increase arises from high immigration rate in coastal northern California. *Wader Study* 124:40–48.

Colwell, M. A., E. J. Feucht, S. E. McAllister, and A. N. Transou. 2017b. Lessons learned from the oldest Snowy Plover. *Wader Study* 124:157–159.

Colwell, M. A., S. J. Hurley, J. N. Hall, and S. J. Dinsmore. 2007. Age-related survival and behavior of Snowy Plover chicks. *Condor* 109:638–647.

Colwell, M. A., J. J. Meyer, M. A. Hardy, S. E. McAllister, A. N. Transou, R. R. Levalley, and S. J. Dinsmore. 2011. Western Snowy Plovers *Charadrius alexandrinus nivosus* select nesting substrates that enhance egg crypsis and improve nest survival. *Ibis* 153:303–311.

Colwell, M. A., C. B. Millett, J. J. Meyer, J. N. Hall, S. J. Hurley, S. E. McAllister, A. N. Transou, and R. R. Levalley. 2005. Snowy Plover reproductive success in beach and river habitats. *Journal of Field Ornithology* 76:373–382.

Colwell, M. A., W. J. Pearson, L. J. Eberhart-Phillips, and S. J. Dinsmore. 2013. Apparent survival of Snowy Plovers (*Charadrius nivosus*) varies with reproductive effort and year and between sexes. *Auk* 130:725–732.

Cormack, R. M. 1964. Estimates of survival from the sighting of marked animals. *Biometrika* 51:429–438.

Davis, A. 1994a. Breeding biology of the New Zealand Shore Plover *Thinornis novaeseelandiae*. *Notornis* 41:195–208.

Davis, A. 1994b. Status, distribution, and population trends of the New Zealand Shore Plover *Thinornis novaeseelandiae*. *Notornis* 41:179–194.

Delaney, S., D. A. Scott, T. Dodman, and D. A. Stroud. 2009. *An Atlas of Wader Populations in Africa and Western Eurasia*. Wetlands International, Wageningen, The Netherlands.

DeRose-Wilson, A., J. D. Fraser, S. M. Karpanty, and D. H. Catlin. 2013. Nest-site selection and demography of Wilson's Plovers on a North Carolina barrier island. *Journal of Field Ornithology* 84:329–344.

DeRose-Wilson, A., J. D. Fraser, S. M. Karpanty, and M. D. Hillman. 2015. Effects of overflights on incubating Wilson's Plover behavior and heart rate. *Journal of Wildlife Management* 79:1246–1254.

Dinsmore, S. J. 2008. Influence of drought on annual survival of the Mountain plover in Montana. *Condor* 110:45–54.

Dinsmore, S. J. 2013. Mountain plover responses to deltamethrin treatments on prairie dog colonies in Montana. *Ecotoxicology* 22:415–424.

Dinsmore, S. J., E. P. Gaines, S. F. Pearson, D. J. Lauten, and K. A. Castelein. 2017. Factors affecting Snowy plover chick survival in a managed population. *Condor* 119:34–43.

Dinsmore, S. J. and D. H. Johnson. 2012. Chapter 15—Population analysis in wildlife biology. In Silvy, N. S. (ed.), *The Wildlife Techniques Manual—Research*, vol. 1. The Wildlife Society: Bethesda, MD, 349–380.

Dinsmore, S. J. and F. L. Knopf. 2005. Differential parental care by adult Mountain plovers, *Charadrius montanus*. *Canadian Field-Naturalist* 119:532–536.

Dinsmore, S. J., D. J. Lauten, K. A. Castelein, E. P. Gaines, and M. A. Stern. 2014. Predator exclosures, predator removal, and habitat improvement increase nest success of Snowy plovers in Oregon, USA. *Condor* 116:619–628.

Dinsmore, S. J. and M. D. Smith. 2010. Mountain plover responses to plague in Montana. *Vector-Borne and Zoonotic Diseases* 10:37–45.

Dinsmore, S. J., G. C. White, and F. L. Knopf. 2002. Advanced techniques for modeling avian nest survival. *Ecology* 83:3476–3488.

Dinsmore, S. J., G. C. White, and F. L. Knopf. 2003. Annual survival and population estimates of Mountain plovers in southern Phillips County, Montana. *Ecological Applications* 13:1013–1026.

Dinsmore, S. J., G. C. White, and F. L. Knopf. 2005. Mountain plover population responses to black-tailed prairie dogs in Montana. *Journal of Wildlife Management* 69:1546–1553.

Dinsmore, S. J., M. B. Wunder, V. J. Dreitz, and F. L. Knopf. 2010. An assessment of factors affecting population growth of the Mountain Plover. *Avian Conservation and Ecology* 5:art5.

Dowding, J. E. and E. C. Murphy. 2001. The impact of predation by introduced mammals on endemic shorebirds in New Zealand: A conservation perspective. *Biological Conservation* 99:47–64.

Drake, K. R., J. E. Thompson, K. L. Drake, and C. Zonick. 2001. Movements, habitat use, and survival of nonbreeding Piping Plovers. *Condor* 103:259–267.

Dreitz, V. 2009. Parental behavior of a precocial species: Implications for juvenile survival. *Journal of Applied Ecology* 46:870–878.

Dreitz, V., R. Conrey, and S. Skagen. 2012. Drought and cooler temperatures are associated with higher nest survival in mountain plovers. *Avian Conservation and Ecology* 7: art6.

Dreitz, V. J. and F. L. Knopf. 2007. Mountain plovers and the politics of research on private lands. *BioScience* 57:681–687.

Dreitz, V. J., P. M. Lukacs, and F. L. Knopf. 2006. Monitoring low density avian populations: An example using Mountain Plovers. *Condor* 108:700–706.

Eberhart-Phillips, L. J. and M. A. Colwell. 2014. Conservation challenges of a sink: The viability of an isolated population of the Snowy Plover. *Bird Conservation International* 24:327–341.

Eberhart-Phillips, L. J., B. R. Hudgens, and M. A. Colwell. 2016. Spatial synchrony of a threatened shorebird: Regional roles of climate, dispersal, and management. *Bird Conservation International* 26:119–135.

Efe, M. A., L. Bugoni, L. V. Mohr, A. Scherer, S. B. Scherer, and O. P. Bairro. 2001. First-known record of breeding for the Black Skimmer (*Rynchops niger*) in a mixed colony in Ibicui River, Rio Grande do Sul state, southern Brazil. *International Journal of Ornithology* 4:103–107.

Edgar, A. T. 1969. Estimated population of the Red-breasted Dotterel. *Notornis* 16:85–100.

Elliott-Smith, E., M. Bidwell, A. E. Holland, and S. M. Haig. 2015. Data from the 2011 International Piping Plover Census. U.S. Geological Survey Data Series 922, Reston, VA.

Elliott-Smith, E., S. M. Haig, and B. M. Powers. 2009. Data from the 2006 International Piping Plover Census. U.S. Geological Survey Data Series 426, Reston, VA, 332.

Evans, P. R. and M. W. Pienkowski. 1984. Population dynamics of shorebirds. Pp. 83–123 in Burger, J. and Olla, B. L. (eds), *Shorebirds: Breeding Behavior and Populations*. Plenum Press, New York.

Foppen, R. P. B., F. A. Majoor, F. J. Willems, P. L. Meininger, G. C. van Houwelingen, and P. A. Wolf. 2006. Survival and emigration rates in Kentish *Charadrius alexandrinus* and Ringed Plovers Ch. *Hiaticula* in the Delta area, SW-Netherlands. *Ardea* 94:159–173.

Forslund, P. and T. Pärt. 1995. Age and reproduction in birds—hypotheses and tests. *Trends in Ecology and Evolution* 10:374–378.

Fransson, T., T. Kolehmainen, C. Kroon, L. Jansson, and T. Wenninger. [online]. 2010. EURING List of Longevity Records for European Birds. <www.euring.org//data-and-codes/longevity-list> (18 July 2016).

Graul, W. D. 1975. Breeding biology of the Mountain Plover. *Wilson Bulletin* 87:6–31.

Graul, W. D. and L. E. Webster. 1976. Breeding status of the Mountain Plover. *Condor* 78:265–267.

Griffith, B., J. M. Scott, J. W. Carpenter, and C. Reed. 1989. Translocation as a species conservation tool: Status and strategy. *Science* 245:477–480.

Groen, N. M., R. Kentie, P. de Goeij, B. Verheijen, J. C. E. W. Hooijmeijer, and T. Piersma. 2012. A modern landscape ecology of Black-tailed Godwits: Habitat selection in southwest Friesland, The Netherlands. *Ardea* 100:19–28.

Haig, S. M., C. L. Ferland, F. J. Cuthbert, J. Dingledine, J. P. Goossen, A. Hecht, and N. McPhillips. 2005. A complete species census and evidence for regional declines in Piping Plovers. *Journal of Wildlife Management* 69:160–173.

Haig, S. M. and L. W. Oring. 1988. Mate, site, and territory fidelity in Piping Plovers. *Auk* 105:268–277.

Haig, S. M. and J. H. Plissner. 1993. Distribution and abundance of Piping Plovers: Results and implications of the 1991 international census. *Condor* 95:145–156.

Hardy, M. A. and M. A. Colwell. 2008. The impact of predator exclosures on Snowy Plover nesting success: A seven-year study. *Wader Study Group Bulletin* 115:161–166.

Harris, W. C., D. C. Duncan, R. J. Franken, D. T. McKinnon, and H. A. Dundas. 2005. Reproductive success of Piping Plovers at Big Quill Lake, Saskatchewan. *Wilson Bulletin* 117:165–171.

Hay, J. R. 1984. The Behavioural Ecology of the Wrybill Plover *Anarhynchus frontalis*. *Ph.D. Thesis*, University of Auckland, Auckland, New Zealand.

Hecht, A. and S. M. Melvin. 2009. Population trends of Atlantic Coast Piping Plovers, 1986–2006. *Waterbirds* 32:64–72.

Herman, D. M. and M. A. Colwell. 2015. Lifetime reproductive success of Snowy Plovers in coastal northern California. *Condor* 117:473–481.

Hillman, M. D., S. M. Karpanty, J. D. Fraser, F. J. Cuthbert, J. M. Altman, T. E. Borneman, and A. DeRose-Wilson. 2012. Evidence for long-distance dispersal and successful interpopulation breeding of the endangered Piping Plover. *Waterbirds* 35:642–644.

Hong, S. and S. Higashi. 2008. Nest site preference and hatching success of the Kentish Plover (*Charadrius alexandrinus*) in the Nakdong Estuary, Busan, Republic of Korea. *Journal of Ecology and Field Biology* 31:201–206.

Hood, S. L. and S. J. Dinsmore. 2007. Abundance of Snowy and Wilson's Plovers in the lower Laguna Madre region of Texas. *Journal of Field Ornithology* 78:362–368.

IUCN. [online]. 2016. IUCN Red List of Threatened Species. Version 2015-4. <www.iucnredlist.org> (30 June 2016).

Jorgensen, J. G., J. P. McCarty, and L. L. Wolfenbarger. 2009. Killdeer *Charadrius vociferous* breeding abundance and habitat use in the Eastern Rainwater Basin, Nebraska. *Wader Study Group Bulletin* 116:65–68.

Kålås, J. A. and I. Byrkjedal. 1984. Breeding chronology and mating system of the Eurasian Dotterel (*Charadrius morinellus*). *Auk* 101:838–847.

Kentie, R., C. Both, J. C. E. W. Hooijmeijer, and T. Piersma. 2015. Management of modern agricultural landscapes increases nest predation rates in Black-tailed Godwits *Limosa limosa*. *Ibis* 157:614–625.

Knopf, F. L. and B. J. Miller. 1994. *Charadrius montanus*-montane, grassland, or bare-ground plover? *Auk* 111:504–506.

Knopf, F. L. and J. R. Rupert. 1995. Habits and habitats of Mountain Plovers in California. *Condor* 97:743–751.

Knopf, F. L. and J. R. Rupert. 1996. Reproduction and movements of Mountain Plovers breeding in Colorado. *Wilson Bulletin* 108:28–35.

Knopf, F. L. and M. B. Wunder. 2006. Mountain Plover (*Charadrius montanus*). In Poole, A. and Gill, F. (eds), *The Birds of North America*, No. 211. The Academy of Natural Sciences, Philadelphia, PA, and The American Ornithologists' Union, Washington, DC.

Knowles, C. J., C. J. Stoner, and S. P. Gieb. 1982. Selective use of black-tailed prairie dog towns by Mountain Plovers. *Condor* 84:71–74.

Kosztolányi, A., S. Javed, C. Küpper, I. C. Cuthill, A. A. Shamsi, and T. Székely. 2009. Breeding ecology of Kentish Plover *Charadrius alexandrinus* in an extremely hot environment. *Bird Study* 56:244–252.

Lafferty, K. D., D. Goodwin, and C. P. Sandoval. 2006. Restoration of breeding by Snowy Plovers following protection from disturbance. *Biodiversity and Conservation* 15:2217–2230.

Larson, M. A., M. R. Ryan, and B. G. Root. 2000. Piping Plover survival in the Great Plains: An updated analysis. *Journal of Field Ornithology* 71:721–729.

Larson, M.A., M. R. Ryan, and R. K. Murphy. 2002. Population viability of Piping Plovers: Effects of predator exclusion. *Journal of Wildlife Management* 66:361–371.

Lei, W. P. 2010. Studies on Migration and Habitat Use of Waterbirds at Typical Wetlands Around Bohei Bay. *M.S. Thesis*, Beijing Normal University, Beijing, China.

Lloyd, P. 2008. Adult survival, dispersal, and mate fidelity in the White-fronted Plover *Charadrius marginatus*. *Ibis* 150:182–187.

Lloyd, P., and É. E. Plagányi. 2002. Correcting observer effect bias in estimates of nesting success of a coastal bird, the White-fronted Plover *Charadrius marginatus*. *Bird Study* 49:124–130.

Loegering, J. P. and J. D. Fraser. 1995. Factors affecting Piping Plover chick survival in different brood-rearing habitats. *Journal of Wildlife Management* 59:646–655.

Long, P. R., S. Zefania, R. H. ffrench-Constant, and T. Székely. 2008. Estimating the population size of an endangered shorebird, the Madagascar Plover, using a habitat suitability model. *Animal Conservation* 11:118–127.

Lukacs, P. M., V. J. Dreitz, F. L. Knopf, and K. P. Burnham. 2004. Estimating survival probabilities of unmarked dependent young when detection is imperfect. *Condor* 106:927–932.

Mabee, T. J. and V. B. Estelle. 2000. Assessing the effectiveness of predator exclosures for plovers. *Wilson Bulletin* 112:14–20.

Mabee, T. J., A. M. Wildman, and C. B. Johnson. 2006. Using egg floatation and eggshell evidence to determine age and fate of Arctic shorebird nests. *Journal of Field Ornithology* 77:163–172.

Maguire, G. S., A. K. Duivenvoorden, M. A. Weston, and R. Adams. 2011. Provision of artificial shelter on beaches is associated with improved shorebird fledging success. *Bird Conservation International* 21:172–185.

Marchant, S. and P. J. Higgins. 1993. *Handbook of Australian, New Zealand, and Antarctic Birds*, vol. 2. Oxford University Press, Melbourne, Australia.

Maslo, B. and J. L. Lockwood. 2009. Evidence-based decisions on the use of predator exclosures in shorebird conservation. *Biological Conservation* 142:3213–3218.

Mayer, P. M. and M. R. Ryan. 1991. Survival rates of artificial Piping Plover nests in American Avocet colonies. *Condor* 93:753–755.

Mayfield, H. R. 1961. Nesting success calculated from exposure. *Wilson Bulletin* 73:255–261.

Mayfield, H. R. 1975. Suggestions for calculating nest success. *Wilson Bulletin* 87:456–466.

McCulloch, N. and K. Norris. 2001. Diagnosing the cause of population changes: Localized habitat change and the decline of the endangered St. Helena Wirebird. *Journal of Applied Ecology* 38:771–783.

McConnell, S., T. J. O'Connell, D. M. Jr. Leslie, and J. S. Shackford. 2009. Mountain Plovers in Oklahoma: Distribution, abundance, and habitat use. *Journal of Field Ornithology* 80:27–34.

McGowan, C. P. 2015. Comparing models of Red Knot population dynamics. *Condor* 117:494–502.

Mettenbrink, C. W., V. J. Dreitz, and F. L. Knopf. 2006. Nest success of Mountain Plovers relative to anthropogenic edges in eastern Colorado. *Southwestern Naturalist* 51:191–196.

Miller, B. J. and F. L. Knopf. 1993. Growth and survival of Mountain Plovers. *Journal of Field Ornithology* 64:500–506.

Mullin, S. M., M. A. Colwell, S. E. McAllister, and S. J. Dinsmore. 2010. Apparent survival and population growth of Snowy Plovers in coastal northern California. *Journal of Wildlife Management* 74:1792–1798.

Neuman, K. K., G. W. Page, L. E. Stenzel, J. C. Warriner, and J. S. Warriner. 2004. Effect of mammalian predator management on Snowy Plover breeding success. *Waterbirds* 27:257–263.

Nguyen, L. P., K. F. Abraham, and E. Nol. 2006. Influence of Arctic Terns on survival of artificial and natural Semipalmated Plover nests. *Waterbirds* 29:100–104.

Nol, E. and M. S. Blanken. 2014. Semipalmated Plover (*Charadrius semipalmatus*) in Poole, A. and Gill, F. (eds), *The Birds of North America*, No. 444. The Academy of Natural Sciences: Philadelphia, PA, and The American Ornithologists' Union, Washington, DC.

Nol, E., S. Williams, and B. K. Sandercock. 2010. Natal philopatry and apparent survival of juvenile Semipalmated Plovers. *Wilson Journal of Ornithology* 122:23–28.

Noszály, G., T. Székely, and J. M. C. Hutchinson. 1995. Brood survival of Kentish Plovers (*Charadrius alexandrinus*) in alkaline grasslands and drained fish-ponds. *Ornis Hungarica* 5:15–21.

Nur, N., G. W. Page, and L. E. Stenzel. 1999. Population Viability Analysis for Pacific Coast Snowy Plovers. Appendix D. in 2007. Recovery plan for the Pacific Coast population of the western Snowy Plover (*Charadrius alexandrines nivosus*). U.S. Fish and Wildlife Service. Sacramento, CA.

Olson, S. L. and D. Edge. 1985. Nest site selection by Mountain Plovers in Northcentral Montana. *Journal of Range Management* 38:280–282.

Olson-Edge, S. L. and W. D. Edge. 1987. Density and distribution of the Mountain Plover on the Charles M. Russell National Wildlife Refuge. *Prairie Naturalist* 19:233–238.

Oyler-McCance, S. J., J. St. John, F. L. Knopf, and T. W. Quinn. 2005. Population genetic analysis of Mountain Plover using mitochondrial DNA sequence data. *Condor* 107:353–362.

Page, G. W., L. E. Stenzel, D. W. Winkler, and C. W. Swarth. 1983. Spacing out at Mono Lake: Breeding success, nest density, and predation in the Snowy Plover. *Auk* 100:13–24.

Pakanen, V. M., S. Lampila, H. Arppe, and J. Valkama. 2015. Estimating sex specific apparent survival and dispersal of Little Ringed Plovers (*Charadrius dubius*). *Ornis Fennica* 92:172–186.

Paton, P. W. C. 1994. Survival estimates for Snowy Plovers breeding at Great Salt Lake, Utah. *Condor* 96:1106–1109.

Paton, P. W. C. 1995. Breeding biology of Snowy Plovers at Great Salt Lake, Utah. *Wilson Bulletin* 107:275–288.

Patterson, M. E., J. D. Fraser, and J. W. Roggenbuck. 1991. Factors affecting Piping Plover productivity on Assateague Island. *Journal of Wildlife Management* 55:525–531.

Pearson, W. J. and M. A. Colwell. 2013. Effects of nest success and mate fidelity on breeding dispersal in a population of Snowy Plovers *Charadrius nivosus*. *Bird Conservation International* 24:342–353.

Pearson, S. F., S. M. Knapp, and C. Sundstrom. 2016. Evaluating the ecological and behavioural factors influencing Snowy Plover *Charadrius nivosus* egg hatching and the potential benefits of predator exclosures. *Bird Conservation International* 26:100–118.

Pienkowski, M. W. 1984. Behaviour of young Ringed Plovers *Charadrius hiaticula* and its relationship to growth and survival to reproductive age. *Ibis* 126:133–155.

Piping Plover Conservation in Nova Scotia. 2017. A 17-Year Old Banded Piping Plover in Cuba. <www.facebook.com/ploverconservation/posts/1860425277300890> (17 February 2017).

Plissner, J. H. and S. M. Haig. 2000. Status of a broadly distributed endangered species: Results and implications of the second International Piping Plover Census. *Canadian Journal of Zoology* 78:128–139.

Plumb, R. E., F. L. Knopf, and S. H. Anderson. 2005. Minimum population size of Mountain Plovers breeding in Wyoming. *Wilson Bulletin* 117:15–22.

Pollock, K. H. 1982. A capture-recapture design robust to unequal probability of capture. *Journal of Wildlife Management* 46:757–760.

Post van der Burg, M., B. Bly, T. VerCauteren, and A. J. Tyre. 2010. Making better sense of monitoring data from low density species using a spatially explicit modelling approach. *Journal of Applied Ecology* 48:47–55.

Post van der Burg, M. P. and A. J. Tyre. 2011. Integrating info-gap decision theory with robust population management: A case study using the Mountain Plover. *Ecological Applications* 21:303–312.

Powell, A. N. 1992. The Effects of Early Experience on the Development, Behavior, and Survival of Shorebirds. *Ph.D. Dissertation*, University of Minnesota, St. Paul, MN.

Powell, A. N. and F. J. Cuthbert. 1993. Augmenting small populations of plovers: An assessment of cross-fostering and captive-rearing. *Conservation Biology* 7:160–168.

Powell, A. N., F. J. Cuthbert, L. C. Wemmer, A. W. Doolittle, and S. T. Feirer. 1997. Captive-rearing Piping Plovers: Developing techniques to augment wild populations. *Zoo Biology* 16:461–477.

Ptacek, J. and M. Schwilling. 1983. Mountain Plover reintroduction in Kansas. *Kansas Ornithological Society Bulletin* 34:21–22.

Que, P., Y. Chang, L. Eberhart-Phillips, Y. Liu, T. Székely, and Z. Zhang. 2015. Low nest survival of a breeding shorebird in Bohei Bay, China. *Journal of Ornithology* 156:297–307.

Riegen, A. C. and J. E. Dowding. 2003. The Wrybill *Anarhynchus frontalis*: A brief review of status, threats and work in progress. *Wader Study Group Bulletin* 100:20–24.

Roche, E. A., J. B. Cohen, D. H. Catlin, D. L. Amirault-Langlais, F. J. Cuthbert, C. L. Gratto-Trevor, J. Felio, and J. D. Fraser. 2010. Range-wide Piping Plover survival: Correlated patterns and temporal declines. *Journal of Wildlife Management* 74:1784–1791.

Root, B. G., M. R. Ryan, and P. M. Mayer. 1992. Piping Plover survival in the Great Plains. *Journal of Field Ornithology* 63:10–15.

Rotella, J. J., S. J. Dinsmore, and T. L. Shaffer. 2004. Modeling nest-survival data: A comparison of recently developed methods that can be implemented in MARK and SAS. *Animal Biodiversity and Conservation* 27:187–205.

Ruhlen, T. D., S. Abbott, L. E. Stenzel, and G. W. Page. 2003. Evidence that human disturbance reduces Snowy Plover chick survival. *Journal of Field Ornithology* 74:300–304.

Saalfield, S. T., W. C. Conway, D. A. Haukos, and W. P. Johnson. 2011. Nest success of Snowy Plovers (*Charadrius nivosus*) in the Southern High Plains of Texas. *Waterbirds* 34:389–399.

Sandercock, B. K. 2003. Estimation of survival rates for wader populations: A review of mark-recapture methods. *Wader Study Group Bulletin* 100:163–174.

Sandercock, B. K., T. Székely, and A. Kosztolányi. 2005. The effects of age and sex on the apparent survival of Kentish Plovers breeding in southern Turkey. *Condor* 107:583–596.

Saunders, S. P., T. W. Arnold, E. A. Roche, and F. J. Cuthbert. 2014. Age-specific survival and recruitment of piping plovers *Charadrius melodus* in the Great Lakes region. *Journal of Avian Biology* 45:437–449.

Saunders, S. P. and F. J. Cuthbert. 2015. Chick mortality leads to male-biased sex ratios in endangered Great Lakes Piping Plovers. *Journal of Field Ornithology* 86:103–114.

Schneider, E. G. and S. R. McWilliams. 2007. Using nest temperature to estimate nest attendance of Piping Plovers. *Journal of Wildlife Management* 76:1998–2006.

Seber, G. A. F. 1982. *The Estimation of Animal Abundance and Related Parameters*, 2nd ed. Macmillan Publishing Company, Inc., New York.

Shackford, J. S., D. M. Jr. Leslie, and W. D. Harden. 1999. Range-wide use of cultivated fields by Mountain Plovers during the breeding season. *Journal of Field Ornithology* 70:114–120.

Shaffer, T. L. 2004. A unified approach to analyzing nest success. *Auk* 121:526–540.

Shaffer, T. L. and F. R. Thompson. 2007. Making meaningful estimates of nest survival with model-based methods. *Studies in Avian Biology* 34:84–95.

Skrade, P. D. B. and S. J. Dinsmore. 2010. Sex-related dispersal in the Mountain Plover (*Charadrius montanus*). *Auk* 127:671–677.

Skrade, P. D. B. and S. J. Dinsmore. 2013. Egg-size investment in a bird with uniparental incubation by both sexes. *Condor* 115:508–514.

Stenzel, L. E., G. W. Page, J. C. Warriner, J. S. Warriner, D. E. George, C. R. Eyster, B. A. Ramer, and K. K. Neuman. 2007. Survival and natal dispersal of juvenile Snowy Plovers (*Charadrius alexandrinus*) in central coastal California. *Auk* 124:1023–1036.

Stenzel, L. E., G. W. Page, J. C. Warriner, J. S. Warriner, K. K. Neuman, D. E. George, C. R. Eyster, and F. C. Bidstrup. 2011. Male-skewed adult sex ratio, survival, mating opportunity, and annual productivity in the Snowy Plover *Charadrius alexandrinus*. *Ibis* 153:312–322.

Stenzel, L. E., J. C. Warriner, J. S. Warriner, K. S. Wilson, F. C. Bidstrup, and G. W. Page. 1994. Long-distance breeding dispersal of Snowy Plovers in Western North America. *Journal of Animal Ecology* 63:887–902.

Studds, C. E., B. E. Kendall, N. J. Murray, H. B. Wilson, D. I. Rogers, R. S. Clemens, K. Gosbell, C. J. Hassell, R. Jessop, D. S. Melville, D. A. Milton, C. D. T. Minton, H. P. Possingham, A. C. Riegen, P. Straw, E. J. Woehler, and R. A. Fuller. 2017. Rapid population decline in migratory shorebirds relying on Yellow Sea tidal mudflats as stopover sites. *Nature Communications* 8:14895.

Summers, R. W. and P. A. R. Hockey. 1980. Breeding biology of the White-fronted Plover (*Charadrius marginatus*) in the southwestern Cape, South Africa. *Journal of Natural History* 14:433–445.

Thomas, S. M., J. E. Lyons, B. A. Andres, E. Elliott-Smith, E. Palacios, J. F. Cavitt, J. A. Royle, S. D. Fellows, K. Maty, W. H. Howe, E. Mellink, S. Melvin, and T. Zimmerman. 2012. Population size of Snowy Plovers breeding in North America. *Waterbirds* 35:1–14.

Tipton, H. C., P. F. Doherty, Jr. and V. J. Dreitz. 2009. Abundance and density of Mountain Plover (*Charadrius montanus*) and Burrowing Owl (*Athene cunicularia*) in eastern Colorado. *Auk* 126:493–499.

Toral, G. M. and J. Figuerola. 2012. Nest success of Black-winged Stilt *Himantopus himantopus* and Kentish Plover *Charadrius alexandrinus* in rice fields, southwest Spain. *Ardea* 100:29–36.

Tulp, I. 1998. Nest success of White-fronted Plover *Charadrius marginatus* and Kittlitz's Plover *Charadrius pecuarius* in a South African dune field. *Wader Study Group Bulletin* 87:51–54.

U.S. Department of the Interior. 2002. Endangered and threatened wildlife and plants: Threatened status and special regulation for the Mountain Plover. *Federal Register* 67 (234):72396–72407.

U.S. Fish and Wildlife Service. 2007. *Recovery Plan for the Pacific Coast Population of the Western Snowy Plover (Charadrius alexandrines nivosus)*. USFWS, Sacramento, CA.

U.S. Geological Survey [online]. 2016. Longevity Records of North American Birds. <www.pwrc.usgs.gov/BBL/longevity/Longevity_main.cfm> (18 July 2016).

Van de Pol, M., Y. Vindenes, B.-E. Sæther, S. Engen, B. J. Ens, K. Oosterbeek, and J. M. Tinbergen. 2010. Effects of climate change and variability on population dynamics in a long-lived shorebird. *Ecology* 91:1192–1204.

Warriner, J. S., J. C. Warriner, G. W. Page, and L. E. Stenzel. 1986. Mating system and reproductive success in a small population of polygamous Snowy Plovers. *Wilson Bulletin* 98:15–37.

Watts, C. M., J. Cao, C. Panza, C. Dugaw, M. Colwell, and E. A. Burroughs. 2012. Modeling the effects of predator exclosures on a Western Snowy Plover population. *Natural Resource Modeling* 25:529–547.

Wemmer, L. C., U. Ozesmi, and F. J. Cuthbert. 2001. A habitat-based population model for the Great Lakes population of the Piping Plover (*Charadrius melodus*). *Biological Conservation* 99:169–181.

Weston, M. A. 2000. The Effect of Human Disturbance on the Breeding Biology of Hooded Plovers. *Ph.D. Thesis*, University of Melbourne, Melbourne, Australia.

Weston, M.A., G. C. Ehmke, and G. S. Maguire. 2011. Nest return times in response to static versus mobile human disturbance. *Journal of Wildlife Management* 75:252–255.

Wilcox, L. 1959. A twenty year banding study of the Piping Plover. *Auk* 76:129–152.

Williams, B. K., J. D. Nichols, and M. J. Conroy. 2002. *Analysis and Management of Animal Populations*. Academic Press: San Diego, CA.

Wunder, M. B., F. L. Knopf, and C. A. Pague. 2003. A high-elevation population of Mountain Plovers in Colorado. *Condor* 105:654–662.

Yasué, M. and P. Dearden. 2006. The potential impact of tourism development on habitat availability and productivity of Malaysian Plovers *Charadrius peronii*. *Journal of Applied Ecology* 43:978–989.

Yu, Y. and X. Pei. 1996. Studies on the breeding ecology of *Charadrius alexandrinus dealbatus*. Pp. 305–308 in *Study on Chinese Ornithology*. China Forestry Publishing House: Beijing, China.

Zefania, S., R. ffrench-Constant, P. R. Long, and T. Székely. 2008. Breeding distribution and ecology of the threatened Madagascar Plover *Charadrius thoracicus*. *Ostrich* 79:43–51.

Ram Papish

CHAPTER ELEVEN

Human Disturbance*

Michael A. Weston

Abstract. Populations of many plovers occupy habitats favored by humans for recreation and other human uses. This chapter describes the responses of shorebirds to the presence of people and the potential impact these responses may have on plovers throughout the annual cycle. Plover responses to people are often frequent, sometimes lengthy, are (1) context-specific, for example, more intense when breeding and less intense in busier areas, and (2) sophisticated, for example, tailored to aspects of human behavior. These responses disrupt time budgets, sometimes displace or alter habitat use of plovers, and variably (mostly negatively) influence reproductive success. Evidence of links between disturbance and plover population viability or trends is sparse, but the few studies available suggest that disturbance can limit population size and mediates habitat quality. Many research gaps exist, and there is a need to reconcile the varying and often contradictory results of available studies.

Keywords: disruption, flight-initiation distance (FID), optimal escape theory, people, recreation, response.

PEOPLE, PLOVERS, AND PROBLEMS

Being ground-dwelling, non-perching, and generally inhabiting open places, plovers are especially heavily exposed to people and their activities. Many plovers inhabit wetlands, marshes, coasts, grasslands, plains, or paddy fields that are also areas used by people for recreation and agriculture (Colwell 2010). Worldwide, increasing human populations and mobility mean that people occur in an ever-increasing proportion of plover habitats. Other processes (e.g., sea-level rise) are likely to increase the co-occurrence of humans and plovers in the future (Seavey et al. 2010).

In many areas, plovers overlap substantially in time and space with people (Figure 11.1). Where people and plovers co-occur, they can interact, with possible detrimental outcomes for plovers. Disturbance is the physiological or behavioral disruption to normal states associated with the proximity of a potential threat ("stimulus") such as a person, vehicle, and aircraft (Box 11.1). Many studies of plover disturbance report habitats with higher densities of people, dogs, and vehicles than plovers. All report overall higher rates of encounters with, and often more intense responses to, anthropogenic stimuli than "natural" stimuli such as predators (e.g., Goldin and Regosin 1998, Lafferty 2001b). While it is possible that

* Michael A. Weston. 2019. Human Disturbance. Pp. 277–308 in Colwell and Haig (editors). The Population Ecology and Conservation of Charadrius Plovers (no. 52), CRC Press, Boca Raton, FL.

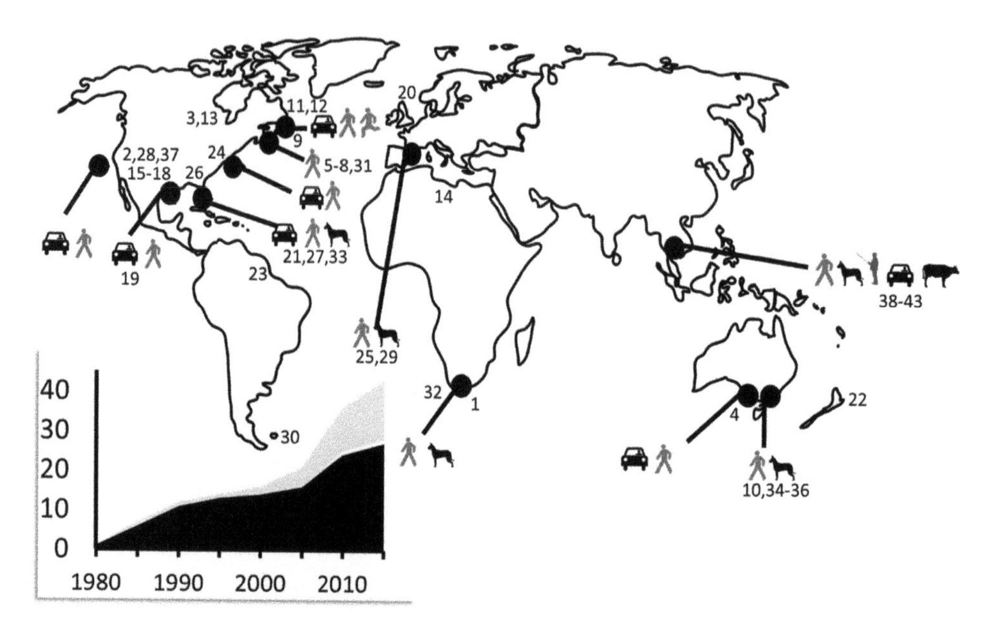

Figure 11.1 A world of disturbance. The main stimuli in plover habitats as reported by major and other studies of plover disturbance. The inset shows the accumulation of studies over time: white, demographic studies; gray, site-based studies; and black, modeling study. Major studies on plover disturbance are numbered: [1]Baudains and Lloyd (2007); [2]Brindock and Colwell (2011); [3]Brunton (1990); [4]Buick and Paton (1989); [5–7]Burger (1990, 1994); [8]Burger et al. (1995); [9]Doherty and Heath (2011); [10]Dowling and Weston (1999); [11]Flemming et al. (1988); [12]Goldin and Regosin (1988); [13]Haffner et al. (2009); [14]Hamza and Selmi (2015); [15–17]Lafferty (2001a–2001c); [18]Lafferty et al. (2006); [19]LeDee et al. (2008); [20]Liley and Sutherland (2007); [21]Loegering and Fraser (1995); [22]Lord et al. (2001); [23]Lunardi and Masedo (2014); [24]MacIvor et al. (1990); [25]Martín et al. (2015); [26]Nicholls and Baldassarre (1990); [27]Patterson et al. (1991); [28]Ruhlen et al. (2003); [29]Gómez-Serrano and López-López (2014); [30]St Claire et al. (2010); [31]Staine and Burger (1994); [32]Summers and Hockey (1981); [33]Webber et al. (2013); [34,35]Weston and Elgar (2005a, 2007); [36]Weston et al. (2011); [37]Wilson and Colwell (2010); [38]Yasué 2006; [39–41]Yasué and Dearden (2006a, 2006b, 2009); and, [42,43]Yasué et al. (2008a, 2008b).

researchers of disturbance may focus on the most human-dominated ecosystems, it is nonetheless evident that many plovers occur in areas where humans are common (Figure 11.1). Geographical gaps exist regarding studies of plover disturbance (i.e., most of Eurasia, Africa, and South America). Some of these areas are the most populous on the planet, and may be areas where disturbance is most impactful, or where habituation or other adaptations are expressed most clearly. This said, in the well-studied temperate areas, peak occurrence of humans often occurs during the plover breeding season, and at times when days are warmest, periods when plovers are thought to be most vulnerable to deleterious effects of disturbance (Lloyd and Plagányi 2000, Baird and Dann 2003; Figure 11.2). Disturbance of plovers also occurs at other times, even by night (Staine

BOX 11.1 Definitions, Glossary, and a Conceptual Model

Colwell (2010) comprehensively reviews the various definitions of disturbance and associated concepts as they apply to shorebirds. However, problems with vague concepts and definitions, and inconsistent usage, persist. One example is the commonly used phrase "response to disturbance," when disturbance itself is the consequence of responses to humans. Another example is the use of terms such as "potential disturbance" to describe a stimulus, an agent, or event which has the capacity to cause a response. I adopt a definition which is: (1) explicit; (2) centers around the consensus definition of "disturbance"; (3) draws on terminology from optimal escape theory, because responses to humans are apparently antipredator in nature; and (4) has some precedence in the literature (Weston et al. 2012). Below explicit definitions are presented; two salient points warrant mention. First, I differentiate "disturbance" from "ecological disturbance," the physical disruption of a habitat (e.g., wetland reclamation). Second, I exclude the direct mechanical destruction of plovers, their eggs, or young (e.g., collisions with vehicles, crushing by people or stock). Destruction co-occurs with, or is sometimes the consequence of, disturbance but these represent distinctly different processes or mechanisms through which humans interact with plovers. Much bird disturbance literature makes this distinction, albeit implicitly. For example, the effects of hunting of birds involve: (1) direct mortality (destruction) and (2) disruption of normal behavior, including displacement. The term "disturbance" is used to describe only the disruption of normal states by hunting (see, for example, Fox and Madsen 1997).

Definitions

Stimulus: An agent/event which has the potential to evoke a response. Stimuli can be "natural" (e.g., raptors) or anthropogenic (humans and their activities), including companion animals. Stimuli are perceived by the senses (i.e., they provide cues), presumably mostly visual (e.g., the appearance of a person) but also auditory (e.g., aircraft noise, barking of dogs), and possibly olfactory cues.

Stimulus attributes: Stimuli may occur in different configurations (e.g., walker with a dog, or a dog or walker alone), numbers (e.g., one walker versus two), frequencies, speeds and direction (e.g., joggers versus walkers), and/or volumes (decibels).

Encounter: An event where a stimulus is within a proximity that has the potential to evoke a response.

Response: A short-term change in behavior or physiology as an immediate result of an encounter. Longer-term or higher-scale consequences of responses are here differentiated from the aforementioned definition.

Disturbance: The disruption of normal activities or states (behavior and physiology) caused by responses.

A Simple Conceptual Model

The mechanistic model of plover disturbance involves a behavioral response which is initiated when a stimulus occurs within a certain proximity and which ceases after the stimulus has moved away (Figure 11.1.1). When encounters with humans are frequent, plovers may exhibit "response discounting" where extended responses occur such that one absence encompasses encounters with multiple stimuli (Weston and Elgar 2005a, 2007).

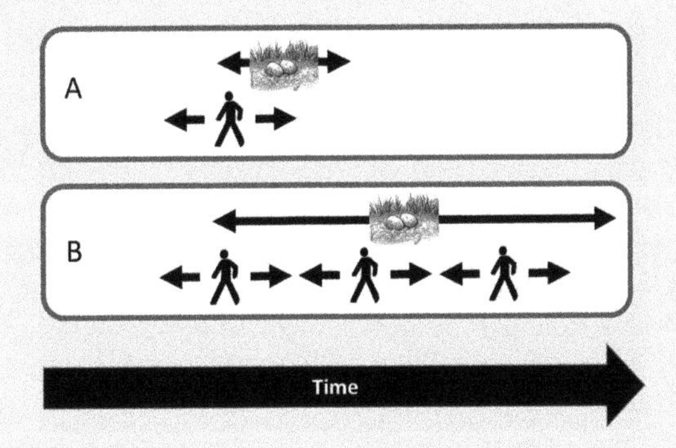

Figure 11.1.1 A conceptual model of plover response to humans. (a) Traditional model and (b) the multiple stimulus circumstance. The example response is an absence from the nest (indicated by arrows).

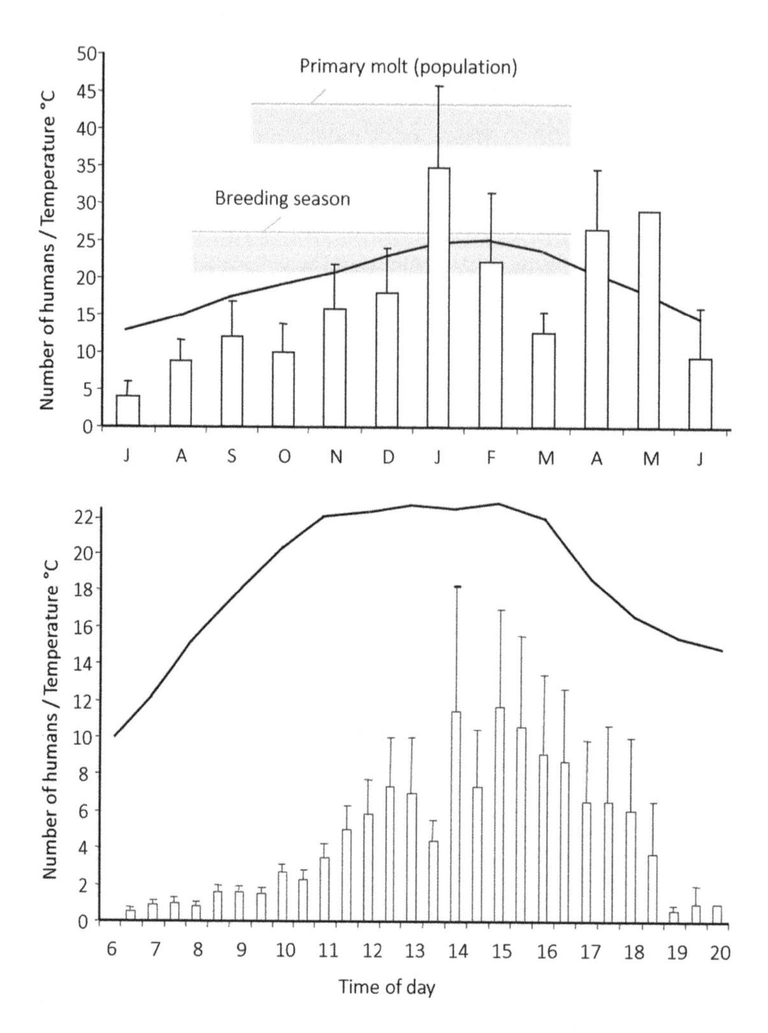

Figure 11.2 An example of virtually complete spatial and temporal overlap between humans and plovers. Any time of year or day the resident beach-nesting Hooded Plover is exposed to people, especially during key life history events such as breeding and primary molt (after Rogers et al. 2014). The average (± SE) number of humans recorded (top) on the beach (5.4 km surveys between Point Lonsdale and Collendina, Victoria, Australia), in each month of the year, and (bottom) during scans of territories every half hour of full day observation periods (n = 27 territories) are shown (after Weston 2000). The solid line is air temperature; top, mean maximum air temperature (at nearby Geelong Airport; www.weatherzone.com.au), bottom, average air temperature experienced during incubation (Weston and Elgar 2005b).

and Burger 1994). Although few studies exist of nocturnal disturbance, some plover responses to humans differ by day and night (Graul 1975), so the impacts of nocturnal disturbance could differ from those during daytime.

Plovers respond to people in the same way that they respond to predators, for example, by hiding or using distraction displays (Brunton 1990, Frid and Dill 2002). It is unsurprising that plovers treat people as predators; while the vast majority of humans intend no harm to plovers, they sometimes inadvertently destroy nests and kill chicks and adults. Indeed, some nonhuman primates can be active plover predators, and human hunting of shorebirds occurs, or has recently occurred, in some areas (Baudains and Lloyd 2007, Glover et al. 2011). The details and intensity of plover responses are often adjusted to a given threat as well as to the costs of responding, so humans often evoke nuanced responses (Flemming et al. 1988, Brunton 1990, Weston and Elgar 2005a, 2007).

To survive, plovers must detect and discriminate an array of benign and dangerous changes in their environment ("agents" or "stimuli"), judge the risk associated with each, and respond to them appropriately. The consequences of inappropriate or inadequate response can be fatal (Ydenberg and Dill 1986). Examples include being mauled by a dog, trampled, or struck by a vehicle (Dowling and Weston 1999, Baudains and Lloyd 2007). Frequent "unnecessary" responses could be maladaptive and cause breeding failure, loss of condition, and other consequences which diminish fitness (see below). Thus, it is unsurprising that plover responses to humans are complex, and generally regarded as nontrivial. Under some circumstances, such as when fitness costs of disturbance are high and refugia are unavailable, plover population viability may theoretically be compromised by disturbance (Gill et al. 2001).

Different mechanisms link responses (and consequent disturbance) with effects on wildlife, and there are many modulating factors (Tablado and Jenni 2016). Evidence is often inconsistent or contradictory when comparisons are made between studies, areas, species, or prevailing human regimes. Here, I summarize and synthesize information on human disturbance (Box 11.2), including some unpublished datasets. I describe plover responses to humans, examine the consequences of these responses, and highlight some key research opportunities.

PLOVER RESPONSES TO HUMANS

Plover responses to stimuli are often discriminatory and sophisticated such that they are tailored to the nature and behavior of the threatening agent. For example, plovers often give more intense responses to humans in comparison with other stimuli (Flemming et al. 1988). White-fronted Plovers bury eggs before departing their nests more frequently when approached by humans than when they leave under other circumstances (Summers and Hockey 1981) and nesting Killdeer give more intense and less variable responses (e.g., distraction displays) to humans than to natural predators (Brunton 1986, 1990). Plover responses are also complex and evidently involve cognition. For example, Piping Plovers defending nests from a human intruder perform distraction displays in a manner that suggests

BOX 11.2 Information Sources on Plover Disturbance

Many of the prominent papers on wildlife disturbance focus on plovers and there has been a steady growth in the number of available studies (Figure 11.1). Major published papers (excluding theses and papers which are not species-specific) are those that focus on, or contain a substantial emphasis on, disturbance (43 studies, all but three of which involved breeding plovers, and all but six of which had a sole focus on plovers). No study focuses on a migratory species breeding in high latitudes, where people are at low densities. Major studies have focused on a small number of plover systems (Piping [35%], Snowy [16%], Malay [12%] and Hooded plovers [12%]; 16 species are studied in some substantive way) and most involved threatened, wetland (mostly coastal) taxa. Most studies are from a few ecosystems.

Studies of plover disturbance can be categorized as: (1) site-based research (35% of major studies); (2) demographic studies quantifying the costs of disturbance in terms of reductions in survival or breeding success, or processes which may affect these (60%); (3) a population perspective focused, for example, on measuring the density-dependent consequences of shifts in distribution resulting from disturbance (<1%); or (4) behavioral studies (<1%) which use standardized human approaches to assess questions which do not involve disturbance *per se* (after Gill 2007). About half of all studies were comparative in nature, with observational and experimental methodologies common.

Several papers which focus at the assemblage level on shorebirds or on important habitat (e.g., waterbird habitat) deal with disturbance at particular sites, usually stopover or nonbreeding sites, and include plovers in their examination of disturbance and its effects (e.g., Yasué et al. 2008). Nonbreeding shorebirds often form mixed-species flocks (in which, for example, various species of plover join other species such as sandpipers), which at least sometimes respond to humans as cohesive social units, making any plover-specific responses or effects difficult to identify (e.g., Lilleyman et al. 2016). These are critical studies because mixed-species flocks may respond to humans differently compared with single-species flocks (Weston et al. 2012). This chapter emphasizes studies in which mechanisms, patterns, and outcomes can be clearly attributed to, or elucidated, in regard to plovers.

purposeful, responsive, goal-oriented attempts to distract, as opposed to purely reflexive, invariant behavior (Ristau 1992). Finally, responses of plovers to humans are context-specific, and vary among individuals and life history stages. For example, Brunton (1990) illustrated the influence of parental investment and roles of nesting Killdeer responding to people. Defense intensity increased during incubation. After hatching, parental defense declined, perhaps because of reduced vulnerability of young during this stage. Overall males took greater risks, remained on the nest longer, defended offspring more intensely, and displayed closer to the approaching human compared with females.

Despite context-specific adjustments and individual variation, the basic form of plover responses to humans (or predators) is generalized within and across species, and involves an escalation of responses after the stimulus detection (Figure 11.3). While physiological responses are to be expected, they remain virtually unstudied (Box 11.3).

Vigilance, Alert, and Alarm

Vigilance is a behavioral state whereby a plover monitors or assesses potential threats and makes decisions regarding risks and responses. Vigilance is an alternative state to that normally exhibited by an individual that is incubating, brooding, or feeding. Vigilant plovers typically raise their head, extend the neck, and sometimes stand upright (Figure 11.4). This posture, coupled with observations that visual obstructions reduce the distance at which plovers respond to people (Gómez-Serrano and López-López 2014, Lomas et al. 2014), reinforces the broadly held idea that much threat detection by plovers is visual in nature. Plovers also have distinct alarm calls, especially evident when breeding or flocking, suggesting social facilitation of states of alarm. For example, brood-rearing

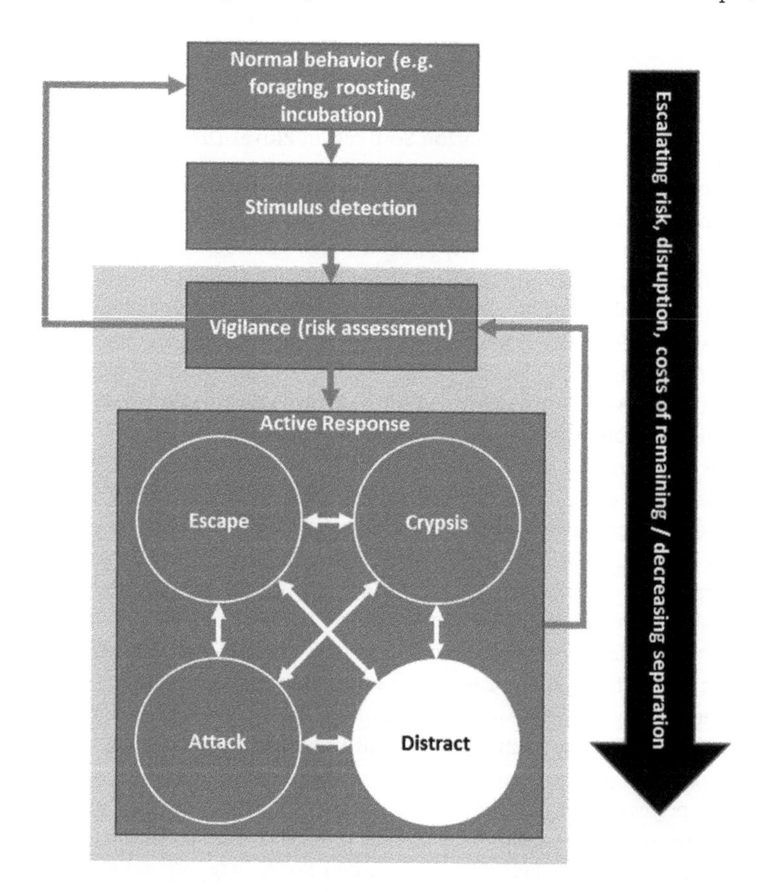

Figure 11.3 A conceptual schematic of plover responses to a stimulus. Dark gray fill indicates aspects which may occur during breeding or nonbreeding, no fill indicates elements which occur during breeding only. Light gray fill indicates the components which are defined here as a "response." "Separation" refers to the distance between the plover and the stimulus.

BOX 11.3 Under the Feathers: Physiological Responses to Humans

Behavioral responses of birds to humans have been the focus of research in plovers to date, partly because they are readily quantified. There is no doubt, however, that such responses are integrated with physiological processes. No studies on the physiological responses to humans have been conducted on plovers, because, as small birds, plovers are more difficult to fit with instrumentation (Colwell 2010). Physiological responses could also conceivably occur in the absence of obvious behavioral responses, and at distances which exceed those detectable behavioral responses (Weston et al. 2012). Despite being virtually unstudied, physiological disturbance is likely to affect plovers. Examples include increased energy expenditure associated with physiological excitement and altered immune-competency associated with stress. For developing embryos or chicks, disturbance-mediated thermal stress could conceivably alter development rates or condition.

Tan et al. (2015) showed that a blood-based measure of chronic stress did not increase in Red-capped Plovers whose nest was fitted with a predator exclusion cage. However, plovers incubating nests in the open (uncovered nests, which were associated with longer parental flight-initiation distances (FIDs); Lomas et al. 2014) exhibited higher stress levels. Thus, the ability to detect more threats (which enables earlier responses), may lead to more physiological stress, although higher temperatures at open nests may also explain this pattern (Tan et al. 2015).

Physiological studies of disturbance face problems associated with intrusive techniques, which themselves can cause physiological disruption. However, innovative, novel, and emerging methodologies (e.g., measuring heart rates via fake eggs containing recording devices, measuring stress hormones from feathers) mean that ecophysiology studies of disturbance represent a major research opportunity. Key questions to be tackled include what degree of physiological disruption is problematic, and how these changes relate to demography (Sutherland 2007).

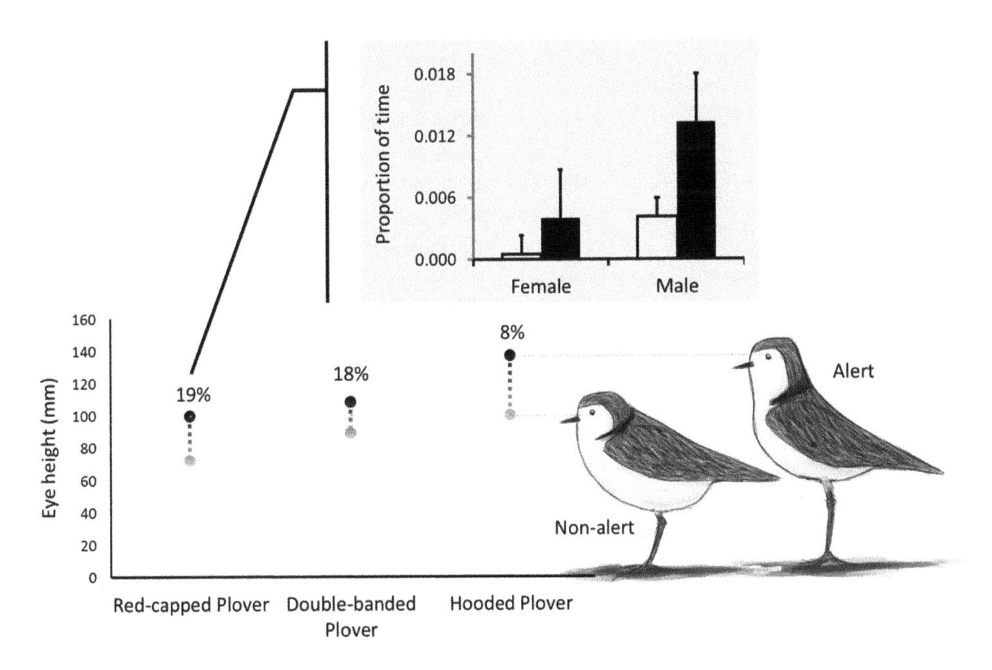

Figure 11.4 The "head up" posture in plovers evidently functions to monitor potential threats. The eye height above ground (mm) is shown for three plover species in roosting (gray dots) and vigilant (black dot) postures. Measurements were made using calipers from clear published images using tarsus length to calibrate height. (Data from H.K. Glover.) Percentages refer to percentage increase in eye heights. The inset shows mean (± SD) "head-up" rate for male and female Red-capped Plovers, singly and in pairs (black bars) versus flocks (open bars). (T. Forge, D. Whisson and M.A. Weston, unpublished data.)

Piping Plovers increased vigilance when one parent became alarmed (Burger 1990). Individual Red-capped Plovers occurring in flocks typically show reduced vigilance (Figure 11.4; but see Lilleyman et al. 2016), in accord with the group size effect that has been shown for many organisms.

Vigilance increases with proximity of a potential threat, permitting "Alert Distances" (ADs) to be measured. As for other birds, plover ADs are positively correlated with escape distances, that is, their FID which is the distance at which "physical" escape is initiated (Table 11.1). While vigilance presumably precedes most escape behavior, except perhaps when plovers are "startled," AD is less reliably measured than FID (Guay et al. 2013) and for plovers is often unmeasurable (of 300 FIDs available for 12 plover species, 49.3% of approaches did not produce a measurable AD; MAW unpublished data). Vigilance represents a disruption to normal activities, and given it occurs at greater distances than escape behavior, documenting ADs can inform management tools like set-back distances. Although no studies are available, vigilance directed at humans may also come at the cost of vigilance for other predators.

Another distinctive behavior possibly associated with plover vigilance is "head-bobbing," where the head is rapidly raised and lowered. The function of head-bobbing remains unknown. It could serve as a non-acoustic signal of perceived risk to conspecifics and predators similar to, for example, tail flicking in the Rallidae (Alvarez 1993), to aid the judgment of distance from eyes that lack binocular convergence (Wilson 1950), or to avoid retinal habituation and maintain maximum neural responsiveness (Waldvogel 1990). Head-bobbing has not yet been definitively linked to the degree of risk or to vigilance and warrants further research.

Plover responses to humans vary in their latency: this is at least partly determined by the latency of the stimulus (Weston et al. 2011). Thus, vigilance is also evidently involved in the decision to end a response and resume normal activities. For example, plovers make complex decisions, such that responses depend on environmental and other factors that balance the benefits and costs of responses, regarding when to return to a nest (Yasué and Dearden 2006b). Some of these decisions clearly involve monitoring the presence and behaviors of humans (Weston et al. 2011).

Distraction Behavior

Distraction essentially attracts the attention of a potential predator by advertising an apparently easy meal. In plovers, it can involve false incubating, rodent runs where the tail is held down and the bird runs haphazardly, or broken-wing displays in which one or both wings are dragged, dropped, or flapped (Cairns 1982, Weston and Elgar 2005a, 2007). These displays are often delivered in close proximity to a threat, and can be vigorous. Distraction displays can also be adjusted in relation to the behavior of the nearby person. Piping Plovers perform displays more intensely in front of, rather than behind, a person near their nest, and birds adjust their behavior when a person does not follow them, which suggests they monitor a person's response to their display (Ristau 1992). Given that distraction displays provide visual cues to potential threats such as humans,

TABLE 11.1

Linear regressions of Alert Distance (AD) on Flight-initiation Distance (FID) for plover populations (where $n \geq 7$).

Species (population)	Equation	r^2 (n)
Double-banded Plover (Australia)	AD = 1.161*FID+5.280	0.676 (10)
Greater Sand-Plover (Africa)	AD = 4.285*FID−55.623	0.478 (11)
Hooded Plover (Australia)	AD = 0.977*FID+8.750	0.580 (23)
Lesser Sand-Plover (Africa)	AD = 1.377*FID+4.066	0.386 (15)
Lesser Sand-Plover (Australia)	AD = 0.819*FID+10.455	0.727 (7)
Red-capped Plover (Australia)	AD = 1.243*FID+2.133	0.664 (47)
Common Ringed Plover (Africa)	AD = 0.629*FID+13.068	0.336 (28)

NOTES: Equations are presented to enable practitioners to predict FIDs from ADs or vice versa. All models were statistically significant. Data from Guay et al. (2016) and unpublished (A.Z. Radkovic, P.J. Guay, W. Van Dongen and M.A. Weston).

it is unsurprising that they are not apparently given at night (Mountain Plovers, Graul 1975; Red-capped Plovers, pers. obs.). People sometimes follow Kentish Plovers giving distraction displays, suggesting they may effectively distract some humans (Gómez-Serrano and López-López 2014).

Apart from the odd, apparently aberrant distraction display, distraction behaviors occur only during defense of eggs or young from nearby threats (Brunton 1990, Weston and Elgar 2005a). Distraction displays are more likely or intense when chicks hatch and are young (Brunton 1990, Lord et al. 2001, Weston and Elgar 2005a).

Crypsis and Deceit

Crypsis or "camouflage" involves concealing the location of the bird itself, its eggs, or young and encompasses a variety of behaviors including hiding and distant or deceitful responses which function to conceal the presence or location of eggs or young. The use of crypsis can profoundly influence aspects of parental care such as incubation scheduling between the sexes (Ekanayake et al. 2015) and cryptic responses have been linked to several mechanisms which may cause reproductive failure (Box 11.4). For flying-age plovers, crypsis can involve remaining still, ducking behind objects, or crouching (e.g., Inland Dotterels; MacLean 1976). Responses of breeding plovers often occur in a manner consistent with concealing the presence or location of nests or young, and involve early and indirect departures from nests or the vicinity of young. Even New Zealand Shore Plovers, which nest under extensive solid cover such as log piles, leave the nest when humans approach (Davis 1994). Some species completely or partially bury their eggs before departure (e.g., White-fronted and Kittlitz's plovers; Summers and Hockey 1981; pers. obs.). Plover chicks routinely crouch, often beside or within cover, and remain motionless until the threat (e.g., person) passes. The use of crypsis is also discriminatory, with some stimuli (e.g., dogs) evoking crypsis more frequently or for

BOX 11.4 Consequences of Disrupted Parental Care

Most studies of disturbance to plovers have focused on breeding, presumably because of the long-held but untested suspicion that disturbance is most impactful and easily linked to population viability through its purported suppression of reproductive success.

Incubation

Human-induced absences from the nest can be frequent and lengthy and are suspected of having a number of negative consequences. First, the number of departures and arrivals from the nest increases with increasing numbers of encounters with humans (Weston 2000), resulting in more plover footprints to and from the nest, potentially compromising crypsis (Colwell 2010). This substrate-dependent effect will be driven by the frequency of responses, interacting with processes that may remove footprints (e.g., wind and rain); it has yet to be definitively demonstrated for any plover.

Second, absences compromise the thermoregulation of the eggs, potentially to the point where embryo death occurs (Yasué and Dearden 2006c, Weston and Elgar 2007, Figure 11.4.1). In hot environments, cooling eggs is a particular challenge, especially given the upper lethal thermal limits of eggs are closer to average incubation temperatures than are lower lethal thermal limits, and that cooling eggs following a disturbance takes longer than warming them (Figure 11.4.2). The degree to which overall egg thermoregulation is compromised by disturbance remains unclear and appears context-dependent. White-fronted Plovers maintained similar egg temperatures in areas of relatively high and low disturbance despite some drastic drops in nest attentiveness (Baudains and Lloyd 2007). However, a false egg in Hooded Plover clutches deviated more from a predicted ideal egg temperature during human-initiated absences (2.3 ± 2.1 [SE] °C cf. 1.6 ± 1.5°C for incubator-initiated absences) and during disturbance eggs experienced temperatures 9.8°C–45.0°C (Weston 2000). The thermal tolerances of embryos, which are likely to change with age, are not understood in detail, though at least these upper temperatures are likely to have been fatal.

BOX 11.4 (*Continued*) Consequences of Disrupted Parental Care

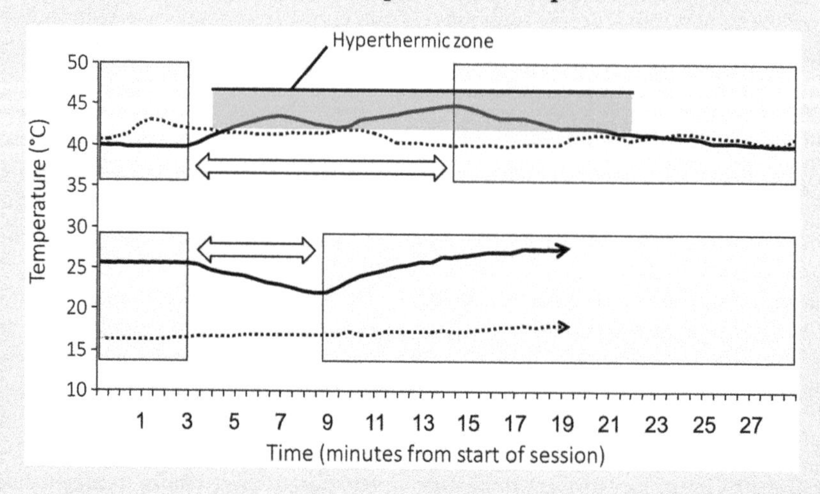

Figure 11.4.1 Examples of temperature change of a model egg incubated by Hooded Plovers when disturbed from its nest on a hot day (upper) and cold day (lower). Egg temperature (black line) and air temperature (dotted line); gray bars indicate incubation and arrows indicate the human-caused absence from the nest (see Weston 2000). The upper lethal thermal limits are indicated (after Yasué et al. 2007).

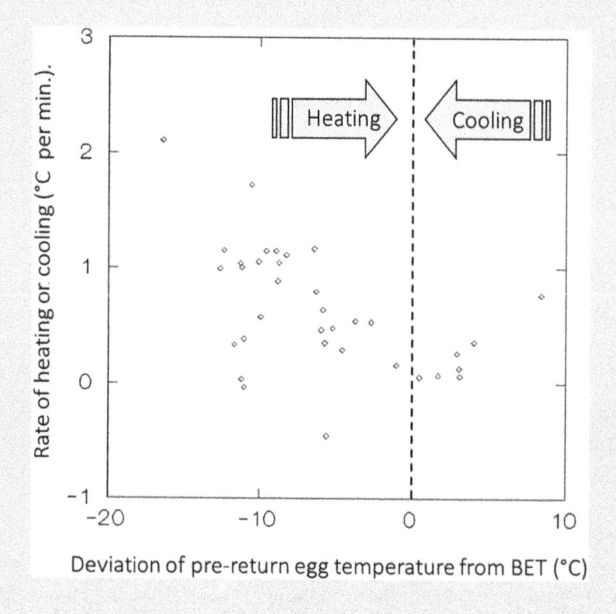

Figure 11.4.2 The restoration of egg temperatures after a human-initiated absence from the nest (Hooded Plover, first 5 min after return, absences ≥10 min). Benchmark Egg Temperature (BET) is the temperature expected at a given air temperature under uninterrupted incubation (Weston 2000).

Reduced attentiveness might result in lengthened incubation durations potentially increasing exposure to threats (e.g., predators), or embryo death may result in non-hatching eggs, and sometimes aberrantly long attempts at incubation (e.g., 60+ days in Hooded Plovers; Weston 2000). Higher rates of non-hatching Hooded Plover eggs occurred in microhabitats where responses to humans are most frequent and longest (Figure 11.4.3). Clutches might be abandoned when embryos die from thermal stress during forced absences, or when incubators perceive their risk

BOX 11.4 (Continued) Consequences of Disrupted Parental Care

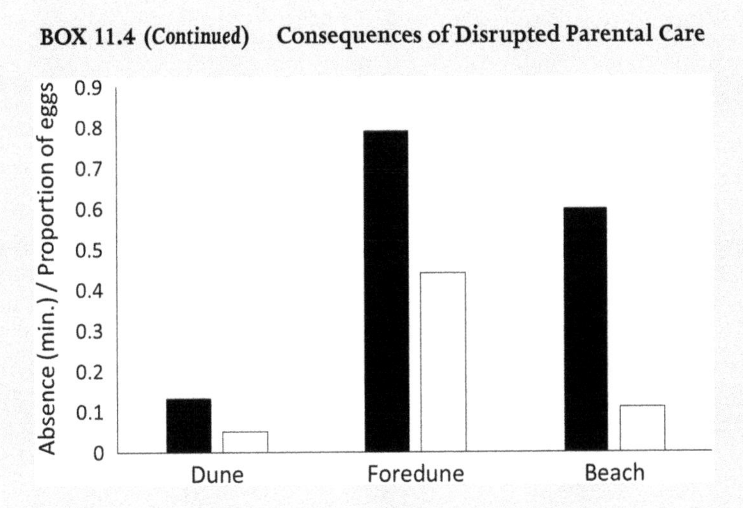

Figure 11.4.3 Disturbance may result in non-hatching plover eggs. Hooded Plovers nesting in different habitats experience different rates of human disruption to incubation, and egg non-hatching rates. The average absence caused by encounters (black bars; the product of response rates and absence durations) and the proportion of all eggs which were abandoned (open bars; n = 323 eggs) (Weston 2000).

is too high, or likelihood of hatching too low, at a particular nest. Abandonment rates can be high (e.g., 23% of Malay Plover eggs; Yasué and Dearden 2006b). The possible consequences of disrupted incubation are shown in Figure 11.4.4.

Brood Rearing

Fewer studies examine disturbance during brood rearing, although, for many plovers, high chick mortality occurs, and disturbance has been reported to lower chick survival. Most studies

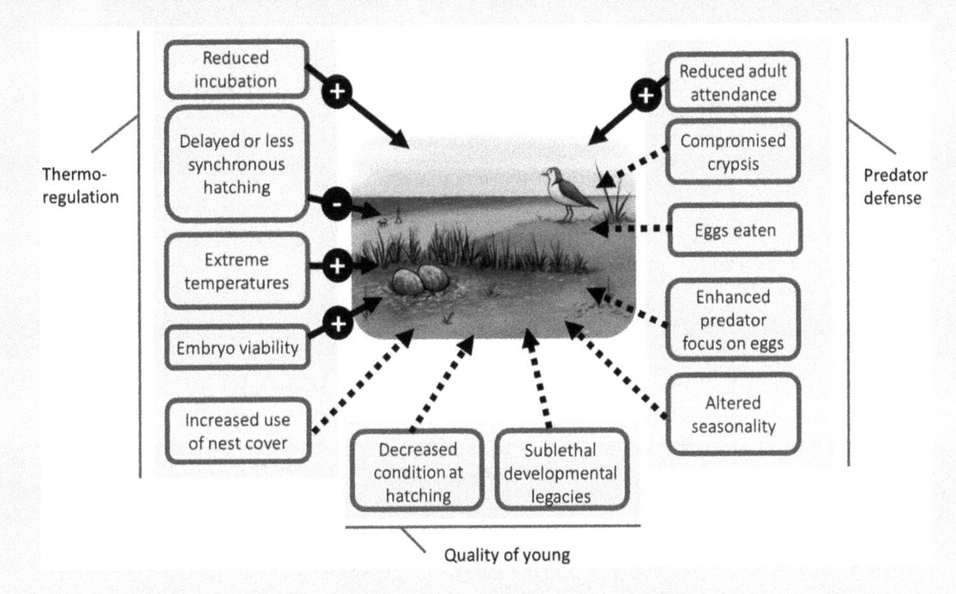

Figure 11.4.4 A summary of known implications of human-initiated nest absences in plovers. Solid arrows indicate the factor has been tested in at least one plover species, dotted lines indicate no test is available. Negative symbols (–) indicate no relationship was present, positive symbols (+) indicate the implication has been demonstrated.

BOX 11.4 (Continued) Consequences of Disrupted Parental Care

examine survival rates to fledging, but few examine the mechanisms through which disturbance could influence survival. Responses to humans during brood rearing potentially compromise chick thermoregulation (lack of brooding), energy balance (lack of, or displaced, foraging or increased energy expenditure), or predator defense (separation from defending parents) (Weston and Elgar 2005a). Humans clearly disrupt plover brooding (Flemming et al. 1988, Baudains and Lloyd 2007), and disruptions can be substantial (Weston and Elgar 2005a). In terms of compromised energy balance, disturbance reduces chick foraging times, fledging success, and microhabitats used (Flemming et al. 1988). Compromised predator defense has not been unequivocally demonstrated, though adult plovers spend more time searching for chicks as encounter rates with humans increase, suggesting they have lost track of the exact location of their young (Weston and Elgar 2005a). No clear mechanistic link between disturbance and reduced chick survival is currently available, partly because of the difficulty of tracking mobile and cryptic broods.

longer than other stimuli (Gómez-Serrano and López-López 2014).

During incubation, the most prominent response to nearby humans is to leave the nest and return when the human departs, thus denying the potential predator cues on the location of the nest containing highly camouflaged eggs (Hoffmann 2005, Weston and Elgar 2007). The process of denying a potential predator cues as to the presence and location of a nest involves a decision to leave and a decision to return by the parent. Most information available is on the decision to leave the nest, which is readily indexed by recording the FID of incubating birds. FIDs are apparently longer for incubating than nonbreeding plovers (Figure 11.5), consistent with the idea that early departures function to deny potential predators cues about the precise location of the nest. This has been termed the LEAD hypothesis: Leave Early to Avoid Detection (Weston et al. 2019).

The decision to leave the nest is influenced by multiple factors. For example, nest habitat, type of stimulus, prevailing temperature, and prevailing human, and predator regimes (the occurrence, frequency, and abundance of humans and different predators). The decision when to return to the nest is also influenced by many factors such as behavior of the stimulus, encounters with other humans, and prevailing temperature (Lord et al. 2001, Baudains and Lloyd 2007, Weston and Elgar 2007, St Claire et al. 2010, Lomas et al. 2014). Several specific examples indicate the context-specificity,

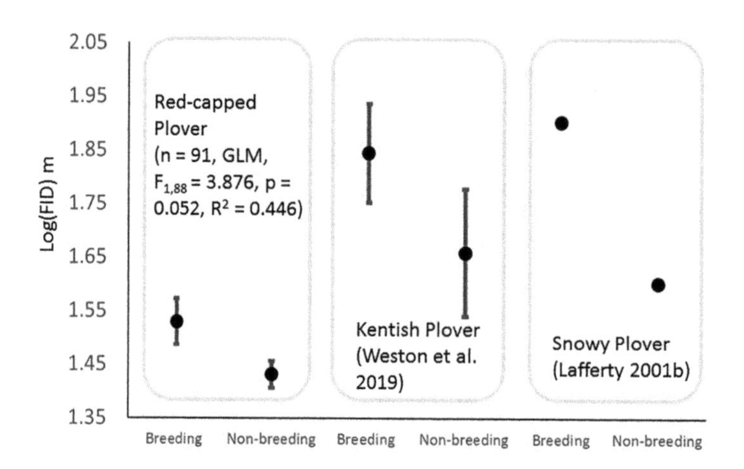

Figure 11.5 Flight-initiation Distance (FID) of breeding and nonbreeding plovers. Mean (±1 SE) FIDs evoked by a standardized approach by a single walker (Red-capped Plover model estimates from a GLM which includes Start Distance are shown). (Sources: Lafferty 2001); Weston et al. 2019, and unpublished data from P. Guay, W. Van Dongen, M.A. Weston, and H.K. Glover.)

and highly adapted nature of parental responses to people. First, incubating plovers discriminate between human and nonhuman stimuli (Weston and Elgar 2007) and among different "human" stimuli (e.g., people with and without dogs; Lord et al. 2001). They also adjust their responses to humans depending on the prevailing risk environment; FIDs of incubating Two-banded Plovers are longer when nonhuman predators are present (St Claire et al. 2010), and FIDs of several species are shorter in areas where humans are more common (Figure 11.4; Box 11.5).

Several studies describe the duration of absences from the nest evoked by encounters with humans, a metric which is affected by: (1) the likelihood and distance at which departure occurs, and (2) the decision on when to return (Figure 11.6). Hooded Plover "voluntary" absences were shorter and less frequent than those caused by people or dogs (Weston and Elgar 2007). The duration of nest absences was positively correlated with incubator FID for Northern Red-breasted Dotterels and was longer when walkers had a dog (Lord et al. 2001). Durations of human-induced nest absence in White-fronted and Snowy plovers were shorter in areas with more people (Baudains and Lloyd 2007, Faillace and Smith 2016). Following human disturbance, Malay Plovers returned to nests faster at higher modeled egg temperatures, in the morning, if they had younger clutches, and when they had high nest attendance prior to the encounter (Yasué and Dearden 2006c).

The result of nest absences is reduced incubation constancy at nests where human disturbance is more common (Baudains and Lloyd 2007, Weston and Elgar 2007). The number of people and dogs was negatively correlated with nest attentiveness for White-fronted Plovers, and humans cause sometimes drastic reductions in attentiveness, from 95% in the absence of disturbance to 22% under highly disturbed conditions (Baudains and Lloyd 2007). Nest absences, when they occur at modest rates, do not necessarily translate into clear-cut relationships with overall rates of nest attentiveness (e.g., Buick and Paton 1989). Absences might be trivial when they are infrequent, of short duration, or both or when incubators may compensate for the disrupted incubation. Many biparentally incubating plovers incubate their eggs for almost 100% of the time, so any absence would reduce incubation constancy to some extent.

Plovers caring for broods also rely on crypsis, which involves adult and chick behavior. When brood-rearing plovers respond to humans, chicks often run and/or hide (often in cover), behavior which is apparently prompted by vocalizations from adults; adults also perform distraction displays or other behaviors. Chick responses to humans change as they grow; young chicks tend to hide whereas older chicks tend to run and hide (Powell et al. 1997, Colwell et al. 2007). Response rates can be high. About half of all encounters between humans and brooding Hooded Plovers caused brooding to cease and 86% of encounters

BOX 11.5 Individual or Local Adaptations of Responses

"Habituation," where responsiveness decreases with increasing exposure to common, usually benign stimuli, has been reported from several species of plover when breeding (Figure 11.5.1). A special case of habituation (attraction) may occur where human presence is associated with heightened plover food availability (Lunardi and Macedo 2014, Hamza and Selmi 2015). "Habituation" generally refers to within-individual learning; however, this type of habituation has not been unambiguously demonstrated for any bird, let alone any plover (Weston et al. 2012). It could equally be applied at the population level, and be explained by other processes such as local selection for bolder individuals (van Dongen et al. 2015). Where habituation is evident, the extent to which it occurs is unknown—responses persist in all shorebirds to at least some extent (Guay et al. 2013). Habituation has the potential to be maladaptive under some circumstances, with "habituated" individuals of several species realizing lower breeding success (Yasué and Dearden 2006b, Baudains and Lloyd 2007). Additionally, habituation may be difficult for migratory or nomadic plovers moving between areas with very different prevailing human environments.

The opposite of habituation ("facilitation") may occur where stimuli are not benign. For example, unleashed dogs which may chase or kill plovers often evoke "escalated" responses among at least some plovers (Weston and Elgar 2007).

BOX 11.5 (Continued) Individual or Local Adaptations of Responses

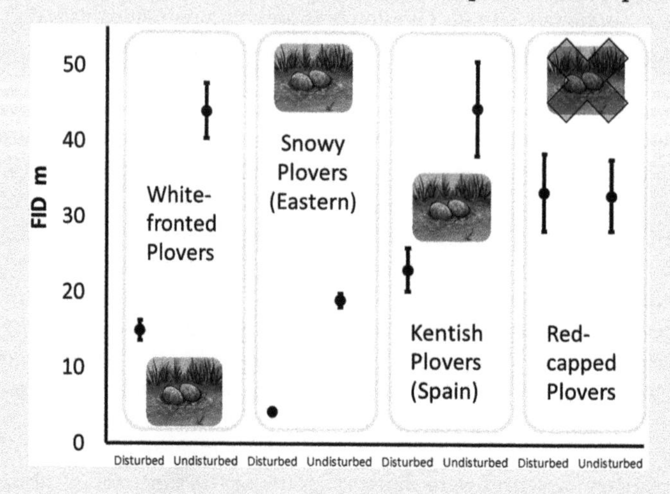

Figure 11.5.1 FID of plovers in areas where humans are common ("disturbed") and uncommon/absent ("undisturbed"). Mean (± SE) FIDs evoked by a standardized approach by a single walker. Sources: Gómez-Serrano and López-López (2014) and unpublished data (P. Guay, W. Van Dongen, M.A. Weston, C. Faillace, H.K. Glover, and P. Lloyd).

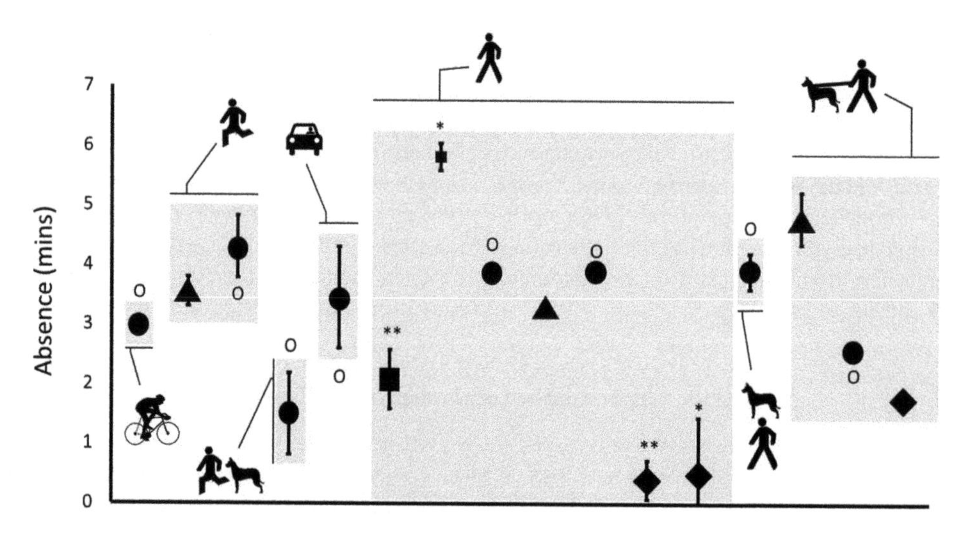

Figure 11.6 Durations of parental absence (mean±SE min) from nests caused by humans for different plovers (Hooded [circles], Northern Red-breasted [triangles], Western Snowy [diamonds] and White-fronted plovers [squares]). Symbols indicate stimulus, "O" indicates observational study, thus stimuli did not necessarily directly approach nests; remainder are experimental studies where the stimulus directly approached nests. "**" represents a relatively intense human regime and "*" represents relatively few humans. (Sources: Baudains and Lloyd (2007), Buick and Paton (1989), Faillace and Smith (2016), Lord et al. (2001) and Weston and Elgar (2007).)

with foraging chicks caused foraging to cease (Weston and Elgar 2005a).

During brood-rearing, plover responses could conceivably be ended by parents or chicks and is generally marked by the cessation of chick hiding or running behavior. Durations of chick disturbance are poorly known, but likely vary with the starting behavior and location of the chicks, distance to refuge, age of the brood, and other factors. Durations of chick disturbance can occasionally be lengthy. For example, disturbed Hooded Plovers left their chicks unbrooded for up to 290 min across ambient temperatures of 10°C–46°C (Weston and Elgar 2005a). Because the response usually entails greater parental separation from chicks than would otherwise occur, delays can occur in the resumption of normal activity. Brood-rearing Hooded Plovers on busy beaches spent 6.1% of their time searching for their hiding chicks while they spent only 2.0% of their time searching under less busy conditions (Weston and Elgar 2005a).

Responses of brood-rearing plovers and their chicks are nuanced. For example, Piping Plover broods in Nova Scotia responded to people at 160 m, but barely responded to vehicles (Flemming et al. 1988). Young White-fronted Plover chicks exhibited longer FIDs and response durations to an investigator at a site with relatively few humans (Baudains and Lloyd 2007).

Given the variety of plover parental care systems, defense of eggs or chicks during encounters with humans may or may not be borne equally by both parents (Chapter Four, this volume), but the degree of equity apparently varies among species. Chick age did not affect the equitable division of brood rearing by Malay Plovers of each sex devoted to responding to stimuli (Yasué and Dearden 2008b). However, male Killdeer defended nests and offspring more intensely when approached by a human (or predator) than females (Brunton 1990). Given that females often perform more of the daylight incubation (Ekanayake et al. 2015), when humans are more likely to be present, females might experience greater disruption from humans.

A special form of crypsis is "deceit," when a plover apparently attempts to mislead a potential threat, such as a person, by attempting to conceal an obvious antipredator or alarm response. For example, several plover species "false feed," pretending to feed at a location away from a nest or concealed brood (Weston and Elgar 2005a). This response presumably denies a predator behavioral cues which might betray the presence of eggs or young.

Escape

Escape ("flight") in plovers involves walking, running, or flying away from the stimulus; some chicks even swim from danger. Escape increases the distance between a stimulus and the bird or provides refuge. Plover escape often ends with the placement of barriers, such as water, between themselves and a person from whom they fled. Escape is perhaps the most obvious and dramatic response of plovers to humans, and the probability of escape increases with decreasing separation between a person and a plover. FIDs vary among plover species, in a manner consistent with that described for other bird groups, in that for example, FID correlates positively with body mass (Weston et al. 2012, Figure 11.7).

A range of internal and environmental factors influence plover FIDs (Figure 11.8). FID may vary given an individual's state such as their sex, age, or breeding status. For foraging, nonbreeding

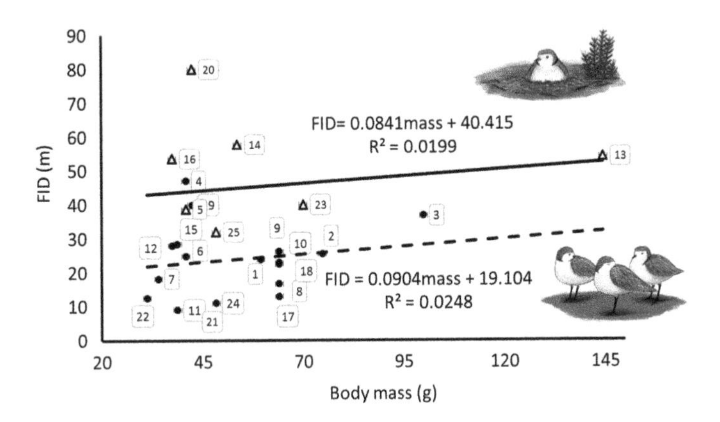

Figure 11.7 Flight-initiation Distances (FID; m) of breeding (solid line; triangles) and nonbreeding (dotted line; circles) plovers against body mass (grams, derived from Dunning 1992 supplemented with Dowding 1994 and Szentirmai et al. 2001). Populations: [1]Double-banded Plover (Australia); [2]Greater Sand-Plover (Africa); [3]Hooded Plover (Eastern); [4]Kentish Plover (Spain); [5]Kentish Plover (Sri Lanka); [6]Kittlitz's Plover; [7]Lesser Sand-Plover (Australia); [8]Lesser Sand-Plover (Sri Lanka); [9]Lesser Sand-Plover (Africa); [10]Little Ringed Plover (Europe); [11]Little Ringed Plover (Sri Lanka); [12]Northern Red-breasted Dotterel; [13]Piping Plover (Western); [14]Red-capped Plover (Australia); [15]Red-capped Plover (Australia); Common Ringed Plover ([16]Africa; [17]Europe); Snowy Plover ([18,19]Eastern USA; [20]Western USA); [21]African Three-banded Plover; [22]Two-banded Plover; and [23,24]White-fronted Plover. Plover FIDs in mixed flocks (Lunardi and Macedo 2014, Lilleyman et al. 2016) are excluded. Analyses are not phylogenetically controlled because of the close relatedness represented by a single genus.

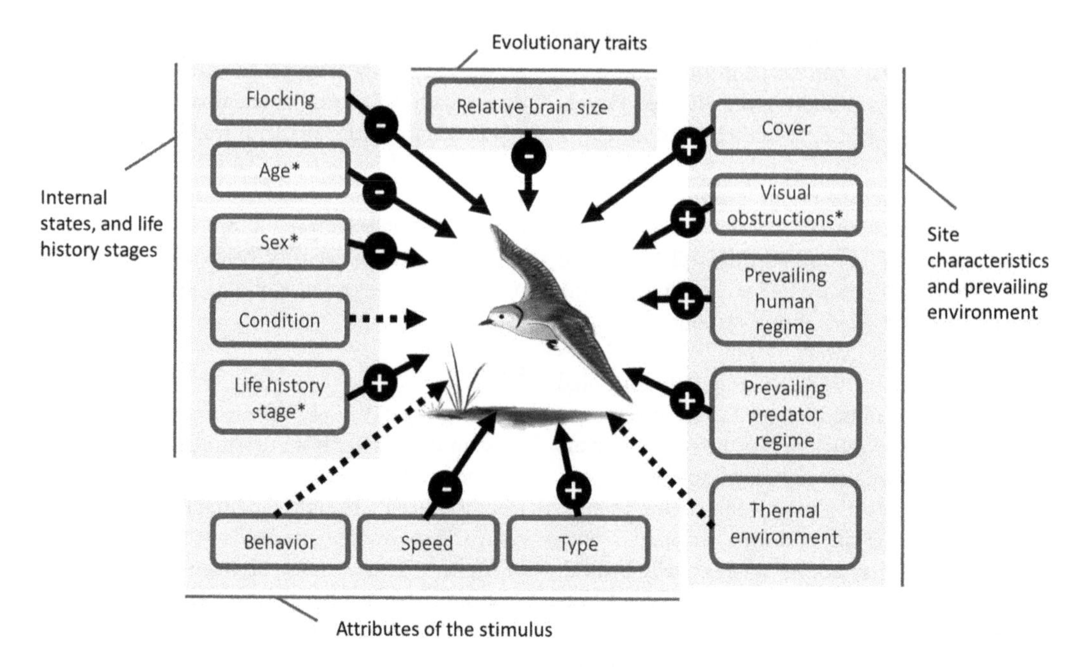

Figure 11.8 A summary of factors which are known to, or might, influence plover Flight-initiation Distances (FIDs). Solid arrows indicate the factor has been tested in at least one plover species, dotted lines indicate no test is available. Negative symbols (−) indicate no relationship was present, positive symbols (+) indicate the factor is known to influence FID. Asterisks indicate analyses presented in this chapter.

Red-capped Plovers, a sexually dimorphic species, FID does not differ between the sexes. Using FIDs (logged) evoked by a single pedestrian in a standardized approach, and including in the model the almost universally significant covariate in such analyses, "Starting Distance" (SD; see below), the effect of sex on FID can be examined. This General Linear Model (GLM) revealed no effect of sex on FID ($F_{1,19}$ = 1.012, P = 0.159, r^2 = 0.614; n = 22). A similar analysis is possible between juvenile and adult Red-capped Plovers, but in this case involves standardized approaches by a walker with a leashed dog. Again, there was no effect (GLM, $F_{1,49}$ = 0.190, P = 0.665, r^2 = 0.337; n = 54). FIDs evoked by walkers tend to be longer for breeding compared to nonbreeding birds (Figure 11.5).

FID varies with attributes of the human stimulus and human environment. Experimental comparisons of FIDs of walkers versus joggers permit examination of whether human speed influences response; for plovers, speed does not appear to alter FID (Lord et al. 2001, Faillace and Smith 2016), although it could conceivably change the modality of escape. For example, Hooded Plovers run from joggers but walk away from walkers (Lethlean et al. 2017). Several experimental studies report the effect of different stimuli. The basic pattern is that vehicles evoke shorter FIDs than walkers, while dogs (usually leashed) evoke relatively long FIDs (Lord et al. 2001, Lafferty 2001a, 2001b, Gómez-Serrano and López-López 2014). Observations suggest that unleashed dogs may elicit even higher responsiveness (Lafferty 2001b, Weston and Elgar 2005a, 2007). FIDs were shorter in areas with more people (Figure 11.5.1).

Two aspects of the human "approach" may influence responses. First, the SD used during experimental approaches to plovers was positively related to FID (e.g., in the GLMs above). While poorly understood, this almost universal finding allows us to infer that fewer visual obstructions at a site will be associated with longer FIDs. For nesting Red-capped and Kentish plovers the amount or use of cover is negatively correlated with FIDs, perhaps because approaching humans cannot be detected earlier (Gómez-Serrano and López-López 2014, Lomas et al. 2014). Second, direct versus tangential approaches, those which might be expected on beaches when people walk past nests in a longshore direction, may influence responses. While no evidence is available for plovers, tangential approaches can result in longer FIDs among some grassland birds (Fernández-Juricic et al. 2005).

Attack or Threat

Plovers rarely give threat displays or attack humans (Gochfeld 1984). Exceptions include breeding Killdeer (Brunton 1990) and Red-capped plovers (pers. obs.), which occasionally give threat displays to humans. In Hooded Plovers, 4% of encounters that evoked aggressive responses from brood-rearing adults involved walkers and dogs (Weston and Elgar 2005a). Injured plovers may also attack humans when pursued (MacLean 1976).

CONSEQUENCES OF DISTURBANCE

A range of consequences of disturbance are conceivable at the level of the individual, including body condition, stress, and compromised immune-competency, which may or may not manifest themselves at the population level. While welfare considerations prevail at the level of the individual, conservation concerns center around the viability of populations. At the population level, the evidence required to demonstrate consequences of disturbance is difficult to collect, and little evidence is available.

Responses Cost Time

Responses cost plovers time, in extreme cases several hours at a time (unpublished data). Time devoted to responses is not available for other activities (e.g., foraging, caring for young), and with few exceptions (e.g., Lafferty 2001a), studies of plovers consistently demonstrate altered time budgets caused by human proximity (Figure 11.9).

Responses Limit or Alter Space Use

The need to maintain a separation distance from humans means that humans displace plovers, thus at least temporarily limiting habitat availability. The strongest evidence for human displacement of plovers is at the local scale and during mobile stages of the life cycle (Table 11.2). Evidence for displacement of foraging plovers is mixed and occasionally, under particular circumstances, some evidence for attraction to humans is available. Kentish Plovers in Tunisia were more common with higher human densities (Hamza and Selmi 2015), and Semipalmated Plovers in Brazil exhibited higher peck and ingestion rates when

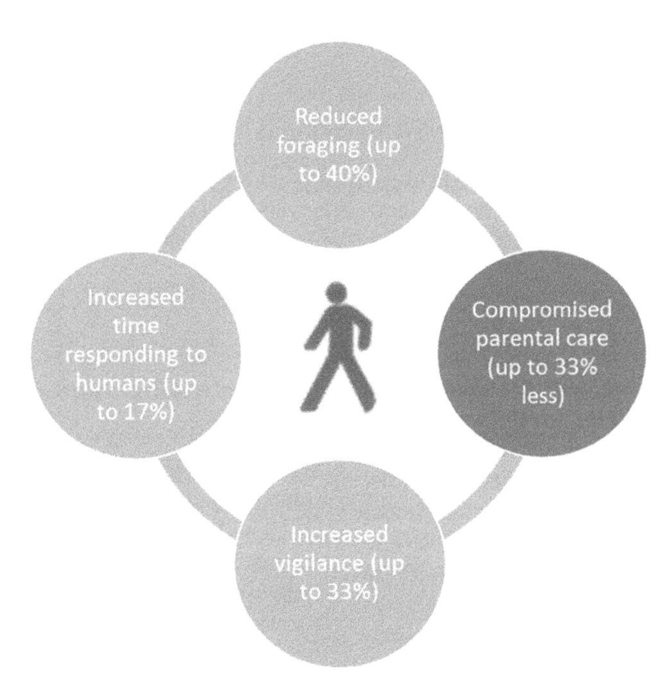

Figure 11.9 Summary of the effects of human disturbance on plover time budgets (across life history stages with maximum recorded changes with close human proximity/activity indicated). Light Gray circles indicate applicable to all life history stages including chicks. (*Sources*: Flemming et al. (1988); Burger (1994); Staine and Burger (1994); Burger et al. (1995); Goldin and Regosin (1998); Blakney (2004); Weston and Elgar (2005a, 2007); Yasué (2006b); Yasué and Dearden (2006b); Yasué et al. (2007, 2008); and Martín et al. (2015).)

TABLE 11.2

Evidence of displacement of plovers by human disturbance. "Site" refers to an area of contiguous habitat.

Within site
Foraging. Displacement between habitats (Piping Plover: Burger 1994, Burger et al. 1995) or levels of habitat (Hooded Plover: Weston and Elgar 2005). Not for Malay or Semipalmated Plovers (Yasué 2006, Yasué et al. 2008). **Breeding.** Malay, Piping and Common Ringed Plover nests were placed in areas with relatively fewer humans (Elias-Gerken 1994, Yasué and Deardon 2006b, Liley and Sutherland 2007), inland with more cover (Kentish Plover, Goméz-Serrano and Lõpez Lõpez 2014). Smaller Piping Plover territories with more people (Haffner et al. 2009), Killdeer and Snowy Plover broods more mobile with human activity (Powell et al. 1997, Wilson and Colwell 2010).

Between sites
Foraging. Between beaches/areas, sometimes inferred from counts at single sites. Collared, Wilson's, Kentish and Common Ringed Plovers (Lunardi and Macedo 2014, Martín et al. 2015) but not for Semipalmated or Snowy Plover (Pfister et al. 1992, Brindock and Colwell 2011, Lunardi and Macedo 2014). **Breeding.** Snowy Plover site occupancy and colonization were negatively associated with human disturbance and site abandonment was positively associated with human disturbance (Webber et al. 2013).

Across land/sea/beachscapes
Non-breeding Piping Plovers in the Gulf of Mexico were less abundant where there were more urbanization and roads, features associated with higher human abundances (among other differences, LeDee et al. 2008). Snowy Plovers were less abundant near trail heads on a California beach (Lafferty 2001b), where people were presumably more common. No association with roads and buildings and Piping Plover chick survival around the Great Lakes (Brudney et al. 2013).

foraging near shellfishes who dig into the substrate, but these rates near recreationists were similar to when people were absent (Lunardi and Macedo 2014). Human disturbance can affect where plovers decide to "settle," that is, initiate the stage of their life history during which they are effectively immobile (nesting) or have limited mobility (brood rearing, Table 11.2). At larger scales, the evidence of displacement is growing, though the surrogates of human presence are less directly relatable to human presence alone, potentially being confounded with factors such as the degree of anthropogenic habitat modification. Nevertheless, at these higher scales, evidence suggests disturbance may alter habitat occupancy.

Responses and Reproductive Success

Disrupted behavior does not translate directly into key demographic processes (Gill 2007). Indeed, responsiveness to humans is not necessarily associated with suppressed reproductive success: Malay Plovers with shorter durations of nest absence in response to a human approach had *lower* hatching success (Yasué and Dearden 2006a) and White-fronted Plovers had *higher* hatching success in busier areas (Baudains and Lloyd 2007). These studies do not control for individual quality or selective processes, which may influence which birds breed where. Unambiguous links between human disturbance and reduced reproductive success in plovers have been demonstrated in only a few cases and some plovers experience enhanced success with disturbance. Nevertheless, several lines of evidence suggest that disturbance can reduce reproductive success in some contexts (Table 11.3).

The link between disturbance and reduced productivity appears inconsistent and complex and varies in time and space. Disturbance may detrimentally affect plovers in one but not all stages of breeding. For example, hatching but not fledging success was decreased by human disturbance for Malay Plovers (Yasué and Dearden 2006b). Even within a species (e.g., Piping Plover chick survival) disturbance may decrease

TABLE 11.3

Evidence linking human disturbance and plover reproductive success.

Study type	Evidence that humans decrease reproductive success	Other effects (examples)
*Space–human substitution studies that describe reproductive success in areas where humans are common in comparison with areas where they are rare or less abundant	Hooded and Piping Plover (Hanisch 1998, Strauss and Dane 1989, Dowling and Weston 1999)	Survival of Piping Plover chicks the same between off-road vehicle (ORV) or non-ORV areas (Patterson et al. 1991), White-fronted Plovers bred more successfully in disturbed areas (Baudains and Lloyd 2007)
*Time–human substitution studies that report changes in survival using presumed temporal variation in human abundances as proxies for disturbance	Snowy Plovers chick mortality: (1) 69%–72% greater on weekend days and holidays over weekdays (Ruhlen et al. 2003); (2) lower during a period when a disturbance refuge was established than beforehand (Wilson and Colwell 2010)	—
Correlative studies that index human disturbance and examine the degree of association of these indices with reproductive success	Piping Plover territories in North Dakota with cattle or motor vehicle disturbance had a lower nest success rate (Gaines and Ryan 1988)	No evidence that variation in human activity near Snowy Plover nests correlated with clutch survival in coastal northern California (Hardy and Colwell 2012). No association with roads and buildings and Piping Plover chick survival, Great Lakes (Brudney et al. 2013)
Studies that experimentally deliver disturbance to nests and examine whether clutch survival is affected	—	Close approaches by researchers to Piping Plover nests reduced the probability that the clutch would be preyed upon MacIvor et al. (1990)
Higher success or survival where or when disturbance refugia exist	Piping Plover broods with access to pondshore (1.6% of time responding to humans) realized higher fledging success than those with access to the beach alone (17.0% of time responding) (Goldin and Regosin 1998). Hooded Plover chick shelters (help manage the effects of disturbance among other things) were associated with improved fledging rates (Maguire et al. 2011). When humans were restricted from an area of Californian beach, Snowy Plovers colonized, bred successfully and increased in abundance (Lafferty et al. 2006)	—

NOTES: Asterisks indicate terminology after Weston et al. (2012).

survival or success in some contexts (Flemming et al. 1988, Strauss 1990, Goldin and Regosin 1998) but not others. At other times and places, especially where efforts to mitigate disturbance occur, reductions in chick survival do not occur (Patterson et al. 1991, Cohen et al. 2009). Within a species and site, disturbance may reduce reproductive success in some but not all years, perhaps due to interannual variation in predators and humans (Yasué and Dearden 2006b). In addition to the anthro-ecological context, the inconsistencies in results could stem from different methodologies, designs, and realized statistical power.

The complexity of the relationship between disturbance and reproductive success is exemplified by White-fronted Plovers breeding at both a relatively disturbed and an undisturbed site in South Africa (Baudains and Lloyd 2007). Daytime nest attentiveness decreased with increasing experimental disturbance at both sites, but incubating birds at the more disturbed site had greater nest attentiveness because they reduced their responsiveness. However, chick mortality was significantly greater at the disturbed site, probably because of predation by domestic dogs resulting from reduced escape responses. Annual fecundity was substantially higher at the more disturbed site, thus apparent reproductive fitness of plovers is not always compromised by human disturbance (assortment of individuals between disturbed and undisturbed sites remains unstudied for plovers). In some contexts, people may protect plovers (Box 11.6).

Responses Might Reduce Condition

Body condition influences the capacity to renest, defend territories, provide care, grow, survive migration, avoid predators, and molt, and good body condition buffers against inclement weather and periods of food limitation. Reduced adult survival, perhaps through compromised energy budgets, could influence plover population trajectories substantially (Plissner and Haig 2000,

BOX 11.6 People and Plovers—Perceived Predators, Protectors, and Paradoxes

Under certain specific circumstances, people and their companion animals may benefit plovers. The following contexts appear to explain these reports:

1. Humans also disturb natural and introduced predators. White-fronted Plovers realized higher hatching success at a busier site, perhaps because humans also excluded baboons which prey upon plover eggs (Baudains and Lloyd 2007). MacIvor et al. (1990) found close approaches to Piping Plover nests reduced the probability of foxes taking eggs. Similarly, dogs accompanying people may discourage mammalian predators (e.g., foxes) from areas of plover habitat (Schneider 2013).
2. Humans initiate conservation management in places where conflict is perceived to be high. Survival of Piping Plover chicks was higher for broods on busy public beaches and there were even hints that those nearer buildings had higher survival, perhaps because management efforts were concentrated there (Brudney et al. 2013). An irony regarding plover–human interactions is that close encounters sometime engender awareness, concern, and conservation action from sympathetic people (Colwell 2010, Maguire et al. 2015).
3. Certain human activities (e.g., shellfishing) permit access to abundant food. For example, Common Ringed, Wilson's, and Semipalmated plovers are attracted to sites frequently used by humans, perhaps because shellfish harvesting activities increased food availability (Lunardi and Macedo 2014, Hamza and Selmi 2015).
4. Humans facilitate the creation of suitable habitat in which they co-occur with plovers. Distance to settlement had no effect on habitat suitability of Madagascan Plovers because human activities (e.g., grazing by Zebus [*Bos indicus*]) help to maintain plover habitat (Long et al. 2007).

Thus, under certain circumstances, people effectively protect plovers by being an accidental or purposeful umbrella species, or they fulfill a role as ecosystem architects.

Wemmer et al. 2001). No direct measurements of adult or chick condition across prevailing human regimes are available for plovers. Responses interfere with plover energy assimilation in two ways, first, via disruption of foraging adults and chicks (Goldin and Regosin 1998, Weston and Elgar 2005a, Baudains and Lloyd 2007). Second, responses entail energy expenditure. Alarm flights in tropical Australia were estimated to increase daily energy expenditure of Lesser and Greater Sand-plovers by 7.5% and 7.8%, respectively (Lilleyman et al. 2016). For eggs and chicks, disturbance may impose energetic costs associated with unaided thermoregulation during parental absences. Given that stress and other physiological responses may occur without obvious behavioral responses, compromised body condition may occur even in apparently "habituated" plovers (Yasué and Dearden 2006b). While the true energetic costs of responses are unknown, they might be proportional to the frequency with which they are used, and for some plovers that frequency is high. For example, there were 115 responses to people per Snowy Plover per week in southern California (Lafferty 2001b).

Disturbance and Populations

Concerns that disturbance is a conservation problem for plovers exist globally, from island and mainland systems and temperate and tropical zones (e.g., Lorenzo and Emmerson 1995, Dowling and Weston 1999). The occurrence of responses to humans does not necessarily translate into negative population effects, and the challenge remains to understand the cumulative effect of responses and to interpret their impact (Sutherland 2007). Even substantial impacts on fitness may not reduce population viability if alternative habitats are available (Gill et al. 2001). Responses may not be linked to fitness in expected ways. For example, longer FIDs may reflect an ability rather than a need to respond (Gill 2007), and lower responsiveness may be maladaptive in some contexts (Baudains and Lloyd 2007). Similarly, evidence of altered demography, such as suppressed reproduction, does not necessarily mean that population viability has been compromised. Plover population modeling shows populations are sensitive to several parameters, for example, adult survival, as well as aspects of reproduction (Eberhart-Phillips and Colwell 2014). Plovers may theoretically be able to tolerate, or adapt to, substantial levels of disturbance and plovers persist in even the most human-dominated habitats (e.g., Dowling and Weston 1999). Clear links between disturbance and altered population trajectories remain elusive and arguably these links have been the major gap for decades in wildlife disturbance research. Several lines of (indirect) evidence infer that at least local plover populations can be impacted by disturbance. First, in some contexts, key demographic parameters (principally reproductive success) appear to be suppressed by disturbance (Figure 11.10). Second, populations in which disturbance (and other) processes are intensively managed have increased in abundance, at least locally (Dowling and Weston 1999, Lafferty et al. 2006, Cohen et al. 2009). These studies often report management, which addresses a variety of key threatening processes (e.g., predation) in conjunction with disturbance, rendering any inference regarding the effect of disturbance ambiguous. These accounts are generally management focused, disturbance-removal "experiments" which are often unreplicated and lack controls. Nevertheless, at the site scale they tell compelling stories. Increases in Hooded Plover reproductive success overtime with management, and higher success rates in more intensively managed areas, combined with population increases infer that disturbance was a process affecting population size (Dowling and Weston 1999, unpublished data). Third, population models predict population recovery or increase if disturbance is managed (Liley and Sutherland 2007, Eberhart-Phillips and Colwell 2014).

The link between individual fitness consequences and population-scale effects will depend upon the scale at which disturbance occurs and, in circumstances where plovers are displaced by disturbance, whether there are available habitats and any secondary consequences for plovers in those less or undisturbed areas (Gill et al. 2001, Gill 2007). Where suitable habitat occurs nearby, the consequences may be less severe than in isolated habitat patches or situations where humans occur across a population's distribution (Gill et al. 2001, Sutherland 2007). The severity of secondary effects will depend on the degree to which density dependence is operating within populations, and any reductions in *per capita* survival or fecundity at the sites to which disturbed plovers

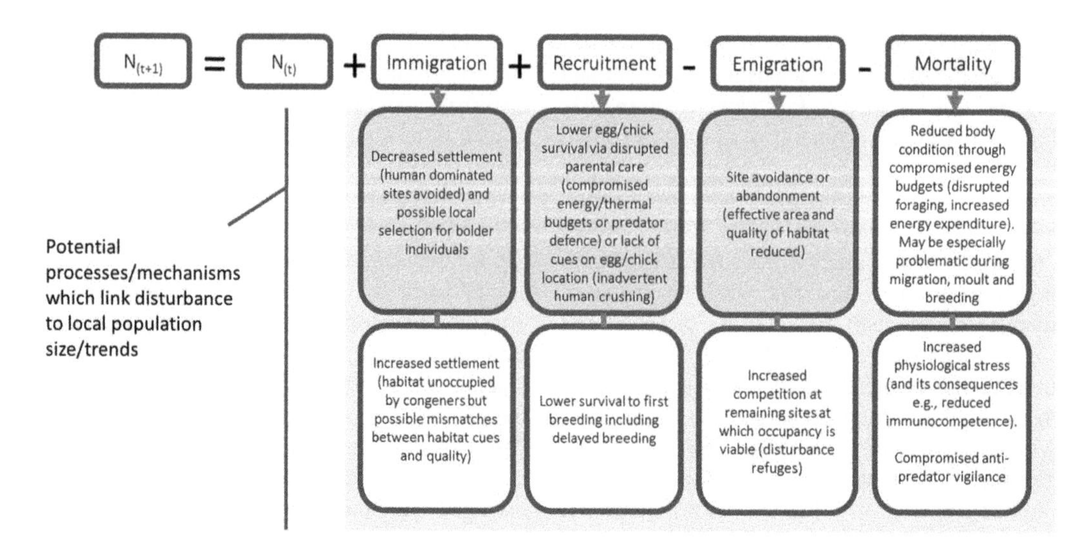

Figure 11.10 A conceptual map of potential consequences of disturbance to local plover populations, based on an assumed, simple geometric population growth equation (where $N_{(t)}$ is the number of plovers at time t). The processes/mechanisms are selected because they are likely candidate processes but do not represent an exhaustive list. Gray tiles indicate the process for which there is confirmatory research for plovers. White tiles represent research gaps and opportunities.

have relocated (Gill 2007). One approach to determining a population response to disturbance is to create population models, incorporate the demographic consequence of disturbance, and assess the consequences for population size (Sutherland 2007). This is currently available for only one plover population (Box 11.7).

Another critical question regarding the impact of disturbance on populations is whether disturbance acts independently, additively, or synergistically with other threatening processes. Most studies of clutch fate in plovers report substantial levels of egg predation in disturbed areas and these are usually higher than losses directly attributable

BOX 11.7 Common Ringed Plover Populations and Disturbance

To date, only a single population model exists which explores disturbance and its population-level consequences for a plover. Liley and Sutherland (2007) use data, theory, and modeling to consider behavioral responses and their population consequences. They present a model that allows predictions of the effect that changes in human numbers, visiting a 9-km-long section of the coastline, may have upon the size of a Common Ringed Plover population at the Wash, England.

Plovers avoided areas of high disturbance. Using the level of human disturbance and habitat variables (which defined territory quality) the model predicted actual and potential areas of beach occupied by plovers. Breeding success, for a given area of beach, was predicted from habitat data. Incorporating known, density-independent, adult mortality allowed the equilibrium population size to be predicted under different, hypothetical, levels of disturbance. If nest loss from human activity was prevented the equilibrium population size was predicted to increase by 8%. A complete absence of human disturbance would cause an equilibrium population increase of 85%. If the numbers of people were to double, the equilibrium population is predicted to decrease by 23%.

Studies such as this are critical to better understanding disturbance (Sutherland 2007), provide useful predictions for future research, and reinforce the idea that disturbance is of conservation importance.

to humans, suggesting that disturbance is one of several prevailing threats. The possibility of disturbance operating synergistically with other threats also exists. For example, nest absences caused by humans may not be especially problematic in the absence of predators, but where abundant human-tolerant predators are attracted to locations to exploit human refuse (e.g., corvids to anglers), then absences from the nest may result in clutch loss by depredation (Rees et al. 2015). In this case, humans encourage the predator into plover habitat and then diminish parental defense by causing disturbance. Humans may detrimentally influence plover energy budgets while simultaneously trampling prey resources (Schlacher et al. 2016). The presence of humans often co-occurs with substantial human-induced changes in habitats. In Spain, Kentish Plover nesting decreased with increases in recreational activities which were associated with seaweed removal (Pietrelli and Biondi 2012), a process which removes a critical ecological subsidy underpinning coastal food webs, as well as directly destroying some nests. Humans thus create a cascade of processes which may detrimentally affect plovers, of which disturbance is but one.

Possible Insidious Impacts of Disturbance

Normal behavioral and physiological states (e.g., the time devoted to different activities) prevail because they are adaptive and optimize lifetime reproductive success, hence, changes to these are expected to have adverse implications. Disturbance is particularly insidious because it is a process which diminishes habitat quality while not altering the physical attributes of a habitat (Colwell 2010). Thus, disturbance may effectively reduce carrying capacity while the habitat appears to remain intact and suitable.

Underutilized habitat may represent an attractive settlement option for immigrating plovers, and this creates the conditions for a potential "disturbance-derived ecological trap" (Sutherland 2007)—where cues used to select habitat no longer reliably reflect habitat quality. Within such traps it is possible that abundance could be misinterpreted as reflecting local population viability (Schlaepfer et al. 2002). For example, plovers may select a suitable beach for breeding well before the breeding season actually starts, perhaps at a time of year when few people occur. The beach may host many thousands of people during the breeding period and represent a poor quality breeding locality. Ecological traps can drive extinctions (Sutherland 2007). Whether plovers consider the prevailing human regime when selecting habitat (or nest sites), as occurs in some other shorebirds (Roche et al. 2016), is currently unknown. Ecological traps and the role of disturbance as a mediator of habitat quality are theoretically likely, but the extent to which they operate for plovers is unclear.

MANAGING DISTURBANCE

Intensely disturbed populations require management intervention (Montalvo and Figuerola 2006). Given the widespread distribution of many plovers, clear targeting of any management is required. This could be to areas of greatest need, importance, feasibility of management, or to avert imminent or growing problems. For plovers, priority areas are likely to include breeding areas in hot or cold climates or those with many predators, and staging sites for migratory species. Once high-priority areas are set, three options exist to manage plover disturbance: change people or change plovers, such that interactions represent coexistence rather than "conflict," or reduce the impacts of disturbance (Figure 11.11).

Perhaps the commonest approach to managing disturbance is to restrict humans through infrastructure (e.g., fences), land use planning (e.g., providing access points away from sensitive or important habitats), or by altering human behavior (e.g., restricting dogs). FIDs (or better, ADs) can be used to establish effective off-limit areas or minimum approach distances (exclusion of zones based around a plover, implemented via codes of conduct) or to buffer designated refuge, critical, or sensitive habitat (Guay et al. 2016). Zonation of recreation which accounts for shorebird usage offers a promising model of coexistence (Stigner et al. 2016). The issue of plover disturbance by humans is almost unique in the degree to which adaptive management occurs, and several published accounts of management success or failure exist. For example, responses to humans by wintering Snowy Plovers were 16 times higher at a public beach than at protected beaches (Lafferty 2001b), and symbolic fencing reduces Hooded Plover disturbance and egg crushing (Weston et al. 2011, 2012). One key issue is how to change

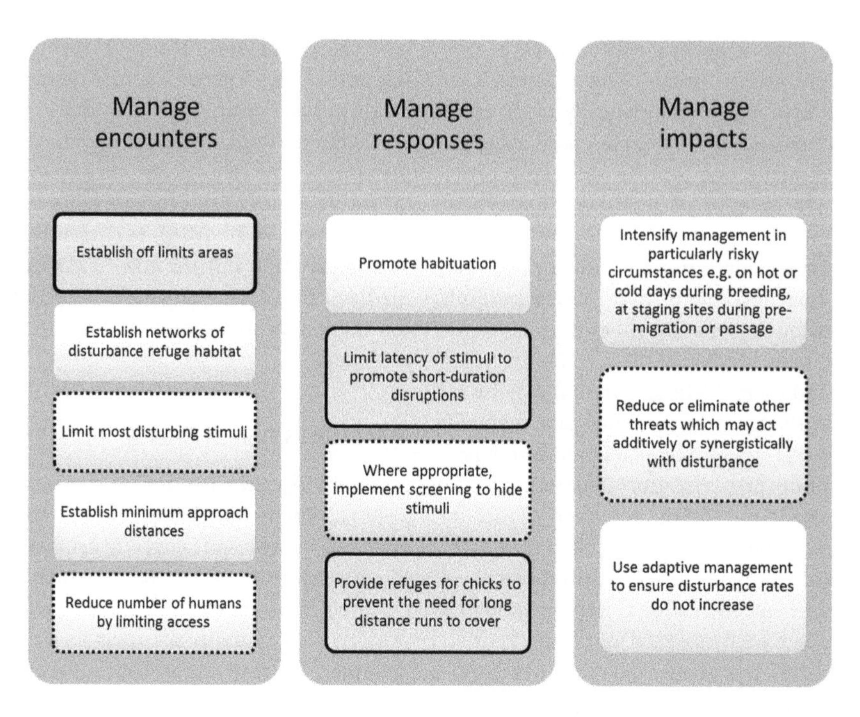

Figure 11.11 Options for managing disturbance to plovers. Gray tiles with solid borders indicate options which have demonstrated efficacy for at least one plover species. Tiles with dotted borders indicate options which have been implemented for at least one plover species but for which efficacy is unknown.

human behavior and the need for social science investigations is critical (Sutherland 2007, Colwell 2010). In relation to plover disturbance, there is an emerging picture of growing human awareness (Maguire et al. 2015) and suitable ways to communicate key messages (Williams et al. 2009, Ormsby and Forys 2010, Rimmer et al. 2013).

Plover behavior might conceivably be modified to reduce any deleterious outcomes of encounters. Decreased responsiveness in busier or less isolated habitat has been reported for a number of plover species (Northern Red-breasted Dotterel, Lord et al. 2001; Two-banded Plover, St Claire et al. 2010; Hooded Plover, Blakney 2004; White-fronted Plover, Baudains and Lloyd 2007; Red-capped Plover, Glover et al. 2011). Promoting habituation is a hypothetical possibility which could desensitize plovers through deliberate repeated benign approaches or through strictly regulating encounters so that they are benign. Such an approach would rely on plover learning lasting for substantial periods of time, and that the new responsiveness does not reduce fitness.

The final approach to managing disturbance is to decouple responses with the way they cause impacts. For example, the provision of refuges (e.g., shelters) may limit the extent and duration of responses and protect responders from exposure to damaging temperatures and/or predators (Maguire et al. 2011). Supplementing food and shelter, perhaps by augmenting seaweed in coastal habitats, may reduce or mitigate the impact of disturbance. Where threats interact with disturbance, a reduction in the impact of disturbance could be achieved by concurrently managing additive or synergistic threats (e.g., predators).

RESEARCH NEEDS

In regard to the impact of disturbance at the population level, a series of interconnected questions exist which, if addressed, will assist in understanding the interactions between disturbance and population viability, and inform management (after Sutherland 2007):

1. To what extent do humans temporally and spatially overlap with sensitive plover populations? One of the substantial information gaps in disturbance research is, perhaps surprisingly, a poor state of knowledge regarding human occurrence

in time and space. Human occurrence is presumably driven by a range of factors such as weather, social constructs such as holidays, cultural differences, accessibility, and infrastructure (Lafferty 2001b). Interannual variation in the occurrence or abundance of humans in plover habitats occurs (e.g., Loegering 1992). Available information tends to be at the site scale, and studies use a variety of noncomparable metrics to describe the prevailing human context, which hampers meta-analyses (Box 11.8). Human behavior within plover habitats is also

BOX 11.8 Understanding Us: Indexing Prevailing Human Regimes

There is remarkable consistency with the type of stimuli involved in major plover disturbance studies around the globe (Figure 11.1). Apart from the most frequent stimuli, most studies report a broad array of less common anthropogenic stimuli which disturb plovers, e.g., jogging, horse-riding, unaccompanied dogs, water-based activities, aircraft, kite-flying, hang gliding, and wind-surfing.

The prevailing human regime can influence evolved and adapted responses of plovers to people (Baudains and Lloyd 2007, Faillace and Smith 2016). To contextualize plover responses to people, it is necessary to quantify the "exposure" plovers have to humans over a variety of temporal and spatial scales. Ironically, in some ways we know more about the distribution of plovers than we do people. Various measures have been used to describe the context within which disturbance occurs. The variety of methods and metrics are understandable given the different aims, methods, and circumstances under which studies are conducted. However, the variety of measures used in plover-disturbance research requires a standard system of classification. Studies have used:

- Direct counts of humans and their companion animals. This can take the form of:
  - Transects in appropriate habitats, usually linear shoreline habitats, occupied by plovers e.g., 6.3 people/km on beaches in southern Australia (Hooded Plover; Weston 2000), and up to 80 people/km and 17 dogs/km in Cape Town, South Africa (White-fronted Plover; Lloyd and Plagányi 2002).
  - Counts within fixed areas resulting in densities, a widely used approach especially on coastal flats and beaches (Yasué et al. 2008). This includes point counts at fixed intervals or over fixed periods, and constrained to, for example, 500 m (Brindock and Colwell 2011).
- Passive track indices (Brindock and Colwell 2011, Webber et al. 2013) or signs (Gaines and Ryan 1988). These approaches are used particularly on soft substrates, though sand-pads can help quantify human occurrence (Weston et al. 2009).
- Human social surrogates of relative human occurrence and abundance, such as the use of weekends and holidays versus weekdays in recreational areas (Ruhlen et al. 2003, Weston and Elgar 2005a). These metrics are relative and assume human abundance and occurrence is relatively higher on nonworking days.
- Encounter rates (per unit time; e.g., Weston et al. 2007), a plover-focused metric which describes the human environment rather than any resultant disturbance (not all encounters cause a response).
- Response rates, another plover-focused metric but in this case "environment" and "response" are assumed to be equivalent. For example, hourly rates of responses (e.g., incubating Malay Plovers spent 70 s/h and 48 s/h responding to anthropogenic and natural disturbance, respectively; Yasué and Dearden 2006c, Yasué et al. 2007) or response rates per plover per hour (e.g., on a public beach in southern California, each Snowy Plover was disturbed, on average, once every 27 weekend-min and once every 43 weekday-min; Lafferty 2001b, Lafferty et al. 2006).

- Indices of urbanization (e.g., density of roads; LeDee et al. 2008). However, human land use change also engenders many other influences, such as altered plover habitat (Yasué and Dearden 2008b) and increased (Huijibers et al. 2015) or decreased (Baudains and Lloyd 2007) number of egg predators near urban areas.

Two newer, spatially explicit, approaches to understanding human behavior offer promise. First, humans and their companion animals can be tracked, and disturbance estimated by overlaying plover FIDs on those tracks. For example, a tracking study of dogs on Australian beaches revealed dogs were highly mobile on beaches, and when their tracks were buffered by the FID of (the resident) Hooded Plover, the spatial extent of habitat impacted by disturbance can be ascertained (Figure 11.8.1). Second, disturbance simulation models such as simulation of disturbance activities (Bennett et al. 2009), offer a useful framework to enhance our understanding of human behavior as it pertains to plover disturbance.

To date, the variety of metrics employed hampers comparability between studies, and for reported data on prevailing human regimes there are almost as many different measures as there are studies. It is, therefore, unclear what represents high versus low levels of human presence in plover habitat, and this may explain some of the inconsistent findings between studies. However, studies of human–plover disturbance describe situations where people are a predominant species in the environment, more abundant than perhaps any other vertebrate species.

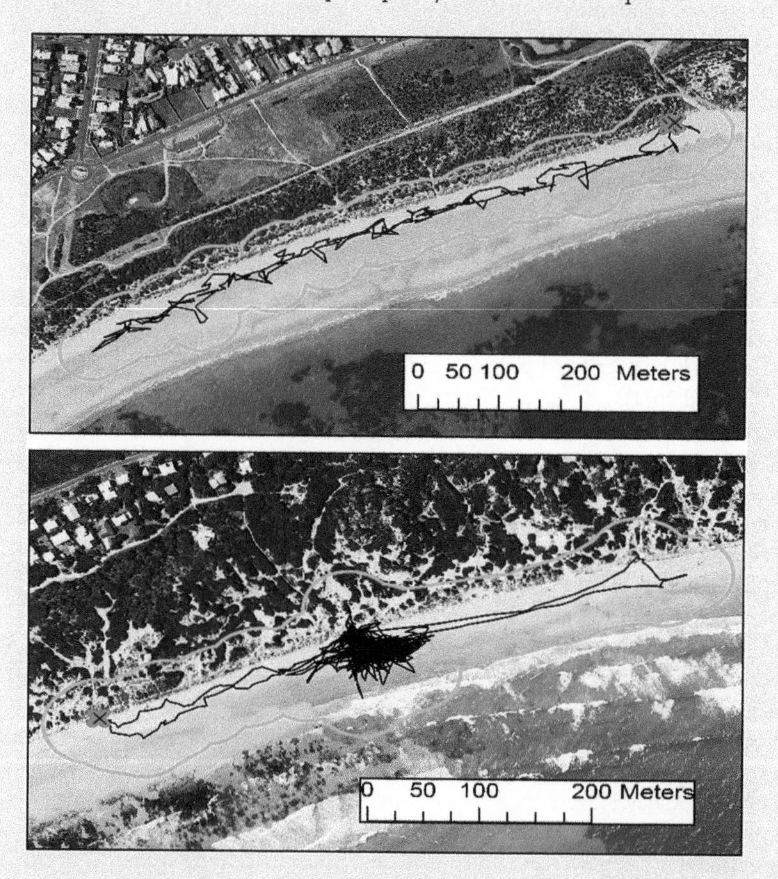

Figure 11.8.1 An overlay of two GPS-tracked dogs (black lines) in Hooded Plover habitat, overlaid with the FID (blue) of Hooded Plovers to dogs (after Schneider 2013). The crosses indicate beach access points.

likely to mediate impacts, yet is poorly documented.

2. Can we unravel general principles linking disturbance and population-level processes, and use these to predict habitats, places, and species where disturbance will be problematic? Are plovers displaced by disturbance, and what are the secondary effects such as density-dependent processes (Gill 2007)?
3. When and where does disturbance mediate habitat quality? Studies that exclude humans from part of a habitat report heightened reproductive success and higher abundances of plovers where humans do not occur (Dowling and Weston 1999). This suggests that disturbance lowers habitat quality in at least some systems, but the generalizability of this, plus the exact nature of the relationship between disturbance intensity and habitat quality remain unknown. Disentangling the direct destruction of eggs and chicks from disturbance effects is also required to interpret these studies.
4. Does disturbance create ecological traps? The operation of ecological traps is notoriously difficult to demonstrate. Long-term studies of disturbed populations will eventually elucidate settlement and habitat selection processes, and may even highlight the need to actively discourage plover settlement from some areas.
5. What tolerance do plovers have for disturbance, and how does this vary across different life history stages? Plovers experience many nonhuman disruptions during their lives and many species appear pre-adapted to low reproductive success (e.g., high renesting and adult survivorship rates). One poorly understood topic is the responsiveness of migratory plovers at various staging sites. Pre- or post-migratory restlessness could lead to high responsiveness (Lilleyman et al. 2016), and the energetic bottlenecks associated with pre-migratory energy assimilation may mean that the consequences of disturbance at these times may alter migratory behavior or survival (Pfister et al. 1992). The relationship between fitness and disturbance is unknown, it may be nonlinear, and could involve thresholds.
6. To what extent do plovers successfully adapt to disturbance, that is, what plasticity do they exhibit? Gradient studies involving multiple sites with different prevailing human disturbance regimes (being otherwise comparable) would elucidate the presence of any thresholds or adaptations regarding disturbance and plover persistence. Any learning (habituation or facilitation), and the degree to which it can "cognitively buffer" plovers from disturbance, is unknown (Guay et al. 2013). Microhabitat preference (e.g., the use of cover), the degree to which any site fidelity can be adjusted, and altered seasonality or diel activity cycles could also represent potential adaptations to disturbance (Gómez-Serrano and López-López 2014).
7. To what extent does disturbance act additively and/or synergistically with other threats? As plover populations come under increasing pressures, interventionist approaches may reduce the effect of additive processes such as predation (e.g., Maguire et al. 2011, Tan et al. 2015). Synergistic interactions of disturbance may include climate change and superabundant or introduced predators, which may exacerbate the effects of any disturbance-caused lapses in egg or chick thermoregulation or defense, respectively. Interactions with other threats may also mask the effects of disturbance (see, for example, Yasué and Dearden 2006b).
8. To what extent can disturbance be managed across large spatial and temporal scales? Optimal use of resources could be guided by specific optimization studies e.g., optimizing ranger patrols aimed at reducing shorebird disturbance (Dhanjal-Adams et al. 2016). While optimization is important, the spatial and temporal scale at which management is required is daunting. Small but widespread gains in coexistence might reap great rewards for plover conservation. Additionally, appropriate planning of reserve networks, within which plover habitat is afforded protection, is critical. Enhanced knowledge of plover

dispersal patterns would underpin reserve system design as well as helping assess any density-dependent effects.

9. What benefits occur to people and plovers by interacting? In a few specific instances, people may aid plovers (Box 11.6). Plovers may also help connect people with the natural world and biodiversity in general, enhancing the way humans value biodiversity and consequently establish legal and other mechanisms for plover conservation.

CONCLUSIONS

Disturbance defies easy characterization of its form, extent, or consequences. Unsurprisingly, the available studies produce a variety of results, and represent a range of prevailing human regimes, species, responses, and methodologies (Colwell 2010). Plovers offer an ideal model for understanding the often subtle ways that humans change wildlife behavior and affect their populations. They also represent an actual or potential flagship group, which can be used to demonstrate how we can manage any welfare or conservation impacts associated with human disturbance.

ACKNOWLEDGMENTS

For unpublished data and images, I thank P. Guay and W. Van Dongen, A. Radkovic, T. Forge, D. Lees, D. Whisson and K. Ekanayake (Deakin University). For illustrations, L. X. Tan. For specifics on published data, C. Faillace, J. St Claire and P. Lloyd. For patience, support, and inspiration, C. and B. Weston, M. Ramondetta and A. Piper. M. Yasué and R. Fuller provided helpful and constructive comments. This work was supported by the Beach Ecology and Conservation Hub (BEACH; Venus Bay).

LITERATURE CITED

Alvarez, F. 1993. Alertness signalling in two rail species. *Animal Behavior* 46:1229–1231.

Baird, B. and P. Dann. 2003. The breeding biology of Hooded Plovers *Thinornis rubricollis* on Phillip Island. Emu 103:323–328.

Baudains, T. P. and P. Lloyd. 2007. Habituation and habitat changes can moderate the impacts of human disturbance on shorebird breeding performance. *Animal Conservation* 10:400–407.

Bennett, V. J., M. Beard, P. A. Zollner, E. Fernández-Juricic, L. Westphal, and C. L. LeBlanc. 2009. Understanding wildlife responses to human disturbance through simulation modelling: A management tool. *Ecological Complexity* 6:113–134.

Blakney, A. H. 2004. Behavioural responses and habituation of the Hooded Plover, *Thinoris rubricollis* (Gmelin 1789), to disturbance stimuli. *Honours dissertation*, University of Tasmania, Hobart, Tasmania.

Brindock, K. M. and M. A. Colwell. 2011. Habitat selection by western Snowy Plovers during the nonbreeding season. *Journal of Wildlife Management* 75:786–793.

Brudney, L. J., T. W. Arnold, S. P. Saunders, and F. J. Cuthbert. 2013. Survival of Piping Plover (*Charadrius melodus*) chicks in the Great Lakes region. *Auk* 130:150–160.

Brunton, D. H. 1986. Fatal antipredator behavior of a Killdeer. *Wilson Bulletin* 98:605–607.

Brunton, D. H. 1990. The effects of nesting stage, sex, and type of predator on parental defense by Killdeer (*Charadrius vociferus*): Testing models of avian parental defense. *Behavioral Ecology and Sociobiology* 26:181–190.

Buick, A. M. and Paton, D. C. 1989. Impact of off-road vehicles on the nesting success of Hooded Plovers (*Charadrius rubricollis*) in the Coorong region of South Australia. Emu 89:159–172.

Burger, J. 1990. Foraging behavior and the effect of human disturbance on the Piping Plover *Charadrius melodus*. *Journal of Coastal Research* 7:39–52.

Burger, J. 1994. The effect of human disturbance on foraging behavior and habitat use in Piping Plover (*Charadrius melodus*). *Estuaries* 17:695–701.

Burger, J., M. Gochfeld, and L. J. Niles. 1995. Ecotourism and birds in coastal New Jersey: Contrasting responses of birds, tourists, and managers. *Environmental Conservation* 22:56–65.

Cairns, W. E. 1982. Biology and behavior of breeding Piping Plovers. *Wilson Bulletin* 94:531–545.

Cohen, J. B., L. M. Houghton, and J. D. Fraser. 2009. Nesting density and reproductive success of Piping Plovers in response to storm- and human-created habitat changes. *Wildlife Monographs* 173:1–24.

Colwell, M. A. 2010. *Shorebirds: Ecology, Conservation and Management*. University of California Press, Oakland, CA.

Davis, A. 1994. Breeding biology of the New Zealand Shore Plover *Thinornis novaeseelandiae*. *Notornis* 41:195–208.

Dhanjal-Adams, K. L., K. Mustin, H. P. Possingham, and R. A. Fuller. 2016. Optimizing disturbance management for wildlife protection: The enforcement allocation problem. *Journal of Applied Ecology* 53:1215–1224.

Doherty, P. J. and J. A. Heath 2011. Factors affecting Piping Plover hatching success on Long Island, New York. *Journal of Wildlife Management* 75:109–115.

Dowding, J. E. 1994. Morphometrics and ecology of the New Zealand Dotterel (*Charadrius obscurus*), with a description of a new subspecies. *Notornis* 41:221–233.

Dowling, B. and M. A. Weston. 1999. Managing the Hooded Plover in a high-use recreational environment. *Bird Conservation International* 9:255–270.

Dunning, J. B. 1992. *CRC Handbook of Avian Body Masses*. CRC Press, Boca Raton, FL.

Eberhart-Phillips, L. J. and M. A. Colwell. 2014. Conservation challenges of a sink: The viability of an isolated population of the Snowy Plover. *Bird Conservation International* 24:327–341.

Ekanayake, K. B., M. A. Weston, D. G. Nimmo, G. S. Maguire, J. A. Endler, and C. Küpper. 2015. The bright incubate at night: Sexual dichromatism and adaptive incubation division in an open-nesting shorebird. *Proceedings of the Royal Society B* 282:20143026.

Elias-Gerken, S.P. 1994. Piping plover habitat suitability on central Long Island, New York barrier islands *PhD dissertation*, Virginia Polytechnic and State University, Blacksburg, VA.

Faillace, C. A. and B. W. Smith. 2016. Incubating Snowy Plovers (*Charadrius nivosus*) exhibit site specific patterns of disturbance from human activities. *Wildlife Research* 43:288–297.

Fernández-Juricic, E., M. P. Vernier, D. Renison, and D. T. Blumstein. 2005. Sensitivity of wildlife to spatial patterns of recreationist behavior: A critical assessment of minimum approaching distances and buffer areas for grassland birds. *Biological Conservation* 125:225–235.

Flemming, S. P., R. D. Chiasson, and P. C. Smith. 1988. Piping Plover status in Nova Scotia related to its reproductive and behavioral responses to human disturbance. *Journal of Field Ornithology* 59:321–330.

Fox, J. D. and J. Madsen. 1997. Behavioral and distributional effects of hunting disturbance on waterbirds in Europe: Implications for refuge design. *Journal of Applied Ecology* 34:1–13.

Frid, A. and L. M. Dill. 2002. Human-caused disturbance stimuli as a form of predation risk. *Conservation Ecology* 6:11.

Gaines, E. P. and M. R. Ryan. 1988. Piping Plover habitat use and reproductive success in North Dakota. *Journal of Wildlife Management* 52:266–273.

Gill, J. A. 2007. Approaches to measuring the effects of human disturbance on birds. *Ibis* 149:9–14.

Gill, J. A., K. Norris, and W. J. Sutherland. 2001. Why behavioural responses may not reflect the population consequences of human disturbance. *Biological Conservation* 97:265–268.

Glover, H. K., M. A. Weston, G. S. Maguire, K. K. Miller, and B. A. Christie. 2011. Towards ecologically meaningful and socially acceptable buffers: Response distances of shorebirds in Victoria, Australia, to human disturbance. *Landscape and Urban Planning* 103:326–334.

Gochfeld, M. 1984. Antipredator behaviour: Aggressive and distraction displays of shorebirds. pp. 289–377. In Burger, J. and Olla, B. L. (eds), *Shorebirds: Breeding Behaviour and Populations*. Plenum Press, New York.

Goldin, M. R. and J. V. Regosin. 1998. Chick behavior, habitat use, and reproductive success of Piping Plovers at Goosewing Beach, Rhode Island. *Journal of Field Ornithology* 69:228–234.

Gómez-Serrano, M. Á. and P. López-López. 2014. Nest site selection by Kentish Plover suggests a trade-off between nest-crypsis and predator detection strategies. *PLoS One* 9:e107121.

Graul, W. D. 1975. Breeding biology of the Mountain Plover. *Wilson Bulletin* 87:6–31.

Guay, P. J., W. W. F. D. van Dongen, R. Robinson, D. T. Blumstein, and M. A. Weston. 2016. AvianBuffer: An interactive tool for characterising and managing wildlife fear responses. *Ambio* 45:841–851.

Guay, P. J., M. A. Weston, M. R. E. Symonds, and H. K. Glover 2013. Brains and bravery: Little evidence of a relationship between brain size and flightiness in shorebirds. *Austral Ecology* 38:516–522.

Haffner, C. D., F. J. Cuthbert, and T. W. Arnold. 2009. Space use by Great Lakes Piping Plovers during the breeding season. *Journal of Field Ornithology* 80:270–279.

Hamza, F. and S. Selmi. 2015. Habitat features and human presence as predictors of the abundance of shorebirds and wading birds wintering in the Gulf of Gabès, Tunisia. *Marine Ecology Progress Series* 540:251–258.

Hanisch, D. 1998. Effects of human disturbance on the reproductive performance of the Hooded Plover. *Honors Thesis*, University of Tasmania, Hobart, Australia.

Hardy, M. A. and M. A. Colwell. 2012. Factors influencing Snowy Plover nest survival on ocean-fronting beaches in coastal Northern California. *Waterbirds* 35:503–656.

Hoffmann, A. 2005. Incubation behavior of female Western Snowy Plovers (*Charadrius alexandrinus nivosus*) on sandy beaches. M.S. thesis, Humboldt State University, Arcata, CA.

Lafferty, K. D. 2001a. Birds at a southern California beach: Seasonality, habitat use and disturbance by human activity. *Biodiversity and Conservation* 10:1949–1962.

Lafferty, K. D. 2001b. Disturbance to wintering Western Snowy Plovers. *Biological Conservation* 101:315–325.

Lafferty, K. D., D. Goodman, and C. P. Sandoval. 2006. Restoration of breeding by Snowy Plovers following protection from disturbance. *Biodiversity and Conservation* 15:2217–2230.

Lilleyman, A., D. C. Franklin, J. K. Szabo, and M. J. Lawes. 2016. Behavioural responses of migratory shorebirds at a high tide roost. *Emu* 116:95–99.

LeDee, O. E., F. J. Cuthbert, and P. V. Bolstad. 2008. A remote sensing analysis of coastal habitat composition for a threatened shorebird, the Piping Plover (*Charadrius melodus*). *Journal of Coastal Research* 24:719–726.

Lethlean, H., W. F. D. van Dongen, K. Kostoglou, P.-J. Guay, and M. A. Weston. 2017. Joggers cause greater avian disturbance than walkers. *Landscape and Urban Planning* 159:42–47.

Liley, D. and W. J. Sutherland. 2007. Predicting the population consequences of human disturbance for Ringed Plover *Charadrius hiaticula*: A game theory approach. *Ibis* 149:82–94.

Lloyd, P. and É. E. Plagányi. 2002. Correcting observer effect bias in estimates of nesting success of a coastal bird, the White-fronted Plover *Charadrius marginatus*: A recently developed observer-effects model gives better estimates than the survival model. *Bird Study* 49:124–130.

Loegering, J. P. 1992. Piping Plover breeding biology, foraging ecology and behavior on Assateague Island national seashore, Maryland. Ph.D. Dissertation, Virginia Polytechnic Institute and State University, Blacksburg, VA.

Loegering, J. P. and J. D. Fraser. 1995. Factors affecting Piping Plover chick survival in different brood-rearing habitats. *Journal of Wildlife Management* 59:646–655.

Lomas, S. C., D. A. Whisson, G. S. Maguire, L. X. Tan, P.-J. Guay, P.J. and M. A. Weston. 2014. The influence of cover on nesting red-capped plovers: a trade-off between thermoregulation and predation risk? *Victorian Naturalist* 131:115–120.

Long, P. R., S. Zefania, R. H. French-Constant, and T. Székely. 2007. Estimating the population size of an endangered shorebird, the Madagascar Plover, using a habitat suitability model. *Animal Conservation* 11:118–127.

Lord, A., J. R. Waas, J. Innes, and M. J. Whittingham. 2001. Effects of human approaches to nests of Northern New Zealand Dotterels. *Biological Conservation* 98:233–240.

Lorenzo, J. A. and K. W. Emmerson. 1995. Recent information on the distribution and status of the breeding population of Kentish Plover *Charadrius alexandrinus* in the Canary Islands. *Wader Study Group Bulletin* 76:43–46.

Lunardi, V. O. and R. H. Macedo. 2014. Shorebirds can adopt foraging strategies that take advantage of human fishing practices. *Emu* 114:50–60.

MacIvor, L. H., S. M. Melvin, and C. R. Griffin. 1990. Effects of research activity on Piping Plover nest predation. *Journal of Wildlife Management* 54:443–447.

MacLean, G. L. 1976. A field study of the Australian Dotterel. *Emu* 76:207–215.

Maguire, G. S., A. K. Duivenvoorden, M. A. Weston, and R. Adams. 2011. Provision of artificial shelter on beaches is associated with improved shorebird fledging success. *Bird Conservation International* 21:172–185.

Maguire, G., J. M. Rimmer, and M. A. Weston. 2015. Stakeholder knowledge of threatened coastal species; the case of beach-goers and the Hooded Plover *Thinornis rubricollis*. *Journal of Coastal Conservation* 19:73–77.

Martín, B., S. Delgado, A. de la Cruz, S. Tirado, and M. Ferrer. 2015. Effects of human presence on the long-term trends of migrant and resident shorebirds: Evidence of local population declines. *Animal Conservation* 18:73–81.

Montalvo, T. and J. Figuerola. 2006. The distribution and conservation of the Kentish Plover *Charadrius alexandrinus* in Catalonia. *Revista Catalana d'Ornitologia* 22:1–8.

Nicholls, J. L. and G. A. Baldassarre. 1990. Winter distribution of Piping Plovers along the Atlantic and Gulf Coasts of the United States. *Wilson Bulletin* 102:400–412.

Ormsby, A. A. and E. A. Forys. 2010. The effects of an education campaign on beach user perceptions of beach-nesting birds in Pinellas County, Florida. *Human Dimensions of Wildlife* 15:119–128.

Patterson, M. E., J. D. Fraser, and J. W. Roggenbuck. 1991. Factors affecting Piping Plover productivity on Assateague Island. *Journal of Wildlife Management* 55:525–531.

Pfister, C., B. A. Harrington, and M. Lavine. 1992. The impact of human disturbance on shorebirds at a migration staging area. *Biological Conservation* 60:115–126.

Pietrelli, L. and M. Biondi. 2012. Long term reproduction data of Kentish Plover *Charadrius alexandrinus* along a Mediterranean coast. *Wader Study Group Bulletin* 119:114–119.

Plissner, J. H. and S. M. Haig. 2000. Viability of Piping Plover *Charadrius melodus* metapopulations. *Biological Conservation* 92:163–173.

Powell, A. N., F. J. Cuthbert, L. C. Wemmer, A. W. Doolittle, and T. Shane. 1997. Captive-rearing Piping Plovers: Developing techniques to augment wild populations. *Zoo Biology* 16:461–477.

Rees, J. D., J. K. Webb, M. S. Crowther, and M. Letnic. 2015. Carrion subsidies provided by fishermen increase predation of beach-nesting bird nests by facultative scavengers. *Animal Conservation* 18:44–49.

Rimmer, J. M., G. S. Maguire, and M. A. Weston. 2013. Perceptions of effectiveness and preferences for design and position of signage on Victorian beaches for the management of Hooded Plovers *Thinornis rubricollis*. *Victorian Naturalist* 130:75–80.

Ristau, C. A. 1992. Cognitive ethology: Past, present and speculations on the future. *Proceedings of the Biennial Meeting of the Philosophy of Science Association* 2:125–136.

Roche, D. V., A. P. Cardilini, D. Lees, G. S. Maguire, P. Dann, C. D. Sherman, and M. A. Weston. 2016. Human residential status and habitat quality affect the likelihood but not the success of lapwing breeding in an urban matrix. *Science of the Total Environment* 556:189–195.

Rogers, K. G., D. I. Rogers, and M. A. Weston. 2014. Prolonged and flexible primary moult overlaps extensively with breeding in beach-nesting Hooded Plovers *Thinornis rubricollis*. Ibis 156:840–849.

Ruhlen, T. D., S. Abbott, L. E. Stenzel, and G. W. Page. 2003. Evidence that human disturbance reduces Snowy Plover chick survival. *Journal of Field Ornithology* 74:300–304.

Schlacher T. A., L. K. Carracher, N. Porch, R. M. Connolly, A. D. Olds, B. L. Gilby, K. B. Ekanayake, B. Maslo, and M. A. Weston. 2016. The early shorebird will catch fewer invertebrates on trampled sandy beaches. *PLoS One* 11:e0161905.

Schlaepfer, M. A., M. C. Runge, and P. W. Sherman. 2002. Ecological and evolutionary traps. *Trends in Ecology and Evolution* 17:474–480.

Schneider, T. J. 2013. The use of Victoria's sandy shores by domestic dogs (*Canis familiaris*). *Honors Thesis*, Deakin University, Melbourne, Australia.

Seavey, J. R., B. Gilmer, and K. M. McGarigal. 2010. Effect of sea-level rise on Piping Plover (*Charadrius melodus*) breeding habitat. *Biological Conservation* 144:393–401.

Staine K. J. and J. Burger. 1994. Nocturnal foraging behavior of breeding Piping Plovers (*Charadrius melodus*) in New Jersey. *Auk* 111:579–587.

St Clair, J. J., G. E. García-Peña, R. W. Woods, and T. Székely. 2010. Presence of mammalian predators decreases tolerance to human disturbance in a breeding shorebird. *Behavioral Ecology* 21:1285–1292.

Stigner, M. G., H. L. Beyer, C. J. Klein, and R. A. Fuller. 2016. Reconciling recreational use and conservation values in a coastal protected area. *Journal of Applied Ecology* 53:1206–1214.

Strauss, E. and B. Dane. 1989. Differential reproductive success in a stressed population of Piping Plovers in areas of high and low human disturbance. *American Zoologist* 29:A42.

Summers, R. W. and P. A. R. Hockey. 1981. Egg-covering behavior of the White-Fronted Plover *Charadrius marginatus*. *Ornis Scandinavica* 12:240–243.

Sutherland, W. J. 2007. Future directions in disturbance research. *Ibis* 149:120–124.

Szentirmai, A., A. Kosztolányi, and T. Székely. 2001. Daily changes in body mass of incubating Kentish Plovers. *Ornis Hungarica* 11:27–32.

Tablado, Z. and L. Jenni. 2016. Determinants of uncertainty in wildlife responses to human disturbance. *Biological Reviews* 92:216–233.

Tan, L. X., K. L. Buchanan, G. S. Maguire, and M. A. Weston. 2015. Cover, not caging, influences chronic physiological stress in a ground-nesting bird. *Journal of Avian Biology*, 46:482–488.

van Dongen, W. F., R. W. Robinson, M.A. Weston, R. A., Mulder and P.-J. Guay. 2015. Variation at the DRD4 locus is associated with wariness and local site selection in urban black swans. *BMC Evolutionary Biology* 15:253

Waldvogel, J. A. 1990. The bird's eye view. *American Scientist* 78:342–353.

Webber, A. F., J. A. Heath, and R. A. Fischer. 2013. Human disturbance and stage-specific habitat requirements influence Snowy Plover site occupancy during the breeding season. *Ecology and Evolution* 3:853–863.

Wemmer, L. C., O. U. Zesmi, and F. J. Cuthbert. 2001. A habitat-based population model for the Great Lakes population of the Piping Plover (*Charadrius melodus*). *Biological Conservation* 99:169–181.

Weston, M. A., M. J. Antos, and C. L. Tzaros. 2009. Sand pads: A promising technique to quantify human visitation into nature conservation areas. *Landscape and Urban Planning* 89:98–104.

Weston, M. A., F. Dodge, A. Bunce, D. G. Nimmo, and K. K. Miller. 2012a. Do temporary beach closures assist in the conservation of breeding shorebirds on recreational beaches? *Pacific Conservation Biology* 18:47–55.

Weston, M. A., G. C. Ehmke, and G. S. Maguire. 2011. Nest return times in response to static versus mobile human disturbance. *Journal of Wildlife Management* 75:252–255.

Weston, M. A. and M. A. Elgar. 2005a. Disturbance to brood-rearing Hooded Plover *Thinornis rubricollis*: Responses and consequences. *Bird Conservation International* 15:193–209.

Weston, M. A. and M. A. Elgar. 2005b. Parental care in Hooded Plovers (*Thinornis rubricollis*). *Emu* 105:283–292.

Weston, M. A. and M. A. Elgar. 2007. Responses of incubating Hooded Plovers (*Thinornis rubricollis*) to disturbance. *Journal of Coastal Research* 23:569–576.

Weston, M. A., E. M. McLeod, D. T. Blumstein, and P.-J. Guay. 2012b. A review of flight initiation distances and their application to managing disturbance to Australian birds. *Emu* 112:269–286.

Weston, M. A., K.-Y. Ju, Guay, P-J., and Naismith, C. 2019. A test of the "Leave Early and Avoid Detection" (LEAD) hypothesis for passive nest defenders. *Wilson Bulletin* 130.

Williams, K. J. H., M. A. Weston, S. Henry, and G. S. Maguire. 2009. Birds and beaches, dogs and leashes: Dog owners' sense of obligation to leash dogs on beaches in Victoria, Australia. *Human Dimensions of Wildlife* 14:89–101.

Wilson, H. 1950. Visual perception among waders: A suggested explanation of the habit of 'bobbing'. *Emu* 50:128–131.

Wilson, C. A. and M. A. Colwell. 2010. Movements and fledging success of Snowy Plover (*Charadrius alexandrinus*) chicks. *Waterbirds* 33:331–340.

Yasue, M. 2006a. Breeding ecology and potential impacts of habitat change on the Malaysian Plover, *Charadrius peronii*, in the Gulf of Thailand *PhD dissertation*, University of Victoria, Victoria, Canada.

Yasué, M. 2006b. Environmental factors and spatial scale influence shorebirds' responses to human disturbance. *Biological Conservation* 128:47–54.

Yasué, M. and P. Dearden. 2006a. The effects of heat stress, predation risk and parental investment on Malaysian Plover nest return times following a human disturbance. *Biological Conservation* 132:472–480.

Yasué, M. and P. Dearden. 2006b. The potential impact of tourism development on habitat availability and productivity of Malaysian Plovers *Charadrius peronii*. *Journal of Applied Ecology* 43:978–989.

Yasué, M. and P. Dearden. 2008a. Parental sex roles of Malaysian Plovers during territory acquisition, incubation and chick-rearing. *Journal of Ethology* 26:99–112.

Yasué, M. and P. Dearden. 2008b. Methods to measure and mitigate the impacts of tourism development on tropical beach-breeding shorebirds: The Malaysian Plover in Thailand. *Tourism in Marine Environments* 5:287–299.

Yasué, M. and P. Dearden 2009. The importance of supratidal habitats for wintering shorebirds and the potential impacts of shrimp aquaculture. *Environmental Management* 43:1108.

Yasué, M., P. Dearden, and A. Moore. 2008. An approach to assess the potential impacts of human disturbance on wintering tropical shorebirds. *Oryx* 42:415–423.

Yasué, M., A. Patterson, and P. Dearden. 2007. Are saltfats suitable supplementary nesting habitats for Malaysian Plovers *Charadrius peronii* threatened by beach habitat loss in Thailand? *Bird Conservation International* 17:211–223.

Ydenberg, R. C. and L. M. Dill. 1986. The economics of fleeing from predators. *Advances in the Study of Behavior* 16:229–249.

Ram Papish

CHAPTER TWELVE

Future Challenges for Charadrius Plovers

Susan M. Haig and Mark A. Colwell

Abstract. The topics covered in this book provide a broad perspective on a group of species that exist at the epicenter of a rapidly changing world. As such, plovers provide critical insight into anthropogenic effects on biodiversity. However, plover research illustrates that few species are studied well enough to guide management of factors that limit their population size and growth. Even so, one-third of the 40 species are listed by IUCN as *Vulnerable* (3), *Threatened* (7), or *Endangered* (3). The remaining species are classified as *Least Concern* despite declining populations. Plover vulnerability often stems from a species' insular, restricted range, exacerbated by anthropogenic factors. Further, some strategies directed at conservation of other shorebird groups may not be as effective for plovers. For example, most sandpipers breed on the arctic tundra and winter on coastal tidal flats. While three plovers breed in the arctic, most of the Charadrius plovers breed on coastal or inland areas. They winter in flocks on agricultural fields and coastal mud or sandflats although some set up territories on beaches. Finally, research efforts with the broadest perspective will now have to be stepped up as the effects of changing climates [e.g., rising sea levels, more intense storms, increased water salinization (inland and oceanic), increasing drought in desert habitats] take a stronger hold on plovers inland, arctic, and coastal systems. Thus, new assessments of species distributions, food availability, predator activity, and human encroachment will contribute to our knowledge of factors that influence their population viability and can be used to develop science-based management options.

Keywords: annual cycle, Charadrius, climate change, conservation, ecology, management, migratory connectivity, plover.

The previous chapters summarized our current knowledge of a group of birds that serve as critical indicators of the status of their human-coveted habitats. Charadrius plovers are great study species for many reasons: they are easy to observe, they can be individually marked, they indicate the status of habitat prized by humans, and, as such, they have been used to address critical conservation issues that other species cannot. And we have come a long way in what we have learned about particular species, such as Piping, Snowy, and Kentish plovers. However, we know surprisingly little about the other 35+ species in the clade. In this chapter, we synthesize what we have learned from the more commonly known species and address future research and conservation issues related to members of the subfamily Charadriinae.

At first blush, the 40 plover species appear to be a rather uniform group; however, many species consist of multiple subspecies (Chapter Two, this volume), and breeding systems run the gamut from polyandry through monogamy to polygyny, with plasticity evident within some species (Chapter Four, this volume). Some populations, notably those breeding in northern latitudes, migrate long distances to reach non-breeding grounds, whereas others are partially migratory or sedentary year-round (Chapter Seven, this volume).

Most plovers occupy habitats that render them vulnerable to human impacts at some point in the annual cycle (Chapter Eleven, this volume). The resulting anthropogenic habitat loss at beaches, sandflats, and agricultural land (Chapter Nine, this volume) is exacerbated by changes in natural or climate-induced ecological factors such as food availability and predation (Chapter Eight, this volume), which influences population stability (Chapter Ten, this volume). Despite a wealth of knowledge on some plover species, we lack high-quality data on population size and trends, as well as the demographic vital rates essential for evaluating effective management actions for recovery of many species. The exceptions are those taxa protected by laws such as the U.S. Endangered Species Act. Here, we offer a prospectus on plover knowledge, status, and conservation in the Anthropocene (Crutzen 2002).

THE ANNUAL CYCLE

We organized the chapters of this volume around the importance of understanding and conserving migratory connectivity throughout the annual cycle (Webster et al. 2002, Haig et al. in review). That is, we consider each phase of the annual cycle as well as the effect of one phase on subsequent phases.

Breeding Biology

Small clutch sizes, precocial chicks, and extended parental care suggest that plovers are conservative breeders. Low annual productivity is coupled with high survival, especially for island taxa or other species that are non-migratory. These life history patterns render plovers vulnerable to natural and anthropogenic factors that chronically compromise reproductive success (Chapter Eleven, this volume). Conversely, many temperate or tropical species have prolonged breeding seasons during which they may successfully produce multiple broods (Chapter Five, this volume). As climates change and phenological mismatch spreads, careful attention must be paid to identifying changing breeding distributions and phenologies in order to diagnose changes in reproductive success (Chapter Three, this volume).

Predation is a major cause of reproductive failure especially for island taxa affected by introduced vertebrates (Chapter Six, this volume). In continental populations, predators, especially intelligent, synanthropic omnivores such as corvids, exert strong negative effects on nest success and chick survival. In coastal and agricultural regions, conservation challenges are exacerbated by human activities in plover habitats. Extensive efforts to manage predators and humans can have positive results, although exceptions and challenges exist such as how circadian sex roles during incubation are adapted to minimize predation risk (Ekanayake et al. 2015; Chapters Six and Eleven, this volume).

Conservation often focuses on the breeding season. Management prescriptions include habitat restoration, predator control, and efforts to manage humans. Each of these approaches aims to improve productivity. However, survival is the most influential vital rate affecting avian population growth (Sæther and Bakke 2000), which suggests that renewed efforts to manage predators of adults and juveniles must include perspectives from throughout the annual cycle.

Migration

Shorebird conservation is steeped in the "staging paradigm." Put succinctly, populations adhering to flyways (Lincoln 1935) are most vulnerable to declines when a large percentage of a population stages at a few wetlands; loss of wetlands, vital to survival during the non-breeding season, may doom global populations to irreversible declines (Morrison 1984, Myers et al. 1987, Robinson and Warnock 1996). This paradigm derives from a rich literature on sandpipers and, while less studied, also applies to a number of plover species (e.g., Piping Plovers, Killdeer, Snowy Plovers; Plissner et al. 2000, Sanzenbacher and Haig 2002, Elliott-Smith et al. 2004, Haig et al. 2005, Taft and Haig 2006), although see Robinson and Warnock (1996). One extreme example of this staging is the Chestnut-banded Plover, which occurs at just three sites in Namibia. Recently, Iwamura et al. (2016) modeled effects of habitat loss owing to sea level rise in estuaries of the East Asian-Australasian Flyway based on the "staging paradigm." Their results, including nine sandpiper species and the Lesser Sand-Plover, indicated negative effects of habitat loss were stronger for all sandpipers than the plover. Potential use of beaches provides plovers with habitat options not always used by other shorebirds.

Non-breeding Ecology

Most plovers, especially arctic and temperate breeders, spend much of the annual cycle in trans-equatorial regions where they forage day and night (Staine and Burger 1994, Plissner et al. 2000, Sanzenbacher and Haig 2002, Elliott-Smith et al. 2004, Haig et al. 2005). Ever vigilant for predators, plovers clearly prefer expansive sand/salt/tidal flat habitats of estuaries, beaches, grasslands, and disturbed habitats, such as fallow or tilled agricultural fields (Taft and Haig 2006). However, in some species, birds set up winter territories on beaches while conspecifics are wintering in flocks on mudflats only a few kilometers away (e.g., Piping, Snowy, and Collared plovers on Gulf of Mexico beaches and inland mudflats in Texas and Mexico; S. Haig, pers. obs.). Further, at least some plover (e.g., Piping Plovers) species exhibit high-site fidelity in winter (Gratto-Trevor et al. 2012, 2016). Thus, permanent loss of habitat or a food source, such as following a hurricane, can result in some individuals trying to occupy suboptimal habitat because it is near their former territory (Aharon-Rotman et al. 2015).

Bottom-up effects of food limitation on plover populations require further study, especially in northern hemisphere wintering areas where prey depletion is more likely than in southern latitudes and the austral summer (Hockey et al. 1992). If habitat loss and degradation affect plover populations, it can be via subtle effects on food availability or intake rates. Climate change and the resulting altered salinity and water levels in coastal and inland areas, will further negatively affect habitat availability (and the food therein). This will likely compromise overwinter survival or decrease fitness for the spring migration.

We would be remiss not to mention that some land use practices may favor plovers owing to their preference for expansive, sparsely vegetated habitats. For example, the St. Helena Plover is reliant on grazing to maintain habitat favorable for breeding (McCulloch and Norris 2001). In North America, Killdeer and Mountain Plover often occupy agricultural settings such as pastures and arable lands (Knopf and Ruppert 1995, Wunder and Knopf 2003, Plissner et al. 2000, Sanzenbacher and Haig 2002, Taft and Haig 2006). Management of these species will benefit from special consideration of land management practices (e.g., farming) to insure optimization of plover needs and agricultural goals. And anthropogenic saltwork habitats provide critical resources for plovers at all stages of the annual cycle (e.g., Laguna Lagartos, Mexico; southern coastal Puerto Rico; Walvis Bay, Namibia; S. Haig, pers. obs.). These areas can be beneficial or toxic depending on the salinity and other elements contained in the production area.

The top-down effects of predation are evident in much of plover evolution, behavior, and ecology: muted plumages, nocturnal foraging, and protective strategies for eggs and young. Despite these observations, we know little of the impacts of predators on plover populations (Chapter Six, this volume). Several studies suggest plovers are taken disproportionately by falcons (e.g., Lank et al. 2003, Lank et al. 2017). This effect, however, is at odds with the observation that, in mixed-species flocks, plovers tend to detect and respond to raptors sooner that sandpipers (Thompson and Thompson 1985), perhaps due to their superior visual abilities.

SPECIES INFORMATION GAPS

The impetus for this book was our collective experiences of several decades of research on plovers, with an interest in applying what is known to effect and improve plover conservation. In compiling this book, we focused on factors that limit population growth. To accomplish this objective, we enlisted experts on plovers, each with a specialization in some facet of their biology. In writing their chapters, we asked authors to tally the number of papers published on their species. Figure 12.1 illustrates our disparate knowledge of the 40 members of the Charadriinae. Three species (Piping, Snowy, and Kentish plovers), two with populations listed under the U.S. Endangered

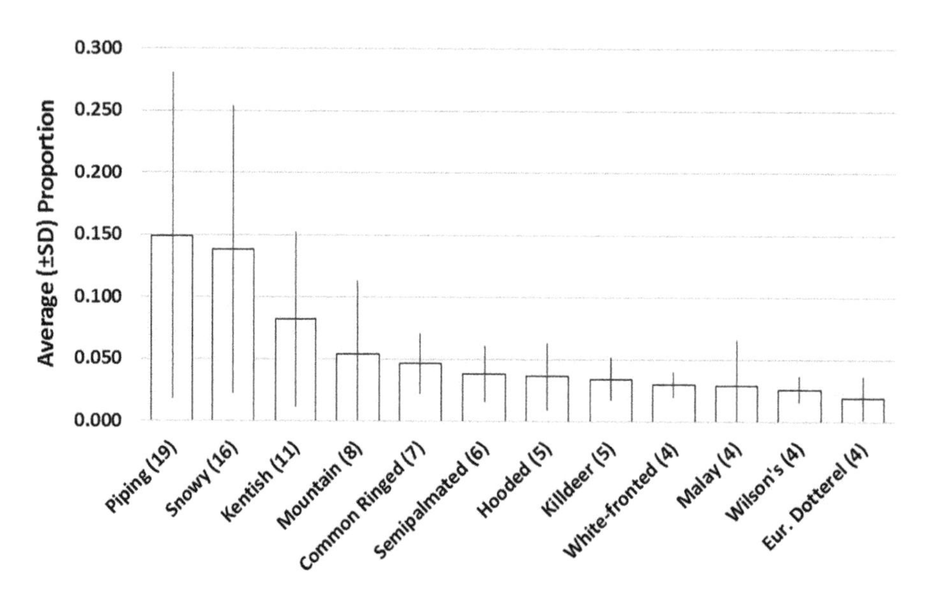

Figure 12.1. The published literature on members of the plover clade is disproportionately represented by a quarter of its taxa. Data are proportion of papers on each species averaged (±SD) across the chapters of this volume. Integers in parentheses are the average number of papers across chapters.

Species Act, represent more than one-third (37%) of research effort; 12 species represent two-thirds (68%) of all publications. Stated another way, most plover species (28) are virtually unstudied, with no taxon surpassing a contribution that exceeds 2% of published papers (on average, across chapter topics). This disparity has a geographical bias whereby underrepresented species primarily breed in southern latitudes. Given the high richness of plover species in Asia and Africa (Figure 7.3), we clearly lack information on the basic ecology of most members of the clade, especially in areas where there are the most pressing conservation needs. It is plausible that this uneven pattern arises, in part, from: (1) legal systems that affect funding of applied research (e.g., ESA mandating data to effect recovery), as evidenced by Piping and Snowy plovers; (2) logistics of research in remote regions; (3) varying funding opportunities across the world; and (4) lack of concern regarding small birds.

SPECIES STATUS ASSESSMENTS

Species status assessments begin with defining the taxon in question, usually via molecular methods (Haig et al. 2011, 2016). The evolutionary relationships of the Charadrius clade have been well resolved with advances in molecular ecology (Chapter Two, this volume). Still, questions with conservation implications remain. For example, many species consist of subspecies; others, such as the White-faced Plover, are of uncertain taxonomic status, which obviates further investigation. From a conservation perspective, distinct taxonomic or demographic segments must be defined, such as the three Piping Plover populations and the Pacific coast population of Snowy Plover that are protected under U.S. law (e.g., U.S. Endangered Species Act; Funk et al. 2007, Miller et al. 2010), in order to provide for accurate status assessments. The recent taxonomic split of the (former) New Zealand Dotterel into southern and northern (red-breasted) species makes their conservation status even more pressing given they are both threatened. Finally, taxonomic and population investigations are warranted for several other relatively unknown species, such as Chestnut-banded Plover with its disjunct populations across Africa (dos Remedios 2013).

The conservation challenge with plovers is dire, as evidenced by their IUCN conservation status (Table 1.2). Twenty-six (65%) species are labeled of "Least Concern" under IUCN standards; one species is "Data Deficient." The remaining 13 taxa (32%) have populations that are low enough to warrant concern (i.e., "Vulnerable," "Near Threatened," "Endangered"), and one

(St. Helena's Plover) is "Critically Endangered." Although population size estimates vary in precision (note order of magnitude estimates for all but some of the most-studied taxa), knowledge of population growth is even more poorly known. Where estimates exist, trends suggest declines for taxa widely considered to be abundant. For example, Breeding Bird Survey data indicate that populations of seemingly common Killdeer have declined across North America (Sanzenbacher and Haig 2002).

One concern in determining plover species status is that while plovers are relatively easy to count, a major issue arises in estimating population sizes if all applicable habitats are not searched uniformly. Detectability measures are designed to statistically estimate population sizes given gaps in sampling. However, most plovers are not evenly distributed across their habitats, especially as populations decline. For example, Long et al. (2008) extrapolated a population estimate for Black-banded (Madagascar) plover of >3,000 based on subsampled habitat and an assumption that the birds are not aggregated. Clearly, there are fewer birds (Table 1.1), demonstrating that overestimating population sizes, particularly small population sizes, can result in a serious loss of conservation efforts for a species when likely more was needed.

ADVANCES IN TECHNOLOGY

Individual identification and the resulting statistical analyses are helpful in census and population analyses even though plover size precludes use of the most powerful tracking devices (Bridge et al. 2011). Traditionally, this tracking has been carried out using small VHF transmitters attached to the bird's back. Unfortunately, the size, weight, and longevity of the batteries are often limiting. Satellite transmitters are too heavy for plovers, but their weight is gradually being reduced to make this a more feasible option. Current efforts to decrease transmitter size to 1 g by P.P. Marra (Smithsonian Migratory Bird Center) and others will uncover a new layer of insight into plover movement.

New technology using miniature archival geolocators (Cooper et al. 2017) is starting to bridge that gap in our ability to understand plover annual cycles. These geolocators have been miniaturized to weigh 0.65 g and will record data for over 1 year. Often the limiting factor in geolocators is that the bird must be recaptured in order to recover the data. Plovers tend to be highly site faithful in breeding and winter locations and relatively easy to catch so the geolocator technology will be important technology used to track annual movements of plovers in the future.

While prices per unit are coming down on all tracking devices, they are still high enough to usually preclude deployment of many devices per population. And birds must be caught at least once to attach the device. Thus, new bioacoustic listening technology might be a way to better identify breeding, migration, or winter sites for plovers, particularly as climate change alters distributions and phenologies (Blumstein et al. 2011). While not generally used to identify individuals, they can be deployed on beaches or other habitats to continuously monitor plover occupancy via recordings of their vocalizations. Automatic recording and determination of species identity as they fly over or land in an area provides a previously untested manner in which to monitor or track groups of individuals (Salamon et al. 2016). This technology could be particularly useful to deploy on vast beaches plovers may or may not be using as a means of determining the importance of the site to the species and help narrow focus on the ground census or survey efforts.

Another non-invasive survey technique involves the use of drones with heat-seeking thermal infrared sensors to record the presence or absence of birds or nests (Gillette et al. 2013, Hanson et al. 2014). Particularly in areas where at some point (often just before dawn) during the 24-h cycle, the bird or nest is the warmest object on the landscape, this technology will provide an efficient means of finding nests, counting birds, or just identifying if an area is being used without drawing predators in via a person's presence. Drones require skilled pilots so this technology might be deployable for only short periods of time but could be quite efficient in surveying new potential nest sites, etc.

Finally, genomic techniques, including eDNA technology, are fast becoming the key to identifying individuals, families, populations, and subspecies (Chapter Two, this volume). While the lab component is complex, major laboratories are taking over processing samples for a nominal fee leaving field biologists with easy sampling techniques (e.g., Vilstrup et al. 2018) and population biologists to deal with data analyses. Thus, use of molecular markers is becoming easier to deploy, more specific in the identifications that can be made, and biologists are able to use orders of magnitude more markers on many more individuals than in the past.

Once some of the technological issues discussed above are resolved, other concerns include addressing broader issues such as examining the effects of drought, flooding, habitat fragmentation, and the need for fresh water. Human development projects such as beach renourishment, agricultural irrigation, contaminants, and river stabilization programs will remain critical issues to address.

SUMMARY

Our understanding of plover population ecology is rich and growing, but the foundation for their conservation is derived from just a few well-studied species. Clearly, much has yet to be deciphered of the basic biology for many taxa. An even greater concern is to understand and mitigate the changes coastal, arctic, and inland plover habitats are increasingly undergoing as a result of climate change. Thus, we hope this synthesis stimulates future generations to study this fascinating group of birds that is readily accessible, easy to study, and so vulnerable to human impacts in nearly all habitats they occupy.

ACKNOWLEDGMENTS

We thank L. Eberhart-Phillips, P. Sanzenbacher, C. Phillips, S. Phillips, and B. Ralston for critical reviews of the manuscript. Funding for this chapter was partially provided by the USGS Forest and Rangeland Ecosystem Science Center. Any use of trade, product, or firm names is for descriptive purposes only and does not imply endorsement by the U.S. Government.

LITERATURE CITED

Aharon-Rotman, Y., M. Soloviev, C. Minton, P. Tomkovich, C. Hassell, and M. Klaassen. 2015. Loss of periodicity in breeding success of waders links to changes in lemming cycles in Arctic ecosystems. *Oikos* 124:861–870.

Blumstein, D.T., D.J. Mennill, P. Clemins, L. Girod, K. Yao, G. Patricelli, J.L. Deppe, A.H. Krakauer, C. Clark, K.A. Cortopassi, S.F. Hanser, B. McCowan, A.M. Ali, and A.N.G. Kirschel. 2011. Acoustic monitoring in terrestrial environments using microphone arrays: Applications, technological considerations and prospectus. *Journal of Applied Ecology* 48:758–767.

Bridge, E.S., K. Thorup, M.S. Bowlin P.B. Chilson, R.H. Diehl, R.W. Fléron, P. Hartl, R. Kays, J.F. Kelly, W.D. Robinson, and M. Wikelski. 2011. Technology on the move: Recent and forthcoming innovations for tracking migratory birds. *BioScience* 61:689–698.

Cooper, N.W., M.T. Hallworth, and P.P. Marra. 2017. Light-level geolocation reveals wintering distribution, migration routes, and primary stopover locations of an endangered long distance migratory songbird. *Journal of Avian Biology* 48:209–219.

Crutzen, P.J. 2002. Geology of mankind: The anthropocene. *Nature* 415:23.

dos Remedios, N. 2013. The Evolutionary History of Plovers, Genus *Charadrius*: Phylogeography and Breeding Systems. *PhD Dissertation*. University of Bath, Bath, UK.

Ekanayake, K.B., M.A. Weston, D.G. Nimmo, G.S. Maguire, J.A. Endler, C. Kupper. 2015. The bright incubate at night: Sexual dichromatism and adaptive incubation division in an open-nesting shorebird. *Proceedings of the Royal Society B* 282:20143026.

Elliott-Smith, E., S.M. Haig, C.L. Ferland, and L. Gorman. 2004. Winter distribution and abundance of Snowy Plovers in eastern North America and the West Indies. *Wader Study Group Bulletin* 104:28–33.

Funk, W.C., T.D. Mullins, and S.M. Haig. 2007. Conservation genetics of Snowy Plovers (*Charadrius alexandrinus*) in the Western Hemisphere: Population genetic structure and delineation of subspecies. *Conservation Genetics* 8:1287–1309.

Gillette, G.L., P.S. Coates, S. Petersen, and J.P. Romero. 2013. Can reliable Sage-grouse lek counts be obtained using aerial infrared technology? *Journal of Fish and Wildlife Management* 4:386–394.

Gratto-Trevor, C.L., D. Amirault-Langlais, D. Catlin, F. Cuthbert, J. Fraser, S. Maddock, E. Roche, and F. Schaffer. 2012. Connectivity in Piping Plovers: Do breeding populations have distinct winter distributions? *Journal of Wildlife Management* 76:348–355.

Gratto-Trevor, C.L., S.M. Haig, M.P. Miller, T.D. Mullins, S. Maddock, E. Roche, and P. Moore. 2016. Breeding sites and winter site fidelity of

Piping Plovers wintering in The Bahamas, a previously unknown major wintering area. *Journal of Field Ornithology* 87:29–41.

Haig, S.M., W. Bronaugh, R. Crowhurst, J. D'Elia, C. Eagles-Smith, C. Epps, B. Knaus, M.P. Miller, M. Moses, S. Oyler-McCance, W.D. Robinson, and B. Sidlauskas. 2011. Perspectives in ornithology: Applications of genetics in avian conservation. *Auk* 128:205–229.

Haig, S.M., C.L. Ferland, D. Amirault, F. Cuthbert, J. Dingledine, P. Goossen, A. Hecht, and N. McPhillips. 2005. The importance of complete species censuses and evidence for regional declines in Piping Plovers. *Journal of Wildlife Management* 69:160–173.

Haig, S.M., M.P. Miller M.R. Bellinger, H.M. Draheim, D.M. Mercer, and T.D. Mullins. 2016. The conservation genetics juggling act: Integrating genetics and ecology, science and policy. *Evolutionary Applications* 9:181–195.

Haig, S.M., S.P. Murphy, J.D. Matthews, I. Arismendi, and M. Safeeq. Climate-altered wetlands challenge waterbird use and migratory connectivity in arid landscapes. Scientific Reports 9:46666.

Hanson, L., C.L. Holmquist-Johnson, and M.L. Cowardin. 2014. Evaluation of the Raven sUAS to Detect and Monitor Greater Sage-Grouse Leks within the Middle Park Population: U.S. Geological Survey Open-File Report 2014–1205, 20.

Hockey, P.A.R., R.A. Navarro, B. Kaljeta, and C.R. Velasquez. 1992. The riddle of the sands: Why are shorebird densities so high in southern estuaries? *American Naturalist* 140:961–979.

Iwamura, T., H.P. Possingham, I. Chadès, C. Minton, N.J. Murray, D.I. Rogers, E.A. Treml, and R.A. Fuller. 2016. Migratory connectivity magnifies the consequences of habitat loss from sea-level rise for shorebird populations. *Proceedings of the Royal Society B* 280:20130325.

Knopf, F.L., and J.R. Rupert. 1995. Habits and habitats of Mountain Plovers in California. *Condor* 97:743–751.

Lank, D.B., R.W. Butler, J. Ireland, and R.C. Ydenberg. 2003. Effects of predation danger on migration strategies of sandpipers. *Oikos* 103:303–319.

Lank, D.B., C. Xu, B.A. Harrington, R.I.G. Morrison, C.L. Gratto-Trevor, P.W. Hicklin, B.K. Sandercock, P.A. Smith, E. Kwon, J. Rausch, L.D. Pirie Dominix, D.J. Hamilton, J. Paquet, S.E. Bliss, S.G. Neima, C. Friis, S.A. Flemming, A.M. Anderson, and R.C. Ydenberg. 2017. Long-term continental changes in wing length, but not bill length, of a long-distance migratory shorebird. *Ecology and Evolution* 7:3243–3256.

Lincoln, F.C. 1935. The Waterfowl Flyways of North America. Circular 342. U.S. Department of Agriculture, Washington, DC.

Long, P.R., S. Zefania, R.H. ffrench-Constant, and T. Szekely. 2008. Estimating the population size of an endangered shorebird, the Madagascar plover, using a habitat suitability model. *Animal Conservation* 11:118–127.

McCulloch, N. and K. Norris. 2001. Diagnosing the cause of population changes: Localized habitat change and the decline of the endangered St. Helena Wirebird. *Journal of Applied Ecology* 38:771–783.

Miller, M.P., S.M. Haig, T.D. Mullins, and C.L. Gratto-Trevor. 2010. Molecular population genetic structure in the Piping Plover. *Auk* 127:57–71.

Morrison, R.I.G. 1984. Migration systems of some New World shorebirds in J. Burger and B.L. Olla (eds). *Shorebirds: Migration and Foraging Behavior*. Plenum Press: New York, 125–202.

Myers, J.P., R.I.G. Morrison, P.Z. Antas, B.A. Harrington, T.E. Lovejoy, M. Sallaberry, S.E. Senner, and A. Tarak. 1987. Conservation strategy for migrating species. *American Scientist* 75:19–26.

Plissner, J.H., L.W. Oring, and S.M. Haig. 2000. Space use of Killdeer at a Great Basin breeding area. *Journal of Wildlife Management* 64:421–429.

Robinson, J.A. and S.E. Warnock. 1996. The staging paradigm and wetland conservation in arid environments: Shorebirds and wetlands of the North American Great Basin. *International Wader Studies* 9:37–44.

Sæther, B.-E. and Ø. Bakke. 2000. Avian life history variation and contribution of demographic traits to the population growth rate. *Ecology* 81:642–653.

Salamon, J., P. Bello, A. Farnsworth, M. Robbins, S. Keen, H. Klinck, and S. Kelling. 2016. Towards the automatic classification of avian flight calls for bioacoustic monitoring. *PLoS One* 11:e0166866.

Sanzenbacher, P.M. and S.M. Haig. 2002. Regional fidelity and movement patterns of wintering Killdeer in an agricultural landscape. *Waterbirds* 25:16–25.

Staine, K.J. and J. Burger. 1994. Nocturnal foraging behavior of breeding Piping Plovers (*Charadrius melodus*) in New Jersey. *Auk* 111:579–587.

Taft, O.W. and S.M. Haig. 2006. Importance of wetland landscape structure to shorebirds wintering in an agricultural valley. *Landscape Ecology* 21:169–184.

Thompson, D.B.A. and M.L.P. Thompson. 1985. Early warning and mixed species associations: The 'Plover's Page' revisited. *Ibis* 127:559–562.

Vilstrup, J., T.D. Mullins, M.P. Miller and S.M. Haig. 2018. Comparing three generic swabs used to sample buccal cavities of Red-cockaded Woodpecker nestlings for mitochondrial and nuclear DNA. *Wilson Journal of Ornithology* 130:326–335.

Webster, M.S., P.P. Marra, S.M. Haig, S. Bensch, and R.T. Holmes. 2002. Links between worlds: Unraveling migratory connectivity. *Trends in Ecology & Evolution* 17:76–83.

Wunder, M.B. and F.L. Knopf. 2003. The Imperial Valley of California is critical to wintering Mountain Plovers. *Journal of Field Ornithology* 74:74–80.

INDEX

A

B

C

D

E

F

G

H

I

J

K

L

M

N

O

P

R

S

T

U

V

W

Z

STUDIES IN AVIAN BIOLOGY

Series Editor: Kathryn P. Huyvaert

http://americanornithology.org

1. Status and Distribution of Alaska Birds. Kessel, B., and D. D. Gibson. 1978.
2. Shorebirds in Marine Environments. Pitelka, F. A., editor. 1979.
3. Bird Community Dynamics in a Ponderosa Pine Forest. Szaro, R. C., and R. P. Balda. 1979.
4. The Avifauna of the South Farallon Islands, California. DeSante, D. F., and D. G. Ainley. 1980.
5. Annual Variation of Daily Energy Expenditure by the Blackbilled Magpie: A Study of Thermal and Behavioral Energetics. Mugaas, J. N., and J. R. King. 1981.
6. Estimating Numbers of Terrestrial Birds. Ralph, C. J., and J. M. Scott, editors. 1981.
7. Population Ecology of the Dipper (Cinclus mexicanus) in the Front Range of Colorado. Price, F. E., and C. E. Bock. 1983.
8. Tropical Seabird Biology. Schreiber, R. W., editor. 1984.
9. Forest Bird Communities of the Hawaiian Islands: Their Dynamics, Ecology, and Conservation. Scott, J. M., S. Mountainspring, F. L. Ramsey, and C. B. Kepler. 1986.
10. Ecology and Behavior of Gulls. Hand, J. L., W. E. Southern, and K. Vermeer, editors. 1987.
11. Bird Communities at Sea off California: 1975 to 1983. Briggs, K. T., W. B. Tyler, D. B. Lewis, and D. R. Carlson. 1987.
12. Biology of the Eared Grebe and Wilson's Phalarope in the Nonbreeding Season: A Study of Adaptations to Saline Lakes. Jehl, J. R., Jr. 1988.
13. Avian Foraging: Theory, Methodology, and Applications. Morrison, M. L., C. J. Ralph, J. Verner, and J. R. Jehl, Jr., editors. 1990.
14. Auks at Sea. Sealy, S. G., editor. 1990.
15. A Century of Avifaunal Change in Western North America. Jehl, J. R., Jr., and N. K. Johnson, editors. 1994.
16. The Northern Goshawk: Ecology and Management. Block, W. M., M. L. Morrison, and M. H. Reiser, editors. 1994.
17. Demography of the Northern Spotted Owl. Forsman, E. D., S. DeStefano, M. G. Raphael, and R. J. Gutiérrez, editors. 1996.
18. Research and Management of the Brown-headed Cowbird in Western Landscapes. Morrison, M. L., L. S. Hall, S. K. Robinson, S. I. Rothstein, D. C. Hahn, and T. D. Rich, editors. 1999.
19. Ecology and Conservation of Grassland Birds of the Western Hemisphere. Vickery, P. D., and J. R. Herkert, editors. 1999.
20. Stopover Ecology of Nearctic–Neotropical Landbird Migrants: Habitat Relations and Conservation Implications. Moore, F. R., editor. 2000.
21. Avian Research at the Savannah River Site: A Model for Integrating Basic Research and Long-Term Management. Dunning, J. B., Jr., and J. C. Kilgo, editors. 2000.
22. Evolution, Ecology, Conservation, and Management of Hawaiian Birds: A Vanishing Avifauna. Scott, J. M., S. Conant, and C. van Riper, II, editors. 2001.

23. Geographic Variation in Size and Shape of Savannah Sparrows (Passerculus sandwichensis). Rising, J. D. 2001.
24. The Mountain White-crowned Sparrow: Migration and Reproduction at High Altitude. Morton, M. L. 2002.
25. Effects of Habitat Fragmentation on Birds in Western Landscapes: Contrasts with Paradigms from the Eastern United States. George, T. L., and D. S. Dobkin, editors. 2002.
26. Ecology and Conservation of the Willow Flycatcher. Sogge, M. K., B. E. Kus, S. J. Sferra, and M. J. Whitfield, editors. 2003.
27. Ecology and Conservation of Birds of the Salton Sink: An Endangered Ecosystem. Shuford, W. D., and K. C. Molina, editors. 2004.
28. Noncooperative Breeding in the California Scrub-Jay. Carmen, W. J. 2004.
29. Monitoring Bird Populations Using Mist Nets. Ralph, C. J., and E. H. Dunn, editors. 2004.
30. Fire and Avian Ecology in North America. Saab, V. A., and H. D. W. Powell, editors. 2005.
31. The Northern Goshawk: A Technical Assessment of its Status, Ecology, and Management. Morrison, M. L., editor. 2006.
32. Terrestrial Vertebrates of Tidal Marshes: Evolution, Ecology, and Conservation. Greenberg, R., J. E. Maldonado, S. Droege, and M. V. McDonald, editors. 2006.
33. At-Sea Distribution and Abundance of Seabirds off Southern California: A 20-Year Comparison. Mason, J. W., G. J. McChesney, W. R. McIver, H. R. Carter, J. Y. Takekawa, R. T. Golightly, J. T. Ackerman, D. L. Orthmeyer, W. M. Perry, J. L. Yee, M. O. Pierson, and M. D. McCrary. 2007.
34. Beyond Mayfield: Measurements of Nest-Survival Data. Jones, S. L., and G. R. Geupel, editors. 2007.
35. Foraging Dynamics of Seabirds in the Eastern Tropical Pacific Ocean. Spear, L. B., D. G. Ainley, and W. A. Walker. 2007.
36. Status of the Red Knot (Calidris canutus rufa) in the Western Hemisphere. Niles, L. J., H. P. Sitters, A. D. Dey, P. W. Atkinson, A. J. Baker, K. A. Bennett, R. Carmona, K. E. Clark, N. A. Clark, C. Espoz, P. M. González, B. A. Harrington, D. E. Hernández, K. S. Kalasz, R. G. Lathrop, R. N. Matus, C. D. T. Minton, R. I. G. Morrison, M. K. Peck, W. Pitts, R. A. Robinson, and I. L. Serrano. 2008.
37. Birds of the US–Mexico Borderland: Distribution, Ecology, and Conservation. Ruth, J. M., T. Brush, and D. J. Krueper, editors. 2008.
38. Greater Sage-Grouse: Ecology and Conservation of a Landscape Species and Its Habitats. Knick, S. T., and J. W. Connelly, editors. 2011.
39. Ecology, Conservation, and Management of Grouse. Sandercock, B. K., K. Martin, and G. Segelbacher, editors. 2011.
40. Population Demography of Northern Spotted Owls. Forsman, E. D. et al. 2011.
41. Boreal Birds of North America: A Hemispheric View of Their Conservation Links and Significance. Wells, J. V., editor. 2011.
42. Emerging Avian Disease. Paul, E., editor. 2012.
43. Video Surveillance of Nesting Birds. Ribic, C. A., F. R. Thompson, III, and P. J. Pietz, editors. 2012.
44. Arctic Shorebirds in North America: A Decade of Monitoring. Bart, J. R., and V. H. Johnston, editors. 2012.
45. Urban Bird Ecology and Conservation. Lepczyk, C. A., and P. S. Warren, editors. 2012.
46. Ecology and Conservation of North American Sea Ducks. Savard, J.-P. L., D. V. Derksen, D. Esler and J. M. Eadie, editors. 2014.
47. Phenological Synchrony and Bird Migration: Changing Climate and Seasonal resources in North America. Wood E. M. and J. L. Kellermann, editors. 2015.
48. Ecology and Conservation of Lesser Prairie-Chickens. Haukos, D.A. and C. Boal, editors. 2016.
49. Golden-winged Warbler Ecology, Conservation, and Habitat Management. Streby, H. M., D. E. Andersen, and D. Buehler, editors. 2016.
50. The Extended Specimen: Emerging Frontiers in Collections based Ornithological Research. Webster, M. S., editor. 2017.
51. Molt in Neotropical Birds: Life History and Aging Criteria. Johnson, E. I. and J. D. Wolfe. 2017.
52. The Population Ecology and Conservation of *Charadrius* Plovers. Colwell, M. A. and S. M. Haig, editors. 2019.

For more information about this series, please visit: https://www.crcpress.com/Studies-in-Avian-Biology/book-series/CRCSTDAVIBIO

Milton Keynes UK
Ingram Content Group UK Ltd.
UKHW050453071024
449327UK00015B/358